Handbook of Mechanical Engineering

Handbook of Mechanical Engineering

Edited by Kimberly Porter

Clanrye International
www.clanryeinternational.com

Clanrye International,
750 Third Avenue, 9th Floor,
New York, NY 10017, USA

ISBN: 978-1-63240-832-7

Cataloging-in-Publication Data

Handbook of mechanical engineering / edited by Kimberly Porter.
p. cm.
Includes bibliographical references and index.
ISBN 978-1-63240-832-7
1. Mechanical engineering. 2. Engineering. I. Porter, Kimberly.
TJ145 .H36 2019
621--dc23

For information on all Clanrye International publications
visit our website at www.clanryeinternational.com

Clanrye International

Contents

Preface

Mechanical engineering is a field of engineering concerned with the design, analysis and manufacture of mechanical systems. It is a multidisciplinary science that applies the principles of engineering mathematics, physics, materials science and engineering. The field focuses on the design of industrial equipment, manufacturing plants, heating and cooling systems, transport systems, etc. Computer-aided engineering is a modern tool in this field. The sub-disciplines of mechanical engineering are structural analysis, mechatronics, robotics, mechanics, etc. Research in mechanical engineering explores the development of safer, more efficient and cheaper machines and mechanical systems. Micro electro-mechanical systems, composites, friction stir welding, finite element analysis and nanotechnology are other significant areas of research under this field. The ever-growing need of advanced technology is the reason that has fueled the research in the field of mechanical engineering in recent times. The objective of this book is to give a general view of the different areas of mechanical engineering and their applications. It aims to equip students and experts with the advanced topics and upcoming concepts in this area.

The information shared in this book is based on empirical researches made by veterans in this field of study. The elaborative information provided in this book will help the readers further their scope of knowledge leading to advancements in this field.

Finally, I would like to thank my fellow researchers who gave constructive feedback and my family members who supported me at every step of my research.

Editor

Fracture Properties Investigation of *Artocarpus odoratissimus* Composite with Polypropylene (PP)

Shah MKM[1], Sapuan SM[2], Al-Fareez Bin-Aslie[1], Irma Wani O[1] and Sarifudin J[1]*

[1]*Faculty of Engineering, University of Malaysia, Sabah 88400, Kota Kinabalu Sabah, Malaysia*
[2]*Department of Mechanical and Manufacturing Engineering, Universiti Putra Malaysia, 43400 UPM Serdang, Selangor, Malaysia*

Abstract

Wood plastic composites (WPC) were done using a matrix of polypropylene (PP) thermoplastic resin with wood fibre from *Artocarpus odoratissimus* as filler. The purpose of this study is to investigate the fracture properties of *Artocarpus odoratissimus* composite with PP. The WPC was manufactured by a hot - press technique with varying formulations which are 10:0 (100% pure PP), 50:50 (40 g of wood fibre and 40 g of PP) and 60:40 (48 g of wood fibre and 32 g of PP). The mechanical properties were investigated. Tensile and flexural were carried out according to ASTM D 638 and ASTM D 790. The results were analysed to calculate the tensile strength. Tensile strength at break is ranged from 13.2 N/mm^2 to 21.7 N/mm^2. While, the flexural strength obtained is varying from 14.7 N/mm^2 to 31.1 N/mm^2. The results of the experiment showed that tensile and flexural properties of the composite increased with the adding of wood fibre material. Finally, the Scanning Electron Microscope (SEM), have been done to study the fracture behaviour of the WPC specimens.

Keywords: WPC; *Artocarpus odoratissimus;* Polypropylene thermoplastic; Wood fibre

Introduction

Wood-plastic composites (WPCs) are emerging as one of the dynamic growth materials in the building industry. WPC is manufactured by dispersing wood particles into molten plastic with coupling agent or additives to form a composite material through various techniques of processing such as extrusion, compression or injection moulding. It was first made commercially from phenol-formaldehyde and wood floor that was used for Rolls-Royce gearshift knob in 1916, and it was reborn as a modern concept in Italy in the 1970s, and popularized in North America in early 1990s [1]. Wood-thermoset composites date back to the early 1900s; however, thermoplastic polymers in WPC are a relatively new innovation. In 1983, an American Woodstock company (Lear Corporation in Sheboygan, WI) began producing automotive interior substrates by using extrusion technology from the mixture of polypropylene (PP) and wood flour [2]. Since then production and market demand for the WPCs have been growing rapidly worldwide. Currently WPCs are mainly used for building products like decking, fencing, siding, garden furniture, exterior windows and doors [3,4], although other applications can also be found in marine structures, railroad crossties, automobile parts and highway structures such as highway signs, guardrail posts, and fence posts [5]. WPCs possess many advantages over the raw materials of polymers and wood filler. WPC had better dimensional stability and durability against bio-deterioration as compared to wood. In addition, WPC also reduces the machine wear and tear of processing equipment, and lower the product cost against inorganic fillers when waste streams such as sawdust are used [4]. As compared to the polymers, WPC had higher mechanical properties, thermal stability, and more resistance to the ultraviolet light and degradation [6-10].

Similarly, large amount of wood waste generated at different stages in the wood processing is mainly destined for landfill. It was reported that waste wood in the form of wood flour, fibres or pulp is suitable as filler for polyolefin's matrix composites. *Artocarpus odoratissimus* wood fibres possessed physical and mechanical properties suitable to the reinforcement of plastics. Hence increased usage of the recycled plastics and the waste wood for WPCs offers the prospect of lessening waste disposal problems and lowering production costs.

Virgin thermoplastics such as HDPE and polypropylene (PP) are widely used for WPCs, and a significant number of papers are available for their mechanical properties, dimensional stability, interface adhesion and durability.

In conclusion, stability and durability performance of WPCs based on post-consumer thermoplastic are not fully understood and the affecting factors are not known, leaving open research opportunities for the optimization of formulation and processing. WPCs performance can be optimized by investigating a wide range of composite formulations and processing techniques. Considering the potentials for applications and resource availability, PP was chosen as the raw materials to produce the WPCs with wood fibre from *Artocarpus odoratissimus* wood through the compression moulding. Mechanical properties need to be investigated. Influence of polymer type and form (virgin) and *Artocarpus odoratissimus* wood fibre was also examined.

Methodology

Wood fibre, as shown in Figure 1 used in this project is obtained from the *Artocarpus odoratissimus* tree at the Kg. Kebayau, Telipok, Sabah. The wood fibres as solid waste were derived from the tree by using a chainsaw and the waste of the wood produce by chainsaw the tree was collected.

The wood fibre was ground to have required 40-80 mesh size of the fibre and then screened to remove impurities. It was then dried in an oven at 103±2°C for 24 hours to a moisture content of 2%. These

***Corresponding author:** Sarifudin J, Faculty of Engineering, University of Malaysia, Sabah 88400, Kota Kinabalu Sabah, Malaysia
E-mail: jumafisabilillah92@gmail.com

fibres are kept in airtight bags to avoid high humidity which would allow moisture ingression into fibres (Table 1).

The type of resin used in this project is polypropylene (PP) thermoplastic resin (Table 2). This resin is obtained from Chemical Laboratory as shown in Figure 2 and was used as a matrix. Table 3 shows the properties of the polypropylene (PP). Besides that, this resin's attributes include light weight (low specific gravity), high temperature resistance, good chemical resistance, excellent rigidity, no stress cracking, high tensile strength, excellent flexural strength, low creep, excellent electrical properties, good barrier properties, impact resistance, wide range of applications, ease of processing and highly environmental friendly.

The wide range of physical properties via modification, ease of processing, and cost advantage of polypropylene, make it an extremely attractive alternative to more expensive resins in a number of applications.

Figure 1: *Artocarpus odoratissimus* wood fiber.

Code	Value	Fiber length (µm)	Fiber diameter (µm)	Lumen diameter (µm)	Wall thickness (µm)
Pith	Average	952.9	20.97	13.23	7.74
	Standard deviation	150.75	5.17	3.85	3.25
Core	Average	1294.28	35	16.21	9.4
	Standard deviation	230.69	5.87	6.05	3.88
Transition	Average	1260	26.21	16.05	5.08
	Standard deviation	203.4	6.29	5.09	1.7
Sapwood	Average	1286.43	28.47	17.1	5.69
	Standard deviation	221.31	5.93	5.28	2.5

Table 1: Dimension value of *Artocarpus odoratissimus* wood fiber.

Code	Value	Runkel Ratio	Power loom	Muhlstep Ratio	Coefficient of Rigidity (CR)	Flexibility Ratio (FR)
Pith	Average	0.67	47.91	37.18	0.19	0.63
	Standard Deviation	0.56	13.38	9.57	0.05	0.10
Core	Average	0.70	37.74	39.32	0.20	0.61
	Standard Deviation	0.34	8.08	9.42	0.05	0.09
Transition	Average	0.69	50.11	39.19	0.20	0.61
	Standard Deviation	0.30	11.89	10.19	0.05	0.10
Sapwood	Average	0.78	47.01	39.67	0.20	0.60
	Standard Deviation	0.57	12.92	14.39	0.07	0.14

Table 2: Properties of *Artocarpus odoratissimus* wood fibers.

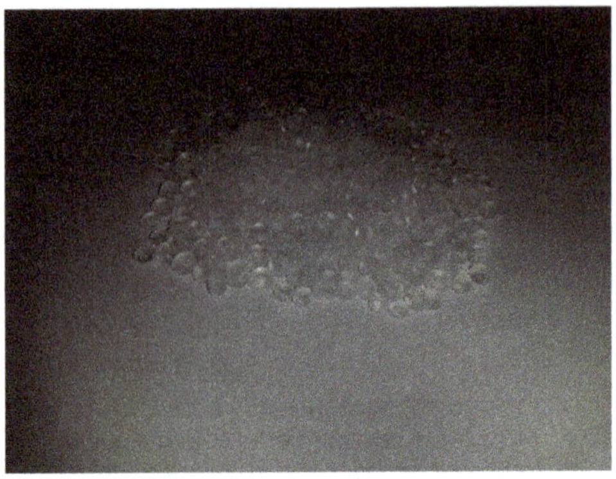

Figure 2: Polypropylene (PP) thermoplastic.

No.	Properties	Unit	Analysis Report
1.	Density	g/cm^3	0.9
2.	Melt flow index	g/10 min. 2.16 kg/190°C	25
3.	Tensile strength	MPa	26.48
4.	Elongation	%	-

Table 3: Properties of Polypropylene (PP) thermoplastic.

The specimen of the WPC was formed by using concrete. The median size is 150 mm × 150 mm × 150 mm. This mould is heated in an oven at temperature of 190°C to 210°C for a day before using. The used of this mould will be explained in the preparation of the Wood Plastic Composite (WPC).

The specimens of WPC were through hot-pressed method. Matrix material which is PP thermoplastic is reinforced with *Artocarpus odoratissimus* wood fibres. These materials are mixed well in a ratio of 71.43:28.57, 57.14:42.86 and 50:50. This ratio was determined by different weight percentage between thermoplastic and wood fibres respectively, and the three panels of the dimensions of 150 mm × 150 mm surface area with different ratios were manufactured.

In the first process of manufacturing the wood, plastic composites, the wood fibres and thermoplastic resin were mixed well in hand by the container. Then, the mixing was put into a mould that has been heated in the oven a day before uniformly. Next, a steel plate is a plate on the composite materials mixing as a cover for the hot - pressed process to obtain flat surfaces. Compressive strength machine was used to compress the composite materials mixing, which is much more similar to actual hot-press method and the pressure applied was around 200-250 kN.

The mould was placed at the centre point position in the compressive machine with the lab technician conduct and helps. Rod was put in the mould for compressing purposes. The mixing was pressed by the machine which is conducted manually around 5 minutes. After the compression done, the campsites were taken out of the mould and left at room temperature for a while. Finally, six of rectangular shapes specimens with 100 mm × 25 mm for mechanical testing were produced.

Tensile test is aimed to obtain tensile properties. Tensile test is commonly used to determine mechanical properties such as strength, toughness and modulus of elasticity. There are 3 specimens for the three different compositions. The tensile test was run by using the GOTECH/AI-7000M Electronic Mechanical Testing. All these samples have been cut according to ASTM D638 and will be used in the tensile test.

All the specimens are tested using the Electronic Mechanical Testing with a speed rate of 1 mm/min. Samples are placed in the grips of the machine at specified grip separation and pulled until failure. The machines will automatically produce a stress versus strain diagram, thus the mechanical behaviour of the composites can be interpreted from the diagram. The specimen will elongate as the tensile test starts. The load value (F) is recorded up to the point where the specimen breaks. The instrument software, which is provided with the machine equipment, will calculate the tensile properties such as the tensile, yield strength and elongation.

Tensile and flexural tests have been performed and the mechanical properties such as tensile strength and tensile modulus were determined. In this project, there were 3 specimens prepared. The test specimens were cut into the dimension of 100 mm × 25 mm (Figure 3).

Result and Discussion

The aim of this paper is to establish mechanical properties such as tensile strength and flexural strength of the wood plastic composites with different formulations. Basically, three main tasks were carried out in order to achieve the objectives of this project. The first task is to prepare the wood, plastic composite by combining wood fibre from *Artocarpus odoratissimus* and polypropylene thermoplastic resin. This is followed by performing the tensile and flexural tests, and lastly, a microstructure analysis is carried out to study the fracture behaviour of the wood plastic composites specimen.

Figure 3: The specimens of wood plastic composites prepared for mechanical properties.

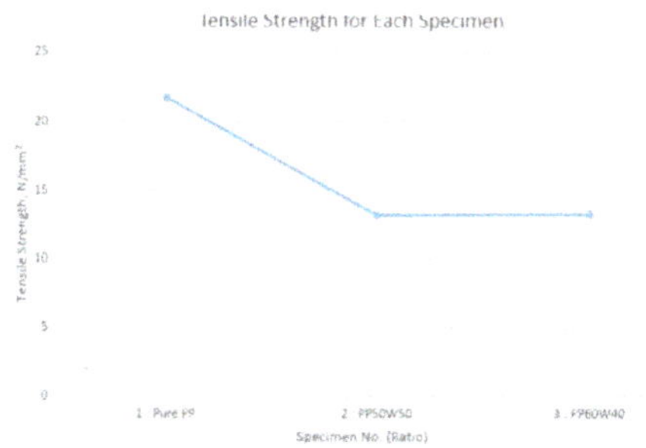

Figure 4: Tensile strength for 3 specimens of different ratio.

The results of the tensile strength and flexural properties of the wood-PP composites are given in Table 3. As it can be seen from the table, the tensile strength of the wood, plastic composites lay in the range of 13.2 N/mm^2 to 21.7 N/mm^2 depending upon composite formulations. The composites based on pure PP exhibited higher tensile strengths, compared to those based on PP for 50 wt. % wood fibre content. For example, with a plastic to wood ratio of 50:50, composite based on VPP was 13.2 MPa. This is because of higher molecular weight of virgin PP (25 g/10 min. 2.16 kg/190°C) as MFI is inversely proportional to MFI. The tensile strength increases with molecular weight due to the effect of better entanglement.

The flexural strength (MOR) and Young's modulus (MOE) were obtained from 3 points bending tests and values are given in Table 4. The flexural MOR exhibited a similar trend as the tensile strength. The composites had a flexural MOR varying from 14.7 Mpa to 31.1 MPa. It was also observed that the MOR increases with decreasing wood content as expected from the rule of mixtures, the addition of wood fibres into the PP matrix significantly increased MOE of the composites and MOE value increased with the wood content. It was observed that for the non-coupled composite formulations, incorporation of wood fibres into the PP matrix did affect the MOR more as compared to MOE. The yield strength of composites followed the similar trend as that of flexural MOR for all composite formulations. The load-displacement curves of the pure PP composites and the wood, plastic composites are shown in Figures 4 and 5, which illustrates the effects of the filler loading. The load-displacement curves for the wood, plastic composites had relatively high slopes initially, but as a failure occurred, the loads dropped off quickly as the material crumbled. However, the pure PP specimen reached ultimate capacity after significant deformation (15 mm in this case), and then the load decreased gradually until the test was concluded, indicating ductile behaviour. The addition of wood fibre in the composites increased the stiffness and brittleness, however, reduced the elongation at break. The stress concentration at the fibre ends and poor interface bonding between wood and PP matrix have been recognized as the main causes for the embrittlement (Figure 4).

However, during loading, fractures occurred at the filler locations, and these fracture locations were more brittle than other parts of the matrix. Hence, these composites showed little change in the specimen appearance in the initial stage of loading until the maximum load was reached when the specimen failed suddenly with extensive breakage at the interface between the wood fibre and the matrix. The elongation at break of the wood, plastic composites was much lower than that of the pure PP panel. This decrease was probably due to the higher degree of brittleness introduced by the incorporation of wood fibres in the PP matrix. As in the case of modulus, improving the adhesion between fibres and PP did not enhance elongation at break.

Specimen code	Flexural Properties				Tensile Strength (MPA)
	MOR (MPA)	MOE (GPa)	Yield Strength (MPa)	Elongation at break (%)	
Pure PP	31.1	1.25	17.3	3.70	21.7
PP50W60	14.7	1.68	7.9	1.55	13.2
PP60W40	22.0	1.71	11.1	2.50	13.3

Table 4: Tensile and flexural properties of the wood plastic composites.

The tensile and flexural properties variation of composites based on wood fibre and PP can be explained differently. It was found that composites exhibited higher tensile and flexural properties (Figure 5).

The morphology of surface structure of wood, plastic composites was investigated by SEM. SEM images of the wood; plastic composite at filler loading of 50 wt. % matrix is shown in Figure 6, in 200x magnification. From these images, it is clearly observed that there were distinct cluster and gaps between thermoplastic polymer matrix and wood. The patterns from wood fibres that were so weakly bonded to the matrix had been released from the matrix during fracture. The failure surface was undulated with clear wood fibre surfaces with visible tragedies, and laymen, indicating the path of weaker part through the wood-wood interface and weakest thermoplastic polymer matrix. This suggests that the interface between the wood fibre and PP matrix was weaker due to the poor dispersion and wettability. In some cases, the part of the wood lumen was filled with plastic that could increase the strength of the composites because of mechanical interlocking. When wood content was increased, the thermoplastic polymer matrix was no longer continuously distributed and many wood fibres were in direct contact with one another, resulting in poor bonding adhesion at the interface.

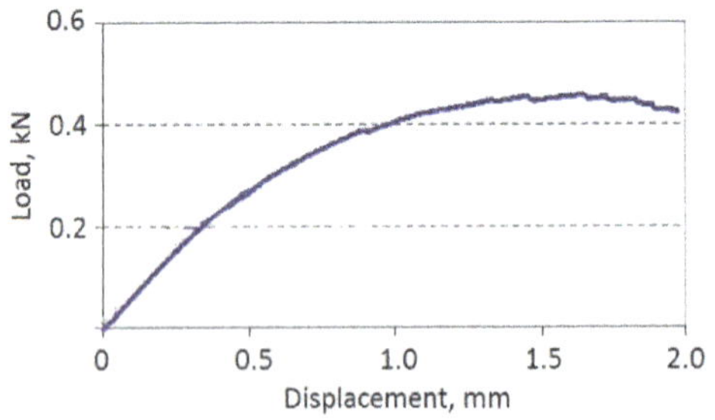

Figure 5: Load displacement curves for the specimen Pure PP.

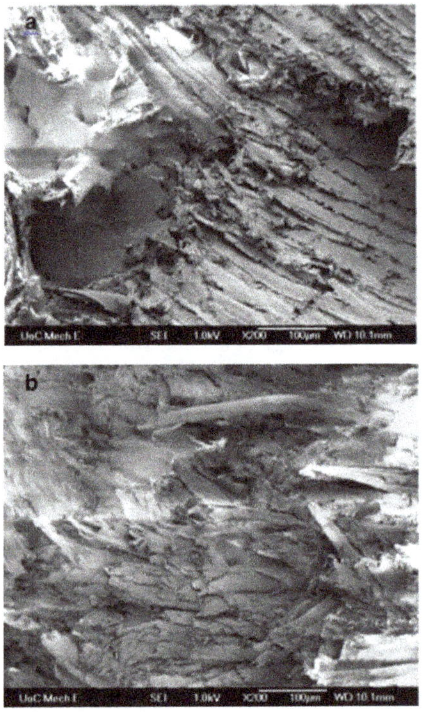

Figure 6: (a) and (b) are SEM images of the specimen PP50W50 with Magnification 200X.

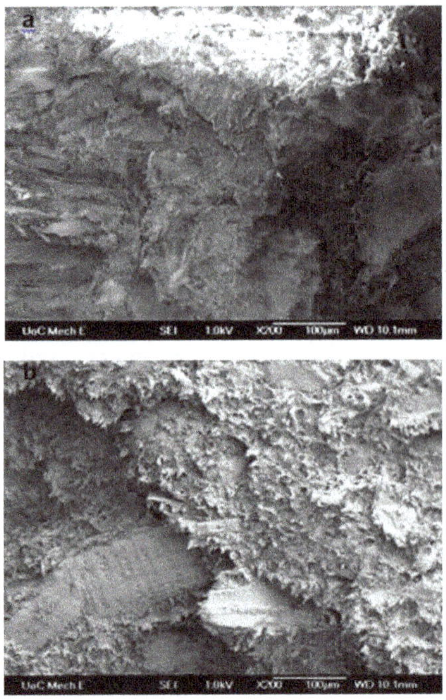

Figure 7: SEM images (x200) of fractured surface of PP60W40.

Mechanical bond is a form of adhesive bonding which adhesive material physically locks into the crevices of the surface. There are two or more separate components of interlocked molecules which are not connected by chemical or covalent bonds. The valleys and crevices of each fibre must filled with matrix and displace trapped air to work well. Matrix and fibre are mechanically interlocked together by adhesion and the overall strength of the bond is dependent upon the quality of this interlocking interface as explained in Bonding Adhesives in the year, 2000 as reported by D. Hull and T. W. Clyne in the year 1996.

However, this compatibility can improve when fibre treated by chemical treatment methods. Some research proved that treatment fibre increases the value of tensile strength by 53% as compared to the composites with untreated fibres. They also stated that the treatment improved the fober-matrix adhesion, allowing an efficient, and stress transfer from the matrix to fibres as studied by Morsyleide et al., and also by Herrera and Valadez. The composite seems to detach from the matrix and have relatively large pilots due to the poor interfacial bonding with the matrix. Hence, the role of treatment is mainly to remove impurities of the natural fibre, thus improving interfacial bonding.

The fracture surface of the composite showed a very limited amount of torn matrix, suggesting that the matrix was more brittle than those composites without coupling agent. It was also seen that a crack running through the wood fibre, and this could be an indication of stress-transfer from the matrix to the wood fibres. The interfacial bonding between the filler and the PP matrix was improved due to the esterification mechanism, and the fracture occurred at the filler itself. This means that the stress was well propagated between the filler and the matrix thermoplastic polymer, resulting in enhanced flexural strength and modulus in response to stress. In addition, the fracture surface showed a very limited amount of torn matrix, suggesting that the campsite was more brittle. In general, coupling agent was randomly distributed in composites and randomly reacted with wood fibres and the matrix to form graft polymerization. Hence, grafting sites was randomly distributed on wood, and a network of coupling agent was formed at the interface (Figure 7).

However, there was a limit for chemical coupling reaction and only part of coupling agent was grafted onto wood surface and even cross-linked at the interface. Further, the fracture surface of the composite containing coupling agent showed a very limited amount of torn matrix, suggesting that the matrix was more brittle when higher weight. % of coupling agent added in the composite. This phenomenon was mainly due to the excessive modification of the base polymer.

It was observed that non-coupled composite samples had a weak interface region and damage mainly occurred along the loose and weak interface between the wood fibre and PP matrix under loading. However, with the coupling agent such as MAPP mix into composites, the wood fibre combined with the PP matrix through the covalent bonding or strong interfacial bonding, and the interfacial fracture usually accompanied with a cross section damage of wood fibre. Hence, after the failure, the fibre surface in the untreated composites was smooth; whereas the wood fibre in the coupling agent treated composites had a rough surface and it was embedded in the matrix with a chemical link. As the interracial bonding in the case of non-coupled composites were from mechanical connection, these composites failed mainly along the direction parallel to fibre length due to shearing failure between fibre bundles under bending.

Conclusion

Wood plastic composite have been successfully developed using *Artocarpus odoratissimus* wood fibre and polypropylene thermoplastic as reinforcements. Overall, the tension and flexural properties of the WPC show increase with increasing *Artocarpus odoratissimus* wood fibre loading. The value of tensile strength and hardness number of this composite material is higher than other existing composite material. This wood plastic composite is widely used for both indoor and outdoor decorations as well as office and high-grade furniture.

Acknowledgment

We would like to take this opportunity to express our deepest appreciation and gratitude to those people who had guided and assisted us doing this research. We thank our colleagues from University Malaysia Sabah who provided insight and expertise that greatly assisted the research, although any errors are our own and should not tarnish the reputations of these esteemed persons.

References

1. Pritchard G (2005) Second-generation wood composites: The US shows Europe the way. Reinforced Plastics 49: 34-35.
2. Schut JH (1999) For compounding, sheet and profile: Wood is good. Plastics Technology 45:46-52.
3. Smith PM, Wolcott MP (2006) Opportunities for wood/natural fiber-plastic composites in residential and industrial applications. Forest Products Journal 56.
4. Optimat Ltd. and MERL Ltd. (2003) Wood plastic composites study - Technology and UK market opportunities. The Waste and Resource Action Programme p: 1- 100.
5. Youngquist JA, Myers GE, Muehl JH, Krzysik AM, Clemens CM, et al. (1994) Composites from recycled wood and plastics: A project summary: US-Environment Protection Agency.
6. Hannequart JP (2004) Good practice guide on waste plastics recycling: A guide by and for local and regional authorities: Association of Cities and Regions for Recycling (ACRR) Belgium.
7. USEPA (2006) Municipal solid waste in the USA: 2005 Facts and figures, Washington D.C.
8. Association of Plastics Manufacturers (2004) Plastics in Europe: An analysis of plastics consumption and recovery in Europe.
9. Plastics (2005) Plastics: New Zealand sustainable end-of-life options for plastics in New Zealand.
10. Panthapulakkal S, Law S, Sain M (1991) Properties of recycled high-density polyethylene from milk bottles. J of Applied Polymer Science 43: 2147-2150.

Identification of Contact Pressures in Two and Three-Dimensional Solid Bodies from Cauchy Data

Kadria ML*

LAMSIN, Ecole Nationale d'Ing´enieurs deTunis, BP no 37-1002Tunis, Tunisia

Abstract

This note deals with the identification of contact pressures in two and three-dimensional elastic bodies via two approaches relying on domain decomposition using electrostatic measurements. These approaches consist in recasting the problem in terms of primal or dual Steklov Poincar´e equations. The numerical performances of these formulations are compared. The proposed methods are applied to some inverse problems: the first application deals with the identification of a Hertizian contact pressures distribution, the second deals with the identification of the indentation pressure of a heterogeneous solid, and the third one with the identification of boundary data at the interface of a bonded structure.

Keywords: Inverse problem; Linear elasticity; Numerical methods; Contact; Indentation

Introduction

Physical phenomena are often governed by partial differential equations, which need an essential set of data to solve them. In linear elasticity, these data are: the geometry of the solid, the mechanical properties of the materials and the boundary conditions. However, in many industrial applications, some of these data are unknown and have to be identified. This leads to an inverse problem whose resolution requires over specified measured data. In this paper we focus on problems of boundary condition identification in linear elasticity. In this case, data measured on part of the easily accessible border are often available. However, contrary to the direct problem, two kinds of boundary conditions (e.g. displacements and tractions) are imposed on the same part of the boundary while no information exists on the remaining part of it. Hence, data completion consists in reconstructing the boundary conditions for the whole boundary of a domain by using the partially over specified measurements. This is the well-known Cauchy problem, which is ill- posed. The ill-posedness of inverse problems may concern the existence and/or the uniqueness of the solution, but their most critical feature is their instability: the solution, whenever a problem has one, is not continuous with respect to the data, i. e. small measurement errors in the data may dramatically amplify the errors in the solution. This is ill-posedness in the Hadamard sense [1]. The Cauchy problem pertains to this kind of inverse problem. Therefore suitable regularizing algorithms that are exempt from this ill-posedness phenomenon are required in order to solve the inverse problem correctly.

For inverse problems in elasticity, we refer to Bonnet et al. overview paper [2] and the huge amount of references therein. The Cauchy problem in linear elasticity was first studied by Yeih et al. [3]. In this paper, the existence and uniqueness of the solution are analyzed as well as the continuity of the solution with respect to the data. Other authors have proposed an alternative regularization procedure, namely the indirect fictitious boundary method, which is based on the simple or the double layer potential theory. The numerical implementation of the aforementioned method has been carried out by Koya et al. [4] who used the BEM and the Nystrom method for discretizing the integrals involved. Marin et al. [5] have determined the approximate solutions of the Cauchy problem in linear elasticity using an alternating iterative BEM that reduces the problem to solving a sequence of well-posed boundary value problems. Marin et al. [6] have used singular value decomposition to solve the same problem numerically. A related inverse problem which allows for interior displacement measurements and inter-facial crack detection has been investigated by Huang and Shih [7]. Weikel et al. [8] have proposed an alternating iterative algorithm in order to reconstruct an internal planar crack laying on an a priori known internal surface inside a three-dimensional elastic body from over determined electrostatic boundary data on the outer surface. Numerical experiments for the identification of internal cracks in a three-dimensional elastic body using the primal and dual formulations of the Steklov-Poincar´e equation are recently carried out and their numerical performances are compared [9]. A general framework for the different approaches (primal, dual and mixed) is presented presentad by Azaiez et al. [10].

In this work, the Steklov-Poincar´e method is applied to the linear elastic data completion problem. In section 2, the Cauchy problem is presented in the context of linear elasticity. In section 3 this problem is recast in condensed form that we will refer to as the Cauchy-Steklov-Poincar´e problem, which leads to the Cauchy-Steklov-Poincar´e equation acting on the boundary of the unknowns. In section 4 we present the Dirichlet-to-Neumann algorithm and we show that it can be interpreted as a preconditioned Richardson procedure for the Cauchy-Steklov-Poincar´e equation. The numerical procedures are presented in sections 3 and 5 and the results obtained by FEM discretization of the problems are presented in sections 7 and 8. The methods are used to solve applications borrowed from engineering mechanics in 2D and 3D frameworks: the identification of contact pressure between two elastic bodies, identification of the indentation pressure in a two layer solid and the identification of boundary data at the interface of a bonded structure.

The Cauchy Problem in Linear Elasticity

Let Ω denote a bounded domain in R^2 *or* R^3 with regular boundary

***Corresponding author:** Kadria ML, LAMSIN, Ecole Nationale d'Ing´enieurs de Tunis, BP no 37-1002 Tunis, Tunisia, E-mail: medlarbi.kadri@lamsin.rnu.tn

$\Gamma=\partial\Omega$. The whole domain is assumed to be filled with a homogeneous linear elastic isotropic medium. It is assumed that Γ is splitted into two open subsets Γ_c and Γ_i, $\Gamma=\Gamma_c\cup\Gamma_i$ where $\Gamma_c, \Gamma_i\neq\emptyset$ and $\Gamma_c\cap\Gamma_i=\emptyset$. In what follows, $u(x)$ denotes the displacements field on Ω.

The local equilibrium equation is given by

$$-div\ \sigma(u(x)) = f\ x\in\Omega, \quad (1)$$

where σ is the stress tensor and f the volume forces. The strain tensor ε is given by

$$\varepsilon(u(x)) = \frac{1}{2}(\nabla u(x) + \nabla u(x)^t). \quad (2)$$

These tensors are related by the Hooke's constitutive law, which is

$$\begin{aligned}\sigma(u(x)) &= 2\mu\varepsilon(u(x)) + \lambda tr\varepsilon(u(x))I \\ &= \mu(\nabla u(x) + \nabla u(x)^t) + \lambda\, div u(x),\end{aligned} \quad (3)$$

where λ and μ are the Lam´e constants of the material and I is the identity tensor.

Let n(x) be the outward normal vector at Γ and $t(x)$ be the traction vector at a point $x\in\Gamma$ defined by

$$t(x)=\sigma(u(x))n(x) \quad x\in\Gamma.$$

In the well-posed direct problem formulation, the knowledge of the displacement on a part of the boundary and traction vectors on another part of the boundary enables us to determine the displacement vector in domain Ω. Then, the strain tensor ε can be calculated from kinematic relation (2) and the stress tensor is determined by constitutive law (3).

If a part of the boundary Γ_i is inaccessible and if it is possible to measure both the displacement and traction vectors on the remaining part of boundary Γ_c, this leads to the mathematical formulation of a direct problem expressed as follows:

$$\begin{cases}\text{-div } \sigma(u(x)) = f & \text{in } \Omega \\ u(x) = \tilde{u}(x) in \Gamma_c \\ t(x) = \tilde{t}(x) in \Gamma_c\end{cases} \quad (4)$$

where $\tilde{u}$ and $\tilde{t}$ are prescribed vector valued functions. This problem is ill-posed because of the formulation of its boundary conditions (4). It can be seen that boundary Γ_c is over specified by prescribing both the displacement $u_{|\Gamma_c}=\tilde{u}$ and the traction $t_{|\Gamma_c}=\tilde{t}$ vectors, while boundary Γ_i is underspecified since both the displacement $u_{|\Gamma_c}=\bar{u}$ and the traction $t_{|\Gamma_c}=\bar{t}$ are unknown and have to be determined. Then, this problem can be stated as follows: find $(\bar{u},\bar{t})$ such that a displacement field $u(x)$ exists that satisfies:

$$\begin{cases}\text{-div } \sigma(u(x)) = f & \text{on } \Omega, \\ u(x) = \tilde{u}(x) & on \Gamma_c, \\ \sigma(u(x))n = \tilde{t}(x) & \text{in } \Gamma_c, \\ u(x) = \tilde{u}(x) & on \Gamma_i, \\ \sigma(u(x))n = \tilde{t}(x) & \text{in } \Gamma_i.\end{cases}$$

$$u_D(\mu) = u^0_D(\mu) + u^*_D \quad (5)$$

$$\sigma(u^0_D(\lambda))\mathbf{n} - \sigma(u^0_N(\lambda))\mathbf{n} = \sigma(u^*_N)\mathbf{n} - \sigma(u^*_D)\mathbf{n}\ on \Gamma_i.$$

$$\begin{cases}\text{-div } \sigma(u^D_0(\mu)) = 0 \text{ in } \Omega \\ u^D_0(\mu) = 0\ on\ \Gamma_c \\ u^D_0(\mu) = \mu\ on\ \Gamma_i\end{cases}$$

This problem, known as the Cauchy problem, is ill-posed in the sense that the dependence of $u(x)$, and consequently of $(\tilde{u},\tilde{t})$ on the data $(\bar{u},\bar{t})$ is not continuous. Although the problem may have a unique solution, it is well-known that this solution is unstable with respect to small perturbations in the data on Γc. In this paper we propose to recover the lacking data by using the Steklov-Poincar´e algorithm introduced by Ben Belgacem et al. in the steady state thermal case in [11]. However, let us first introduce an operator acting on the boundary where data are unknown: the Steklov-Poincar´e operator which is very familiar in domain decomposition and recently introduced for the Cauchy problem of the Laplace equation by Andrieux et al. [13] and by Ben Belgacem et al. [11,12].

The Cauchy-Steklov-Poincar´e equation

To keep the notational complexity to a minimum let us remove x from the notations. Let λ denote the unknown displacement vector on Γ_i. We consider both Dirichlet and Neumann elliptic problems obtained by duplicating the solution u into a couple of vectors (u_N, u_D). The Cauchy problem (5) is then split into:

$$\begin{cases}\text{-div } \sigma(u_D) = f \text{ in } \Omega \\ u_D = \tilde{u}\ on\ \Gamma_c \\ u_D = \lambda\ on\ \Gamma_i\end{cases} \quad \text{and} \quad \begin{cases}\text{-div } \sigma(u_N) = f \text{ in } \Omega \\ \sigma(u_N)n = \tilde{u}\ on\ \Gamma_c \\ u_N = \lambda\ on\ \Gamma_i\end{cases}$$

If the pair $(\tilde{u},\tilde{t})$ is compatible (i.e. a vectors field exists that verifies (1) for which $(\tilde{u},\tilde{t})$ are the Cauchy data on Γ_c), the solution of the Cauchy problem is recovered, i.e. $u=u_N=u_D$ in Ω, if and only if

$$\sigma(u_D(\lambda))n=\sigma(u_N(\lambda))n\ on\ \Gamma_i. \quad (6)$$

Now for μ, a displacements vector defined on Γ_i, the linear parts of $u_N(\mu)$ and $u_D(\mu)$ are denoted $u^N_0(\mu)$ and which $u^D_0(\mu)$ solve respectively:

$$\begin{cases}\text{-div } \sigma(u^D_0(\mu)) = 0 \text{ in } \Omega \\ u^D_0(\mu) = 0\ on\ \Gamma_c \\ u^D_0(\mu) = \mu\ on\ \Gamma_i\end{cases} \quad \text{and} \quad \begin{cases}\text{-div } \sigma(u^N_0(\mu)) = 0 \text{ in } \Omega \\ \sigma(u^N_0(\mu))n = 0\ on\ \Gamma_c \\ u^N_0(\mu) = \mu\ on\ \Gamma_i\end{cases}$$

We consider also u^*_D and u^*_N such that

$$\begin{cases}\text{-div } \sigma(u^*_D) = f \text{ in } \Omega \\ u^*_D = \tilde{\mu}\ on\ \Gamma_c \\ u^*_D = 0\ on\ \Gamma_i\end{cases} \quad \text{and} \quad \begin{cases}\text{-div } \sigma(u^*_N) = f \text{ in } \Omega \\ u^*_N = \tilde{t}\ on\ \Gamma_c \\ u^*_N = 0\ on\ \Gamma_i\end{cases}$$

By superposition, we have $u_N(\mu) = u^0_N(\mu) + u^*_N$ and $u_D(\mu) = u^0_D(\mu) + u^*_D$. With this partition, condition (6) is written as

$$\sigma(u^0_D(\lambda))\mathbf{n} - \sigma(u^0_N(\lambda))\mathbf{n} = \sigma(u^*_N)\mathbf{n} - \sigma(u^*_D)\mathbf{n}\ on \Gamma_i. \quad (7)$$

Using the following notations:

$$S_D(\lambda) = \sigma(u^0_D(\lambda))\mathbf{n},$$

$$S_N(\lambda) = \sigma(u^0_N(\lambda))\mathbf{n},$$

$$\chi = (\sigma(u^*_N) - \sigma(u^*_D))\mathbf{n},$$

equation (7) becomes:

$$S(\lambda)=S_D(\lambda)-S_N(\lambda)=\chi on \Gamma_i. \qquad (8)$$

Equation (8) is called the Steklov-Poincar´e interface equation and S is the Steklov- Poincar´e operator. It is familiar in the domain decomposition framework [14] for the direct boundary value problem. More precisely, things happen as if vectors u_D *and* u_N were defined on two different domains with common boundary Γ_i. In this case, the equation (8) expresses the Neumann transmission condition, but the (-) sign in S would be (+) in the domain decomposition formulation [14]. The (-) sign which appears in S is at the origin of the ill-posedness of the Cauchy problem. From the discrete point of view, the finite element discretization of S leads to the Schur complement matrix [14]. It corresponds to having all interior nodes eliminated by static condensation [15]. A numerical study of the Cauchy-Poisson problem, based on the Steklov-Poincar´e formulation is performed by Andrieux and Baranger [16].

We now continue with the analogy with domain decomposition and show how the Cauchy-Steklov-Poincar´e equation can be expressed, as in domain decomposition, in terms of the Dirichlet-to-Neumann problem.

Remark

A one dimensional example [17]: To illustrate how the ill-posedness of the Cauchy problem occurs in the Steklov-Poincare equation, we consider the problem of reconvering the end conditions *(u,f)* of a pre-tensioned string lying on a Winkler-type foundation, the end conditions at the other extremity $(\tilde{u},\tilde{t})$ being given (Figure 1). Denoting by *F* the tension of the string and by *K* the spring stiffness density of the foundation, the Cauchy problem can be written: find *(u,f)* such that there exists a vertical displacement field *v* verifying :

$$-v''+k^2v=0 \text{ on }]0,L[\; and \; v(0)=\tilde{u},\; v'(0)=-\frac{\tilde{t}}{F}$$

where $k^2=\frac{K}{F}$. In this case *S* is a simple scalar function of *k and L*. It is easy to show that $S=\frac{shkL}{chkL}-\frac{chkL}{shkL}$ which vanishes monotonically with respect to kL. This means that, as expected, the ill-posedness of the problem becomes more and stronger when the length of the string or stiffness of the foundation increases and when the pretension of the string decreases [17].

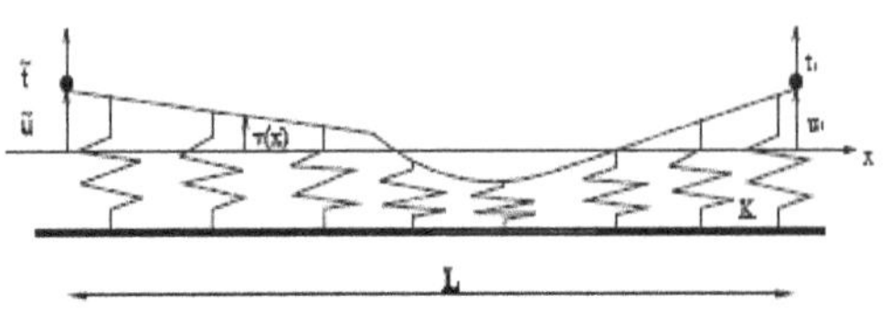

Figure 1: Pre-tensioned string lying on a Winkler-type foundation.

Interpretation of the steklov-Poincar´e equation: Solving the steklov-Poincar´e equation is equivalent to the optimality condition of the first order associated to the energy-like error functional introduced in [18]. The proof is analogous to that done by Koslov and Maz'ya [19] for the Cauchy-Stokes problem.

The Dirichlet-to-Neumann algorithm, also borrowed from domain decomposition and introduced first by Ben Fatma et al. [20] to solve the Cauchy-Poisson problem, can be interpreted as a Richardson procedure applied to the Steklov-Poincar´e equation preconditioned by SD. The proof is the same than that used in domain decomposition [14] and that used [21] for Cauchy-Helmoltz problem.

The Dirichlet-to-Neumann solver for the Cauchy problem

When describing the Dirichlet-to-Neumann approach it should be noted that when the complete data are available on *Γ*, we have an over specied boundary value problem

$$\begin{cases} -\text{div }\sigma(\text{u})=\text{f} & \text{in } \Omega, \\ \sigma(\text{u})n=\tilde{t},\; u=\tilde{u} & on\ \Gamma_c, \\ \sigma(\text{u})n=\overline{t},\; u=\overline{u} & on\ \Gamma_i. \end{cases}$$

This problem can be split into two well-posed sub problems with different boundary conditions. For one of them (Neumann/Dirichlet) conditions are imposed on (Γ_c, Γ_i)

$$\begin{cases} -\text{div }\sigma(\hat{\text{u}})=\text{f} & \text{in } \Omega, \\ \sigma(\hat{\text{u}})n=\tilde{t}, & on\ \Gamma_c, \\ \hat{\text{u}}=\overline{u} & on\ \Gamma_i. \end{cases}$$

while this is reversed for the other and (Dirichlet/Neumann) conditions are imposed on (Γ_c, Γ_i)

$$\begin{cases} -\text{div }\ (\text{u})=\text{f} & \text{in }\ , \\ \text{u}\ \ u & on\ \ , \\ (\text{u})n=\overline{t}, & on\ \Gamma\ . \end{cases}$$

Solving the Cauchy system (5) is achieved when extension $(\overline{t},\overline{\text{u}})$ makes û and ŭ coincide so the solution is u = û = ŭ .

Basically, the iterative method proposed for the Cauchy-Poisson problem and studied by Koslov et al. [20], is derived from these observations: starting from an arbitrary prediction of the Dirichlet condition (here the displacement vector $\overline{\text{u}}$) on Γ_i , we add several corrections by solving alternately a Dirichlet on Γ_c/Neumann on Γ_i problem and a Neumann on Γ_c/Dirichlet on Γ_i problem, where at each iteration the appropriate boundary data are inferred from the solution computed in the previous step. More specifically, we construct a sequence of a pair of vectors $(u_N^{(k)},(u_D^{(k)})k$ from the following recurrence: given $u_D^{(0)}$, the following systems are solved for each $k \geq 0$:

$$\begin{cases} -\text{div }\sigma(u_N^{(k+1)})=\text{f} & \text{in }\Omega \\ \sigma(u_N^{(k+1)})=\tilde{\text{t}} & on\ \Gamma_e \\ u_N^{(k+1)}=u_D^{(k)} & on\ \Gamma_i. \end{cases} \quad \text{and}$$

$$\begin{cases} -\text{div }\sigma(u_D^{(k+1)})=\text{f} & \text{in }\Omega \\ (u_D^{(k+1)})=\tilde{\text{u}} & on\ \Gamma_e \\ \sigma(u_D^{(k+1)})\text{n}=\sigma(u_N^{(k+1)})n & on\ \Gamma_i. \end{cases} \qquad (9)$$

The convergence of the alternating method toward the solution of the Cauchy problem and its stabilizing properties are established by Koslov et al. [20] for the steady state thermal case. In the linear elastic framework, no convergence result has been proved till now but the result of convergence established by Koslov et al. may be applied for any elliptic operator. When convergence is achieved, we may obtain $(\overline{t},\overline{\text{u}})$

$=(\sigma(u)n,\ u)$ on Γ_i. By using straightforward computations, it can be established that the Dirichlet-to-Neumann scheme can be interpreted as a pre- conditioned Richardson procedure for the Cauchy-Steklov-Poincar´e equation. For this purpose, the Dirichlet-to-Neumann algorithm is rewritten, using the previous notations, as follows: Given λ_0, the following systems are solved for each $k \geq 0$

$$\begin{cases} -\mathrm{div}\ \sigma(u_N^{(k+1)}) = f & \text{in } \Omega, \\ (u_N^{(k+1)}) = \tilde{u} & on\ \Gamma_e, \\ u_N^{(k+1)} = \lambda^{(k)} & on\ \Gamma_i, \end{cases} \quad \text{and}$$

$$\begin{cases} -\mathrm{div}\ \sigma(u_D^{(k+1)}) = f & \text{in } \Omega, \\ \sigma(u_D^{(k+1)})n = \tilde{t} & on\ \Gamma_e, \\ \sigma(u_D^{(k+1)})n = \sigma(u_N^{(k+1)})n & on\ \Gamma_i. \end{cases}$$

The last equality $\sigma(u_D^{(k+1)})n=\sigma(u_N^{(k+1)})n$ on Γ_i can be written as

$$\sigma(u_D^{0(k+1)})n + \sigma(u_D^{*})n = \sigma(u_N^{0(k+1)})n + \sigma(u_N^{*})n.$$

Since $\sigma(u_D^{(k+1)})n=S_D\lambda^{(k+1)}$ and $\sigma(u_N^{0(k+1)})n=S_N\lambda^{(k)}$ on Γ_i it follows that $S_D\lambda^{(k+1)} = S_N\lambda^{(k)} + \chi,$ and therefore

$$\lambda^{(k+1)} = \lambda^{(k)} - S_D^{-1}(S\lambda^{(k)} - \chi).$$

We are thus left with a Richardson procedure for the Cauchy-Steklov-Poincare equation (8) with the operator SD as a preconditioner.

The Dual Cauchy-Steklov-Poincare equation

The Dual Steklov-Poincar´e (DSP) problem consists in introducing $\lambda=(\sigma(u)\ n)_{|\Gamma i}$ as the unknown rather than $u_{|\Gamma i}$. To write the Dual Steklov-Poincar´e equation (DSP) on Γi we use the same process which is used for the primal formulation. We consider again both Dirichlet and Neumann elliptic problems obtained by duplicating the solution u into a couple of vectors (u_N, u_D). The Cauchy problem (4) is then split into:

$$\begin{cases} -\mathrm{div}\ \sigma(u_D) = f & \text{in } \Omega \\ u_D = \tilde{u} & on\ \Gamma_e, \\ \sigma(u_D)n = \lambda & on\ \Gamma_i \end{cases} \quad \text{and} \quad \begin{cases} -\mathrm{div}\ \sigma(u_N) = f & \text{in } \Omega \\ t = \sigma(u_N)n = \tilde{t} & on\ \Gamma_e \\ \sigma(u_N) = \lambda & on\ \Gamma_i \end{cases} \quad (10)$$

The DSP equation is obtained by writing $u_D(\lambda)=u_N(\lambda)$ on Γ_i which is now the necessary and sufficient condition to solve the Cauchy problem. By splitting (u_D, u_N) in linear parts $(\overset{0}{u}_D, \overset{0}{u}_N)$ and remaining parts $(\overset{*}{u}_D, \overset{*}{u}_N)$, the relation above leads to

$$P(\lambda)=P_D(\lambda) - P_N(\lambda)=\psi \ on\ \Gamma_i \quad (11)$$

With $P_D(\lambda)=\overset{0}{u}_D(\lambda), P_N(\lambda)=\overset{0}{u}_N(\lambda)$, and $\psi=\overset{*}{u}_N-\overset{*}{u}_D$. We call the operator P Dual Steklov-Poincare operator sometimes called Poincare e-Steklov operator by the domain decomposition community. It is clearly equal to $S_D^{-1}-S_N^{-1}$.

Remark

When implementing this approach, one has to keep in mind that the solution of the second PDE system in (10) is defined up to a rigid body motion. In order to relieve this problem, one can devote a well-chosen part of Γ_c to set the Dirichlet boundary condition $\tilde{u}$. This approach does not alter the generality of the work presented.

Algorithmic Strategy

The approximation of the problems 8 and 11 leads to the ill-conditioned linear systems: $S_h\lambda_h=\chi_h$ for the SPP algorithm and $D_h\mu_h=\Psi_h$ for the SPD one. S_h and D_h are the discrete Schur complement matrices. These linear systems are ill-posed since they are inverse problems. The vectors λ_h and μ_h are respectively the discretized unknown displacement and stress on Γ_i. The Schur complement is too expensive to compute and ill-conditioned, an iterative procedure is henceforth in order. We use the GMRes algorithm, which is a popular iterative method for the solution of large linear systems. When starting vector λ_0 is zero, it generates a sequence of iterates $\lambda_1, \lambda_2, \ldots$ such that λ_k is in the Krylov subspaces $K_m(S_h, r_0)=span\{r_0, \ldots, S_{m-1}r_0\}$, where $r_0\chi_h - S_h\lambda_0$. Applying a Krylov method to $S_h\lambda_h=\chi_h$ has been shown to have a regularizing effect [22]. In fact the Krylov subspace K_m gives an approximation of the subspace generated by the m eigenvectors associated to the largest eigenvalues of S_h [22], thus GMRes iterate λ_k can be considered as an approximation of the truncated singular value decomposition solution [23].

The advantage of using the GMRES Algorithm is that it does not calculate the approximation of the solution at each iteration; only the final approximation λ_m is computed.

To terminate the iterations of the GMRES algorithm, we use the L-curve criterion. The L-curve is the graph obtained by connecting consecutive points in the sequence [22,23]:

$p_k=(ln(\lambda_k), ln(r_k)),\ k=1, 2, 3\ldots$

In [23] Hansen shows that for discrete ill-posed problems it turns out that the L-curve always has a characteristic L-shaped appearance with a distinct corner separating the vertical and the horizontal parts of the curve. Calvetti et al. [24] shows that the optimal stopping criterion for the GMRes algorithm applied to liner ill-conditoned system corresponds to the vertex of the L-curve.

In this way, the L-curve clearly displays the trade-off between minimizing the residual norm and the solution norm.

Contact Pressures Identification

The example concerns contact identification on an inaccessible contact area. The data of the inverse problem are generated by solving the finite elements discretization of the direct problem. The thicknesses of meshes used for that are finer then that we used for solving the inverse problem.

Domain Ω is a square plate $(1. \times 1.)$ with a circular hole $(R_1=0.20225)$, where a fixed rigid disc $(R_2=0.2)$ is placed. Figure 2 shows the geometry

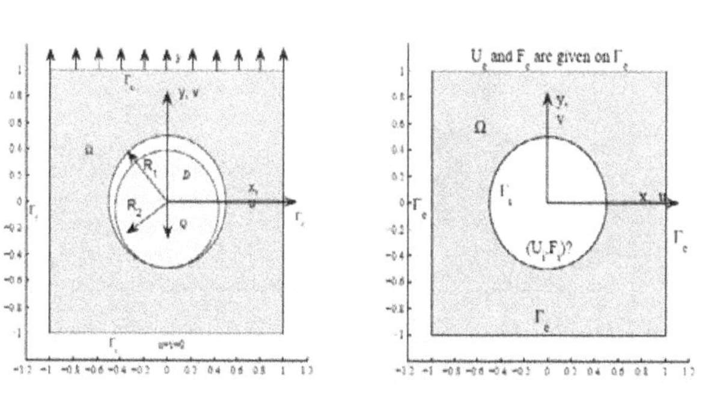

Figure 2: Geometry of the problem studied.

and boundary conditions applied to the plate. The mechanical properties of the plate are given in Table 1. When tractions are applied on the plate, it comes into contact with the lower part of the disc (Figure 2). The problem is to identify the contact pressure distribution and the displacements on the interface between the plate and the disc, by using over specified data provided for the external boundary. It should be noted here that the contact problem is nonlinear. However the data completion problem must be posed for that part of the domain where linear elasticity exists. The over specified data are generated by solving a direct problem using Hertz's analytical contact law. Here, we consider a frictionless contact so that only normal pressure is taken into account. Moreover, plane strain hypothesis is assumed (Figure 3).

Disc and plate	Aluminium
Modulus of elasticity	E=70000 M P A
Poison coefficient	v=0.31
Friction coefficient	μ=0
Load applied to the plate	*F=1e + 7 N/m*

Table 1: Mechanical characteristics of the plate and the disc.

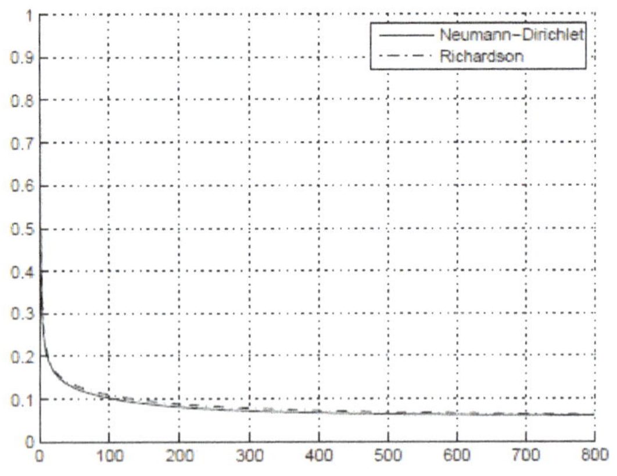

Figure 3: Illustration of the equivalence between Dirichlet-to-Neumann and the preconditioned Richardson procedure for the Cauchy-Steklov-Poincare equation.

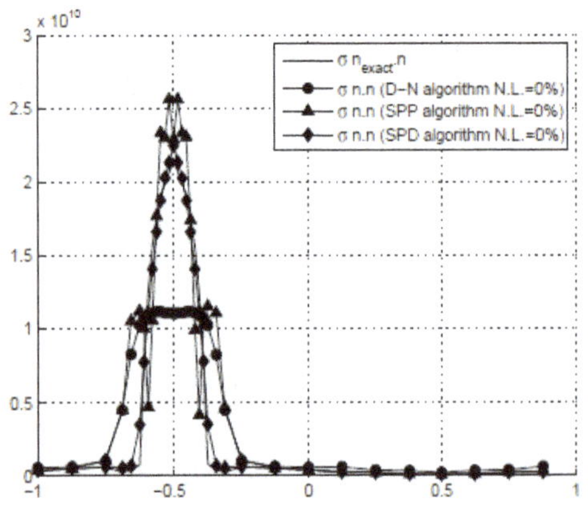

Figure 4: Reconstruction of the normal stress component distribution on the internal boundary of the plate for noise free (N. L.=0%)) data using DN, SPP and SPD algorithms.

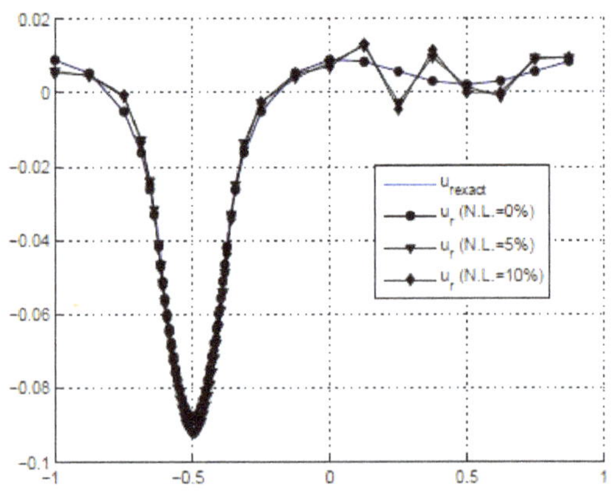

Figure 5: Dual Steklov-Poincare Solver: Reconstruction of the normal displacement component distribution on the internal boundary of the plate for noise free (N. L.=0%)) data and noisy (N. L.=5%, N.L;=10%) data.

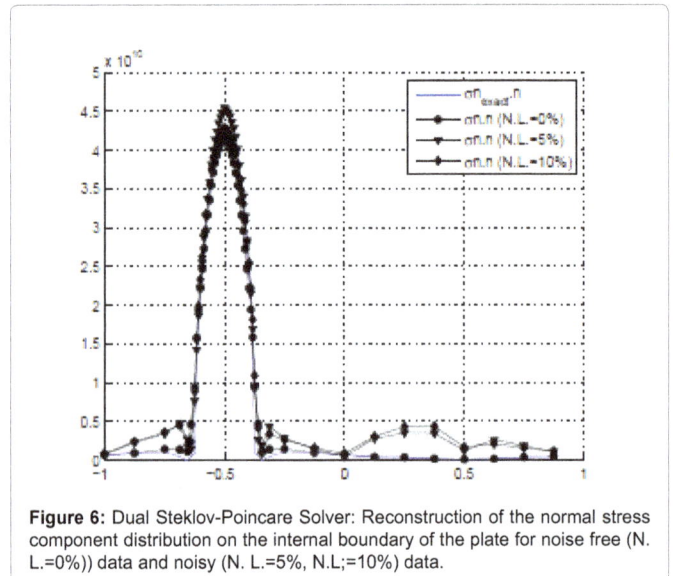

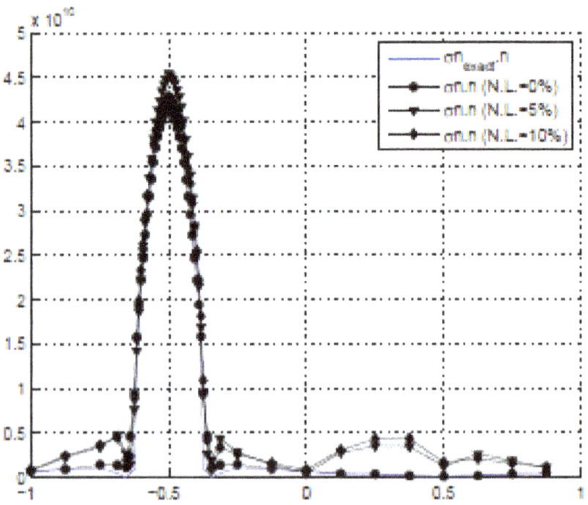

Figure 6: Dual Steklov-Poincare Solver: Reconstruction of the normal stress component distribution on the internal boundary of the plate for noise free (N. L.=0%)) data and noisy (N. L.=5%, N.L;=10%) data.

The finite elements discretization of the equations (8) and (11) leads to ill-posed linear systems. This was expected as they are inverse problems. Since the Schur complement is too expensive to compute, we use the GMRes algorithm described above.

The results obtained by solving the corresponding Cauchy problem are the normal stress components and the displacements field on Γ_i. Hence, the contact zone is the part of the boundary where the normal stress components are not null. When carrying out identification based on measurements, it must be kept in mind that measured data are subject to noise whose effects have to be studied. In this case, the data are synthetic, and therefore suffer from some errors (approximation error, roundoff error . . . etc). We added a noise generated by a MATLAB routine (randn) to the computational noise. The displacement measurements are polluted by a noise level at 5% and 10%.

Figures 4-6 depict the radial displacements u_r and normal stress σ.n reconstruction on the internal boundary of the plate with ND, SPP and SPD algorithms.

For the ND and SPP algorithms, the identification was carried out using over-specified data on the upper and lateral boundaries. It can be seen that the reconstructed stress does not approximate well the exact one: this is a consequence of the derivation operation. Although not presented here, we obtained an accurate numerical solution when the over-determined boundary Γ_c *and* Γ_i are complementary over $\partial\Omega$ *(i.e* $\partial\Omega=\Gamma_c \cup \Gamma_i$*)*. However, in practice, we cannot have stress measurements on the dirichlet boundary real-world problems.

Figure 3 shows how the Dirichlet-to-Neumann algorithm is equivalent to a pre-conditioned Richardson procedure for the Cauchy-Steklov-Poincar´e equation as we have mentioned in section 4.

For SPD algorithm, the identification was carried out using over specified data only on the upper boundary which was discretized onto 65 nodes. We note that the numerical solution presented in Figures 5 and 6 approaches the exact solution as the noise level decreases and that the numerical results obtained by the SPD method are accurate and convergent with respect to decreasing the levels of noise added into the input data. We note that in order to preserve the stability of the method it is necessary to use a stopping criterion to cease the iterative process before the point where the errors start to increase due to the added noise. Figure 7 shows the error $e_2 = \frac{\|(u_{calc} - u_{exact})\|}{\|u_{exact}\|}$ as a function of the number of iterations. It can be seen that the error e_2 decreases rapidly over the first few iterations but the rate of convergence decreases as the number of iterations increases. Thus the iterative process has to be stopped at a point where the error e_2, obtained by comparing the calculated solution with the exact solution, stops decreasing. Moreover, if the number of iterations is large, due to the accumulation of the numerical noise, the error e2 starts to increase, this property is called semi-convergence.

As expected, displacements reconstruction is better than that for the stresses, particularly when the data are noisy. The reason is that the stresses are homogeneous with the displacements gradient and it is well known that the derivation is an ill-posed operation (the influence of noise is considerable). The identification is very satisfactory for free noisy data. For noisy data, the contact zone is well localized and the contact pressures are recovered correctly. However, some fairly significant oscillations appear on the free boundary. The table 2 provides relative error on displacements and number of iterations as

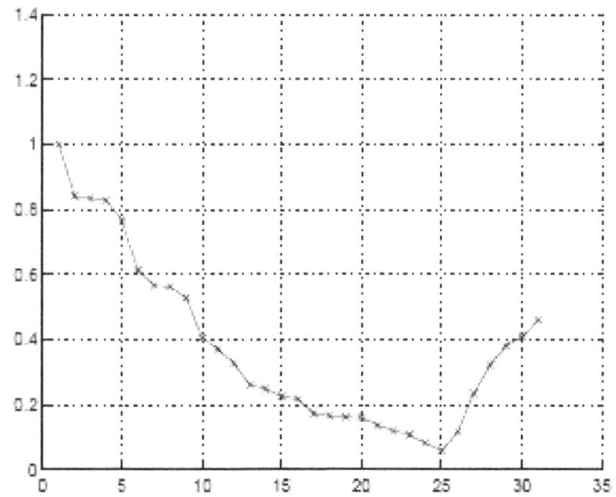

Figure 7: Dual Steklov-Poincare Solver: Rate of convergence for noisy (N. L.=5%) data.

dof on Γ_c	*kiter*	$e_2(u)$
17	34	0.0109
33	34	0.0109
65	37	0.0104
129	37	0.0105
257	35	0.0108

Table 2: Effect of the number of degrees of freedom on Γ_c.

Γ_c	γ_1	γ_2	γ_3	γ_4	γ_5
kiter	34	26	34	26	26
$e_2(u)$	0.0093	0.0882	0.0109	0.0692	0.0698

Table 3: The accuracy and the speed of the SPD algorithm with respect to length and position of Γ_c.

a function of the number of degrees of freedom on Γ_c. Γ_c is the upper boundary of the plate (boundary subjected to the traction load).

We discretize Γ_c respectively to 17, 33, 65, 129 and 257 nodes, the table 2 provides for each of the selected meshes, the number of iterations necessary to reach the solution for a free noisy data. The main conclusion we draw to is that provided that the data information on Γ_c is sufficiently rich, the performances of the SPD algorithm are weakly sensitive to the amount of that information. Indeed, the quality of reconstruction process remains almost unchanged; no improvement is achieved by adding finite element nodes on Γ_c.

The following point to investigate is how the SPD solver depends on the geometrical characteristics of Γ_c: length and position. For this purpose we run five numerical experiments for different types of over specified boundary's (γ_1: the upper and lateral boundaries of the plate, γ_2: the upper boundary, γ_3: the middle half of the upper boundary, γ_4: the right half of the upper boundary and γ_5: the left half of the upper boundary).

Two relevant indicators are recoding in the table 3: the number of iterations to reach the solution and the relative error on displacements. We note that the quality of reconstruction clearly suffer from decreasing the measure of Γ_c, besides keeping the size of Γ_c unchanged the position of it is very important regarding the quality of the information that can contain; we can see that γ_3 and γ_5 lead to nearly the same identification but better than that obtained with γ_2 (although of the same size) this is due to the information concerning the singularity in the stress tensor at the corner which is directly contained in γ_4 and γ_5. The SPD method appears to be powerful and economic in comparison to the SPP one. In fact the primal schur complement matrix results from the static condensation on Γ_i of the rigidity matrix, so it has the same distribution of singular values and thus inherits its initial ill-posedness, whereas the dual complement schur matrix can be seen as the inverse of the primal one so its condition number is obviously better. Therefore SPD is suitable for dealing with more complex data completion problems (e.g. 3D).

Dual Steklov-Poincare Algorithm for the resolution of the Cauchy problem in 3D linear elasticity

We investigate now the performance and accuracy of The SPD method through numerical three-dimensional examples where the measured data are extracted from the results of the direct problems: the identification of the indentation pressure of a heterogeneous solid and boundary data completion at the interface between two bonded elastic bodies.

Identification of indentation pressure on a two-layer solid

The identification of the indentation pressure on a composite solid and the stress on the interface is a major challenge in the design of composites. The example to be dealt with concerns the indentation of a two-layer solid composed of two different materials (steel and titanium) (Figure 8). By symmetry, the problem is reduced to that of a quarter of the solid. The problem is stated as follows:

- Over specified data: displacements and surface traction are known on the rectangular area denoted by Γ_m.
- Boundary Γ_b gathering the four side faces: two have Dirichlet boundary conditions of symmetry, the two other are free from surface tractions.
- Boundary Γ_u where the data are unknown includes the remaining faces (the bottom face and the remaining area of the top face).
- Elastic parameters: Layer 1 is steel with $E_{s=}2.1 \times 1011P$ a, ν_s=0.34; Layer 2 is titanium with $E_{ti=}1.05 \times 1011P$ a, $\nu_{ti=}0.29$.

Our aim is to identify the indentation area and the pressure distribution on the solid boundaries including the interface and the cross section by using the over specified data measured on Γ_m. The measured data are generated numerically by solving the direct problem where the pressurized area (disk with radius R_p) and the expression of the pressure distribution as a function of the radius r are given by:

$$\begin{cases} p=\dfrac{-8\mu\sqrt{R_p^2-r^2}}{1+x_i} \quad with\ x_i=\dfrac{\lambda+3\mu}{\lambda+\mu}, \\ \lambda=\dfrac{V_a E_a}{(1-2a)(1+V_a)}, and\ \mu=\dfrac{E_a}{2(1+V_a)}, \end{cases}$$

The direct and inverse problems have been carried out using MATLAB software [25]. The geometry and mesh of the problem are displayed on Figure 8. The number of nodes on Γ_u is 347, whereas the number of nodes on Γ_m is 142. Hence, there are 345 × 6=2082 variables to identify for only 142 × 6=852 over specified known data. The measured data used are altered with noise that is a function of the mesh refinement, since it is extracted from finite element results of the direct indentation problem. But, in order to assess the efficiency of the method we add also some noise to the measured data.

Figures 9-17 show various fields obtained by the exact and the identified solutions. Figure 9 show the map of the displacement norm || u ||L_2 obtained from the exact and the identified solutions with noisy data. Figure 10 show the map of the Von Mises equivalent stress obtained from the exact and the identified solutions. Figures 11 and 12 show the profiles of the Von Mises equivalent stress σ_{eq} and the stress component σ_{zz} on edges 1 and 2 and on the diagonal line; good reconstruction of the exact fields can be seen.

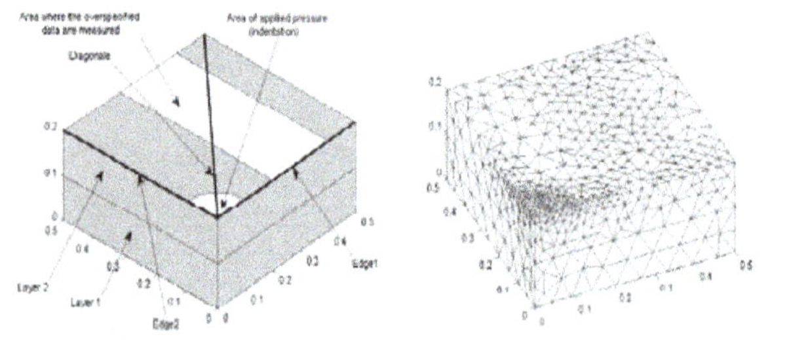

Figure 8: The geometry and the mesh of the solid.

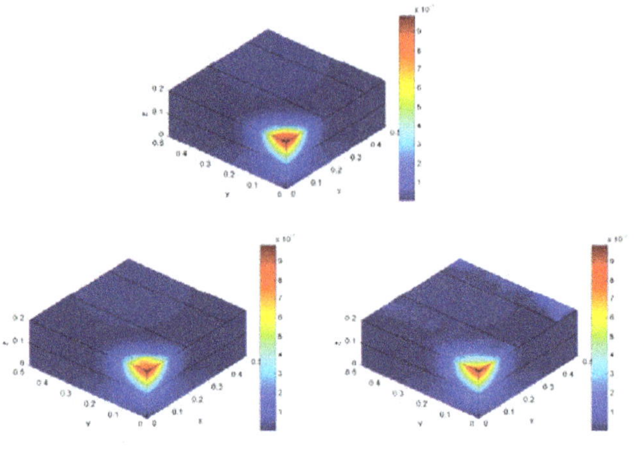

Figure 9: Top: exact displacement field-Left: Identified displacement field. Noise 0%. Bruit 0%-Right: Identified displacement field. Noise 5%.

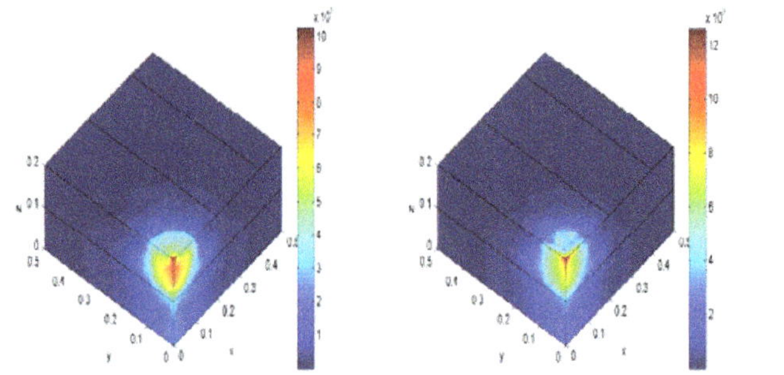

Figure 10: Left: Exact Von Mises equivalent stress. Right: Identified Von Mises equivalent stress. Noise 0%.

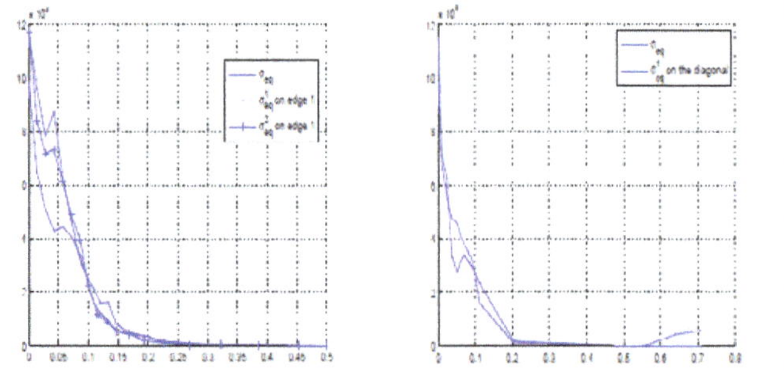

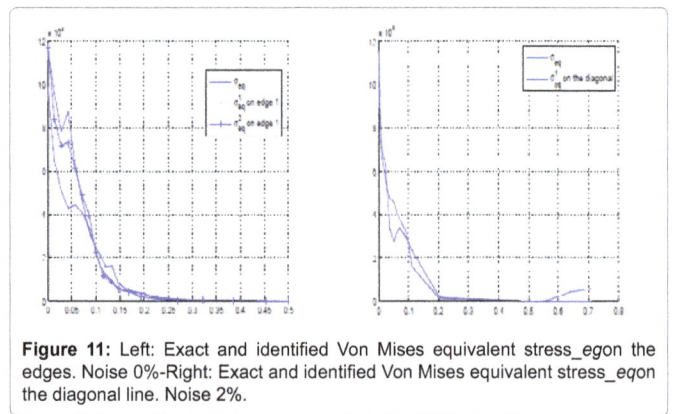

Figure 11: Left: Exact and identified Von Mises equivalent stress_*eq*on the edges. Noise 0%-Right: Exact and identified Von Mises equivalent stress_*eq*on the diagonal line. Noise 2%.

Figures 13 and 14 show the exact and identified displacement component u_z and Von Mises equivalent stress distributions on a diagonal section of the solid. These show good agreement with the exact fields for the displacement component u_z and Von Mises equivalent stress. The stress discontinuity at the interface of the two materials is well recovered. Figure 15 show maps of the exact and identified displacement component u_z distributed on the top face of the solid.

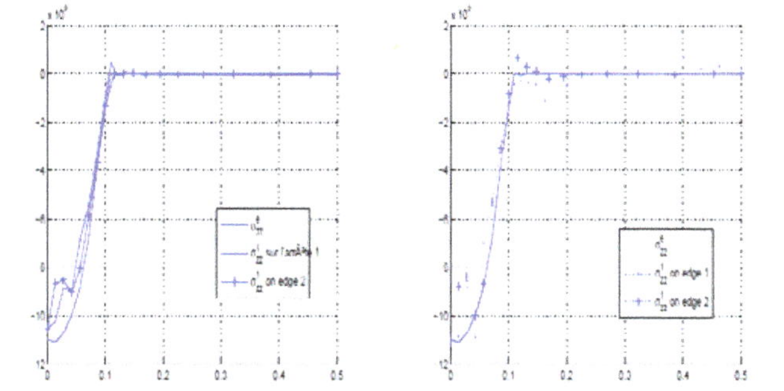

Figure 12: Left: Exact and identified stress_zzon the edges. Noise 0%-Right: Exact and identified Stress_zzon the edges. Noise 5%.

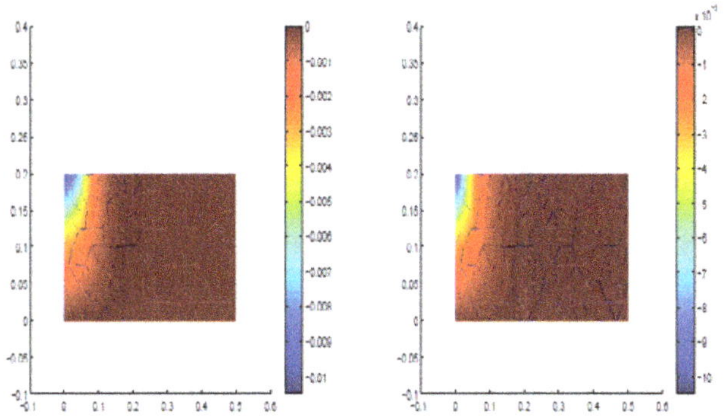

Figure 13: Left: Exact displacement *uz*on the diagonal section. Right: Identified displacement *uz*on the diagonal section noise 5%.

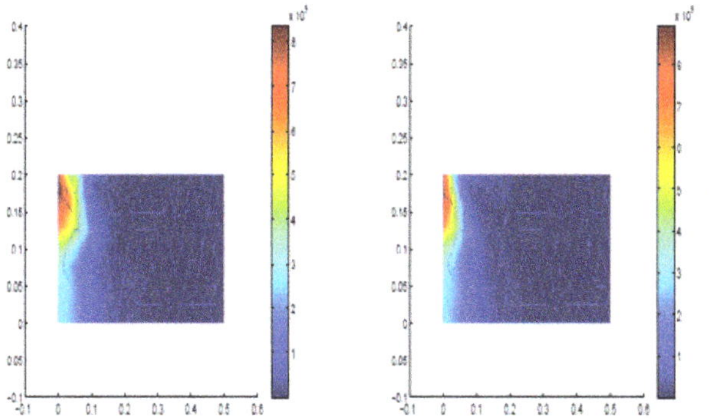

Figure 14: Left: Exact Von Mises equivalent stress on the diagonal section. Right: Identified Von Mises equivalent stress on the diagonal section. Noise 5%.

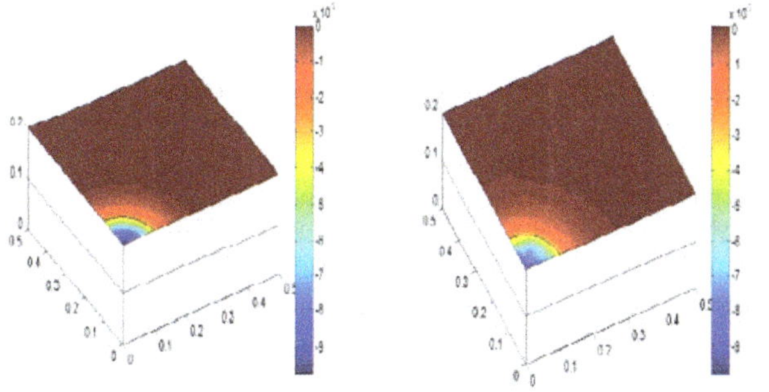

Figure 15: Left: Exact displacement uzon the topface. Right: Identified displacement uzon the top face. noise 0%.

The displacement is well recovered.

Figures 16 and 17 show the maps of the exact and identified stress component σ_{zz} and the norm of the tangential stress vector σ_t defined by the equation $\sigma_t = \sqrt{(\sigma_{xz}^2 + \sigma_{yz}^2)}$, distributed on the interface between the two materials of the solid. Here, too, a good correlation exists between the exact and identified results with reasonable accuracy, even for the interface tangential stress.

Boundary data identification of three-dimensional bonded structure

This application considers a bonded structure i.e. two bodies bonded along their common interface, by a thin adhesive layer. In the simplified models, the adhesive disappears, replaced by an interface transmission condition [26]. The problem has two planes of symmetry. Hence, only one quarter is modeled. Figure 18 shows the deformed shape and the distribution of the displacement field issued from the direct problem, defined as follows:

- The first solid is a cylinder with radius r=0.248 m and a length L=3 m. The material is steel with a Young's Modulus E_s=2.1 × 1011Pa and a Poisson coefficient ν_s=0.34.
- The second solid is a rectangular box of dimension 1 × 1 × 2 m, which contains a cylindrical hole of radius 0.250 m. The

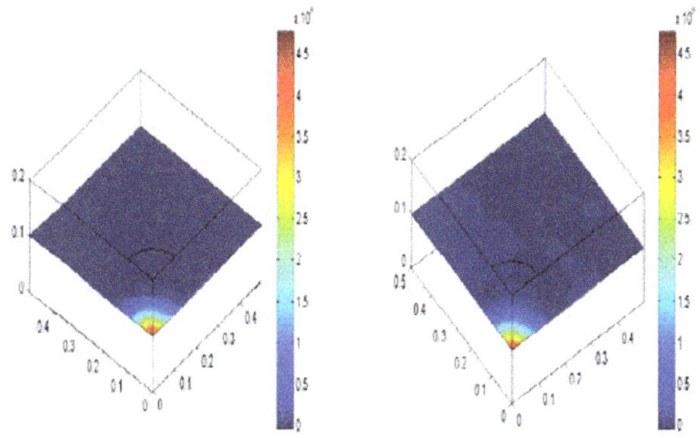

Figure 16: σ_{zz} on the interface.-Right :Identied stress-σ_{zz} on the interface. noise 5%.

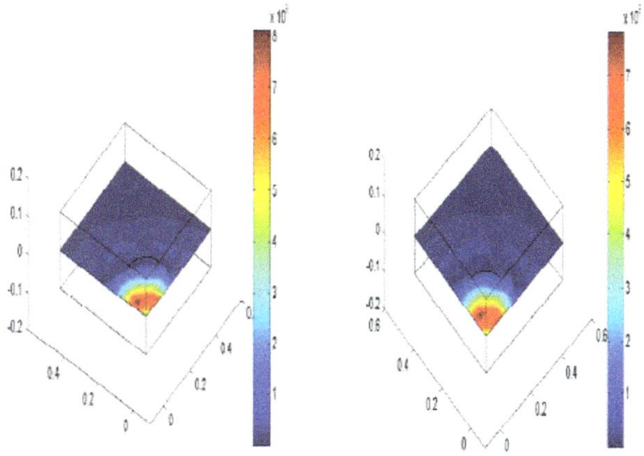

Figure 17: Left: Exact tangential stresson the interface-Right: Identified tangential stresson the interface. Noise 0%.

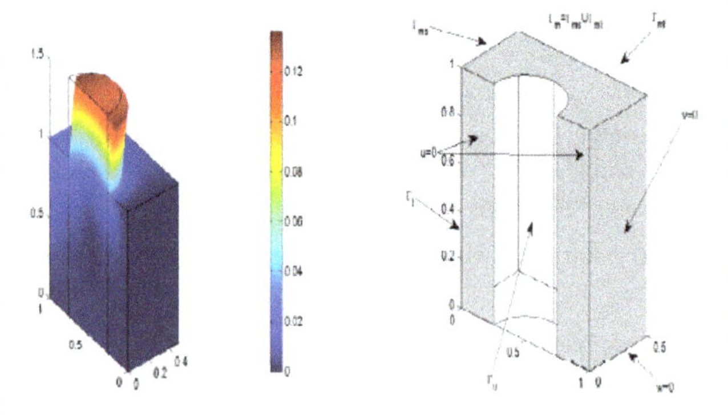

Figure 18: Left: the geometry used in the identification process. Right: displacement field distribution in the direct problem.

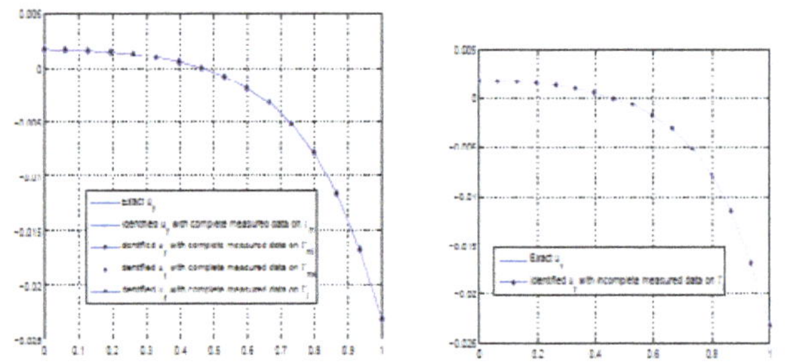

Figure 19: Sensitivity of the identified displacement ***uy*** along the edge $x=0$ and $y=0.25$, Left: to the localization of the over specified data. Right: to the amount of over specified **data**.

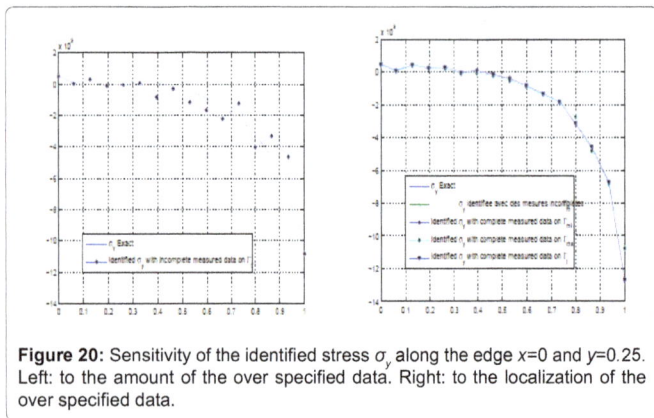

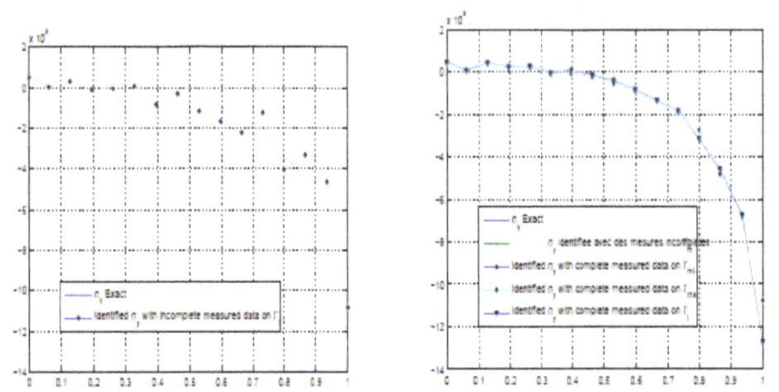

Figure 20: Sensitivity of the identified stress σ_y along the edge $x=0$ and $y=0.25$. Left: to the amount of the over specified data. Right: to the localization of the over specified data.

material is aluminum with a Young's Modulus $E_a=7 \times 1010$Pa and a Poisson coefficient $\nu_a=0.27$.

- The face y=0 is clamped: u=v=w=0.
- The face x=0 is a plane of symmetry and is fixed in the x-direction: u=0.
- The face z=0 is a plane of symmetry and is fixed in the z-direction: w=0.
- On the circular face of the cylinder at z=1.5 m, a displacement is prescribed in the y-direction: $D_{imp}=-0.1$ m.

A static finite element analysis was carried out using finite elements with the MATLAB Software. The result is displayed in Figure 18. Measured data (displacements and forces) are extracted on Γ_m for use in the identification problem. Figure 18 shows the geometry used in the identification problem: the cylinder is ignored. The mesh used in this case has 1203 nodes, 4514 elements, 313 nodes on Γ_u. We identify the stress and displacements using different localizations of the over-specified data Figure 18:

- Γ_{ms}: the top face of the solid with 82 nodes.
- Γ_{ml}: the lateral side of the solid with 225 nodes.
- Γ_l: the left lateral side of the solid with 118 nodes.
- Γ_{ml}: the boundary defined by $\Gamma_m=\Gamma_{ms} \cup \Gamma_{ml}$ with 293 nodes.

To stress the efficiency of the SPD algorithm we use also incomplete data: only tangential displacements on the over specified boundaries are used for the identification problem, the normal displacement is left unknown.

Figures 19 and 20 show the identified displacement u_y and stress σ_{yy} along the edge defined by x=0 and y=0.25. We can see that the displacements are perfectly identified even when we use few and incomplete over specified data. The identified stresses are as well sensitive to the quantity of over specified data as to their localization; however, an acceptable identification remains possible.

Conclusion

In this work we presented three numerical methods for solving the Cauchy problem in the framework of linear elasticity. The methods proposed were applied in practical situations taken from engineering mechanics: contact pressure recovery, identification of the indentation pressure of a heterogeneous solid and boundary data completion in a bonded structure. The numerical results also suggest that the SPD algorithm is an accurate and reliable numerical technique for the identification of variables in elasticity in both two and three-dimensional domains even if the area of the measured data is smaller than the area where the data are identified. The application of the methods to identify material properties and shape of the boundaries will be the subject of a forthcoming paper.

References

1. Hadamard J (1953), Lectures on Cauchy's problem in linear partial differential equation. Dover, New York.
2. Bonnet M, Constantinescu A (2005) Inverse problems in elasticity, Inverse problems 21: R1-R50.
3. Koya T, Yeih W, Mura T (1993). An inverse problem in elasticity with partially over specified boundary conditions. Transactions ASME Journal of Applied Mechanics 60: 601-606.
4. Marin L, Elliot L, Ingham D, Lesnic D (2001) Boundary element method for the Cauchy problem in linear elasticity. Engineering Analysis with Boundary Elements 25: 783-793.
5. Marin L, Lesnic D (2002) Boundary element solution for the Cauchy problem in linear elasticity using singular value decomposition. Comput.Methods Appl Mech Engrg 19: 3257-3270.
6. Huang C, Shih W (1999) An inverse problem in estimating interfacial crack in biomaterials by boundary element technique, International Journal for Numerical Methods in Engineering 45(11): 1547-1567.
7. Weikl W, Andra H, Schnack E (2001) An alternating iterative algorithm for the reconstruction of internal cracks in 3D solid body. Inverse Problems 17: 1957-1975.
8. Kadri ML, Ben Abdallah J, Baranger TN (2011) Identification of internal cracks in a three-dimentional body via steklov-Poincar approaches, Comptes Rendus M´ecanique 339: 674-681.

9. Baranger TN, Andrieux S (2010) Constitutive law gap functionals to solve Cauchy problem for a linear elliptic PDE: a review.

10. Azaiez M, Ben Belgacem F, El Fekih H (2005) On Cauchy's problem: II Completion, regularization and approximation. Inverse Problems 22: 1307-1336.

11. Ben Belgacem F, El Fekih H (2006) On Cauchy's problem: I A variational Stecklov-Poincar´e theory. Inverse Problems 21: 1915-1936.

12. Andrieux S, Baranger TN (2006) Solving Cauchy problems by minimizing an energy-like functional. Inverse Problems 22: 115-133.

13. Quarteroni A, Valli A (1999) Domain Decomposition Methods for Partial Differential Equations. Oxford University Press, Oxford.

14. Przemieniecki J (1985) Theory of Matrix Structural Analysis. Dover, New York.

15. Ben Abdallah J (2007) A conjugate gradient type method for the Stecklov-Poincar´e formulation of the Cauchy-Poisson Problem. International Journal of Applied Mathematics and Mechanics 3: 27-40.

16. Andrieux S, Baranger TN (2008) An energy error-based method for the resolution of the Cauchy problem in 3D linear elasticity. Comput Methods Appl Mech Engrg 197: 902-920.

17. Andrieux S, Baranger T, Ben Abda A (2005) Data completion via an energy error functional. C. M´ecanique 333: 171-177.

18. Ben Abda A, Ben Saad I, Hassine M (2009) Data completion for the Stokes system. C. R. M´ecanique 337: 703-708.

19. Koslov V, Maz'ya VA (1992) Fomin, An iterative method for solving the Cauchy problem for elliptic Equations. Comput Meth Math Phys 31: 45-52.

20. Ben Fatma R, Aza¨ıez M, Ben Abda A, Gmati N (2007) Missing boundary data recovering for the Hel-moltz problem. C. R. M´ecanique 335: 787-792.

21. Hanke M (1995) Conjugate Gradient type method for ill-posed problems. Pitman Research Notes in Math-ematics. 327.

22. Hansen P (1998) Rank-deficient and discrete ill-posed problems. SIAM, Philadelphia.

23. Calvetti D, Lewis B, Reichel L (2002) GMRes, L-curves, and Discrete ill-posed problems. BIT 42: 44-65.

24. Koko J (2008) Convergence analysis of optimization-based domain decomposition methods for a bonded structure. Applied Numerical Mathematics 58: 69-87.

Determination of Temperature Distribution for Porous Fin with Temperature-Dependent Heat Generation by Homotopy Analysis Method

Hoshyar HA[1,2]*, Ganji DD[2] and Abbasi M[2]

[1]*Department of Mechanical Engineering, Bbabol Branch, Emam Sadegh University, Babol, Iran*
[2]*Department of Mechanical Engineering, Sari Branch, Islamic Azad University, Sari, Iran*

Abstract

In this study, highly accurate analytical methods, Homotopy Analysis Method (HAM), is applied for predicting the temperature distribution in a porous fin with temperature dependent internal heat generation. The heat transfer through porous media is simulated using passage velocity from the Darcy's model. It has been attempted to show the capabilities and wide-range applications of the Homotopy Analysis Method in comparison with a type of numerical analysis as Boundary Value Problem (BVP) in solving this problem. The results show that the HAM is an attractive method in solving this problem.

Keywords: Homotopy analysis method; Porous fin; Temperature-dependent heat generation

Introduction

Fins are frequently used in many heat transfer applications to improve performance. In the other hand, for many years, High rate of heat transfer with reduced size and cost of fins are main targets for a number of engineering applications such as heat exchangers, economizers, super heaters, conventional furnaces, gas turbines, etc. Some engineering applications such as airplane and motorcycle also require lighter fin with higher rate of heat transfer. Increasing the heat transfer mainly depend on heat transfer coefficient (h), surface area available and the temperature difference between surface and surrounding fluid. However, this requirement is often justified by the high cost of the high-thermal-conductivity metals, that cost of high thermal conductivity metals is also high. fin is porous to allow the flow of infiltrate through it. Extensive research has been done in this area and many references are available especially for heat transfer in porous fins. Described below are a few papers relevant to the study described herein. The theoretical study of MHD has been a subject of great interest due to its widespread applications, such as plasma studies, petroleum industries, MHD power generators, cooling of nuclear reactors, the boundary layer control in aerodynamics, and crystal growth. For instance, MHD induced in rockets can improve heat transfer through porous fins, located on rocket surface. On the effect of MHD flow, although there are many studies regarding the free convection regime, there are only a few regarding the mixed convection regime. Chamkha et al. [1] studied the effects of localized heating (cooling), suction (injection), buoyancy forces, and magnetic field for the mixed convection flow on a heated vertical plate. Aldoss et al. [2] investigated the effect of MHD on heat transfer from a circular cylinder. Nonlinear problems and phenomena play an important role in applied mathematics, physics, engineering and other branches of science specially some heat transfer equations. Except for a limited number of these problems, most of them do not have precise analytical solutions. Therefore, these nonlinear equations should be solved using approximation methods. Perturbation techniques are too strongly dependent upon the so-called "small parameters" [3]. Other many different methods have introduced to solve nonlinear equation such as the δ-expansion method [4], Adomian's decomposition method [4], Homotopy Perturbation Method (HPM) [5-11] and Variational Iteration Method (VIM) [12-21]. Homotopy analysis method is another techniques), which was introduced by Liao. This method has been successfully applied to solve many types of nonlinear problem [22-28].

In this work, we have applied Homotopy Analysis Method to find the approximate solutions of nonlinear differential equations governing on porous fin with temperature dependent internal heat generation. Results demonstrate that HAM is simple and accuracy compared with the BVP as a numerical method. Also, it is found that this method is powerful mathematical tools and that they can be applied to a large class of linear and nonlinear problems arising in different fields of science and engineering.

Analysis

As shown in Figure 1, a rectangular porous fin profile is considered. The dimensions of this fin are length L, width wand thickness t. The cross section area of the fin is constant and the fin has temperature-dependent internal heat generation. Also, the heat loss from the tip of the fin compared with the top and bottom surfaces of the fin is assumed to be negligible. Since the transverse Biot number should be small for the fin to be effective [29], the temperature variation in the transverse direction are neglected. Thus heat conduction is assumed to occur solely in the longitudinal direction [30]. Energy balance can be written as:

$$q(x)-q(x+\Delta x)+q^{*}.A\,\Delta x=\dot{m}\,c_{p}\left[T(x)-T_{\infty}\right]+h(p.\Delta x)\left[T(x)-T_{\infty}\right] \quad (1)$$

The mass flow rate of the fluid passing through the porous material can be written as:

$$\dot{m}=\rho\,\overline{\vartheta_{w}}\,\Delta x\,w \quad (2)$$

The value of $\overline{\vartheta}_w$ should be estimated from the consideration of the flow in the porous medium. From the Darcy's model we have:

***Corresponding author:** Hoshyar HA, Department of Mechanical Engineering, Sari Branch, Islamic Azad University, Sari, Iran
E-mail: hoshyarali@ymail.com

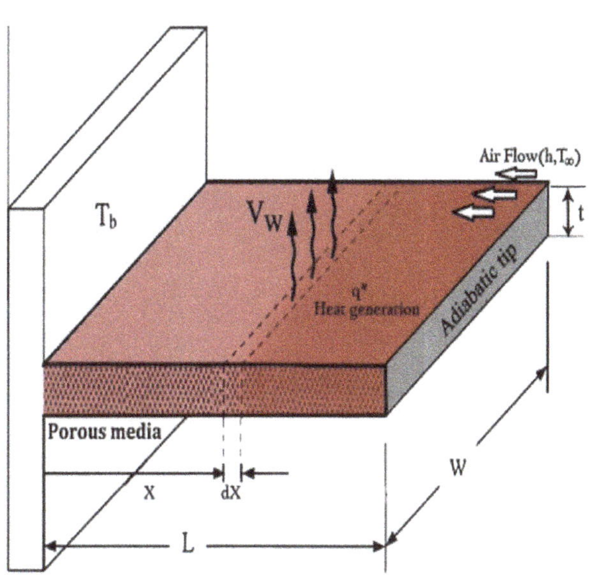

Figure 1: Schematic diagram for the problem under consideration.

$$\overline{\vartheta_w} = \frac{gk\beta}{\upsilon}\left[T(x)\text{-}T\infty\right] \tag{3}$$

Substitutions of Eqs. (2) and (3)into Eq. (1) yields:

$$\frac{q(x)-q(x+\Delta x)}{\Delta x}+q^*.A=\frac{\rho c_p g k\beta w}{\Delta x}\left[T(x)-T_\infty\right]^2+hp\left[T(x)-T_\infty\right] \tag{4}$$

As, $\Delta x \to 0$ Eq.(4)becomes

$$\frac{dq}{dx}+q^*.A=\frac{\rho c_p g k\beta w}{\Delta x}\left[T(x)-T_\infty\right]^2+hp\left[T(x)-T_\infty\right] \tag{5}$$

Also from Fourier‘s Law of conduction: $q=-k_{eff}\,A\dfrac{dT}{dx}$ (6)

Where A is the cross-sectional area of the fin A=$w.t$ and k_{eff} is the effective thermal conductivity of the porous fin that can be obtained from following equation:

$$k_{eff}=\varphi.k_f+(1-\varphi)k_s \tag{7}$$

Where φ is the porosity of the porous fin. Substitution Eq. (6) into Eq. (5) leads to:

$$\frac{d^2T}{dx^2}-\frac{\rho c_p g k\beta w}{t\,k_{eff}\upsilon}\left[T(x)-T_\infty\right]^2+\frac{hp}{k_{eff}\,A}\left[T(x)-T_\infty\right]+\frac{q^*}{k_{eff}}=0 \tag{8}$$

It is assumed that heat generation in the fin varies with temperature as Eq.(9) [15]:

$$q^*=q^*_\infty\left[1+\varepsilon(T-T_\infty)\right] \tag{9}$$

Where q^*_∞ is the internal heat generation at temperature T_∞. For simplifying the above equations some dimensionless parameters are introduced as follows:

$$\theta=\frac{(T-T_\infty)}{(T_b-T_\infty)},\quad X=\frac{x}{L},\quad M^2=\frac{hpL^2}{k_0A},\quad Sh=\frac{Da\,x\,Ra}{kr}\left(\frac{L}{t}\right)^2$$
$$G=\frac{q^*_\infty}{h\,p(T_b-T_\infty)},\quad \varepsilon g=\varepsilon(T_b-T_\infty) \tag{10}$$

Where Sh is a porous parameter that indicates the effect of the permeability of the porous medium as well as buoyancy effect so higher value of Sh indicates higher permeability of the porous medium or higher buoyancy forces. M is a convection parameter that indicates the effect of surface connecting of the fin. Finally, Eq.(8) can be rewritten as:

$$\frac{d^2\theta}{dX^2}-M^2\theta+M^2G(1+\varepsilon g\,\theta)-Sh\theta^2=0 \tag{11}$$

In this research we study finite-length fin with insulated tip. For this case, the fin tip is insulated so that there will not be any heat transfer at the insulated tip and boundary condition will be,

$$\theta(1)=1\quad,\quad\theta'(0)=0 \tag{12}$$

Homotopy Analysis Method (HAM)

For HAM solutions, we choose the initial guess and auxiliary linear operator in the following form:

$$\theta_0(x)=1 \tag{13}$$

$$L(\theta)=\theta'' \tag{14}$$

$$L(c_1x+c_2)=0 \tag{15}$$

Where $c_i(i=1,2)$ are constants. Let $P\in[0,1]$ denotes the embedding parameter and η indicates non-zero auxiliary parameters. We then construct the following equations:

Zeroth-order deformation equations

$$(1-P)L[\theta(x,p)-\theta_0(x)]=p\hbar H(x)N[\theta(x,p)] \tag{16}$$

$$\theta(0;p)=1;\qquad \theta'(1;p)=0 \tag{17}$$

For p=0 and p=1 we have: $\theta(x;0)=\theta_0(x)\quad \theta(x;1)=\theta(x)$

When p increases from 0 to 1 then θ(x;p) varies from $\theta_0(x)$ to $\theta(x)$. By Taylor's theorem and using Eqs. (19), θ(x;p) can be expanded in a power series of p as follows:

$$\theta(x;p)=\theta_0(x)+\sum_{m-1}^{\infty}\theta_m(x)p^m\quad,\quad \theta_m(x)=\frac{1}{m!}\frac{\partial^m(\theta(x;p))}{\partial p^m}\bigg|_{p=0} \tag{20}$$

In which η is chosen in such a way that this series is convergent at p=1; therefore we have through Eq. (20) that,

$$\theta(x)=\theta_0(x)+\sum_{m-1}^{\infty}\theta_m(x), \tag{21}$$

mth -order deformation equations

$$L\left[\theta_n(X)-\chi_n\theta_{n-1}(X)\right]=\hbar H(X)R_n(X) \tag{22}$$

$$\theta(0;p)=0;\qquad \theta'(1;p)=0 \tag{23}$$

Where

$$R_n(X)==\frac{d^2\theta(X;p)}{dX^2}-M^2\theta_{n-1}+M^2G(1+\varepsilon g\,\theta_{n-1})-Sh\sum_{k=0}^{n-1}\theta_{n-1-k}\,\theta_k=0 \tag{24}$$

Now we determine the convergence of the result, the differential equation, and the auxiliary function according to the solution expression. So let us assume:

$$H(X)=1 \tag{25}$$

We have found the answer by maple analytic solution device. For three deformation of the solution are presented below

$$\theta_1(X)=\frac{1}{2}\hbar\left(-Sh-M^2+M^2G(1+\varepsilon g)\right)X^2+\frac{1}{2}\hbar Sh+\frac{1}{2}\hbar M^2-\frac{1}{2}\hbar M^2G-\frac{1}{2}\hbar M^2G\,\varepsilon g \qquad (26)$$

$$\begin{aligned}\theta_2(X)=\;&0.041667\hbar^2X^4\left(-Sh-M^2+M^2G+M^2G\varepsilon g\right)\left(-2Sh-M^2+M^2G\varepsilon g\right)\\&+0.5\big[-\hbar Sh-\hbar M^2+\hbar M^2A+1.5Sh\hbar^2M^2A\varepsilon g-Sh\hbar^2-\hbar^2M^2-\hbar^2Sh^2-0.5M^4\hbar^2\\&+\hbar^2M^2A-1.5Sh\hbar^2M^2+0.5M^4\hbar^2A+Sh\hbar^2M^2G+M^4\hbar^2A\varepsilon g+\hbar M^2G\varepsilon g\\&-0.5M^4A^2\varepsilon g\hbar^2-0.5M^4G^2\varepsilon g^2\hbar^2+\hbar^2M^2G\varepsilon g\big]X^2+0.20833M^4\hbar^2\\&-0.5\hbar^2M^2G+0.625Sh\hbar^2M^2-0.20833M^4\hbar^2G+0.5\hbar M^2+0.20833M^4G^2\varepsilon g^2\hbar^2\\&-0.41667SH\hbar^2M^2G-0.41667M^4\hbar^2G\varepsilon g+0.20833M^4G^2\varepsilon g\hbar^2-0.5\hbar^2M^2G\varepsilon g\\&-0.5\hbar M^2G\varepsilon g-0.5\hbar M^2G-0.62500Sh\hbar^2M^2G\varepsilon g+0.5Sh\hbar^2+0.5\hbar^2M^2\\&+0.5\hbar Sh+0.41667\hbar^2Sh^2\end{aligned} \qquad (27)$$

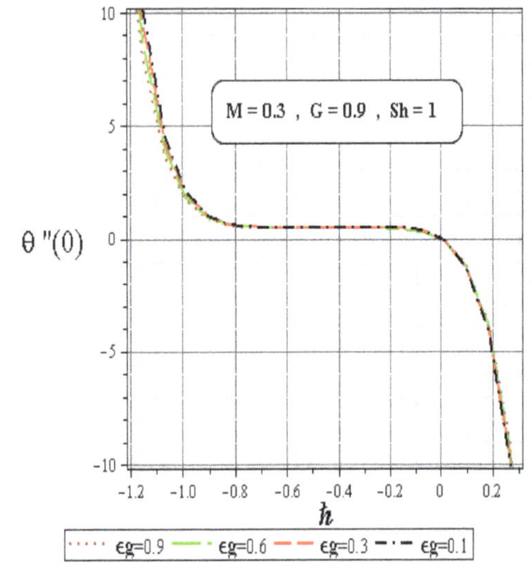

Figure 2: The $\hbar$ -validity for X=0 and different value of *H*.

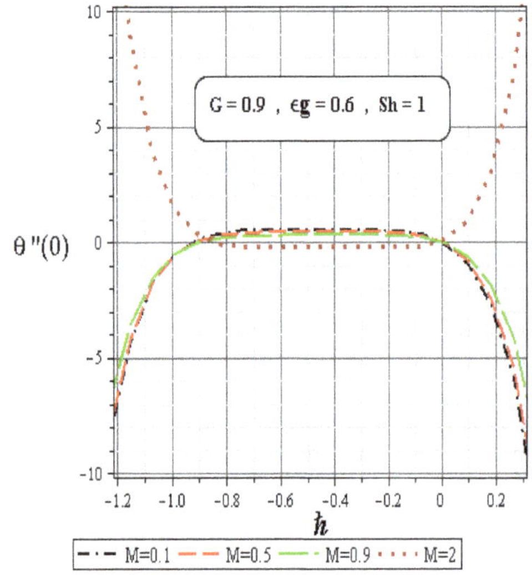

Figure 3: The $\hbar$ - validity for X=0 and different value of *M*.

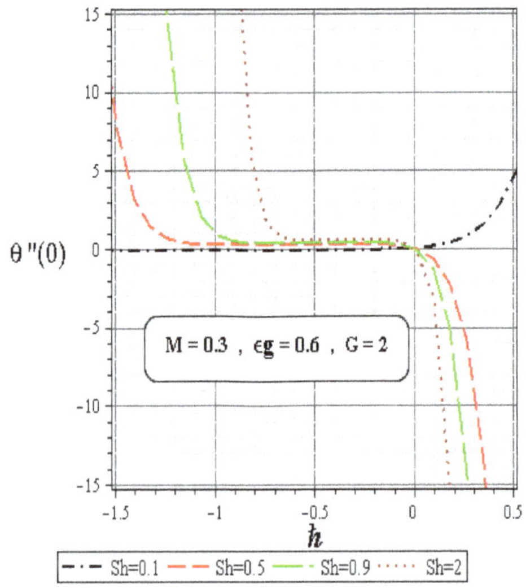

Figure 4: The h - validity for X=0 and different value of *Sh*.

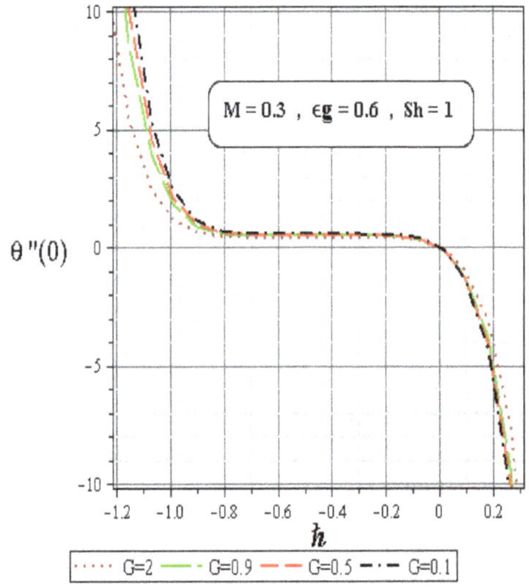

Figure 5: The $\hbar$ - validity for X=0 and different value of*G*.

The solutions $\theta(x)$ were too long to be mentioned here, therefore, they are shown graphically

Convergence of the HAM solution

The series solutions contain the auxiliary parameter h. The validity of the method is based on such an assumption that the series (20) converges at p=1. It is the auxiliary parameter h which ensures that this assumption can be satisfied. As pointed out by Liao [23], in general, by means of the so-called h-curve, it is straightforward to choose a proper value of $\hbar$ which ensures that the solution series is convergent. In this way, we choose h=-0.2 in following computational works (Figures 2-5).

Results and Discussion

In this manuscript, the Homotopy Analysis Method such as

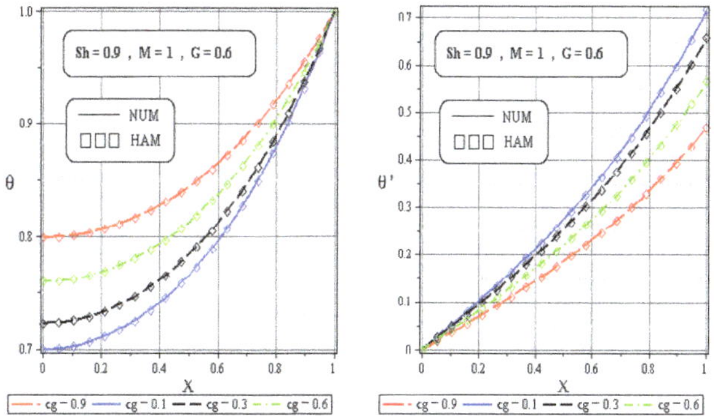

Figure 6: The comparison between the Numerical and HAM solution for θ and θ' atdifferent value of εg.

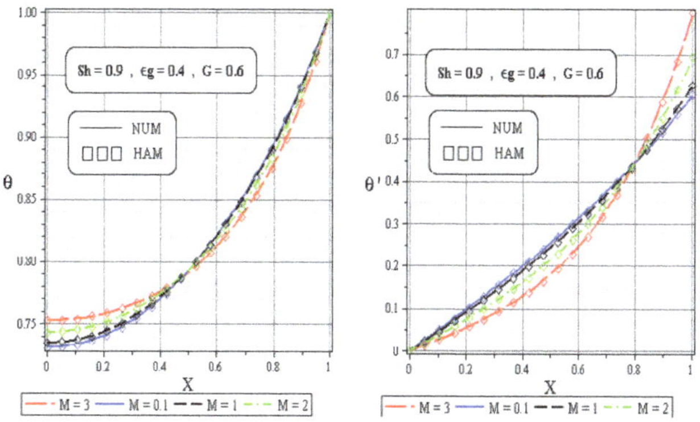

Figure 8: The comparison between the Numerical and HAM solution for θ and θ' atdifferent value of *M*.

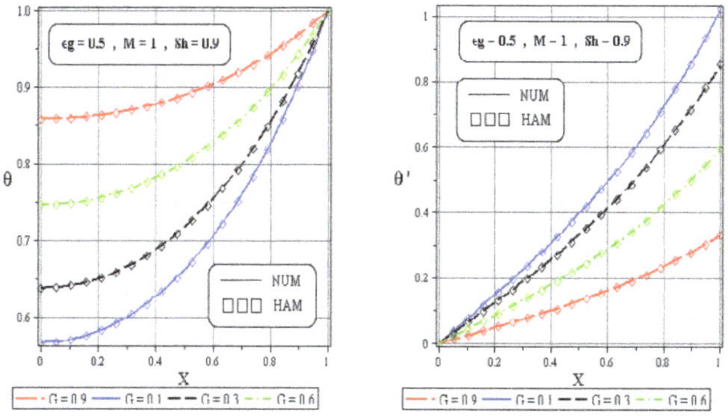

Figure 7: The comparison between the Numerical and HAM solution for θ and θ' atdifferent value of *G*.

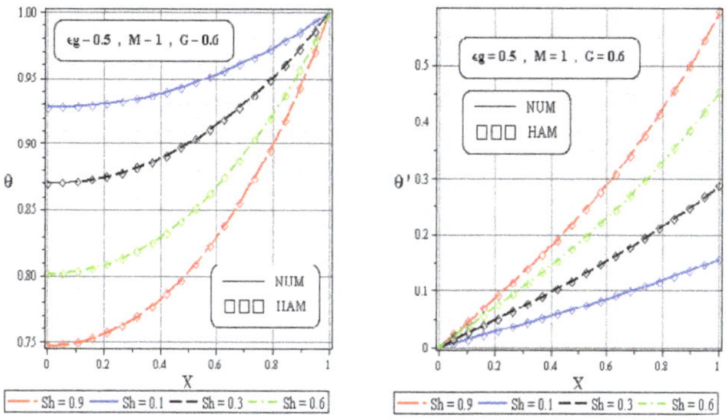

Figure 9: The comparison between the Numerical and HAM solution for θ and θ' atdifferent value of *Sh*.

X	$\theta(X)$			$\theta'(X)$		
	HAM	**NUM**	**Error**	**HAM**	**NUM**	**Error**
0.00	0.452726862	0.452787880	0.0000610676	0.000000000	0.000000000	0.000000000
0.05	0.453655038	0.453715480	0.0000604418	0.037163422	0.037139216	0.000024205
0.10	0.456450499	0.456509076	0.0000585769	0.074764710	0.0747149290	0.000049781
0.15	0.461146198	0.461201593	0.0000553947	0.113247727	0.113169914	0.000077812
0.20	0.467797557	0.467848300	0.0000507434	0.153068448	0.152959108	0.00010934
0.25	0.476483231	0.476527631	0.0000443994	0.194701325	0.194556142	0.000145183
0.30	0.487306202	0.487342342	0.000036140	0.238646034	0.238460303	0.00018573
0.35	0.500395212	0.500420955	0.0000257432	0.285434739	0.285203912	0.000230827
0.40	0.515906592	0.515919591	0.0000129992	0.335640062	0.335360431	0.000279631
0.45	0.534026494	0.534024250	0.0000022448	0.389883917	0.389553555	0.000330363
0.50	0.554973606	0.554953590	0.0000200162	0.448847455	0.448467480	0.000379975
0.55	0.579002383	0.578962239	0.0000401438	0.513282354	0.512858616	0.000423738
0.60	0.606406896	0.606344722	0.0000621731	0.584023786	0.583569072	0.000454714
0.65	0.637525367	0.637440131	0.0000852359	0.662005412	0.661542398	0.000463014
0.70	0.672745525	0.672637657	0.000107868	0.748276866	0.747842077	0.000434789
0.75	0.712510897	0.712383109	0.000127788	0.844024269	0.843673377	0.000350892
0.80	0.757328227	0.757186633	0.000141594	0.950594447	0.950409414	0.000185033
0.85	0.807776202	0.807631870	0.000144332	1.069523667	1.069622444	0.000098777
0.90	0.864515764	0.864386826	0.000128938	1.202571916	1.203121649	0.000549733
0.95	0.928302283	0.928216816	0.0000854667	1.351763967	1.352999072	0.001235105
1.00	1.000000000	1.000000000	0.000000000	1.519438801	1.521685895	0.002247094

Table 1: The results of HAM and Numerical methods for $\theta(X)$ and $\theta'(X)$ for $M = 2,\ \varepsilon g = 0.1, Sh = 0.9\ \ and\ G = 0.3$.

analytical technique is employed to find an analytical solution of the temperature distribution in a porous fin. Figure 6 show comparison between the numerical solution and HAM solution for θ and θ' when Sh=0.9, M=1, G=0.6 and different value of εg.

Figure 7 illustrate the accuracy of HAM solution compare to numerical solution when, εg=0.5, M=1, Sh=0.9 and different value of G. Figure 8 show comparison between numerical solution and HAM solution when εg=0.4, G=0.6, Sh=0.9 and different value of M.

Figure 9 show comparison between numerical solution and HAM solution when M=1, G=0.6, εg=0.4 and different value of Sh. According to Table 1 and Figures 6-9, clearly show that the results by HAM are in excellent agreement with the exact solutions. Also, the auxiliary parameter h provides us with a convenient way to adjust and control the convergence and its rate for the solutions series.

Conclusion

In this study, the MHD on a Porous Fin to a Vertical Isothermal Surface with temperature dependent internal heat generation was analyzed using HAM. The found that the approximations obtained by HAM are valid when compared with the exact solutions. It should be emphasized that the HAM provides us with a convenient way to control the convergence of approximation series. Finally, it has been attempted to show the capabilities and wide-range applications of the HAM in engineering.

References

1. Chamkha AJ, Takhar HS, Nat G (2004) Mixed convection flow over a vertical plate with localized heating (cooling), magnetic field and suction (injection), Heat Mass Transfer 40: 835-841.
2. Aldoss TK, Ali YD, Al-Nimr MA (1996) MHD mixed convection from a horizontal circular cylinder, Numer. Heat Transfer 30(4): 379-396.
3. Nayfeh AH (2000) Perturbation Methods, Wiley, New York, USA.
4. Ganji DD, Hashemi Kachapi, Seyed H (2011) Analytical and numerical method in Engineering and applied Science, progress in nonlinear science 3: 1-579.
5. He, JH (2000) A coupling method of homotopy technique and perturbation technique for nonlinear problems, Internat. J Non-Linear Mech 35(1): 37-43.
6. He JH (2005) Homotopy perturbation method for bifurcation of nonlinear problems, Int. J Nonlinear Sci Numer Simul 6: 207-208.
7. He JH (2005) Application of homotopy perturbation method to nonlinear wave equations, Chaos Solitons Fractals 26: 695-700.
8. Mohsen T, Hessameddin Y, Seyfolah S (2011) assessment of Homotopy Perturbation Method in nonlinear convective-radiative nonfourier conduction heat transfer equation with variable coefficient, Thermal science 15(2): S263-S274.
9. Esmaeilpour M, Ganji DD, Mohseni E (2009) Application of homotopy perturbation method to micropolar flow in a porous channel, J Porous Media 12(5): 451-459.
10. Ganji DD, Rostamiyan Y, Rahimi Petroudi I, Khazayi Nejad M, (in press) analytical investigation of nonlinear model arising in heat transfer through the porous fin, Thermal sciences.
11. Ganji DD (2006) The application of He's homotopy perturbation method to nonlinear equations arising in heat transfer, Physics Letters A 355: 337-341.
12. Ganji DD, Sadighi A (2007) Application of homotopy-perturbation and variational iteration methods to nonlinear heat transfer and porous media equations, J Comput Appl Math 207(1): 24-34.
13. Jitendra S, Praveen Kumar G, Kabindra Nath R (2011) Variation Iteration Method to solve moving boundary problem with temperature dependent physical properties, Thermal sciences 15(2): S229-S239.
14. He JH (2007) Variational iteration method – some recent results and new interpretations, Journal of Computational and Applied Mathematics 207: 3-17.
15. He JH, Wu XH (2006) Construction of solitary solution and compaction-like solution by variational iteration method, Chaos Solitons and Fractals, 29(1): 108-113.
16. Momani S, Abuasad S (2006) Application of He's variational iteration method to Helmholtz equation, Chaos Solitons and Fractals 27: 1119-1123.
17. Ganji DD, Jamshidi N, Ganji ZZ (2009) HPM and VIM methods for finding the exact solutions of the nonlinear dispersive equations and seventh-order Sawada–Kotera equation, International Journal of Modern Physics B 23: 39-52.
18. Ganji DD, Hafez T, Bakhshi Jooybari M (2007) Variational iteration method and homotopy perturbation method for nonlinear evolution equations, Computers and Mathematics with Applications 54: 1018-1027.
19. Ganji DD, Afrouzi GA, Talarposhti RA (2007) Application of variational iteration method and homotopy-perturbation method for nonlinear heat diffusion and heat transfer equations, Physics Letters A, 368: 450-457.
20. He JH (1999) Variational iteration method—a kind of nonlinear analytical technique: Some examples, International Journal of Non-linear Mechanics 34: 699-708.
21. He JH (1998) Approximate analytical solution for seepage with fractional derivatives in porous media, Computational Methods in Applied Mechanics and Engineering, 167: 57-68.
22. Davood Domairy G, Ehsan Mohseni L (2010) Mathematical Methods in Nonlinear Heat transfer, Xlibris Corporation, USA.
23. Liao SJ (1992) The proposed homotopy analysis technique for the solution of nonlinear problems, PhD thesis, Shanghai Jiao Tong University.
24. Liao SJ (1995) An approximate solution technique not depending on small parameters: a special example, Int J Non-Linear Mech 303: 371-80.
25. Liao SJ (1997) Boundary element method for general nonlinear differential operators. Eng Anal Bound Elem 202: 91-9.
26. Liao SJ (2003) Beyond perturbation: introduction to the homotopy analysis method. Boca Raton: Chapman and Hall, CRC Press.
27. Liao SJ, Cheung KF (2003) Homotopy analysis of nonlinear progressive waves in deep water, J Eng Math 45: 103-16.
28. Liao SJ (2004) On the homotopy analysis method for nonlinear problems, Appl Math Comput 47: 499-513.
29. Aziz A, Bouaziz MN (2011) A least squares method for a longitudinal fin with temperature dependent internal heat generation and thermal conductivity, Energy Convers Manage 52: 2876-82.
30. Hatami M, Hasanpour A, Ganji DD (2013) Heat transfer study through porous fins (Si3N4and AL) with temperature-dependent heat generation, Energy Conversion and Management 74: 9-16.

Contact Pressures and Cracks Identification by using the Dirichlet-to-Neumann Solver in Elasticity

Mohamed Larbi Kadri*

Ecole Nationale d'Ingénieurs deTunis (ENIT), Lamsin, Campus Universitaire, B.P. 37, 1002Tunis, Tunisia

Abstract

In this work we present a numerical data completion method based on the Dirichlet-to-Neumann algorithm, by working in a linear elasticity framework. We begin by recasting the problem in terms of the Steklov-Poincaré operator which is commonly used in domain decomposition. Then we present the Dirichlet-to-Neumann algorithm and state the equivalence between both formulations. The proposed method is applied to identify a contact pressures distribution and interfacial cracks.

Keywords: Inverse problem; Data completion; Cauchy problem; Identification; Dirichlet-to-Neumann algorithm

Introduction

Physical phenomena are often governed by partial differential equations, which need an essential set of data to solve them. In linear elasticity, these data are: the geometry of the solid, the mechanical properties of the materials and the boundary conditions. However, in many industrial applications, some of these data are unknown and have to be identified. This leads to an inverse problem whose resolution requires over-specified measured data. In this paper we focus on a problem of boundary condition identification in linear elasticity. In this case, data measured on part of the easily accessible border are often available. However, contrary to the direct problem, two kinds of boundary conditions (e.g. displacements and tractions) are imposed on the same part of the boundary while no information exists on the remaining part of it. Hence, data completion consists in reconstructing the boundary conditions for the whole boundary of a domain by using the partially overspecified measurements. This is the well-known Cauchy problem, which is ill-posed.

The ill-posedness of inverse problems may concern the existence and/or the uniqueness of the solution, but their most critical feature is their instability: the solution, whenever a problem has one, is not continuous with respect to the data, i. e. small measurement errors in the data may dramatically amplify the errors in the solution. This is ill-posedness in the Hadamard sense [1]. The Cauchy problem pertains to this kind of inverse problem. Therefore suitable regularizing algorithms that are exempt from this ill-posedness phenomenon, are required in order to solve the inverse problem correctly.

The Cauchy problem in linear elasticity was first studied by Yeih et al. [2]. In this paper, the existence and uniqueness of the solution are analyzed as well as the continuity of the solution with respect to the data. Others authors have proposed an alternative regularization procedure, namely the indirect fictitious boundary method, which is based on the simple or the double layer potential theory. The numerical implementation of the aforementioned method has been carried out by Koya et al. [3] who used the BEM and the Nystrom method for discretizing the integrals involved. Marin et al. [4] have determined the approximate solutions of the Cauchy problem in linear elasticity using an alternating iterative BEM that reduces the problem to solving a sequence of well-posed boundary value problems [5]. In Marin and Lesnic have used singular value decomposition to solve the same problem numerically. A related inverse problem which allows for interior displacement measurements and inter-facial crack has been investigated by Huang and Shih [6]. In Weikel et al. [7] have proposed an alternating iterative algorithm in order to reconstruct an internal planar crack laying on an a priori known internal surface inside a three-dimensional elastic body from over determined elastostatic boundary data on the outer surface. Furthermore, Koslov and his co-authors adapted the iterative Dirichlet-to-Neumann method to approximate the solution of the Cauchy-Poisson problem, governed by the Laplace equation, and they provide proof of its convergence and its regularizing properties [8]. In [9] the iterative Dirichlet-to-Neumann method was applied to recover the missing boundary data for the Cauchy Helmholtz problem. More recently kadri et al. [10] have used the SteklovPoincaré approach relying on domain decomposition for the identification of internal planar cracks inside a three-dimensional using elastostatic measurements.

In this work, the iterative Dirichlet-to-Neumann method is applied to the linear elastic data completion problem. In section 2, the Cauchy problem is presented in the context of linear elasticity. In section 3 this problem is recast in condensed form that we will refer to as the Cauchy-Steklov-Poincaré problem, which leads to the Cauchy-Steklov-Poincaré equation acting on the boundary of the unknowns. In section 4 we present the Dirichlet-to-Neumann algorithm and we show that it can be interpreted as a preconditioned Richardson procedure for the Cauchy-Steklov-Poincaré equation. The numerical procedure and the results obtained by FEM discretization of the problem are presented in section 5. The method is used to solve two applications borrowed from engineering mechanics: the identification of contact pressures and coating defect in a double layered composite domain.

The Cauchy Problem in Linear Elasticity

Let denote a bounded domain in $\mathbb{R}^2$ or $\mathbb{R}^3$ with regular boundary $\Gamma=\partial\Omega$. The whole domain is assumed to be filled with a homogeneous linear elastic isotropic medium. It is assumed th at Γ is splitted into two

***Corresponding author:** Mohamed Larbi Kadri, Ecole Nationale d'Ingénieurs de Tunis (ENIT), Lamsin, Campus Universitaire, B.P. 37, 1002 Tunis, Tunisia
E-mail: medlarbi.kadri@lamsin.rnu

open subsets Γ_c, and Γ_i, $\Gamma = \Gamma_c \cup \Gamma_i$, where Γ_c, Γ_i, $\Gamma_i=\phi$ and $\Gamma_c \cap \Gamma_i = \varnothing$. In what follows, u(X) denotes the displacements field on Ω.

The local equilibrium equation is given by

$$-\text{div}\ \sigma(\mathbf{u}(\mathrm{x})) = \mathbf{f} \qquad \mathrm{x} \in \Omega, \tag{1}$$

where **σ** is the stress tensor and **f** the volume forces. The strain tensor ε is given by

$$\varepsilon(\mathbf{u}(\mathrm{x})) = \frac{1}{2}(\nabla \mathbf{u}(\mathrm{x}) + \nabla \mathbf{u}(\mathrm{x}))$$

These tensors are related by the Hooke's constitutive law, which is

$$\sigma(\mathbf{u}(\mathrm{x})) = 2\mu\varepsilon(\mathbf{u}(\mathrm{x})) + \lambda \mathrm{tr}\varepsilon(\mathbf{u}(\mathrm{x}))\mathrm{I} = \mu(\nabla \mathbf{u}(\mathrm{x}) + \nabla \mathbf{u}(\mathrm{x})) + \lambda \text{div}\ \mathbf{u}(\mathrm{x})$$

where λ and μ are the Lamé constants of the material and I is the identity tensor.

Let n(x) be the outward normal vector at Γ and t(x) be the traction vector at a point x ϵ Γ defined by

$$\mathbf{t}(\mathrm{x}) = \sigma(\mathbf{u}(\mathrm{x}))\sigma\mathbf{n}(\mathrm{x})\, \mathrm{x} \in \Gamma$$

In the well-posed direct problem formulation, the knowledge of the displacement on a part of the boundary and traction vectors on another part of the boundary enables us to determine the displacement vector in domain **Ω**. Then, the strain tensor **ε** can be calculated from kinematic relation (2) and the stress tensor is determined by constitutive law (3).

If a part of the boundary Γ_i is inaccessible and if it is possible to measure both the displacement and traction vectors on the remaining part of boundary Γ_c, this leads to the mathematical formulation of a direct problem expressed as follows:

$$\begin{cases} -div\sigma(\mathbf{u}(\mathrm{x})) = f & in\boldsymbol{\Omega} \\ u(\mathrm{x}) = \tilde{\upsilon}(\mathrm{x}) & on\Gamma_c \\ t(\mathrm{x}) = \tilde{\mathbf{t}}(\mathrm{x}) & on\ \Gamma_c \end{cases}$$

Where $\tilde{\mathbf{u}}$ and $\tilde{\mathbf{t}}$ are prescribed vector valued functions. This problem is ill-posed because of the formulation of its boundary conditions (5). It can be seen that boundary Γ_c and the traction tis overs pecified by prescribing both the displacement $\upsilon_{|\Gamma_c} = \tilde{\upsilon}$ and the tractions $\mathbf{t}_{|\Gamma_c} = \tilde{\mathbf{t}}$ vectors, while boundary Γ_i underspecified since both the displacement $\mathbf{u}_{|\Gamma_i} = \bar{\mathbf{u}}$ and the traction $\mathbf{t}_{|\Gamma_i} = \bar{\mathbf{t}}$ are unknown and have to be determined. Then, this problem can be stated as follows: find $(\bar{\mathbf{u}}, \bar{\mathbf{t}})$ that a displacement field u(x) exists that satisfies:

$$\begin{cases} -\text{div}\sigma(\mathbf{u}(\mathrm{x})) = \mathbf{f} & \text{on } \Omega \\ \mathbf{u}(\mathrm{x}) = \tilde{\mathbf{u}}(\mathrm{x}) & \text{in } \Gamma_c \\ \sigma(\mathbf{u}(\mathrm{x}))\mathrm{n} = \tilde{\mathbf{t}}(\mathrm{x}) & \text{in } \Gamma_c \\ \mathbf{u}(\mathrm{x}) = \bar{\mathbf{u}}(\mathrm{x}) & \text{in } \Gamma_i \\ \sigma(\mathbf{u}(\mathrm{x}))\mathbf{n} = \bar{\mathbf{t}}(\mathrm{x}) & \text{in } \Gamma_i \end{cases}$$

This problem, known as the Cauchy problem, is ill-posed in the sense that the dependence of u(x), and consequently of $(\bar{\mathbf{u}}, \bar{\mathbf{t}})$, on the data $(\tilde{\mathbf{u}}, \tilde{\mathbf{t}})$ is not continuous. Although the problem may have a unique solution, it is well-known that this solution is unstable with respect to small parturbation in the data on Γ_c. In this paper we propose to recover the lacking data by using the Dirichlet-to-Neumann algorithm introduced by Kozlov et al. in the steady state thermal case [8]. However, let us first introduce an operator acting on the boundary where data are unknown: the Steklov-Poincaré operator which is very familiar in domain decomposition and recently introduced for the Cauchy problem of the Laplace equation by Andrieux et al. in [11] and by Ben Belgacem et al. in [12,13].

The Cauchy-Steklov-Poincare Equation

To keep the notational complexity to a minimum let us remove **x** from the notations. Let **λ** denote the unknown displacement vector on Γ_i. We consider both Dirichlet and Neumann elliptic problems obtained by duplicating the solution u into a couple of vectors $\mathbf{u}_N$, $\mathbf{u}_D$. The Cauchy problem (6) is then split into:

$$\begin{cases} -div\sigma(\mathbf{u}_N(\mathrm{x})) = f & in\Omega \\ t = \sigma(\mathbf{u}_N(\mathrm{x}))n = \tilde{\mathbf{t}}(\mathrm{x}) & in\Gamma_c \\ \mathbf{u}_N = \lambda & in\Gamma_i \end{cases} \qquad \begin{cases} -div\sigma(\mathbf{u}_D(\mathrm{x})) = f & in\Omega \\ \mathbf{u}_D = \tilde{\mathbf{u}}(\mathrm{x}) & in\Gamma_c \\ \mathbf{u}_D = \lambda & in\Gamma_i \end{cases}$$

If the pair $(\tilde{\mathbf{u}}, \tilde{\mathbf{t}})$ is compatible (i.e. a vectors field exists that verifies (1) for

Which $(\tilde{\mathbf{u}}, \tilde{\mathbf{t}})$ are the Cauchy data on Γ_c, the solution of the Cauchy problem (1-5) is recovered, i.e. $\mathbf{u}=\mathbf{u}_D=\mathbf{u}_N$ in Ω, if and only if $\sigma(\mathbf{u}_D(\lambda))\mathbf{n} = \sigma(\mathbf{u}_N(\lambda))\mathbf{n}$ on Γ_i

Now for **μ**, a displacements vector defined on Γ_i, the linear parts of $\mathbf{u}_N$ (**μ**) and $\mathbf{u}_D$ (**μ**) are denoted $\mathbf{u}_N^0(\mu)$ and $\mathbf{u}_D^0(\mu)$ which solve respectively:

$$\begin{cases} -div\sigma(\mathbf{u}_N^0(\mu)) = 0 & in\Omega \\ t = \sigma(\boldsymbol{u}_N^0(\mu))n = 0 & in\Gamma_c \\ \boldsymbol{u}_N^0(\mu) = \mu & in\Gamma_i \end{cases}$$

$$\begin{cases} -div\sigma(\boldsymbol{u}_D^0(\ \ )) = f & in\Omega \\ \boldsymbol{u}_D^0(\mu)n = 0 & in\Gamma_c \\ \boldsymbol{u}_D^0(\mu) = \mu & in\Gamma_i \end{cases}$$

We consider also $\boldsymbol{u}_N^*$ and $\boldsymbol{u}_D^*$ such that:

$$\begin{cases} -div\sigma(\boldsymbol{u}_N^*) = f & in\Omega \\ t = \sigma(\boldsymbol{u}_N^*)n = \tilde{t} & in\Gamma_c\ f \\ \boldsymbol{u}_N^* = 0 & in\Gamma_i \end{cases} \qquad \begin{cases} -div\sigma(\boldsymbol{u}_D^*) = f\ in\Omega \\ \boldsymbol{u}_D^* = 0 \quad in\Gamma_c \\ \boldsymbol{u}_D^* = 0 \quad in\Gamma_i \end{cases}$$

And, by superposition, we obtain $\boldsymbol{u}_N(\mu) = \boldsymbol{u}_N^0(\mu) + \boldsymbol{u}_N^*$ and $\boldsymbol{u}_D(\mu) = \boldsymbol{u}_D^0(\mu) + \boldsymbol{u}_D^*$. With this partition, condition (8) is written as

$$\sigma(\boldsymbol{u}_D^0(\lambda))\boldsymbol{n} - \sigma(\boldsymbol{u}_N^0(\lambda))\boldsymbol{n} = \sigma(\boldsymbol{u}_N^*)\boldsymbol{n} - \sigma(\boldsymbol{u}_D^*)\boldsymbol{n} \text{ on } \Gamma_i$$

Using the following notations:

$S_D(\lambda) = \lambda(\lambda u^0\lambda_D(\lambda))n$, $S_N(\lambda) = \sigma(u^0{}_D(\lambda))n$ and $\chi = (\sigma(u^*{}_N) - \sigma(u^*{}_D))n$

Equation (11) becomes : $S(\lambda)=S_D(\lambda)-S_N(\lambda)=\chi$ on Γ_i

Equation (12) is called the Steklov-Poincar′e interface equation and S is the Steklov-Poincar′e operator. It is familiar in the domain decomposition framework [14] for the direct boundary value problem. More precisely, things happen as if vectors u_D and u_N were defined on two different domains with common boundary Γ_i. In this case, the equation (12) expresses the Neumann transmission condition, but the (-) sign in S would be (+) in the domain decomposition formulation [14]. The (-) sign which appears in S is at the origin of ill-posedness of the Cauchy problem. From the discrete point of view, the finite element discretization of S leads to the Schur complement matrix [14]. It corresponds to having all interior nodes eliminated by static condensation [15]. In Ben Abdallah [16] a numerical study of the Shur complement matrix is performed for the Cauchy-Poisson problem. We will propose a study of the Cauchy problem in elasticity based on the Stecklov-Poincar´e equation in a forthcoming paper.

We now continue with the analogy with domain decomposition and show how the Cauchy-Steklov-Poincar´e equation can be expressed, as in domain decomposition, in terms of the Dirichlet-to-Neumann problem.

The Dirichlet-to-Neumann Solver for the Cauchy Problem

When describing the Dirichlet-to-Neumann approach it should be noted that when the complete data are available on Γ, we have an overspecified boundary value problem $-div\sigma(u)=f \quad in\,\Omega$

$$\sigma(u)n=\tilde{t},\ u=\tilde{u} \quad on\ \Gamma_c;\ \sigma(u)n=\overline{t},\ u=\overline{u} \quad on\ \Gamma_i;$$

This problem can be split into two well-posed subproblems with different boundary conditions. For one of them (Neumann/Dirichlet) conditions are Imposed on (Γ_c/Γ_i)

$$-div\sigma(\hat{u})=f \quad in\ \Omega$$

$$\sigma(\hat{u})n=\tilde{t} \quad on\ \Gamma_c$$

$$\hat{u}=\overline{u} \quad on\ \Gamma_i$$

$$-div\sigma\left(\overset{\text{i}}{u}\right)=f \quad in\ \Omega$$

$$\overset{\text{i}}{u}=\tilde{u} \quad on\ \Gamma_c$$

$$\sigma\left(\overset{\text{i}}{u}\right)n=\overline{t} \quad on\ \Gamma_i c$$

Solving the Cauchy system (1)-(5) is achieved when extension $(\overline{t},\overline{u})$ makes $\hat{u}$ and $\overset{\text{i}}{u}$ coincide so the solution is then $u=\hat{u}=\overset{\text{i}}{u}$.

Basically, the iterative method proposed for the Cauchy-Poisson problem and studied in Kozlov et al. [8], is derived from these observations: starting from an arbitrary prediction of the Dirichlet condition (here the displacement vector $\hat{u}$) on Γ_i, we add several corrections by solving alternately a Dirichlet on Γ_c/Neumann on Γ_i problem, where at each iteration the appropriate boundary data are inferred from the solution computed in the previous step. More specifically, we construct a sequence of a pair of vectors $\left(u_N^{(k)},u_D^{(k)}\right)_k$ from the following recurrence: given $u_D^{(0)}$, the following systems are solved for each :$k\geq 0$

$$\begin{cases}-div\sigma\left(u_N^{(k+1)}\right)=f & in\Omega\\ \sigma\left(u_N^{(k+1)}\right)n=\tilde{t} & on\Gamma_c\\ u_N^{(k+1)}=u_D^{(k)} & on\Gamma_i\end{cases} \quad \begin{cases}-div\sigma\left(u_D^{(k+1)}\right)=f & in\Omega\\ u_D^{(k+1)}=\tilde{u} & on\Gamma_c\\ \sigma\left(u_D^{(k+1)}\right)n=\sigma\left(u_N^{(k+1)}\right)n & on\Gamma_i\end{cases}$$

The convergence of the alternating method toward the solution of the Cauchy problem and its stabilizing properties are established by Kozlov et al. [8] for the steady state thermal case. In the linear elastic framework, no convergence result has been proved till now but the result of convergence established by Koslov et al. may be applied for any elliptic operators. When convergence is achieved, we may obtain $(\overline{t},\overline{u})=(\sigma(u)n,u)$ on Γ_i. By using straightforward computations, it can be established that the Dirichlet-to-Neumann scheme can be interpreted as a preconditioned Richardson procedure for the CauchySteklov-Poincaré equation. For this purpose, the Dirichlet-to-Neumann algorithm is rewritten, using the previous notations, as follows: Given λ^0,

$$\begin{cases}-div\sigma\left(u_N^{(k+1)}\right)=f & in\Omega\\ u_N^{(k+1)}=\tilde{u} & on\Gamma_c\\ u_N^{(k+1)}=\lambda^{(k)} & on\Gamma_i\end{cases} \quad \begin{cases}-div\sigma\left(u_D^{(k+1)}\right)=f & in\Omega\\ \sigma(\mathrm{u}_D^{(k+1)})n=\tilde{\mathrm{t}} & on\Gamma_c\\ \sigma\left(\mathrm{u}_D^{(k+1)}\right)n=\sigma\left(\mathrm{u}_N^{(k+1)}\right)n & on\Gamma_i\end{cases}$$

The last equality $\sigma\left(\mathrm{u}_D^{(k+1)}\right)\mathrm{n}=\sigma\left(\mathrm{u}_N^{(k+1)}\right)\mathrm{n}$ on Γ_i, can be written as

$$\sigma\left(\mathrm{u}_D^{0(k+1)}\right)\mathrm{n}+\sigma\left(\mathrm{u}_D^{*}\right)\mathrm{n}=\sigma\left(\mathrm{u}_N^{0(k+1)}\right)\mathrm{n}+\sigma\left(\mathrm{u}_N^{*}\right)\mathrm{n}\cdot$$

Since $\sigma\left(\mathrm{u}_D^{0(k+1)}\right)\mathrm{n}=\mathrm{S}_D\lambda^{(k+1)}$ and $\sigma\left(\mathrm{u}_N^{0(k+1)}\right)\mathrm{n}=\mathrm{S}_N\lambda^{(k)}$ on Γ_i it follows that $\mathrm{S}_D\lambda^{(k+1)}=\mathrm{S}_N\lambda^{(k)}+\chi$, and therefore $\lambda^{(k+1)}=\lambda^{(k)}-\mathrm{S}_D^{-1}\left(\mathrm{S}\lambda^{k}-\chi\right)$ We are thus left with a Richardson procedure for the Cauchy-Steklov-Poincaré equation (12) with the operator S_D as a preconditioner. In the following section we will discuss the efficiency of the Dirichlet-to-Neumann algorithm presented above as a numerical solver for several particular Cauchy problems in linear elasticity.

Applications

This section is devoted to the presented method in two situations taken from engineering mechanics. The first example concerns contact identification on an inaccessible contact area. The second deals with coating defect identification.

Contact pressures identification

Domain Ω is a square plate (1.*1.) with a circular hole (R=0.20225), where a fixed rigid disc R=0.2 is placed. Figure 1 shows the geometry and boundary conditions applied to the plate. The mechanical properties of the plate are given in Table 1. When tractions are applied on the plate, it comes into contact with the lower part of the disc (Figure 1).

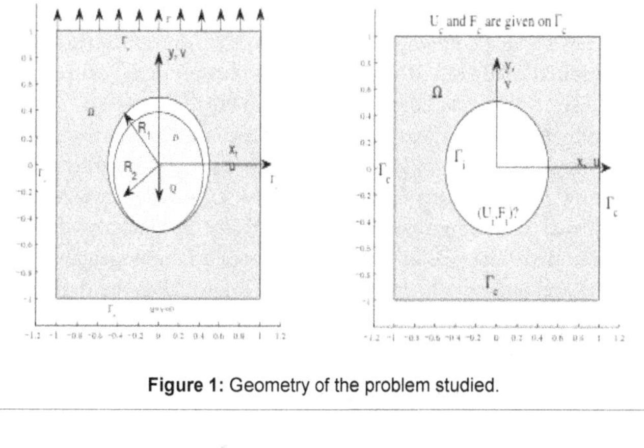

Figure 1: Geometry of the problem studied.

Disc and plate	Aluminium
Modulus of elasticity	*E*=70000 *MPA*
Poison coefficient	*υ*=0.31
Friction coefficient	*μ*=0
Load applied to the plate	F=1^{+7} *N/m*

Table 1: Mechanical characteristic of the plate and the disc.

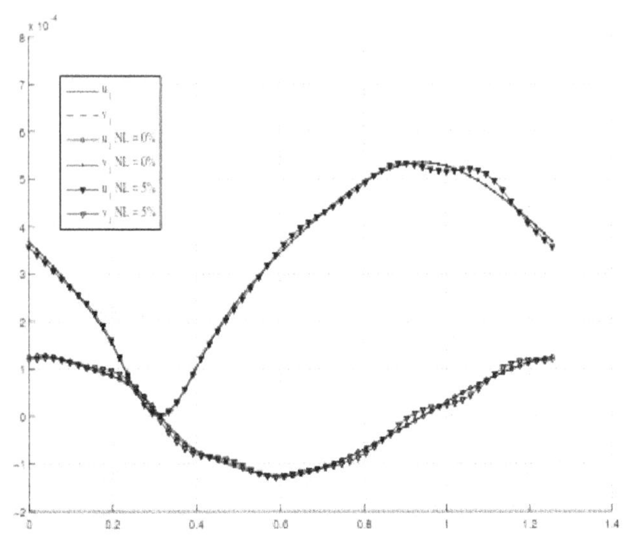

Figure 2: Reconstruction of horizontal (U) and vertical (V) displacements on the internal boundary of the plate for noise free (N. L.=0%) and noisy (N. L.=5%) data.

The problem is to identify the contact pressure distribution and the displacements on the interface between the plate and the disc, by using overs pecified data provided for the external boundary. These overs pecified data are generated by solving a direct problem using Hertz's analytical contact law. Here, we consider a frictionless contact so that only normal pressure is taken in account. Moreover, plane strain hypothesis is assumed.

The results obtained by solving the corresponding Cauchy problem are the normal stress components and the displacements field on Γ_i. Hence, the contact zone is the part of the boundary where the normal stress components are not null.

When carrying out an identification based on measurements, it must be kept in mind that measured data are subject to noise whose effects have to be studied. In this case, the data are synthetic, and therefore suffer from some errors (approximation error, roundoff error, . . . etc). We added a noise generated by a MATLAB routine (randn) to the computational noise. The displacement measurements are polluted by a noise level at 5%. Figure 2 depicts the horizontal (resp. vertical) displacements U (resp.) reconstruction on the internal boundary of the plate. Figure 3 shows the identification of the normal stress distribution on the internal boundary of the plate. As expected, displacements reconstruction is better than that for the stresses, particularly when the data are noisy. The reason is that the stresses are homogeneous with the displacements gradient and it is well known that the derivation is an ill-posed operation (the influence of noise is considerable). The identification is very satisfactory free noisy data. For noise-free data.

For noisy data, the contact zone is well localized and the contact pressures are recovered correctly. However, some fairly significant oscillations appear on the free boundary.

Coating defect identification

The identification of inter-facial cracks is a crucial issue in detecting coating defects or delamination in composite material. Our second experiment focuses on the detection of curved inter-facial cracks. We consider a double-layered annular domain centered at the origin with an inner-radius 0.6 (Γ_{c-}), middleradius 0.8 (Γ_i) and an outer-radius 1 (Γ_{c+}) as shown on Figure 4. The coating defect lies at Γ_i. The simulation is run using synthetic data generated by a finite element resolution of the direct problem. The direct problem is solved with prescribed displacements on the inner boundary and with prescribed surface tractions acting on the outer boundary. The cracks are approximated by two thin cavities on which a homogeneous Neumann condition is prescribed. In order to detect the coating defect, two Cauchy problems are solved. The first, P_+, is defined on subdomain Ω_+ where the overspecified data are given on the external boundary Γ_{c+} and the unknowns are identified on the boundary Γ_i. The second Cauchy problem P_ is defined on subdomain

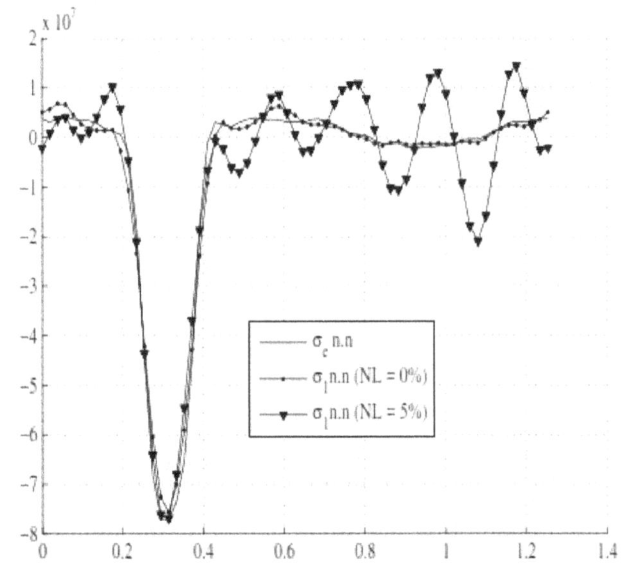

Figure 3: Reconstruction of the normal stress distribution on the internal boundary of the platefor noise free (N.L.=0%) and noisy (N.L.=5%) data.

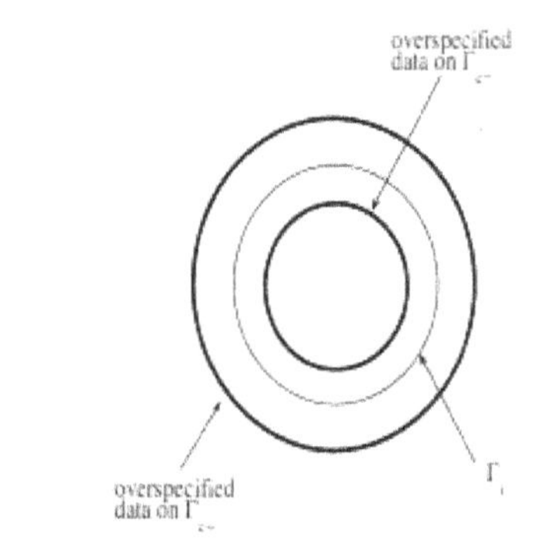

Figure 4: Geometry of the domain with a coating defect.

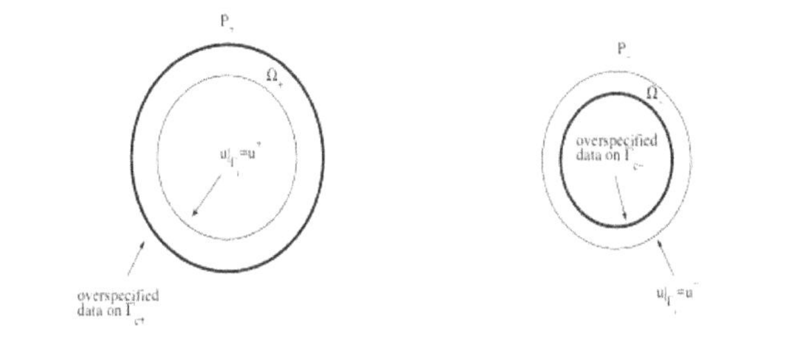

Figure 5: Geometries of subdomains on which Cauchy problems are solved.

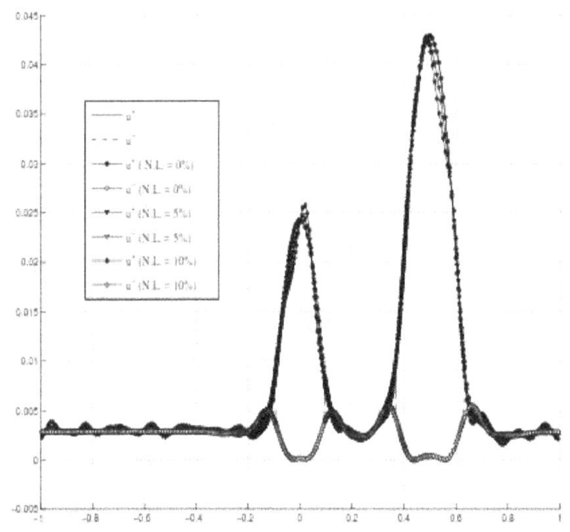

Figure 6: Interfacial displacement identification.

Ω_- where the overspecified data are given on internal boundary Γ_{c-} and the unknowns are identified on boundary (Figure 5). Among the unknowns we are only interested in the displacements. In fact, if we use u^+ (resp. u^-) to denote the displacements field on Γ_i provided by P_+ (resp.P_-), the cracks will appear as the part of Γ_i where the jump of the displacements vector $[u^+ - u^-]$ does not vanish.

Two interfacial cracks with different widths are simulated. In this case also we tested the reconstruction algorithm in the case of noise free and noisy data. The displacements were polluted with noise at 5% and 10% level. The reconstructed u^- and u^+ and the reconstruction of the jump $[u^+ - u^-]$ across the interface are plotted in Figures 6 and 7. It can be seen again that good agreement is achieved with the exact solution, even for noisy cases. It seems that the width of the crack has no influence on the accuracy of the reconstructed procedure: both cracks are well recovered.

Conclusion

In this work we presented a numerical method for solving the Cauchy problem in the framework of linear elasticity. The method proposed was applied in two practical situations taken from engineering mechanics: contact pressure recovery and coating defect identification. We also presented an alternative formulation of the Cauchy problem which lead to an operator acting on the boundary of the unknown: the

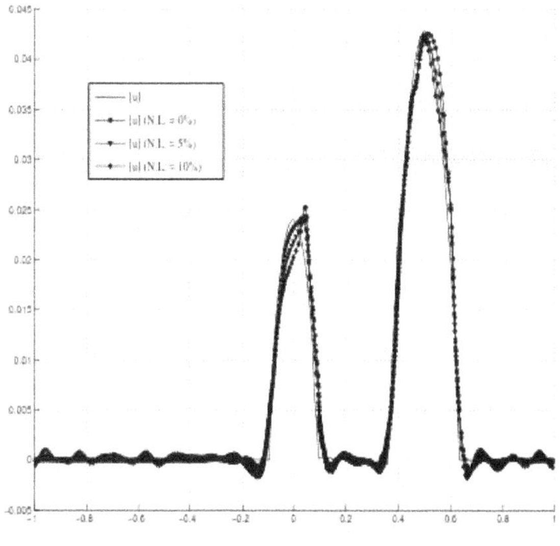

Figure 7: Interfacial jump displacement identification.

Steklov-Poincaré operator. The study of the properties of this operator for Neumann and Dirichlet variable, comparison with the energy approach recently presented by Baranger et al. [17,18] and its use in practical situations will be subject of a forthcoming paper [19-22].

Acknowledgment

This work was funded by the Ministère de la Recherche Scientifique, de la Technologie et du Développement des Compétences (MRSTDC, Tunisia) within the LAB-STI-02 program.

References

1. Hadamard J (1953) Lectures on Cauchy's problem in linear partial differential equation, Dover.
2. Yeih WC, Koya T, Mura T (1993) An inverse problem in elasticity with partially overspecified boundary conditions, Theoretical approach, Transactions. ASME Journal of Applied Mechanics 60: 595-600.
3. Koya T, Yeih WC, Mura T (1993) An inverse problem in elasticity with partially over specified boundary conditions. II. Numerical details. Transactions ASME Journal of Applied Mechanics 60: 601-606.
4. Marin L, Elliot L, Ingham DB, Lesnic D (2001) Boundary element method for the Cauchy problem in linear elasticity. Engineering Analysis with Boundary Elements 25(9): 783-793.
5. Marin L, Lesnic D (2002) Boundary element solution for the Cauchy problem in linear elasticity using singular value decomposition. Comput Methods Appl Mech Engrg, 191: 3257-3270.
6. Huang CH, Shih WY (1999) An inverse problem in estimating interfacial crack in bimaterials by boundary element technique. International Journal for Numerical Methods in Engineering 45: 1547-1567.
7. Weikl W, Andra H, Schnack E (2001) An alternating iterative algorithm for the reconstruction of internal cracks in 3D solid body. Inverse Problems 17: 1957-1975.
8. Koslov VA, Maz'ya VG, Fomin AV (1991) An iterative method for solving the Cauchy problem for elliptic equations. Comput Meth Math Phys 31: 45-52.
9. Riadh Ben Fatma, Mejdi Azaez, Amel Ben Abda, Nabil Gmati (2007) a Missing boundary data recovering for the Helmholtz problem. Comptes Rendus M'ecanique 335: 787-792.
10. Mohamed Larbi Kadri, Jalel Ben Abdallah, Thouraya Nouri Baranger (2011) Identification of internal cracks in a three-dimensional solid body via Steklov Poincar approaches. Comptes Rendus M'ecanique 339: 674-681.

11. Andrieux S, Baranger TN, Ben Abda A (2006) Solving Cauchy problems by minimizing an energy-like functional. Inverse Problem 22: 115-133.

12. Ben Belgacem F, El Fekih (2006) On Cauchy's problem: I A variational Stecklov-Poincar´e theory. Inverse Problem 21: 1915-1936.

13. Azaiez M, Ben Belgacem F, El Fekih (2005) On Cauchy's problem: II Completion, regularization and approximation. Inverse Problem 22: 1307-1336.

14. Quarteroni A, Valli A (1999) Domain Decomposition Methods for Partial Differential Equations. Oxford University Press.

15. Przemieniecki JS (1985) Theory of Matrix Structural Analysis, Dover: New York.

16. Ben Abdallah J (2007) A conjugate gradient type method for the Stecklov Poincar´e formulation of the Cauchy-Poisson Problem. International Journal of Applied Mathematics and Mechanics 3: 27-40.

17. Baranger TN, Andrieux S (2006) An energy approach to solve a Cauchy problem in elasticity, III European Conference on Computational Mechanics: Solids, Structures and Coupled Problems in Engineering, ECCM-2006, Lisbon, Portugal, 5-8 June, 2006.

18. Baranger TN, Andrieux S (2007) An optimization approach to solve Cauchy problem in linear elasticity. Journal of Structural and Multidisciplinary Optimization to appear.

19. Andrieux S, Ben Abda A, Baranger TN (2005) Data completion via an energy like error functional. Comptes Rendus M´ecanique 333: 171-177.

20. Azaiez M, Ben Abda A, Ben Abdallah J (2005) Revisiting the Dirichlet-to Neumann solver for data completion and application to some inverse problems. International Journal of Applied Mathematics and Mechanics 1: 106-121.

21. Johnson KL (1985) Contact mechanics. Cambridge: Cambridge, University Press.

22. Tikhonov AN, Arsenin VY (1977) Solution to Ill-posed Problems, Winston Wiley.

Stress Analysis of Steam Piping System

Yogita B Shinger* and Thakur AG

Mechanical Department, Sir Visvesvaraya Institute ofTechnology, Kopargoan, Maharashtra 422102, India

Abstract

This is about the design of steam piping and its stress analysis of a given process flow diagram. The prime objective of this project is to design the piping system and then to analyze its main components. Wall thicknesses are calculated for all pipes which were found very safe for the operating pressure. For header pipe the calculated wall thickness is 0.114 inch and the standard minimum wall thickness is 0.282 inch which is greater than the calculated one by more than 2.4 times. Different loads such as static loads, thermal loads of all pipes were also calculated. After load calculations, spacing supports carried out. Thermal and static analysis of main system pipe has been done and results were compared with ASME Power Piping Code B31.1. After calculation of all applied loads, standard circular column of 4 inch nominal size were designed and analyzed both manually and on ANSYS software. The results obtained from both methods were compared and found safe under available applied loads.

Keywords: Code B31.1; ANSYS; Steam piping

Introduction

Piping System design and analysis is a very important field in any process and power industry. Piping system is analogous to blood circulating system in human body and is necessary for the life of the plant. The steam piping system, mentioned in the project will be used for supplying steam to different locations at designed temperature and pressure. This piping system is one of the major requirements of the plant to be installed [1]. The goal of quantification and analysis of pipe stresses is to provide safe design. There could be several designs that could be safe. A piping engineer would have a lot of scope to choose from such alternatives, the one which is most economical, or most suitable etc. Good piping system design is always a mixture of sound knowledge base in the basics and a lot of ingenuity. The aim of the project is to design and analyze piping system according to standard piping Codes. The design should prevent failure of piping system against over stresses due to: Sustained loadings which act on the piping system during its operating time. While piping stress analysis is used to ensure:

1) Safety of piping and piping components

2) Safety of the supporting structures

Basically the sizing of this steam piping has already done and contained nearly on $750 \times 300m^2$ area, including 48 pipes and 52 junctions [2]. The detail of the piping system e.g. length of each pipe, Nominal Pipe Size (NPS) with pipe no. starting from 208 and ending on pipe no. 256 are shown from the following Figure 1. The rest of the data e.g. inlet and out let velocities of each pipe, inlet and out let pressure of each pipe and inlet and out let temperature of each and every pipe are arranged which will be used in further calculations.

This project is about the design of steam piping and its stress analysis of a given process flow diagram. The prime objective of this project is to design the piping system and then to analyze its main components (Figure 2).

This project includes the following tasks:

a) Process design of the complete piping system

b) Structural design of the pipes manually

c) Stress analysis of the pipes using ANSYS or compatible CAE software

d) Structural design of supports manually

Piping Standard and Codes

Before the selection of codes for the steam piping, a little detail about codes, standards and its historical background is given below.

Piping code development

The increase in operating temperatures and pressures led to the development of the ASA (now ANSI) B31 Code for pressure piping. During the 1950s, the code was segmented to meet the individual requirements of the various developing piping industries, with codes being published for the power, petrochemical and gas Transmission industries among others. The 1960s and 1970s encompassed a period of development of standard concepts, requirements and methodologies. The Development and use of the computerized mathematical models of piping system have brought analysis, design and drafting to new levels of sophistication. Codes and standards were established to provide methods of manufacturing, listing and reporting design data [3]. "A standard is a set of specifications for parts, materials or processes intended to achieve uniformity, efficiency and a specified quality". Basic purpose of the Standards is to place a limit on the number of items in the specifications, so as to provide a reasonable inventory of tooling, sizes and shapes and verities [4]. Some of the important document related to piping are:

I. American Society of Mechanical Engineers (ASME)

II. American National Standards Institute (ANSI)

III. American Society of Testing and Materials (ASTM)

IV. Pipe Fabrication Institute (PFI)

***Corresponding author:** Yogita B Shinger, Mechanical Department, Sir Visvesvaraya Institute of Technology, Kopargoan, Maharashtra 422102, India
E-mail: yogitashinger@gmail.com

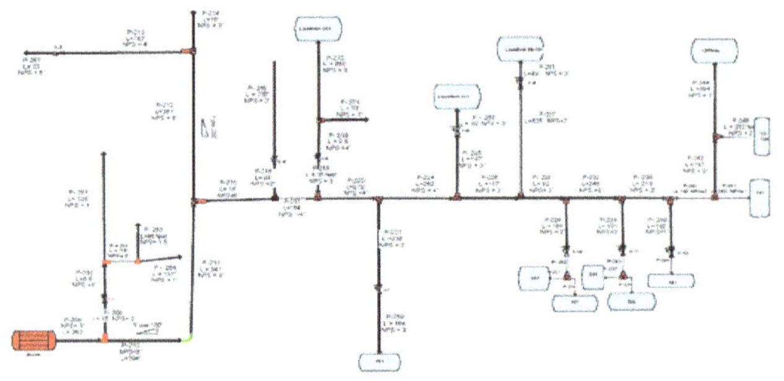

Figure 1: A conceptual model based on health, safety and ergonomic measures for sustainable workplace improvement that increase working capacity and decrease workload.

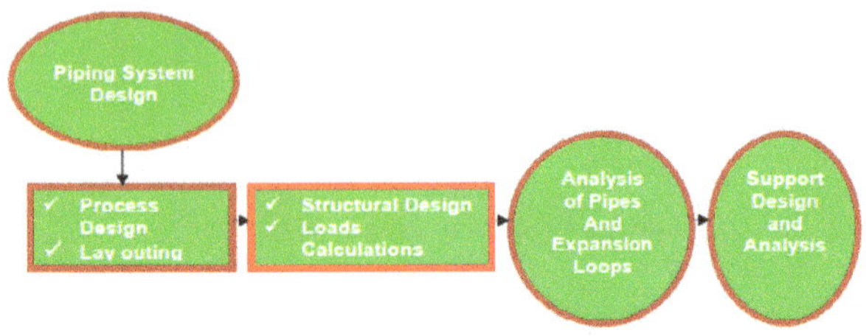

Figure 2: Complete stage designing of piping system.

V. American Welding Institute (AWS)

VI. Nuclear Regulatory Commission (NRC)

On the other side "A code is a set of specifications for analysis, design, Manufacture and construction of something". The basic purpose of code is to provide design criterion such as permissible material of construction, allowable working stresses and loads sets [4]. ASME Boiler and Pressure vessel codeB31 used for the design of commercial power and industrial piping system. This section has the following sub section [1].

B31.1: For Power Piping.

B31.3: For Chemical plant and Petroleum Refinery Piping.

B31.4: Liquid transportation system for Hydrocarbons, liquid petroleum gas, and Alcohols.

B31.5: Refrigeration Piping.

B31.8: Gas transportation and distribution piping system.

B31.1 Power piping code concerns mononuclear piping such as that found in the turbine building of a nuclear plant or in a fossil-fueled power plant.

B31.3 code governs all piping within limits of facilities engaged in the processing or handling of chemical, petroleum, or related products. Examples are a chemical plant compounding plant, bulk plant, and tank farm. B31.4 governs piping transporting liquids such as crude oil, condensate, natural gasoline, natural gas liquids, liquefied petroleum gas, liquid alcohol, and liquid anhydrous ammonia. These are auxiliary piping with an internal gauge pressure at or below 15 psi regardless of temperature. B31.5 covers refrigerants and secondary coolant piping for temperatures as low as 320°F. B31.8 governs most of the pipe lines in gas transmission and distribution system up to the outlet of the customer's meter set assembly. Excluded from this code with metal temperature above 450°F or below - 20°F. As for as the steam piping is concerned, B31.1 Power piping is used because of its temperature and pressure limitations which is discussed below in detail.

B31.1 power piping

This code covers the minimum requirements for the design, materials, fabrication, erection, testing, and inspection of power and auxiliary service piping systems for electric generation stations, industrial institutional plants, and central and district heating plants. The code also covers external piping for power boilers and high temperature, high-pressure water boilers in which steam or vapor is generated at a pressure of more than 15 psig and high-temperature water is generated at pressures exceeding 160 psig or temperatures exceeding 250°F. This code is typically used for the transportation of steam or water under elevated temperatures and pressure so this is the reason that why this code is selected for the steam piping system which is external to the boiler [5].

ASME code requirements

Boiler outlet section of the steam system comes under the category of ASME Code B31.1 Power. In order to ensure the safety of the piping system, code requirements should be fully satisfied. For different loads this code incorporates different relationships for stress level as given below [6].

Stresses due to sustained loadings:

The effects of the pressure, weight, and other sustained loads must meet the requirements of the following equation [1].

$$S_L=(PD_O/4t)+(0.75i\times M_A/Z) \quad (1)$$

Where,

P=internal pressure, psi

D_0=outside diameter of pipe, in

T=nominal wall thickness, in

Z=section modulus of pipe, in^3

M_A=resultant moment due to loading on cross section due to weight and other sustained load, in-lb

S_L=basic material allowable stress at design pressure, psi.

Stresses due to thermal loadings

The effect of thermal expansions must meet the following equations [1],

$$iM_c/Z\leq S_A+f(S_h-S_l) \quad (2)$$

Where,

f=stress range reduction factor,

Mc=range of resultant moment due to thermal expansions, lb-in

S_A=allowable stress range for expansions

The rest of term are same as above.

Piping Design Procedure

Process design

This process is based on the requirement of the process variables. It defines the required length and cross sectional area of pipe, the properties of fluid inside the pipe, nature and rate of flow in it. These variables affect the positioning and placements of equipments during lay outing and routing. The operating and design working conditions are clearly defined. The end of Process Plan Design is the creation of a Process Flow Diagram (PFD) and Process and Instrumental diagram (PID), which are used in the designing and lay outing of the Pipe [7].

Piping structural design

In piping structural design, according to pressure in pipelines, the design and minimum allowable thicknesses are calculated; according to the required codes and standards. ASME codes for various standards are available, for process fluid flow, ASME B31.1 is used [8].

Pipe thickness calculations

Piping codes ASME B31.1 require that the minimum thickness t_m including the allowance for mechanical strength, shall not be less than the thickness [2].

$$t_m=(P\times D_O/2\times S\times E_q+P\times Y)+A \quad (3)$$

Where,

t_m=minimum required thickness, inches

t=pressure design thickness, inches

p=internal pressure, psig

Do = outside diameter of pipe, inches

S=Allowable stress at design temperature(know as hot stress), psi

A=allowance additional thickness to provide for material removed in threading corrosion or erosion allowance ,manufacturing tolerance should also be consider

Y = coefficient that takes material properties and design temperature into account

For temperature below 900°F, 0.4 may be assumed

E_q=quality factor.

Allowable working pressure

The allowable working pressure of a pipe can be determined by Equation [2]

$$P= 2(S\times E_q)\times t/(D_{o-}2Yt) \quad (4)$$

Where,

t=specified wall thickness or actual wall thickness in inches.

For bends the minimum wall thickness after bending should not be less than the minimum required for straight pipe [9].

Sustained load calculations

Sustained loads are those loads which are caused by mechanical forces and these loads are present through out the normal operation of the piping system. These loads include both weight and pressure loadings. The support must be capable of holding the entire weight of the system, including that of that of the pipe, insulation, fluid components, and the support themselves [2].

$$\text{Pipe weight} =(\pi/4)\times\rho_{steel}\times(D_O^2-Di^2)\times(g/g_c)$$

$$\text{Fluidweight}=(\pi/4)\times\rho_{fluid}\times(Di)^2\times(g/g_c) \quad (5)$$

$$\text{Insulation weight}=\text{Insulation factor}\times\rho_{insulation}\times(g/g_c)$$

D_0=outside diameter of pipe, in

Di=inside diameter ofpipe, in

t=insulation thickness depends upon NPS, in

g=acceleration due to gravity, ft/sec^2

g_c=gravitionalconstant, lbm-ft/ft sec^2

ρ_{steel}=density of steel, lb/in^3

ρ_{fluid}=density of water, lb/in^3

$\rho_{insulation}$=density of insulation, lb/in^3

Insulation factor depends on the thickness of the insulation of the pipe.

Static Loads Calculations

For Static loads calculation, considering a pipe no. 208 and taking its section up to first vertical leg of the expansion loop. This pipe is to be considering as a straight beam with uniformly distributed load [10].

Design specifications

NPS (Nominal Pipe Size)=8 in=200 mm

Pipe outer Diameter=8.625 in

Pipe thickness=0.322 in

Total (metal +Insulation +Fluid) distributed weight of pipe=50 lb/ft=4.167 lb/in Section Modulus, Z=16.8 in3

Moment of Inertia, I=72.5 in4

Modulus of Elasticity, E=27.5 Mpsi

For segment A-B as shown in Figures 3-5 below, taking the weight, shear force (Refer Figure 6) and

Moment equation and then using length L_1=22.22 ft

$$W(x)=-M_o(x)^{-2} + R_0(x)^{-1}-w(x)^1 -R_1(x-a)^{-1}-M_1(x-L)^{-2}$$

$$V(x)=-M_o(x)^{-1}+ R_0(x)^0-w(x)^1 -R_1(x-a)^{-1}-M_1(x-L)^{-2} \quad (6)$$

$$M(x)= -M_o(x)^0 + R_0(x)^1 -w(x)^2 -R_1(x-a)^1-M_1(x-L)^0$$

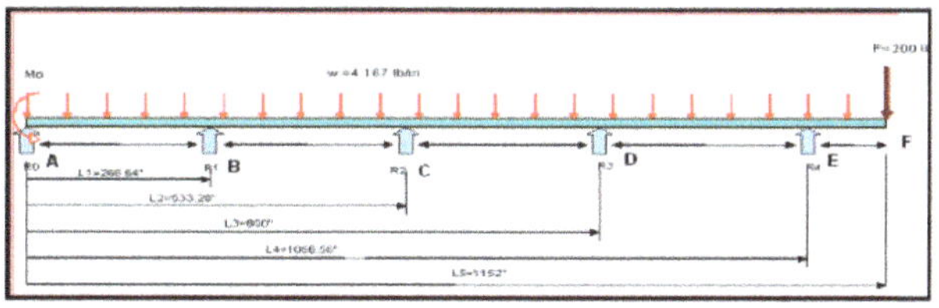

Figure 3: Symmetry of header pipe considering as a beam Solving Segment A-B.

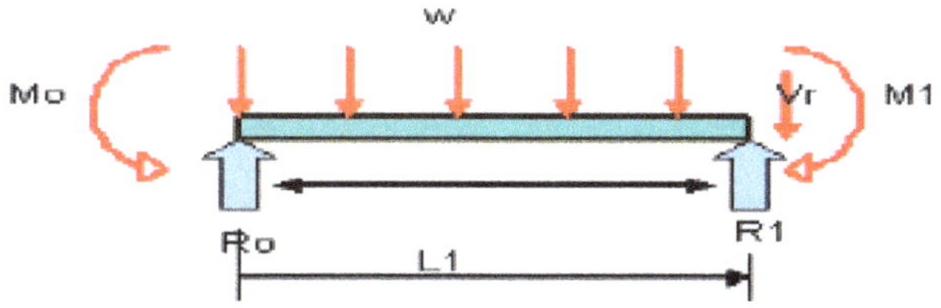

Figure 4: Segement A-B.

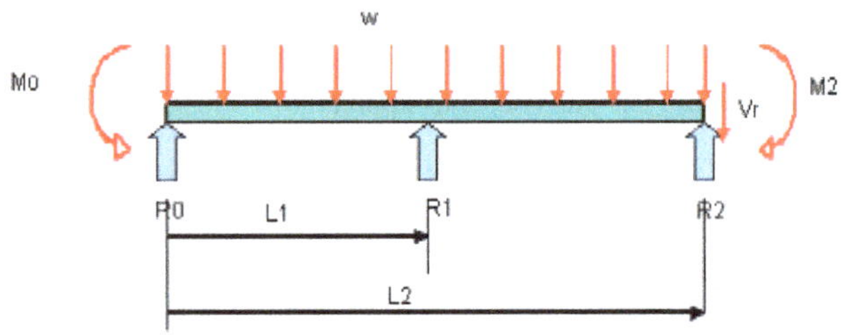

Figure 5: Segment A-B-C.

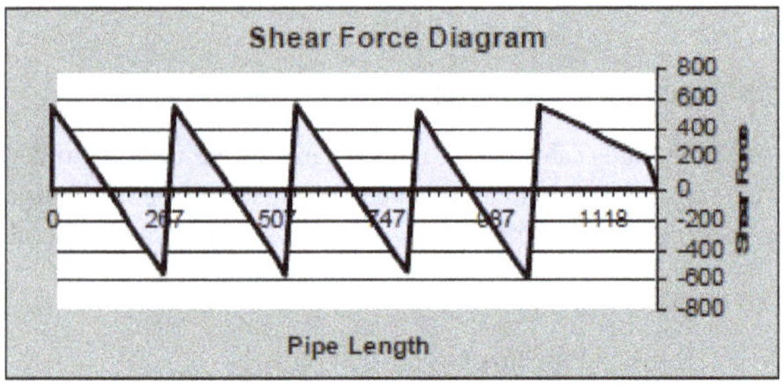

Figure 6: Shear force diagram.

Integrating the moment equation twice and putting boundary conditions we get,

$EIy(x) = -M_0 (x)^2/2 + R_0(x)^3/6 - w(x)^4/24 = 0$ (7)

As the segment AB,x=L_1=266.64 in

$-M_0(l_1)^2/2 + R_0(l_1)^3/6 - w(l_1)^4/24 = 0$

$-35548.44M_0 + 315945.77R_0 - 877634043.8 = 0$

$EIy(l_2) = - M_0(l_2)^2 + R_0(l_2)^3/6 + R_1(l_2-l_1)^3/6 - w(l_2)^4/24 = 0$ (8)

$-142193.78M_0 + 252766366.2R_0 + 3159545.78R_1 - 1.404_e^{10} = 0$

Similarly for segment C-D:

$EIy(l_3) = -M_0(l_3)^2 + R_0(l_3)^3/6 + R_1(l_3-l_1)^3/6 - R2(l_3-l_2)^3/6 - w(l_3)^4/24 = 0$ (9)

$-320000M_0 + 85333333.33R_0 + 25287743.4R_1 + 3159545.78R_2 - 7.11e^{11} = 0$

For segment D-E:

$EIy(l_4) = -M_0(l_4)^2 + R_0(l_4)^3/6 + R_1(l_4-l_1)^3/6 - R_2(l_4-l_2)^3/6 + R_3(l_4-l_3)^3/6 - w(l_3)^4/24 = 0$ (10)

$568775.12M_0 + 202210929R_0 + 85307735.9R_1 + 252776366.2R_2 + 3156702R_3 - 2.24e^{11} = 0$

Now taking summation of moment at left end of right end support

$M_0 + R_0l_4 + R_1(l_4-l_1) + R_2(l_4-l_2) + R_3(l_4-l_3) - wl_1(l_4-a) - w(l_2-l_1)(l_4-b) - w(l_3-l_2)(l_4-c) - w(l_3-l_2)(l_4-c) - w(l_4-l_3)(l_4-d) - wx^2/2 - P\times x = 0$ (11)

$M_0 + 1066.56R_0 + 800R_1 + 533.28R_2 + 266.64R_3 - 2369952.574 = 0$

Solving the above (7),(8),(9),(10),(11) equation we get

M_o = -24401 lb.in, R_o = 552 lb, R1=1123 lb R_2=1067 lb,

R_3= 1266 lb

For R_4,taking

$R_0 + R_1 + R_2 + R_3 + R_4 = wL + 800$

R_4 = 159 1 lb. Maximum Bending Moment (Figure 7) M_{max} = -32741.44533 lb-in at x =799.92 in

Verification from code

The effects of the pressure, weight, and other sustained loads must meet the requirements of the following equation [1].

$$SL = \frac{PD_0}{4t} + \frac{0.75ixM_A}{Z} \le 1.0S_h$$

Where,

P = Internal Pressure, psi

Do = Out Side diameter of Pipe, in

t = nominal wall thickness, in

Z=Section modulus of pipe, in^3

M_A=Resultant moment due to weight and other sustained loads, lb-in

S_L=Allowable stress at design hot pressure, psi

i =stress intensification factor

$193.7\times8.625 / (4\times0.322) + 0.75 \text{ x } 1\times32700/ (16.8) \le 1.0\times14400$

$2756.92 \le 14400$

$2.75\ 10^3 \le 14.4\times10^3$

Piping Analysis on ANSYS

Analysis was performed for the pipe in ANSYS for using the following data.

Element type=Beam 3

Material properties

Modulus of Elasticity=27.5 Mpsi

Poison's Ratio=0.283

Density=0.283 lb/in^3

Type of Loads

Four Vertical constraints in the middle and one all degree of Freedom constrained at the start.

Gravity=9.81(386.22 in/sec^2)

Final Meshing=96 elements for total length of the beam (Figures 8-10) (22 elements for first four each sections and 8 elements for the last section. Refining the mesh from 24 elements up to 96 elements but there is no change found in deformation values and bending moment values).

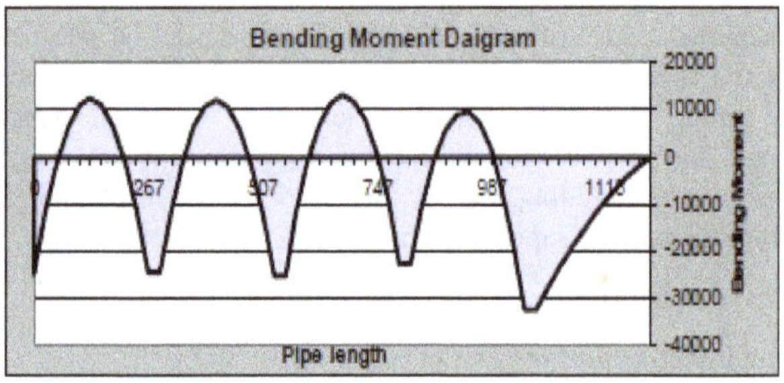

Figure 7: Bending moment diagram.

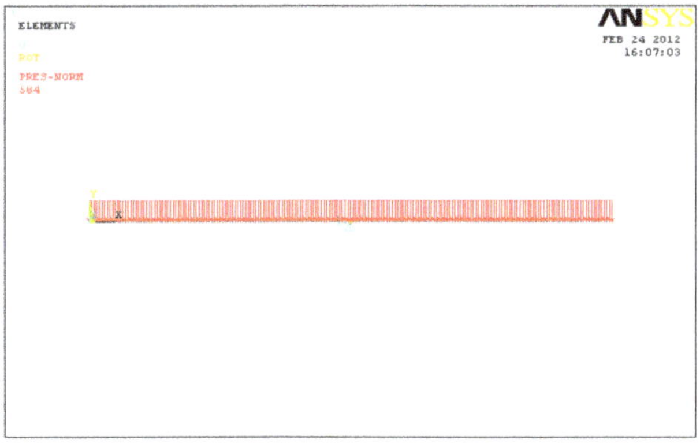

Figure 8: Loaded view of meshed beam.

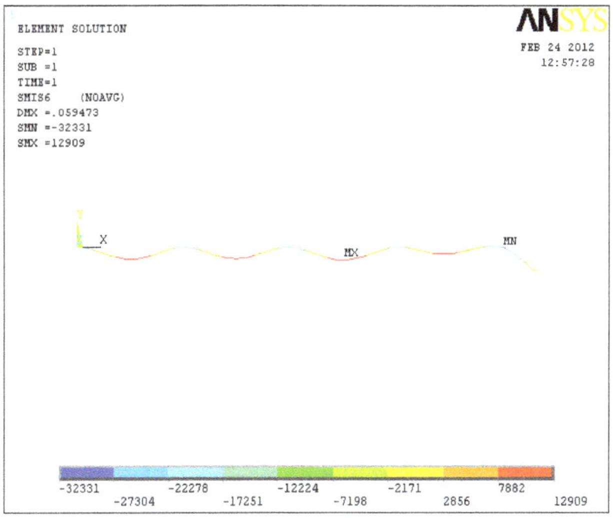

Figure 9: Deflection (inch) in pipe.

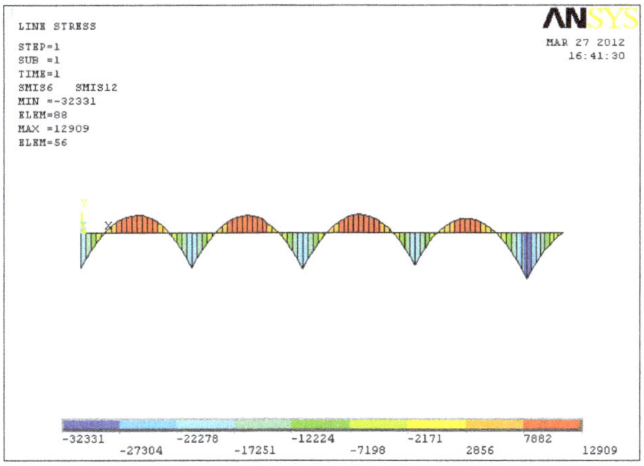

Figure 10: Bending stress (psi) in pipe.

Method	Max. Deflection (in)	Max. Bending (lb-in)
Manual	0.064	32741.445
ANSYS Results	0.0596	32921.00

Table 1: Comparision of analysis of beam.

Comparison of analysis

The maximum deflections and bending moment values obtained from both methods are arranged in Table 1.

Results and Discussion

From the results obtained both manually and on ANSYS, the difference in maximum deflection is 6.4% where on the other hand the difference in the max. Bending moment is 1.35%. Deformation is less than 0.1 inch and also the maximum bending stress is 1947.55 psi which is quite less than the allowable stress of the pipe.

Conclusion

Following conclusions are made from the analysis of the designed system.

1. The designed pipe verified all the conditions defined by the ASME Boiler and Pressure Vessel code B31.1. Thickness and working pressure calculated are in the safe limit. Thermal and Sustained analysis results obtained are in the safe limits defined by the Code.
2. Supporting Assembly confirms to the safety requirements of AISC standards.
3. The analysis shows that the complete system is safe and the results are verified by manual calculations and ANSYS software.
4. On the positive side of the manual calculations lays the fact that it gives fully basic concept of the piping system. While the assumptions made during manual calculations make the results slightly differ from the software results.

References

1. Li M, Manohar LA (2011) Stress analysis of non-uniform thickness piping system with general piping analysis software, Nuclear Engineering & Design 241: 555-56.
2. De Melo FJMQ, Oliveira CAM, Fonseca EMM (2009) The thermal and mechanical behavior of structural steel piping, International Journal of Pressure Vessels and Piping 82: 145-15.
3. Mathan G, Siva Prasad N (2011) Study of dynamic response of piping system with gasketed flanged joints using finite element analysis, International Journal of Pressure Vessels and Piping.
4. Martin M. Schwarz (2004) Flexibility analysis of the vessel-piping interface, International Journal of Pressure Vessels and Piping 81: 181-189.
5. Fu-Zhen X, Pei-Ning L (2004) Finite element-based limit load of piping branch junctions under combined loadings, Nuclear Engineering and Design 231:141-150.
6. The American Society of Mechanical engineers, ASME B31.1-2001 Power piping, revised edition.
7. Sam K (2006) Introduction to pipe stress analysis, John Wiley & Sons, USA.
8. Shigley JE, Mischke CR (2000) Mechanical Engineering Design, 5th edition, McGraw-Hill Book Co.2000.
9. Spirax Sarco Company Ltd. (2008) Supports and Expansion Loops, International Site for spirax sarco.
10. TPC Training system, Piping system, A Dun & Brad Stress Comp. 1999.

6

Transient Thermoelectroelastic Response of a Functionally Graded Piezoelectric Material Strip with Two Parallel Cracks in Arbitrary Positions

Ueda S* and Kishimoto T

Department of Mechanical Engineering, Osaka Institute of Technology, Japan

Abstract

In this article, the problem of two parallel cracks in arbitrary positions of a functionally graded piezoelectric material (FGPM) strip is analyzed under transient thermal loading conditions. It is assumed that the thermoelectroelastic properties of the strip vary continuously along the thickness of the strip, and that the crack faces are supposed to be insulated thermally and electrically. By using both the Laplace transform and the Fourier transform, the thermal and electromechanical problems are reduced to two systems of singular integral equations. The singular integral equations are solved numerically, and a numerical method is then employed to obtain the time dependent solutions by way of a Laplace inversion technique. The intensity factors versus time for various geometric and material parameters are calculated and presented in graphical forms. Temperature change, the stress and electric displacement distributions in a transient state are also included.

Keywords: Thermo electroelasticity; Fracture mechanics; Functionally graded piezoelectric material; Arbitrary positions; Two parallel cracks; Integral transform; Transient response

Introduction

Piezoelectric materials widely have been used as sensors and actuators in smart or intelligent systems to sense thermally induced distortions and to adjust for adverse thermomechanical conditions [1,2]. The requirements of structural strength, reliability and lifetime of these structures call for a better understanding of the mechanics of fracture in piezoelectric materials under thermal loading.

Recently, functionally graded piezoelectric materials (FGPMs) have been developed to improve their reliability [3], and the electromechanical fracture of the FGPM under mechanical and electrical loadings has received much attention [4-6]. Thus, it is also important to investigate the fracture behavior of FGPMs under thermal load, and some interesting results have been reported [7-12].

While the fact that piezoelectric materials involve multiple cracks, most of the existing contributions are concerned with the fracture behavior of a single crack. Then some thermal fracture problems of homogeneous piezoelectric strips with two dimensional cracks, such as two coplanar cracks [13], two parallel cracks [14], parallel multi cracks [15] and a T-shaped crack [16], have been treated. Moreover, the over shooting phenomenon of intensity factors is observed in a piezoelectric plate under the thermal shock loading [17,18]. So, in this type of research, it is important to investigate the transient thermal fracture behavior of piezoelectric materials with multiple cracks. Although the present authors investigated the thermoelectromechanical interaction between two parallel axisymmetric cracks in an FGMP strip [19,20], one of the remaining problems that need to be fully understood is that of interaction between cracks in arbitrary positions of FGPMs under thermal shock loading.

In this study, the problem of two parallel cracks in arbitrary positions in a plate of an FGPM strip is analyzed under transient thermal loading conditions. It is assumed that the thermoelectroelastic properties of the strip vary continuously along the thickness of the strip, and that the crack faces are supposed to be insulated thermally and electrically [5,9]. By using both the Laplace and Fourier transform techniques [21,22], the thermal and electromechanical problems are reduced to two systems of singular integral equations. The singular integral equations are solved numerically [23], and a numerical method is employed to obtain time-dependent solutions by way of a Laplace inversion technique [24]. The intensity factors versus time for various geometric and material parameters are calculated.

Formulation of the Problem

Consider an infinite FGPM strip of thickness $h = h_1 + h_2$ containing two parallel through cracks of different length $2C_k$ (k = 1, 2) being spaced at distances $2d$ in the x-direction and $2h_0$ in the Z -direction as shown in Figure 1. The rectangular coordinates x, y and z are

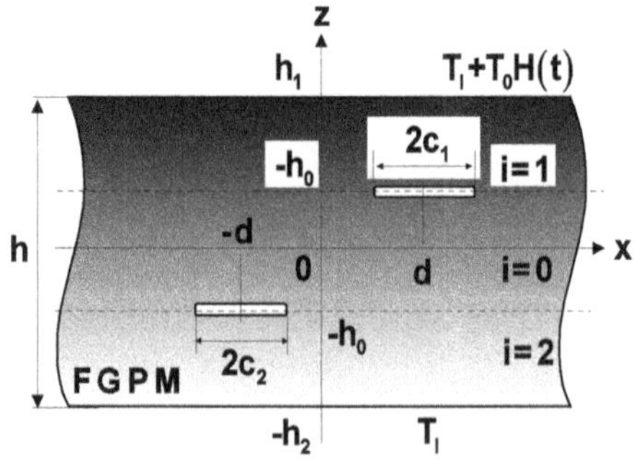

Figure 1: Geometry of the crack problem in a functionally graded piezoelectric strip.

***Corresponding author:** Ueda S, Department of Mechanical Engineering, Osaka Institute of Technology, 5-16-1 Omiya, Asahi-ku, Osaka, 535-8585, Japan E-mail: ueda@med.oit.ac.jp

coincident with the principal axes of the material. The piezoelectric strip is poled in the z-direction and is in the plane strain conditions perpendicular to the y -axis. It is assumed that initially the medium is at the uniform temperature T_I and is suddenly subjected to a uniform temperature rise $T_0\, H(t)$ along the boundary $Z = h_1$, where $H(t)$ is the Heaviside step function and t denotes time. The temperature along the boundary $Z = -h_2$ is maintained at T_I. The crack faces remain thermally and electrically insulated [5,9]. In the following, the subscripts x, y, z will be used to refer to the direction of coordinates.

The material properties, such as the elastic stiffness constants C_{kl}, the piezoelectric constants e_{kl}, the dielectric constants ε_{kk}, the stress-temperature coefficients λ_{kk}, the coefficients of heat conduction, K_x, K_z and the pyroelectric constant P_z are one-dimensionally dependent as

$$\left.\begin{aligned} (c_{kl}, e_{kl}, \varepsilon_{kk}) &= (c_{kl0}, e_{kl0}, \varepsilon_{kk0})\exp(\beta z) \\ (\lambda_{kk}, p_z) &= (\lambda_{kk0}, p_{z0})\exp[(\beta+\omega)z] \\ (\kappa_x, \kappa_z) &= (\kappa_{x0}, \kappa_{z0})\exp(\delta z) \end{aligned}\right\} \tag{1}$$

where β, ω and δ, are positive or negative constants, and the subscript 0 indicates the properties at the plane $Z = 0$. For some materials, the thermal diffusivity λ_0 indeed doesn't vary dramatically, and then λ_0 is assumed to be a constant.

The constitutive equations for the electroelastic fields are,

$$\left.\begin{aligned} \sigma_{xxi} &= c_{11}\frac{\partial u_{xi}}{\partial x} + c_{13}\frac{\partial u_{zi}}{\partial z} + e_{31}\frac{\partial \phi_i}{\partial z} - \lambda_{11}T_i \\ \sigma_{zzi} &= c_{13}\frac{\partial u_{xi}}{\partial x} + c_{33}\frac{\partial u_{zi}}{\partial z} + e_{33}\frac{\partial \phi_i}{\partial z} - \lambda_{33}T \\ \sigma_{zxi} &= c_{44}\left(\frac{\partial u_{xi}}{\partial z} + \frac{\partial u_{zi}}{\partial x}\right) + e_{15}\frac{\partial \phi_i}{\partial x} \end{aligned}\right\} \quad (i = 0,1,2) \tag{2}$$

where $T_i \equiv T_i\,(x,z,t)$ is the temperature, $\varphi_i \equiv \varphi_i\,(x,z,t)$ is the electric potential, $u_{xi} \equiv u_{xi}\,(_{zxi})$, $u_{zi} \equiv u_{zi}(x,z,t)$, are the displacement components, $\sigma_{xxi} \equiv \sigma_{xxi}\,(x,z,t)$, $\sigma_{zzi} \equiv \sigma_{zzi}(x,z,t)$, $_{zxi} \equiv \sigma_{zxi}\,(x,z,t)$ $(i = 0,1,2)$ are the stress components. The subscript i = 0,1,2 denotes the thermoelectroelastic fields in $-h_0 \le Z \le h_0$, $h_0 \le Z \le h_1$, $-h_2 \le Z \le -h_1$ respectively. For the electric field, the constitutive relations are

$$\left.\begin{aligned} D_{xi} &= e_{15}\left(\frac{\partial u_{xi}}{\partial z} + \frac{\partial u_{zi}}{\partial x}\right) - \varepsilon_{11}\frac{\partial \phi_i}{\partial x} \\ D_{zi} &= e_{31}\frac{\partial u_{xi}}{\partial x} + e_{33}\frac{\partial u_{zi}}{\partial z} - \varepsilon_{33}\frac{\partial \phi_i}{\partial z} + p_z T \end{aligned}\right\} \quad (i = 0,1,2) \tag{3}$$

Where $D_{xi} \equiv D_{xi}\,(x,z,t)$, $D_{zi} \equiv D_{zi}\,(x,z,t)$ $(i = 0,1,2)$ are the electric displacement components.

The temperature is assumed to satisfy the Fourier heat conduction equations:

$$\kappa^2\frac{\partial^2 T_i}{\partial x^2} + \frac{\partial^2 T_i}{\partial z^2} + \delta\frac{\partial T_i}{\partial z} = \frac{1}{\lambda_0}\frac{\partial T_i}{\partial t} \quad (i = 0,1,2) \tag{4}$$

where $\kappa^2 = \kappa_x/\kappa_z$. The equations of equilibrium and electrostatics are,

$$\left.\begin{aligned} & c_{110}\frac{\partial^2 u_{xi}}{\partial x^2} + c_{440}\frac{\partial^2 u_{xi}}{\partial z^2} + (c_{130}+c_{440})\frac{\partial^2 u_{zi}}{\partial x\partial z} + (e_{310}+e_{150})\frac{\partial^2 \phi_i}{\partial x\partial z} \\ & +\beta\left[c_{440}\left(\frac{\partial u_{xi}}{\partial z} + \frac{\partial u_{zi}}{\partial x}\right) + e_{150}\frac{\partial \phi_i}{\partial z}\right] = \lambda_{110}\exp(\omega z)\frac{\partial T_i}{\partial x} \\ & c_{440}\frac{\partial^2 u_{zi}}{\partial x^2} + c_{330}\frac{\partial^2 u_{zi}}{\partial z^2} + (c_{130}+c_{440})\frac{\partial^2 u_{xi}}{\partial x\partial z} + e_{150}\frac{\partial^2 \phi_i}{\partial x^2} + e_{330}\frac{\partial^2 \phi_i}{\partial z^2} \\ & +\beta\left(c_{130}\frac{\partial u_{xi}}{\partial x} + c_{330}\frac{\partial u_{zi}}{\partial z} + e_{330}\frac{\partial \phi_i}{\partial z}\right) = \lambda_{330}\exp(\omega z)\left[\frac{\partial T_i}{\partial z} + (\beta+\omega)T_i\right] \\ & e_{150}\frac{\partial^2 u_{zi}}{\partial x^2} + e_{330}\frac{\partial^2 u_{zi}}{\partial z^2} + (e_{150}+e_{310})\frac{\partial^2 u_{xi}}{\partial x\partial z} - \varepsilon_{110}\frac{\partial^2 \phi_i}{\partial x^2} - \varepsilon_{330}\frac{\partial^2 \phi_i}{\partial z^2} \\ & +\beta\left(e_{310}\frac{\partial u_{xi}}{\partial x} + e_{330}\frac{\partial u_{zi}}{\partial z} - \varepsilon_{330}\frac{\partial \phi_i}{\partial z}\right) = -p_{z0}\exp(\omega z)\left[\frac{\partial T_i}{\partial z} + (\beta+\omega)T_i\right] \end{aligned}\right\} (i = 0,1,2) \tag{5}$$

The initial and boundary conditions for the temperature field can be written as

$$T_i(x,z,0) = T_I \quad (i = 0,1,2) \tag{6}$$

$$\left.\begin{aligned} & \frac{\partial T_0(x,\theta_{0i},t)}{\partial z} = 0 && (a_i < x < b_i) \\ & T_0(x,\theta_{0i},t) = T_i(x,\theta_{0i},t) && (-\infty < x \le a_i, b_i \le x < \infty) \end{aligned}\right\} \quad (i = 1,2) \tag{7}$$

$$\left.\begin{aligned} & \frac{\partial T_0(x,\theta_{0i},t)}{\partial z} = \frac{\partial T_i(x,\theta_{0i},t)}{\partial z} && (-\infty < x < \infty) \\ & T_i(x,\theta_{1i},t) = \begin{cases} T_I + T_0 H(t) & (i = 1) \\ T_I & (i = 2) \end{cases} && (-\infty < x < \infty) \end{aligned}\right\} \quad (i = 1,2) \tag{8}$$

If the electrically impermeable boundary is chosen as an idealized crack face electric boundary condition [9,25], the boundary conditions of this problem can be written as

$$\left.\begin{aligned} & \sigma_{zz0}(x,\theta_{0i},t) = 0 && (a_i < x < b_i) \\ & u_{z0}(x,\theta_{0i},t) = u_{zi}(x,\theta_{0i},t) && (-\infty < x \le a_i, b_i \le x < \infty) \end{aligned}\right\} \quad (i = 1,2) \tag{9}$$

$$\left.\begin{aligned} & \sigma_{zx0}(x,\theta_{0i},t) = 0 && (a_i < x < b_i) \\ & u_{x0}(x,\theta_{0i},t) = u_{xi}(x,\theta_{0i},t) && (-\infty < x \le a_i, b_i \le x < \infty) \end{aligned}\right\} \quad (i = 1,2) \tag{10}$$

$$\left.\begin{aligned} & D_{z0}(x,\theta_{0i},t) = 0 && (a_i < x < b_i) \\ & \phi_0(x,\theta_{0i},t) = \phi_i(x,\theta_{0i},t) && (-\infty < x \le a_i, b_i \le x < \infty) \end{aligned}\right\} \quad (i = 1,2) \tag{11}$$

$$\left.\begin{aligned} & \sigma_{zz0}(x,\theta_{0i},t) = \sigma_{zzi}(x,\theta_{0i},t) && (-\infty < x < \infty) \\ & \sigma_{zx0}(x,\theta_{0i},t) = \sigma_{zxi}(x,\theta_{0i},t) && (-\infty < x < \infty) \\ & D_{z0}(x,\theta_{0i},t) = D_{zi}(x,\theta_{0i},t) && (-\infty < x < \infty) \end{aligned}\right\} \quad (i = 1,2) \tag{12}$$

$$\left.\begin{aligned} & \sigma_{zzi}(x,\theta_{1i},t) = 0 && (-\infty < x < \infty) \\ & \sigma_{zxi}(x,\theta_{1i},t) = 0 && (-\infty < x < \infty) \\ & D_{zi}(x,\theta_{1i},t) = 0 && (-\infty < x < \infty) \end{aligned}\right\} \quad (i = 1,2) \tag{13}$$

In Eqs. (7)-(13), θ_{0i}, θ_{1i}, α_i and b_i $(i = 1,2)$ are given by

$$(\theta_{0i}, \theta_{1i}, a_i, b_i) = \left\{\begin{matrix} (h_0, h_1, d-c_1, d+c_1) & (i = 1) \\ (-h_0, -h_2, -d-c_2, -d+c_2) & (i = 2) \end{matrix}\right\} \tag{14}$$

Temperature Field

For the problem considered here, it is convenient to represent the temperature $T_i\,(x,z,t)$ $(i = 0,1,2)$ as the sum of the uniform temperature T_I and two functions.

$$T_i(x,z,t) = T_I + T^{(1)}(z,t) + T_i^{(2)}(x,z,t) \quad (i = 0,1,2) \tag{15}$$

where the non-disturbed temperature $T^{(1)} \equiv T^{(1)}(z,t)$ satisfies the following equation accompanied by initial and boundary conditions:

$$\frac{\partial^2 T^{(1)}}{\partial z^2} + \delta \frac{\partial T^{(1)}}{\partial z} = \frac{1}{\lambda_0}\frac{\partial T^{(1)}}{\partial t} \tag{16}$$

$$T^{(1)}(z,0) = 0 \tag{17}$$

$$\left.\begin{array}{l} T^{(1)}(h_1,t) = T_0 H(t) \\ T^{(1)}(-h_2,t) = 0 \end{array}\right\} \tag{18}$$

and the disturbed temperatures $T_i^{(2)} \equiv T_i^{(2)}(x,z,t)\ (i = 0,1,2)$ are subjected to the relations:

$$\kappa^2 \frac{\partial^2 T_i^{(2)}}{\partial x^2} + \frac{\partial^2 T_i^{(2)}}{\partial z^2} + \delta \frac{\partial T_i^{(2)}}{\partial z} = \frac{1}{\lambda_0}\frac{\partial T_i^{(2)}}{\partial t} \quad (i = 0,1,2) \tag{19}$$

$$T_i^{(2)}(x,z,0) = 0 \quad (i = 0,1,2) \tag{20}$$

$$\left.\begin{array}{ll} \dfrac{\partial T_0^{(2)}(x,\theta_{0i},t)}{\partial z} = -\dfrac{\partial T^{(1)}(\theta_{0i},t)}{\partial z} & (a_i < x < b_i) \\ T_0^{(2)}(x,\theta_{0i},t) = T_i^{(2)}(x,\theta_{0i},t) & (-\infty < x \le a_i, b_i \le x < \infty) \end{array}\right\} \quad (i = 1,2) \tag{21}$$

$$\left.\begin{array}{ll} \dfrac{\partial T_0^{(2)}(x,\theta_{0i},t)}{\partial z} = \dfrac{\partial T_i^{(2)}(x,\theta_{0i},t)}{\partial z} & (-\infty < x < \infty) \\ T_i^{(2)}(x,\theta_{1i},t) = 0 & (-\infty < x < \infty) \end{array}\right\} \quad (i = 1,2) \tag{22}$$

Define a Laplace transform pair by

$$f^*(p) = \int_0^\infty f(t)\exp(-pt)dt, \quad f(t) = \frac{1}{2\pi i}\int_{Br} f^*(p)\exp(pt)dp \tag{23}$$

where Br denotes the Bromwich path in pertinent complex planes, and applying the Laplace transform, it is easy to find from Eqs. (16)-(18) that

$$T^{(1)*}(z,p) = \frac{T_0}{p[1-\exp(-2\mu_0 h)]}\{\exp[\mu_2(h_1 - z)] - \exp[-2\mu_0 h + \mu_1(h_1 - z)]\} \tag{24}$$

with

$$\left.\begin{array}{l} \mu_0 = \left(\dfrac{\delta^2}{4} + \dfrac{p}{\lambda_0}\right)^{1/2} \\ \mu_1 = \dfrac{\delta}{2} + \mu_0, \quad \mu_2 = \dfrac{\delta}{2} - \mu_0 \end{array}\right\} \tag{25}$$

In the following, the superscript * is used to refer to the physical quantities in the Laplace transform plane.

The general solutions of the governing Eq. (19) can be obtained by using the Laplace-Fourier integral transform techniques [21]:

$$T_i^{(2)*}(x,z,p) = \frac{1}{2\pi}\sum_{j=1}^{2}\int_{-\infty}^{\infty} D_{ij}(s,p)\exp(|s|\tau_{ij}z)\exp(-isx)ds \quad (i = 0,1,2) \tag{26}$$

In the above expressions, $D_{ij}(s,p)(i = 0,1,2,\ j = 1,2)$ are unknown functions to be solved and

τ_{ij} $\tau_{ij}(s,p)$ $(i = 0,1,2, j = 1,2)$ are defined as

$$\left.\begin{array}{l} \tau_{i1}(s,p) = -\tau - \dfrac{\delta}{2|s|}, \quad \tau_{i2}(s,p) = \tau - \dfrac{\delta}{2|s|} \\ \tau \equiv \tau(s,p) = \left[\kappa^2 + \left(\dfrac{\delta}{2|s|}\right)^2 + \dfrac{p}{\lambda_0 s^2}\right]^{1/2} \end{array}\right\} \quad (i = 0,1,2) \tag{27}$$

The problem may be reduced to a system of singular integral equations by defining the following new unknown functions $G_{k0}(x,p)$ $(k = 1,2)$ [22]:

$$G_{k0}(x,p) = \left\{\begin{array}{ll} \dfrac{\partial}{\partial x}\left[T_0^{(2)*}(x,\theta_{0k},p) - T_k^{(2)*}(x,\theta_{0k},p)\right] & (a_k < x < b_k) \\ 0 & (-\infty < x \le a_k, b_k \le x < \infty) \end{array}\right\} \quad (k = 1,2) \tag{28}$$

Making use of the first boundary conditions (21) with Eqs. (22), we have the following system of the singular integral equations for the determination of the unknown functions $G_{k0}(\xi, p)(k = 1,2)$:

$$\frac{\kappa}{2\pi}\int_{a_1}^{b_1}\left[\frac{1}{\xi - x} + M_{011}(\xi,x,p)\right]G_{10}(\xi,p)d\xi$$

$$-\int_{a_2}^{b_2} M_{012}(\xi,x,p)G_{20}(\xi,p)d\xi = \frac{d}{dz}T^{(1)*}(h_0,p) \quad (a_1 < x < b_1) \tag{29}$$

$$\frac{\kappa}{2\pi}\int_{a_1}^{b_1} M_{021}(\xi,x,p)G_{10}(\xi,p)d\xi$$

$$-\int_{a_2}^{b_2}\left[\frac{1}{\xi - x} + M_{022}(\xi,x,p)\right]G_{20}(\xi,p)d\xi = \frac{d}{dz}T^{(1)*}(-h_0,p) \quad (a_2 < x < b_2) \tag{30}$$

In Eqs. (29) and (30), the kernel functions $M_{0nk}(\xi,x,p)(n,k = 1,2)$ are given by

$$\left.\begin{array}{l} M_{011}(\xi,x,p) = \displaystyle\int_0^\infty \left\{\left(\frac{\tau_{02}\rho_1}{\kappa\tau\rho_0} - 1\right) - \frac{\tau_{01}\rho_1}{\kappa\tau\rho_0}\exp[-2s\tau(h_2 + h_0)]\right\}\sin[s(\xi - x)]ds \\ M_{012}(\xi,x,p) = -\displaystyle\int_0^\infty \frac{\rho_2}{\kappa\tau\rho_0}\{\tau_{01}\exp(2s\tau_{01}h_0) \\ \qquad -\tau_{02}\exp[-s(2\tau h_1 - \tau_{01}h_0 - \tau_{02}h_0)]\}\sin[s(\xi - x)]ds \\ M_{021}(\xi,x,p) = \displaystyle\int_0^\infty \frac{\rho_2}{\kappa\tau\rho_0}\{\tau_{02}\exp(2s\tau_{02}h_0) \\ \qquad -\tau_{01}\exp[-s(2\tau h_2 - \tau_{02}h_0 - \tau_{01}h_0)]\}\sin[s(\xi - x)]ds \\ M_{022}(\xi,x,p) = -\displaystyle\int_0^\infty \left\{\left(\frac{\tau_{01}\rho_2}{\kappa\tau\rho_0} - 1\right) - \frac{\tau_{02}\rho_2}{\kappa\tau\rho_0}\exp[-2s\tau(h_1 + h_0)]\right\}\sin[s(\xi - x)]ds \end{array}\right\} \tag{31}$$

Where

$$\left.\begin{array}{l} \rho_1 \equiv \rho_1(s,p) = -\tau_{11} + \tau_{12}\exp[-2s\tau(h_1 - h_0)] \\ \rho_2 \equiv \rho_2(s,p) = \tau_{22} - \tau_{21}\exp[-2s\tau(h_2 - h_0)] \\ \rho_0 \equiv \rho_0(s,p) = 1 - \exp(-2s\tau h) \end{array}\right\} \tag{32}$$

It is noted that the kernel functions M_{011} (,x,p), $M_{022}(\xi,x,p)$ are semi-infinite integrals which have rather slow rate of convergence. To simplify the numerical analysis, the kernels are evaluated as follows:

$$M_{011}(\xi,x,p) = M_{011}^{\infty}(\xi,x,p) + \int_0^\infty \frac{\tau}{\kappa\rho_0}\{\exp[-2|s|\tau(h_1 - h_0)] + \exp[-2|s|\tau(h_2 + h_0)]$$

$$+2\exp[-2|s|\tau(h_1 + h_2)]\}\sin[s(\xi - x)]ds$$

$$+\int_0^\infty \frac{\delta}{\kappa\rho_0|s|}\{\exp[-2|s|\tau(h_2 + h_0)] + \exp[-2|s|\tau(h_1 - h_0)]\}\sin[s(\xi - x)]ds$$

$$-\int_0^\infty \frac{1}{\kappa\tau\rho_0}\left(\frac{\delta}{2|s|}\right)^2\{1 + \exp(-2|s|\tau h) - \exp[-2|s|\tau(h_2 + h_0)]$$

$$-\exp[-2|s|\tau(h_1 - h_0)]\}\sin[s(\xi - x)]ds \tag{33}$$

$$M_{022}(\xi,x,p) = M_{022}^{\infty}(\xi,x,p) + \int_0^\infty \frac{\tau}{\kappa\rho_0}\{\exp[-2|s|\tau(h_1 + h_0)] + \exp[-2|s|\tau(h_2 - h_0)]$$

$$+2\exp[-2|s|\tau(h_1 + h_2)]\}\sin[s(\xi - x)]ds$$

$$+\int_0^\infty \frac{\delta}{\kappa\rho_0|s|}\{\exp[-2|s|\tau(h_2 - h_0)] - \exp[-2|s|\tau(h_1 + h_0)]\}\sin[s(\xi - x)]ds$$

$$-\int_0^\infty \frac{1}{\kappa\tau\rho_0}\left(\frac{\delta}{2|s|}\right)^2\{1 + \exp(-2|s|\tau h) - \exp[-2|s|\tau(h_1 + h_0)]$$

$$-\exp\left[-2|s|\tau(h_2-h_0)\right]\}\sin\left[s(\xi-x)\right]ds \tag{34}$$

Where the kernel functions $M_{011}^{\infty}(\xi,x,p)$, $M_{022}^{\infty}(\xi,x,p)$ are given in Appendix A.

The system of the singular integral equations (29) and (30) is to be solved with the following subsidiary conditions obtained from the second boundary conditions of Eqs. (21).

$$\int_{a_k}^{b_k} G_{k0}(\xi,p)d\xi = 0 \quad (k=1,2) \tag{35}$$

The solution procedure of the system of the singular integral equations will be explained lately.

Once $G_{k0}(\xi, p)(k = 1,2)$ are obtained from Eqs. (29), (30) and (35), the temperature field in the Laplace transform plane can be easily calculated as follows:

$$T_i^{(2)*}(x,z,p) = -\frac{1}{\pi}\sum_{j=1}^{2}\sum_{k=1}^{2}\int_0^{\infty}\frac{1}{s}R_{ijk}\exp\left[s(\tau_{ij}z+\tau h_{ij})\right]ds$$

$$\times\int_{a_k}^{b_k} G_{k0}(\xi,p)\sin\left[s(\xi-x)\right]d\xi \quad (i=0,1,2) \tag{36}$$

where the functions $R_{ijk} \equiv R_{ijk}(s,p)$ are given in Appendix B, and constants h_{ij} $(i = 0,1,2, j,k = 1,2)$ are

$$\left.\begin{aligned} h_{01} = h_{02} = -h_0, \quad h_{11} = h_{22} = h_0 \\ h_{12} = -2h_1 + h_0, \quad h_{21} = -2h_2 + h_0 \end{aligned}\right\} \tag{37}$$

The temperature fields $T^{(1)}(z,t)$ and $T_i^{(2)}(x,z,t)$ $(i = 0,1,2)$ in the time-domain can be obtained from $T^{(1)*}(z,p)$ and $T^{(2)*}(x,z,p)(i = 0,1,2)$ by use of the numerical inversion technique of the Laplace transform [24].

Thermally Induced Singular Elastic and Electric Fields

The non-disturbed temperature field $T^{(1)*}(z,p)$ given by Eq. (24) does not induce the stress and electric displacement components affecting the singular field. Thus, we consider the elastic and electric fields due to the disturbed temperature distribution $T_i^{(2)*}(x,z,p)(i=0,1,2)$ only in this section. It is convenient to represent the solutions $u_{zi}^*(x,z,p)$, $u_{xi}^*(x,z,p)$ and $\phi_i^*(x,z,p)(i=0,1,2)$ in the Laplace transform plane as the sum of two functions, respectively.

$$\left.\begin{aligned} u_{zi}^*(x,z,p) = u_{zi}^{(1)*}(x,z,p) + u_{zi}^{(2)*}(x,z,p) \\ u_{xi}^*(x,z,p) = u_{xi}^{(1)*}(x,z,p) + u_{xi}^{(2)*}(x,z,p) \\ \phi_i^*(x,z,p) = \phi_i^{(1)*}(x,z,p) + \phi_i^{(2)*}(x,z,p) \end{aligned}\right\} \quad (i=0,1,2) \tag{38}$$

where $u_{zi}^{(1)*}(x,z,p)$, $u_{xi}^{(1)*}(x,z,p)$, $\phi_i^{(1)*}(x,z,p)$ are the particular solutions of Eq. (5) replaced $T_i^*(x,z,p)$ by $T_i^{(2)*}(x,z,p)$, and $u_{zi}^{(2)*}(x,z,p)$, $u_{xi}^{(2)*}(x,z,p)$, $\phi_i^{(2)*}(x,z,p)$ are the general solutions of homogeneous equations obtained by setting $T_i^*(x,z,p)=0\,(i=0,1,2)$ in Eq. (5). In the following, the superscripts (1) and (2) indicate the particular and general solutions of Eq. (5). Substituting Eq. (38) into Eqs. (2) and (3), one can obtain the stress $\sigma_{xxi}^*(x,z,p)$, $\sigma_{zzi}^*(x,z,p)$, $\sigma_{zxi}^*(x,z,p)$ and electric displacement $D_{xi}^*(x,z,p)$, $D_{zi}^*(x,z,p)$ $(i=0,1,2)$ expressions in the Laplace transform plane.

Using the displacement potential function method and the Fourier integral transform techniques [21], the particular and general solutions can be obtained as follows:

$$\left.\begin{aligned} u_{zi}^{(1)*}(x,z,p) &= \frac{1}{2\pi}\sum_{j=1}^{2}\int_{-\infty}^{\infty}\frac{1}{s|s|}p_{4ij}^{(1)}F_{ij}(s,p)\exp\left[|s|(f_{ij}z+\tau h_{ij})\right]\exp(-\mathrm{i}sx)ds \\ u_{xi}^{(1)*}(x,z,p) &= \frac{1}{2\pi}\sum_{j=1}^{2}\int_{-\infty}^{\infty}\frac{1}{s^2}p_{5ij}^{(1)}F_{ij}(s,p)\exp\left[|s|(f_{ij}z+\tau h_{ij})\right]\exp(-\mathrm{i}sx)ds \\ \phi_i^{(1)*}(x,z,p) &= -\frac{1}{2\pi}\sum_{j=1}^{2}\int_{-\infty}^{\infty}\frac{1}{s|s|}p_{6ij}^{(1)}F_{ij}(s,p)\exp\left[|s|(f_{ij}z+\tau h_{ij})\right]\exp(-\mathrm{i}sx)ds \end{aligned}\right\} \quad (i=0,1,2) \tag{39}$$

$$\left.\begin{aligned} u_{zi}^{(2)*}(x,z,p) &= \frac{\mathrm{i}}{2\pi}\sum_{j=1}^{6}\int_{-\infty}^{\infty}p_{4ij}^{(2)}A_{ij}(s,p)\exp(|s|\gamma_{ij}z)\exp(-\mathrm{i}sx)ds \\ u_{xi}^{(2)*}(x,z,p) &= \frac{\mathrm{i}}{2\pi}\sum_{j=1}^{6}\int_{-\infty}^{\infty}\frac{|s|}{s}p_{5ij}^{(2)}A_{ij}(s,p)\exp(|s|\gamma_{ij}z)\exp(-\mathrm{i}sx)ds \\ \phi_i^{(2)*}(x,z,p) &= -\frac{\mathrm{i}}{2\pi}\sum_{j=1}^{6}\int_{-\infty}^{\infty}p_{6ij}^{(2)}A_{ij}(s,p)\exp(|s|\gamma_{ij}z)\exp(-\mathrm{i}sx)ds \end{aligned}\right\} \quad (i=0,1,2) \tag{40}$$

where $A_{ij}(s,p)$ are the unknown functions to be solved. The functions $p_{mij}^{(1)} \equiv p_{mij}^{(1)}(s,p)$, $f_{ij} \equiv f_{ij}(s, p)$ $(m = 4,5,6,\ i = 0,1,2,\ j = 1,2)$ are given in Appendix C, and $p_{mij}^{(2)} \equiv p_{mij}^{(2)}(s)$, $\gamma ij \equiv \gamma ij\ (S)$ $(m = 4,5,6,\ i = 0,1,2,\ j = 1,2,\ldots,6)$ are given in Appendix A of the previous paper [19]. The functions $F_{ij}(s,p)$ $(i = 0,1,2, j = 1,2)$ are

$$F_{ij}(s,p) = \sum_{k=1}^{2}R_{ijk}\int_{a_k}^{b_k}G_{k0}(\xi,p)\exp(\mathrm{i}s\xi)d\xi \quad (i=0,1,2,\ j=1,2) \tag{41}$$

Similar to the temperature analysis, the problem may be reduced to a system of singular integral equations by defining the following new unknown functions $G_{km}(x,p)$ $(k = 1, 2, m = 1, 2, 3)$ [21]:

$$G_{k1}(x,p) = \begin{cases} \dfrac{\partial}{\partial x}\left[u_{z0}^*(x,\theta_{0k},p) - u_{zk}^*(x,\theta_{0k},p)\right] & (a_k < x < b_k) \\ 0 & (-\infty < x \le a_k,\ b_k \le x < \infty) \end{cases} \quad (k=1,2) \tag{42}$$

$$G_{k2}(x,p) = \begin{cases} \dfrac{\partial}{\partial x}\left[u_{x0}^*(x,\theta_{0k},p) - u_{xk}^*(x,\theta_{0k},p)\right] & (a_k < x < b_k) \\ 0 & (-\infty < x \le a_k,\ b_k \le x < \infty) \end{cases} \quad (k=1,2) \tag{43}$$

$$G_{k3}(x,p) = \begin{cases} -\dfrac{\partial}{\partial x}\left[\phi_0^*(x,\theta_{0k},p) - \phi_k^*(x,\theta_{0k},p)\right] & (a_k < x < b_k) \\ 0 & (-\infty < x \le a_k,\ b_k \le x < \infty) \end{cases} \quad (k=1,2) \tag{44}$$

Making use of the first boundary conditions (9)-(11) with Eqs. (12) and (13), we have the following system of six singular integral equations for the determination of the unknown functions $G_{km}(\xi, p)$ $(k = 1, 2, m = 1, 2, 3)$

$$\int_{a_1}^{b_1}\left\{\left[\frac{Z_{1111}^{(1)\infty}}{\xi-x}+M_{111}^{(1)}(\xi,x)\right]G_{11}(\xi,p)+M_{112}^{(1)}(\xi,x)G_{12}(\xi,p)+\left[\frac{Z_{1131}^{(1)\infty}}{\xi-x}+M_{113}^{(1)}(\xi,x)\right]G_{13}(\xi,p)\right\}d\xi$$

$$+\int_{a_2}^{b_2}\sum_{m=1}^{3}M_{12m}^{(1)}(\xi,x)G_{2m}(\xi,p)d\xi = \pi\sigma_{zz0}^{T*}(x,h_0,p) \quad (a_1<x<b_1) \tag{45}$$

$$\int_{a_1}^{b_1}\left\{M_{211}^{(1)}(\xi,x)G_{11}(\xi,p)+\left[\frac{Z_{2121}^{(1)\infty}}{\xi-x}+M_{212}^{(1)}(\xi,x)\right]G_{12}(\xi,p)+M_{213}^{(1)}(\xi,x)G_{13}(\xi,p)\right\}d\xi$$

$$+\int_{a_2}^{b_2}\sum_{m=1}^{3}M_{22m}^{(1)}(\xi,x)G_{2m}(\xi,p)d\xi = \pi\sigma_{zx0}^{T*}(x,h_0,p) \quad (a_1<x<b_1) \tag{46}$$

$$\int_{a_1}^{b_1}\left\{\left[\frac{Z_{3111}^{(1)\infty}}{\xi-x}+M_{311}^{(1)}(\xi,x)\right]G_{11}(\xi,p)+M_{312}^{(1)}(\xi,x)G_{12}(\xi,p)+\left[\frac{Z_{3131}^{(1)\infty}}{\xi-x}+M_{313}^{(1)}(\xi,x)\right]G_{13}(\xi,p)\right\}d\xi$$

$$+\int_{a_2}^{b_2}\sum_{m=1}^{3}M_{32m}^{(1)}(\xi,x)G_{2m}(\xi,p)d\xi = \pi D_{z0}^{T*}(x,h_0,p) \quad (a_1<x<b_1) \tag{47}$$

$$\int_{a_2}^{b_2}\left\{\left[\frac{Z_{1211}^{(2)\infty}}{\xi-x}+M_{121}^{(2)}(\xi,x)\right]G_{21}(\xi,p)+M_{122}^{(2)}(\xi,x)G_{22}(\xi,p)+\left[\frac{Z_{1231}^{(2)\infty}}{\xi-x}+M_{123}^{(2)}(\xi,x)\right]G_{23}(\xi,p)\right\}d\xi$$

$$+\int_{a_1}^{b_1}\sum_{m=1}^{3}M_{11m}^{(2)}(\xi,x)G_{1m}(\xi,p)d\xi = \pi\sigma_{zz0}^{T*}(x,-h_0,p) \quad (a_2<x<b_2) \tag{48}$$

$$\int_{a_2}^{b_2}\left\{M_{221}^{(2)}(\xi,x)G_{21}(\xi,p)+\left[\frac{Z_{2221}^{(2)\infty}}{\xi-x}+M_{222}^{(2)}(\xi,x)\right]G_{22}(\xi,p)+M_{223}^{(2)}(\xi,x)G_{23}(\xi,p)\right\}d\xi$$

$$+\int_{a_1}^{b_1}\sum_{m=1}^{3}M_{21m}^{(2)}(\xi,x)G_{1m}(\xi,p)d\xi = \pi\sigma_{zx0}^{T*}(x,-h_0,p) \quad (a_2<x<b_2) \tag{49}$$

$$\int_{a_2}^{b_2}\left\{\left[\frac{Z_{3211}^{(2)\infty}}{\xi-x}+M_{321}^{(2)}(\xi,x)\right]G_{21}(\xi,p)+M_{322}^{(2)}(\xi,x)G_{22}(\xi,p)+\left[\frac{Z_{3231}^{(2)\infty}}{\xi-x}+M_{323}^{(2)}(\xi,x)\right]G_{23}(\xi,p)\right\}d\xi$$

$$+\int_{a_1}^{b_1}\sum_{m=1}^{3} M_{31m}^{(2)}(\xi,x)G_{1m}(\xi,p)d\xi = \pi D_{z0}^{T*}(x,-h_0,p) \quad (a_2 < x < b_2) \tag{50}$$

In the above equations, the kernel functions $M_{jkm}^{(n)}(\xi,x)(n,k=1,2, j,m=1,2,3)$ are given by

$$M_{jkm}^{(n)}(\xi,x) = \left\{\begin{array}{l} \left.\begin{array}{ll} \int_{-\infty}^{\infty}\sum_{l=1}^{2}\left[Z_{jkml}^{(n)}(s)-Z_{jkml}^{(n)\infty}\right]\sin\left[s(\xi-x)\right]ds & (m=1,3) \\ -\int_{-\infty}^{\infty}\sum_{l=1}^{2}\left[Z_{jkml}^{(n)}(s)-Z_{jkml}^{(n)\infty}\right]\cos\left[s(\xi-x)\right]ds & (m=2) \end{array}\right\} (j=1,3) \\ \left.\begin{array}{ll} \int_{-\infty}^{\infty}\sum_{l=1}^{2}\left[Z_{jkml}^{(n)}(s)-Z_{jkml}^{(n)\infty}\right]\cos\left[s(\xi-x)\right]ds & (m=1,3) \\ \int_{-\infty}^{\infty}\sum_{l=1}^{2}\left[Z_{jkml}^{(n)}(s)-Z_{jkml}^{(n)\infty}\right]\sin\left[s(\xi-x)\right]ds & (m=2) \end{array}\right\} (j=2) \end{array}\right\} (n,k=1,2) \tag{51}$$

where the functions $Z_{jkml}^{(n)}(s)$ and the constants $Z_{jkml}^{(n)\infty}$ $(n,k,l=1,2, j,m=1,2,3)$ are given in Appendix D. The functions $\sigma_{zz0}^{T*}(x,\pm h_0,p)$, $\sigma_{zx0}^{T*}(x,\pm h_0,p)$ and $D_{z0}^{T*}(x,\pm h_0,p)$, which correspond to the stress and electric displacement components induced by the disturbed temperatures $T_i^{(2)*}(x,z,p)\,(i=0,1,2)$ on the $z = \pm h_0$ planes in the absence of the crack, are obtained as follows:

$$\begin{aligned}\sigma_{zz0}^{T*}(x,\pm h_0,p) &= \lim_{z\to\pm h_0}\frac{1}{2\pi}\sum_{j=1}^{3}\int_{-\infty}^{\infty}|s|\,p_{10j}^{(2)}d_{0j}^{T}(s,p)\exp\left[|s|\gamma_{oj}(h_0+z)\right]\exp(-\mathrm{i}sx)ds \\ &+\lim_{z\to\pm h_0}\frac{1}{2\pi}\sum_{j=4}^{6}\int_{-\infty}^{\infty}|s|\,p_{10j}^{(2)}d_{0j}^{T}(s,p)\exp\left[-|s|\gamma_{0j}(h_0-z)\right]\exp(-\mathrm{i}sx)ds \\ &-\lim_{z\to\pm h_0}\frac{\mathrm{i}}{2\pi}\sum_{j=1}^{2}\int_{-\infty}^{\infty}\frac{1}{s}p_{10j}^{(1)}F_{0j}(s,p)\exp\left[|s|(\tau h_{0j}+f_{0j}z)\right]\exp(\mathrm{i}sx)ds\end{aligned} \tag{52}$$

$$\begin{aligned}\sigma_{zx0}^{T*}(x,\pm h_0,p) &= \lim_{z\to\pm h_0}\frac{1}{2\pi}\sum_{j=1}^{3}\int_{-\infty}^{\infty}s\,p_{20j}^{(2)}d_{0j}^{T}(s,p)\exp\left[|s|\gamma_{oj}(h_0+z)\right]\exp(-\mathrm{i}sx)ds \\ &+\lim_{z\to\pm h_0}\frac{\mathrm{i}}{2\pi}\sum_{j=4}^{6}\int_{-\infty}^{\infty}s\,p_{20j}^{(2)}d_{0j}^{T}(s,p)\exp\left[-|s|\gamma_{0j}(h_0-z)\right]\exp(-\mathrm{i}sx)ds \\ &-\lim_{z\to\pm h_0}\frac{\mathrm{i}}{2\pi}\sum_{j=1}^{2}\int_{-\infty}^{\infty}\frac{1}{|s|}p_{20j}^{(1)}F_{0j}(s,p)\exp\left[|s|(\tau h_{0j}+f_{0j}z)\right]\exp(\mathrm{i}sx)ds\end{aligned} \tag{53}$$

$$\begin{aligned}D_{z0}^{T*}(x,\pm h_0,p) &= \lim_{z\to\pm h_0}\frac{1}{2\pi}\sum_{j=1}^{3}\int_{-\infty}^{\infty}|s|\,p_{30j}^{(2)}d_{0j}^{T}(s,p)\exp\left[|s|\gamma_{oj}(h_0+z)\right]\exp(-\mathrm{i}sx)ds \\ &+\lim_{z\to\pm h_0}\frac{1}{2\pi}\sum_{j=4}^{6}\int_{-\infty}^{\infty}|s|\,p_{30j}^{(2)}d_{0j}^{T}(s,p)\exp\left[-|s|\gamma_{0j}(h_0-z)\right]\exp(-\mathrm{i}sx)ds \\ &-\lim_{z\to\pm h_0}\frac{\mathrm{i}}{2\pi}\sum_{j=1}^{2}\int_{-\infty}^{\infty}\frac{1}{s}p_{30j}^{(1)}F_{0j}(s,p)\exp\left[|s|(\tau h_{0j}+f_{0j}z)\right]\exp(\mathrm{i}sx)ds\end{aligned} \tag{54}$$

In Eqs. (52)-(54), the functions $d_{0j}^{T}(s,p)\,(j=1,2,\ldots,6)$ are also given in Appendix D. Of course, these components are superficial quantities and have no physical meaning in this analysis. However, they are equivalent to the crack face tractions in solving the crack problem by a proper superposition. The singular integral equations (45)-(50) are to be solved with the following subsidiary conditions obtained from the second boundary conditions (9)-(11).

$$\int_{a_k}^{b_k} G_{km}(\xi,p)d\xi = 0 \quad (k=1,2, m=1,2,3) \tag{55}$$

To solve the system of the singular integral equations (29), (30) and (45)-(50) with the subsidiary conditions (35), (55), we introduce the following functions $\Phi_{km}(\xi,p)\,(k=1,2, m=0,1,2,3)$:

$$G_{km}(\xi,p) = \frac{c_k}{[(a_k-|\xi|)(|\xi|-b_k)]^{1/2}}\Phi_{km}(\xi,p) \quad (k=1,2, m=0,1,2,3) \tag{56}$$

Using the Gauss-Jacobi integration formula [22], the functions $\Phi_{km}(\xi,p)$ (k = 1,2, m = 0,1,2,3) can be determined by solving the integral equations.

The stress intensity factors $K_{\mathrm{IA}}^{(k)*}(p)$, $K_{\mathrm{IB}}^{(k)*}(p)$, $K_{\mathrm{IIA}}^{(k)*}(p)$, $K_{\mathrm{IIB}}^{(k)*}(p)$ and the electric displacement intensity factors $K_{\mathrm{DA}}^{(k)*}(p)$, $K_{\mathrm{DB}}^{(k)*}(p)$ at the crack tips $x = a_k, b_k$ on the $z = \theta_{0k}$ (k = 1,2) planes in the Laplace transform plane may be, respectively defined and evaluated as:

$$\left.\begin{aligned}K_{\mathrm{IA}}^{(k)*}(p) &= -\lim_{x\to a_k^-}[2\pi(a_k-x)]^{1/2}\sigma_{zz0}^{*}(x,\theta_{0k},p) \\ &= -(\pi c_k)^{1/2}\left[Z_{1k11}^{\infty}\Phi_{k1}(a_k,p)+Z_{1k31}^{\infty}\Phi_{k3}(a_k,p)\right] \\ K_{\mathrm{IIA}}^{(k)*}(p) &= -\lim_{x\to a_k^-}[2\pi(a_k-x)]^{1/2}\sigma_{zx0}^{*}(x,\theta_{0k},p) \\ &= -(\pi c_k)^{1/2}Z_{2k21}^{\infty}\Phi_{k2}(a_k,p) \\ K_{\mathrm{DA}}^{(k)*}(p) &= -\lim_{x\to a_k^-}[2\pi(a_k-x)]^{1/2}D_{z0}^{*}(x,\theta_{0k},p) \\ &= -(\pi c_k)^{1/2}\left[Z_{3k11}^{\infty}\Phi_{k1}(a_k,p)+Z_{3k31}^{\infty}\Phi_{k3}(a_k,p)\right]\end{aligned}\right\} (k=1,2) \tag{57}$$

$$\left.\begin{aligned}K_{\mathrm{IB}}^{(k)*}(p) &= \lim_{x\to b_k^+}[2\pi(x-b_k)]^{1/2}\sigma_{zz0}^{*}(x,\theta_{0k},p) \\ &= (\pi c_k)^{1/2}\left[Z_{1k11}^{\infty}\Phi_{k1}(b_k,p)+Z_{1k31}^{\infty}\Phi_{k3}(b_k,p)\right] \\ K_{\mathrm{IIB}}^{(k)*}(p) &= \lim_{x\to b_k^+}[2\pi(x-b_k)]^{1/2}\sigma_{zx0}^{*}(x,\theta_{0k},p) \\ &= (\pi c_k)^{1/2}Z_{2k21}^{\infty}\Phi_{k2}(b_k,p) \\ K_{\mathrm{DB}}^{(k)*}(p) &= \lim_{x\to b_k^+}[2\pi(x-b_k)]^{1/2}D_{z0}^{*}(x,\theta_{0k},p) \\ &= (\pi c_k)^{1/2}\left[Z_{3k11}^{\infty}\Phi_{k1}(b_k,p)+Z_{3k31}^{\infty}\Phi_{k3}(b_k,p)\right]\end{aligned}\right\} (k=1,2) \tag{58}$$

Thus, the stress intensity factors $K_{\mathrm{IA}}^{(k)}(t)$, $K_{\mathrm{IB}}^{(k)}(t)$, $K_{\mathrm{IIA}}^{(k)}(t)$, $K_{\mathrm{IIB}}^{(k)}(t)$ and the electric displacement intensity factors $K_{\mathrm{DA}}^{(k)}(t)$, $K_{\mathrm{DB}}^{(k)}(t)$ (k = 1, 2) in the time-domain are

$$\left.\begin{aligned}K_{\mathrm{IA}}^{(k)}(t) &= -(\pi c_k)^{1/2}\frac{1}{2\pi\mathrm{i}}\int_{Br}\left[Z_{1k11}^{\infty}\Phi_{k1}(a_k,p)+Z_{1k31}^{\infty}\Phi_{k3}(a_k,p)\right]\exp(pt)dp \\ K_{\mathrm{IIA}}^{(k)}(t) &= -(\pi c_k)^{1/2}\frac{1}{2\pi\mathrm{i}}\int_{Br}Z_{2k21}^{\infty}\Phi_{k2}(a_k,p)\exp(pt)dp \\ K_{\mathrm{DA}}^{(k)}(t) &= -(\pi c_k)^{1/2}\frac{1}{2\pi\mathrm{i}}\int_{Br}\left[Z_{3k11}^{\infty}\Phi_{k1}(a_k,p)+Z_{3k31}^{\infty}\Phi_{k3}(a_k,p)\right]\exp(pt)dp\end{aligned}\right\} (k=1,2) \tag{59}$$

$$\left.\begin{aligned}K_{\mathrm{IB}}^{(k)}(t) &= (\pi c_k)^{1/2}\frac{1}{2\pi\mathrm{i}}\int_{Br}\left[Z_{1k11}^{\infty}\Phi_{k1}(b_k,p)+Z_{1k31}^{\infty}\Phi_{k3}(b_k,p)\right]\exp(pt)dp \\ K_{\mathrm{IIB}}^{(k)}(t) &= (\pi c_k)^{1/2}\frac{1}{2\pi\mathrm{i}}\int_{Br}Z_{2k21}^{\infty}\Phi_{k2}(b_k,p)\exp(pt)dp \\ K_{\mathrm{DB}}^{(k)}(t) &= (\pi c_k)^{1/2}\frac{1}{2\pi\mathrm{i}}\int_{Br}\left[Z_{3k11}^{\infty}\Phi_{k1}(b_k,p)+Z_{3k31}^{\infty}\Phi_{k3}(b_k,p)\right]\exp(pt)dp\end{aligned}\right\} (k=1,2) \tag{60}$$

The values of them at $t \to \infty$ are given by

$$\left.\begin{aligned}K_{\mathrm{IA}}^{(k)}(\infty) &\equiv \lim_{t\to\infty}K_{\mathrm{IA}}^{(k)}(t) = \lim_{p\to 0}pK_{\mathrm{IA}}^{(k)*}(p) \\ K_{\mathrm{IIA}}^{(k)}(\infty) &\equiv \lim_{t\to\infty}K_{\mathrm{IIA}}^{(k)}(t) = \lim_{p\to 0}pK_{\mathrm{IIA}}^{(k)*}(p) \\ K_{\mathrm{DA}}^{(k)}(\infty) &\equiv \lim_{t\to\infty}K_{\mathrm{DA}}^{(k)}(t) = \lim_{p\to 0}pK_{\mathrm{DA}}^{(k)*}(p)\end{aligned}\right\} (k=1,2) \tag{61}$$

$$\left.\begin{aligned}K_{\mathrm{IB}}^{(k)}(\infty) &\equiv \lim_{t\to\infty}K_{\mathrm{IB}}^{(k)}(t) = \lim_{p\to}pK_{\mathrm{IB}}^{(k)}(p) \\ K_{\mathrm{IIB}}^{(k)}(\infty) &\equiv \lim_{t\to\infty}K_{\mathrm{IIB}}^{(k)}(t) = \lim_{p\to}pK_{\mathrm{IIB}}^{(k)}(p) \\ K_{\mathrm{DB}}^{(k)}(\infty) &\equiv \lim_{t\to\infty}K_{\mathrm{DB}}^{(k)}(t) = \lim_{p\to}pK_{\mathrm{DB}}^{(k)}(p)\end{aligned}\right\} (k=1,2) \tag{62}$$

Numerical Results and Discussion

For the numerical calculations, the material is considered to be cadmium selenide, with the following properties [2] are used properties of the FGPM plate at the plane z = 0:

$$\left.\begin{array}{ll} c_{110} = 7.41\times10^{10}[\mathrm{N/m^2}], & c_{130} = 3.93\times10^{10}[\mathrm{N/m^2}], \\ c_{330} = 8.36\times10^{10}[\mathrm{N/m^2}], & c_{440} = 1.32\times10^{10}[\mathrm{N/m^2}], \\ e_{310} = -0.16[\mathrm{C/m^2}], & e_{330} = 0.347[\mathrm{C/m^2}], \\ e_{150} = -0.138[\mathrm{C/m^2}], & \\ \varepsilon_{110} = 0.825\times10^{-10}[\mathrm{C/Vm}], & \varepsilon_{330} = 0.903\times10^{-10}[\mathrm{C/Vm}], \\ \lambda_{110} = 0.621\times10^{6}[\mathrm{N/Km^2}], & \lambda_{330} = 0.551\times10^{6}[\mathrm{N/Km^2}], \\ p_{z0} = -2.94\times10^{-6}[\mathrm{C/Km^2}] & \end{array}\right\} \quad (63)$$

Since the values of the coefficients of heat conduction for cadmium selenide could not be found in the literature, the value $\kappa^2 = \kappa_x/_z = 1/1.5$ is assumed.

To examine the effects of the crack geometry and material parameters on $K_{\eta A}^{(k)}$, $K_{\eta B}^{(k)}$ $(\eta = \mathrm{I,II,D}, k = 1,2)$, the normalized parameters ($c_1/h, c_2/h, h_0/h, h_1/h, h_2/h, d/h$ and (βh, δh, ωh) are used. Because there are many geometric parameters, we focus on the influence of the crack distance parameter d/h and the material non-homogeneity on the fracture behavior. Thus it is supposed to be the crack location parameters $h_1/h = h_2/h = 0.5$ and the crack length parameters $c_1/h = c_2/h$. And the normalized non-homogeneous parameters βh, δh and ωh are assumed to be $\beta h = \delta h = \omega h$.

The electroelastic fields without cracks

Figures 2a-2c indicate the normalized stress components $(\sigma_{zz0}^T(x,\pm h_0,t),\ \sigma_{zx0}^T(x,\pm h_0,t))/\lambda_{330}T_0$ and the normalized electric displacement component $D_{z0}^T(x,\pm h_0,t)/p_{z0}T_0$ on the $z/h = \pm 0.2$ planes in the strip without crack at various normalized time $F = t\lambda_0/h^2$ for the crack distance parameter d/h = 1.5, the crack spacing parameter h_0/h = 0.2, the crack length parameter $c_1/h = c_2/h = 1.0$ and βh = 1.0. Above mentioned before, these components are given by Eqs. (52)-(54) and are superficial quantities. The maximum values of $\sigma_{zz0}^T(x,h_0,t)$, $\sigma_{zx0}^T(x,h_0,t)$ and $D_{z0}^T(x,h_0,t)$ are seen to occur at about $F \approx 0.5$, whereas the maximum values of $\sigma_{zz0}^T(x,-h_0,t)$, $\sigma_{zx0}^T(x,-h_0,t)$ and $D_{z0}^T(x,-h_0,t)$ occur at $F\to\infty$.

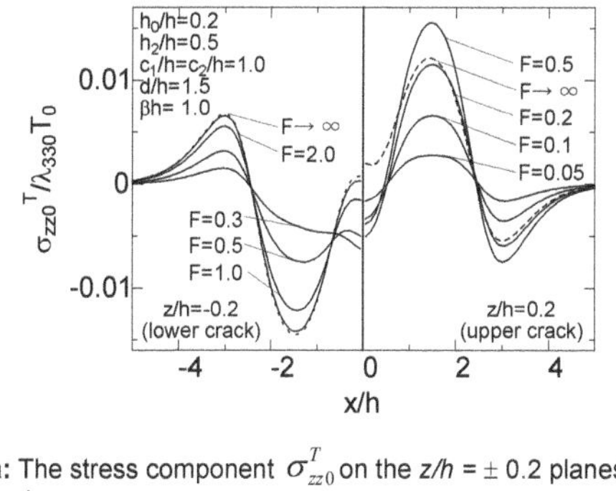

Figure 2a: The stress component σ_{zz0}^T on the *z/h* = ± 0.2 planes in the strip without crack.

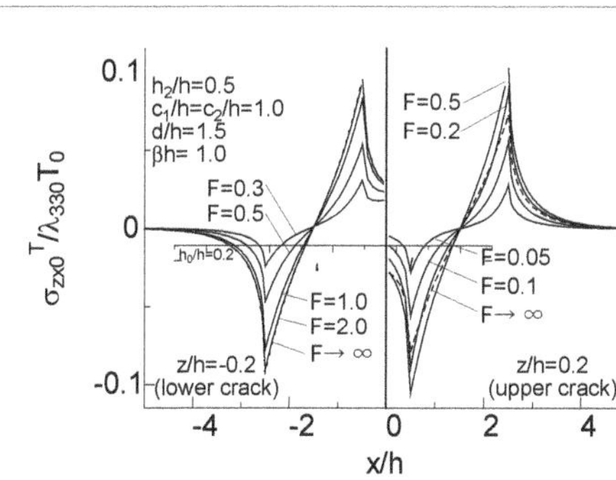

Figure 2b: The stress component ${}_{zx}$ on the *z/h* = ± 0.2 planes in the strip without crack.

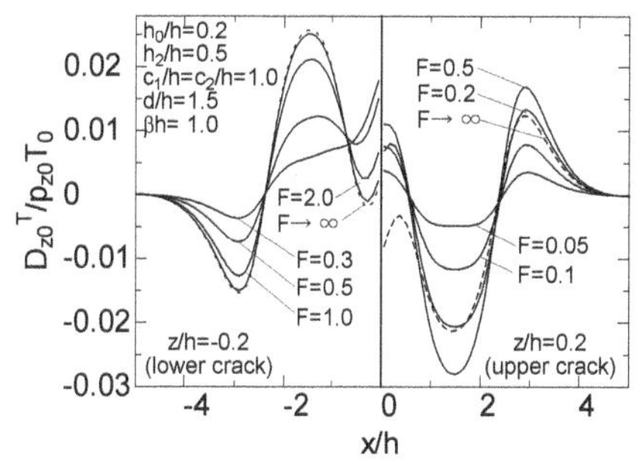

Figure 2c: The electric displacement component D_{z0}^T on the *z/h* = ± 0.2 planes in the strip without crack.

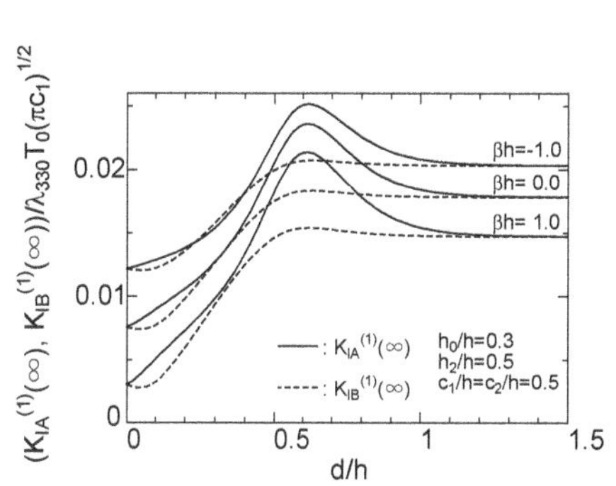

Figure 3a: The effects of the material non-homogeneity and the crack distancedon the static values of the stress intensity factors $K_{\mathrm{IA}}^{(1)}(\infty)$ and $K_{\mathrm{IB}}^{(1)}(\infty)$.

The static behavior of the stress and electric displacement intensity factors

Due to above discussion, the intensity factors of the upper crack would be larger than those of the lower crack, thus only the results for the upper crack will be shown. Figures 3a-3c show the effects of the material non-homogeneity βh and the crack distance d/h on the static values of the normalized stress intensity factors $(K_{\eta A}^{(1)}(\infty), K_{\eta B}^{(1)}(\infty))/\lambda_{330}T_0(\pi c)^{1/2}$ $(\eta = \mathrm{I,II})$ and the static values of the normalized electric displacement intensity factors $(K_{\mathrm{DA}}^{(1)}(\infty), K_{\mathrm{DB}}^{(1)}(\infty))/p_{z0}T_0(\pi c)^{1/2}$ for βh = -1.0,0.0,1.0 with h_0/h = 0.3 and $c_1/h = c_2/h$ = 0.5. The results for the cases of $d/h\to\infty$ and βh = 0.0 coincide with the results of single parallel crack [8] and with the results for the homogeneous case [14], respectively. The values of the intensity factors tend to increase/decrease at first, reach maximum/minimal values and then decrease/increase with increasing d/h. The absolute maximum values of the intensity factors tend to occur at about $c_1/h = d/h$, and the interaction between the two cracks may vanish for the range $3c_1/h<d/h$. Moreover, it is evident that the intensity factors can be reduced by increasing the material property gradient of functionally graded piezoelectric materials.

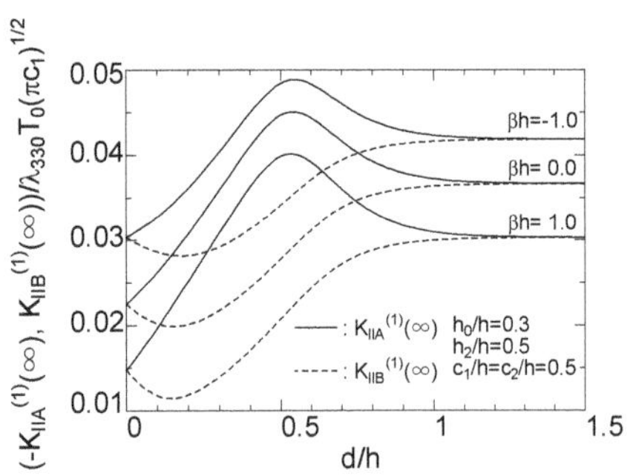

Figure 3b: The effects of the material non-homogeneity and the crack distance *d* on the static values of the stress intensity factors $K_{\mathrm{IIA}}^{(1)}(\infty)$ and $K_{\mathrm{IIB}}^{(1)}(\infty)$.

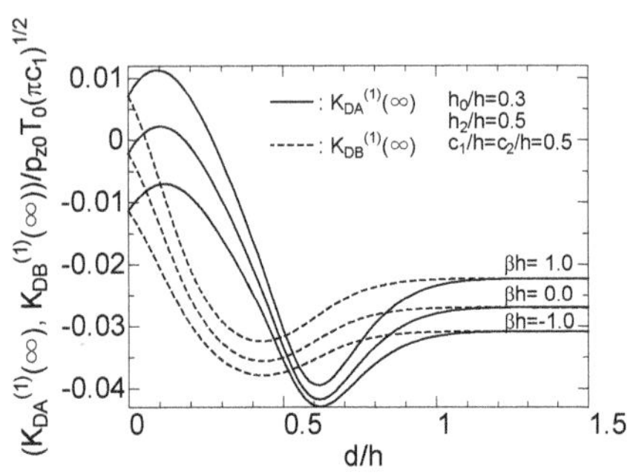

Figure 3c: The effects of the material non-homogeneity and the crack distance *d* on the static values of the electric displacement intensity factors $K_{\mathrm{DA}}^{(1)}(\infty)$ and $K_{\mathrm{DB}}^{(1)}(\infty)$.

The transient behavior of the stress and electric displacement intensity factors

Figures 4a-4c show the effect of the crack distance d/h on the transient behavior of the normalized stress intensity factors $(K_{\eta\mathrm{A}}^{(1)}, K_{\eta\mathrm{B}}^{(1)})/\lambda_{330}T_0(\pi c)^{1/2}(\eta=\mathrm{I,II})$ and the normalized electric displacement intensity factors $(K_{\mathrm{DA}}^{(1)}, K_{\mathrm{DB}}^{(1)})/p_{z0}T_0(\pi c)^{1/2}$ are plotted versus F for βh = 1.0 with h_0/h = 0.2 and $c_1/h = c_2/h$ = 1.0. In these figures, the dashed, solid and dotted lines indicate the results for d/h = 0.5,1.0 and 3.0, respectively.

Similar to the static values of the normalized stress and electric displacement intensity factors, the interaction between the two cracks may vanish for $3c_1/h = d/h$, and the absolute values of the intensity factors become $|K_{\eta\mathrm{A}}^{(1)}| = |K_{\eta\mathrm{B}}^{(1)}|\ (\eta=\mathrm{I,II,D})$. The absolute values of the intensity factors $|K_{\eta\mathrm{A}}^{(1)}|$, $|K_{\eta\mathrm{B}}^{(1)}|\ (\eta=\mathrm{I,II,D})$ increase at first, have the peak values $|K_{\eta\mathrm{A}}^{(1)peak}|$, $|K_{\eta\mathrm{B}}^{(1)peak}|(\eta=\mathrm{I,II,D})$, then decrease and approach the static values $|K_{\eta\mathrm{A}}^{(1)}(\infty)|$, $|K_{\eta\mathrm{B}}^{(1)}(\infty)|\ (\eta=\mathrm{I,II,D})$ with increasing F. The value of $K_{\mathrm{IA}}^{(1)}$ (d/h = 0.5) becomes negative so that the contact of the crack faces would occur, and these results for F >1.3 have no physical meaning. As shown in the previous paper [18], the results presented here without considering this effect may not be exactly correct but would be more conservative, since the contact of the crack faces will increase the friction between the faces and make heat and electric transfer across the crack faces easier. Thus the intensity factors would be lowered by these two factors.

Figures 5a-5c show the effect of the material non-homogeneity βh on $(K_{\eta\mathrm{A}}^{(1)}, K_{\eta\mathrm{B}}^{(1)})/\lambda_{330}T_0(\pi c)^{1/2}(\eta=\mathrm{I,II})$ and $(K_{\mathrm{DA}}^{(1)}, K_{\mathrm{DB}}^{(1)})/p_{z0}T_0(\pi c)^{1/2}$ are plotted versus F for d/h = 0.0 with $c_1/h = c_2/h$ = 1.0. In these figures, the solid, dotted and dashed lines indicate the results for βh = 2.0, 0.0 and -2.0, respectively. The results for the case of βh = 0.0 coincident with two parallel cracks [17] and with the results for the homogeneous case [18]. Because of symmetry, the values of $K_{\eta\mathrm{A}}^{(1)}$, $K_{\eta\mathrm{B}}^{(1)}\ (\eta=\mathrm{I,D})$ for βh = 0.0 approach zero and the values of the intensity factors are $K_{\eta\mathrm{A}}^{(1)} = K_{\eta\mathrm{B}}^{(1)}\ (\eta=\mathrm{I,D})$ and $K_{\mathrm{IIA}}^{(1)} = -K_{\mathrm{IIB}}^{(1)}$. The value of $K_{\mathrm{IA}}^{(1)}$ for βh = 2.0 also becomes negative so that the contact of the crack faces would occur.

Figures 6a-6c are the same as Figures 5a-5c for d/h = 1.0. With

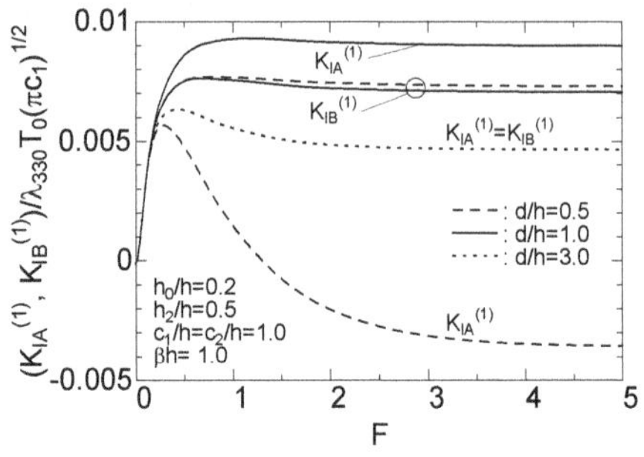

Figure 4a: The effect of the crack distance *d* on the stress intensity factors $K_{\mathrm{IA}}^{(1)}$ and $K_{\mathrm{IB}}^{(1)}$.

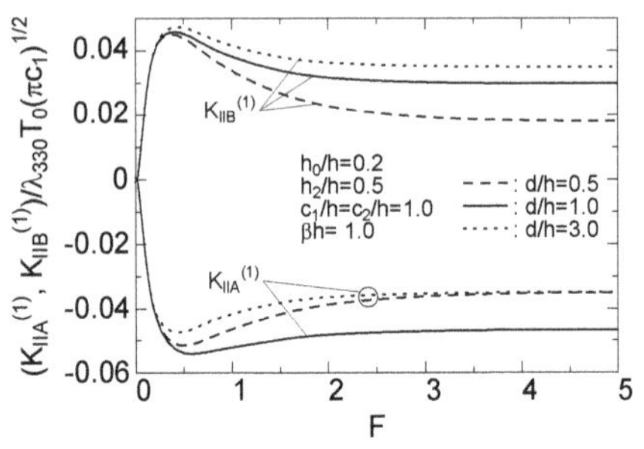

Figure 4b: The effect of the crack distance *d* on the stress intensity factors $K_{\mathrm{IIA}}^{(1)}$ and $K_{\mathrm{IIB}}^{(1)}$.

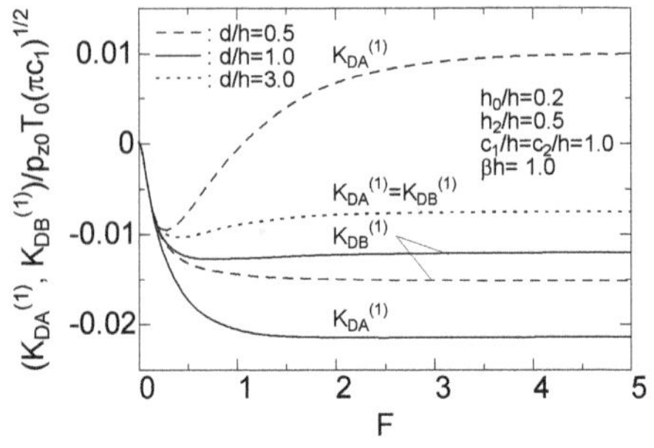

Figure 4c: The effect of the crack distance *d* on the electric displacement intensity factors $K_{\mathrm{DA}}^{(1)}$ and $K_{\mathrm{DB}}^{(1)}$.

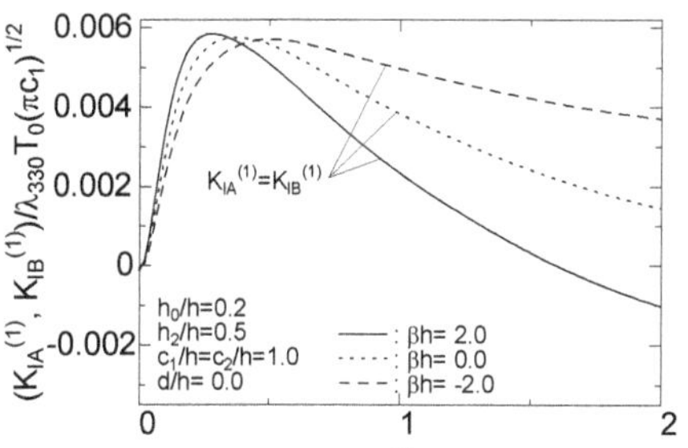

Figure 5a: The effect of the material non-homogeneity on the stress intensity factors $K_{IA}^{(1)}$ and $K_{IB}^{(1)}$ for *d*/*h* = 0.0 and *βh* = 2.0,0.0, -2.0.

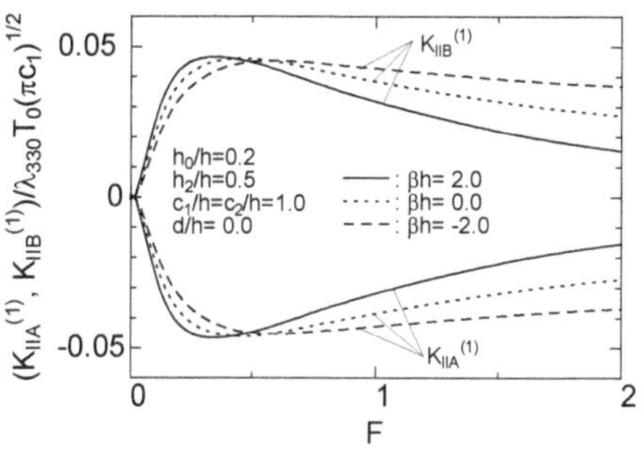

Figure 5b: The effect of the material non-homogeneity on the stress intensity factors $K_{IIA}^{(1)}$ and $K_{IIB}^{(1)}$ for *d*/*h* = 0.0 and *βh* = 2.0,0.0, -2.0

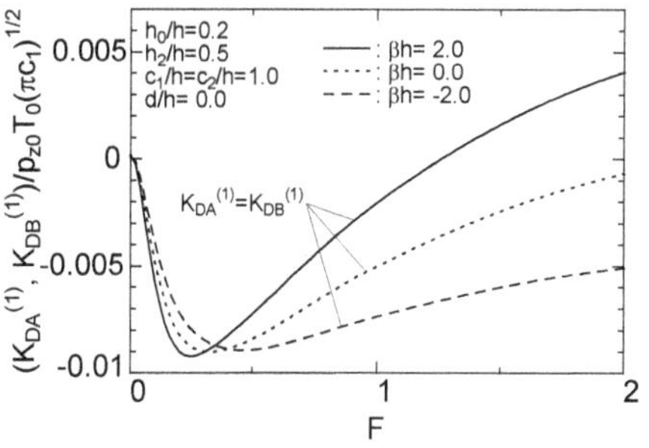

Figure 5c: The effect of the material non-homogeneity on the electric displacement intensity factors $K_{DA}^{(1)}$ and $K_{DB}^{(1)}$ *d*/*h* = 0.0 and *βh* = 2.0,0.0, -2.0.

increasing *F*, the intensity factors $K_{\eta A}^{(1)}$, $K_{\eta B}^{(1)} (\eta = \mathrm{I, D})$ and $K_{IIA}^{(1)}$ for *βh* = -2.0 increase monotonically, and then approach static values $K_{\eta A}^{(1)}(\infty)$, $K_{\eta B}^{(1)}(\infty)(\eta = \mathrm{I, D})$ and $K_{IIA}^{(1)}(\infty)$. On the other hand, the intensity factors $K_{\eta A}^{(1)}$, $K_{\eta B}^{(1)} (\eta = \mathrm{I, II})$ and $K_{DB}^{(1)}$ for *βh* = 2.0 have clear peak values, and $K_{IA}^{(1)}$, $K_{IB}^{(1)}$ for *βh* = 0.0 have slight peak values. In addition, these peak values, static values and the interesting values $(|K_{\eta A}^{(1)peak} - K_{\eta A}^{(1)}(\infty)|)/|K_{\eta A}^{(1)}(\infty)|$, $(|K_{\eta B}^{(1)peak} - K_{\eta B}^{(1)}(\infty)|)/|K_{\eta B}^{(1)}(\infty)| (\eta = \mathrm{I, II, D})$, which mean the overshooting effect, are presented in Tables 1-3. It is found that the peak values of the intensity factors and the overshooting effects increase with increasing *βh*.

Conclusion

The transient mixed-mode thermoelectroelastic fracture problem of a functionally graded piezoelectric material strip with two parallel cracks in arbitrary positions is studied theoretically. For the special cases of symmetrical geometry ($h_1/h = h_2/h = 0.5$ and $c_1/h = c_2/h$), the effects of the crack distance and material non-homogeneity on the stress and electric displacement intensity factors are clarified. The following facts can be found from the numerical results.

For the case of the static behavior

1. The increase of the material parameter is beneficial for reducing the static values of the intensity factors.

2. The absolute maximum values of the intensity factors tend to occur at about $c_1/h = d/h$, and the interaction between the two cracks becomes 0 at about $3c_1/h = d/h$.

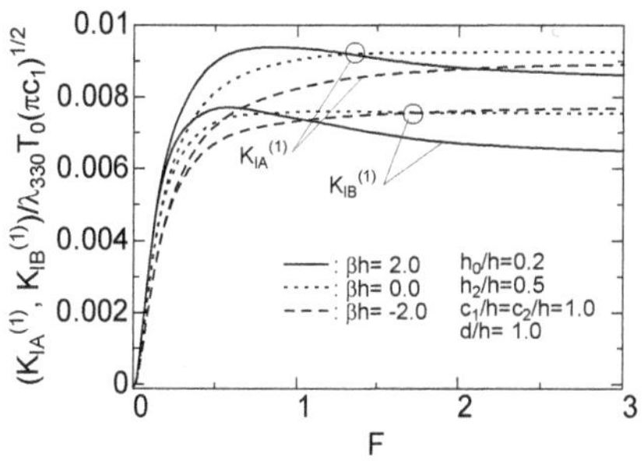

Figure 6a: The effect of the material non-homogeneity on the stress intensity factors $K_{IA}^{(1)}$ and $K_{IB}^{(1)}$ for *d*/*h* = 1.0 and *βh* = 2.0,0.0, -2.0.

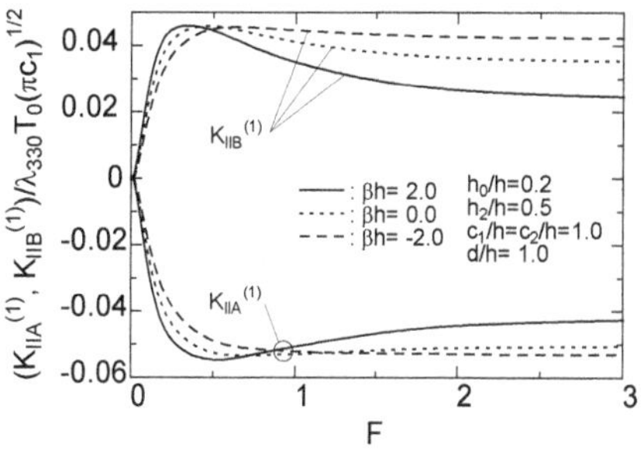

Figure 6b: The effect of the material non-homogeneity on the stress intensity factors $K_{IIA}^{(1)}$ and $K_{IIB}^{(1)}$ for *d*/*h* = 1.0 and *βh* = 2.0,0.0, -2.0.

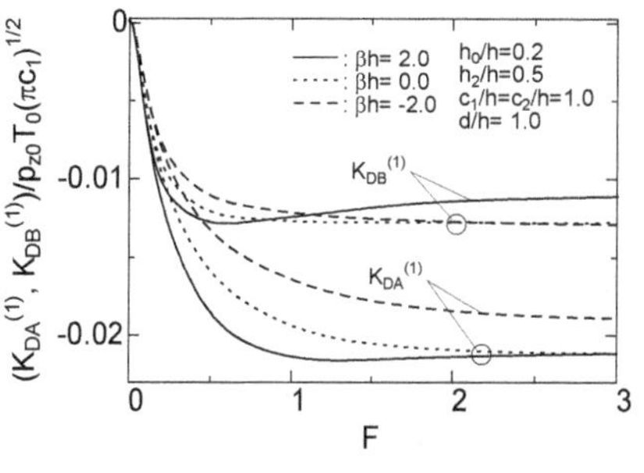

Figure 6c: The effect of the material non-homogeneity on the electric displacement intensity factors $K_{DA}^{(1)}$ and $K_{DB}^{(1)}$ for *d*/*h* = 1.0 and *βh* = 2.0,0.0, -2.0.

Figure	k	d/h	βh	$\frac{\lvert K_{Ik}^{peak} \rvert}{\lambda_{330}T_0(\pi c_1)^{1/2}}$	$\frac{\lvert K_{Ik}(\infty) \rvert}{\lambda_{330}T_0(\pi c_1)^{1/2}}$	$\frac{\lvert K_{Ik}^{peak} - K_{Ik}(\infty) \rvert}{\lvert K_{Ik}(\infty) \rvert}$
			2.0	9.383×10^{-3}	8.616×10^{-3}	0.089
Figure 6a	A	1.0	0.0	9.254×10^{-3}	9.239×10^{-3}	0.002
			-2.0	8.906×10^{-3}	8.906×10^{-3}	0.000
			2.0	7.718×10^{-3}	6.498×10^{-3}	0.188
Figure 6a	B	1.0	0.0	7.611×10^{-3}	7.547×10^{-3}	0.008
			-2.0	7.691×10^{-3}	7.691×10^{-3}	0.000

Table 1: The values of $|K_{Ik}^{peak}|/\lambda_{330}T_0(\pi c_1)^{1/2}$, $|K_{Ik}(\infty)|/\lambda_{330}T_0(\pi c_1)^{1/2}$ and $(|K_{Ik}^{peak} - K_{Ik}(\infty)|)/|K_{Ik}(\infty)|$

Figure	k	d/h	βh	$\frac{\lvert K_{IIk}^{peak} \rvert}{\lambda_{330}T_0(\pi c_1)^{1/2}}$	$\frac{\lvert K_{IIk}(\infty) \rvert}{\lambda_{330}T_0(\pi c_1)^{1/2}}$	$\frac{\lvert K_{IIk}^{peak} - K_{IIk}(\infty) \rvert}{\lvert K_{IIk}(\infty) \rvert}$
			2.0	5.466×10^{-2}	4.263×10^{-2}	0.282
Figure 6b	A	1.0	0.0	5.340×10^{-2}	5.055×10^{-2}	0.056
			-2.0	5.308×10^{-2}	5.308×10^{-2}	0.000
			2.0	4.591×10^{-2}	2.455×10^{-2}	0.870
Figure 6b	B	1.0	0.0	4.582×10^{-2}	3.536×10^{-2}	0.296
			-2.0	4.553×10^{-2}	4.224×10^{-2}	0.078

Table 2: The values of $|K_{IIk}^{peak}|/\lambda_{330}T_0(\pi c_1)^{1/2}$, $|K_{IIk}(\infty)|/\lambda_{330}T_0(\pi c_1)^{1/2}$ and $(|K_{IIk}^{peak} - K_{IIk}(\infty)|)/|K_{IIk}(\infty)|$.

Figure	k	d/h	βh	$\frac{\lvert K_{Ik}^{peak} \rvert}{\lambda_{330}T_0(\pi c_1)^{1/2}}$	$\frac{\lvert K_{Ik}(\infty) \rvert}{\lambda_{330}T_0(\pi c_1)^{1/2}}$	$\frac{\lvert K_{Ik}^{peak} - K_{Ik}(\infty) \rvert}{\lvert K_{Ik}(\infty) \rvert}$
			2.0	2.161×10^{-2}	2.116×10^{-2}	0.021
Figure 6c	A	1.0	0.0	2.116×10^{-2}	2.116×10^{-2}	0.000
			-2.0	1.892×10^{-2}	1.892×10^{-2}	0.000
			2.0	1.283×10^{-2}	1.110×10^{-2}	0.156
Figure 6c	B	1.0	0.0	1.277×10^{-2}	1.277×10^{-2}	0.000
			-2.0	1.287×10^{-2}	1.287×10^{-2}	0.000

Table 3: The values of $|K_{Dk}^{peak}|/\lambda_{330}T_0(\pi c_1)^{1/2}$, $|K_{Dk}(\infty)|/\lambda_{330}T_0(\pi c_1)^{1/2}$ and $(|K_{Dk}^{peak} - K_{Dk}(\infty)|)/|K_{Dk}(\infty)|$.

For the case of the transient behavior

1. The distinct overshooting phenomenon can be observed and this fact may suggest the importance of these transient analyses.

2. The peak values of the intensity factors increase with increasing βh.

3. The overshooting effect depends on the crack distance and material non-homogeneity. The large βh induces the large overshooting effect.

References

1. Rao SS, Sunar M (1994) Piezoelectricity and its use in disturbance sensing and control of flexible structures: A survey. Applied Mechanics Review 47: 113-123.
2. Ashida F, Tauchert TR (1998) Transient response of a piezothermoelastic circular disk under axisymmetric heating. Acta Mechanica 128: 1-14.
3. Wu CM, Kahn M, Moy W (1996) Piezoelectric ceramics with functionally gradients: A new application in material design. J American Ceramics Society 79: 809-812.
4. Chen J, Liu ZX, Zou ZZ (2003) Electromechanical impact of a crack in a functionally graded piezoelectric medium theoretical and applied fracture mechanics 39: 47-60.
5. Wang BL, Zhang XH (2004) A mode III crack in functionally graded piezoelectric material strip transactions of the ASME. J Applied Mechanics 71: 327-333.
6. Ueda S (2006) A Finite crack in a semi-infinite strip of a grade piezoelectric material under electric loading. European J Mechanics A/Solids 25: 250-259.
7. Ueda S (2008) A cracked functionally graded piezoelectric material strip under transient thermal loading. Acta Mechanica 199: 53-70.
8. Ueda S (2007) Thermal intensity factors for a parallel crack in a functionally graded piezoelectric strip. J Thermal Stresses 30: 321-342.
9. Ueda S (2007) Effects of crack surface conductance on intensity factors for a cracked functionally graded piezoelectric material under thermal load. J Thermal Stresses 30: 731-752.
10. Ueda S (2007) A penny-shaped crack in a functionally graded piezoelectric strip under thermal loading. Engineering Fracture Mechanics 74: 1255-1273.
11. Ueda S (2008) Transient thermoelectroelastic response of a functionally graded piezoelectric strip with a penny-shaped crack. Engineering Fracture Mechanics 75: 1204-1222.
12. Ueda S, Nishimura N (2008) An annular crack in a functionally graded piezoelectric strip under thermoelectric loadings. J Thermal Stresses 31: 1079-1098.
13. Ueda S, Tani Y (2008) Thermal stress intensity factors for two coplanar cracks in a piezoelectric strip. Journal of Thermal Stresses 31: 403-415.
14. Ueda S, Nishikohri H (2013) Two parallel cracks in arbitrary positions of a piezoelectric material strip under thermo-electric loadings. J Thermal Stresses 36: 480-500.
15. Ueda S, Uemura Y (2009) Thermoelectromechanical interaction among multi parallel cracks in a piezoelectric Material. J Thermal Stresses 32: 1005-1023.
16. Ueda S, Hatano H (2012) T-Shaped crack in a piezoelectric material thermo-electro-mechanical loadings. J Thermal Stresses 35: 12-29.

17. Ueda S, Ikeda Y, Ishii A (2012) Transient thermoelectromechanical response of a piezoelectric strip with two parallel cracks of different lengths. J Thermal Stresses 35: 534-549.

18. Ueda S, Nishikohri H (2016) Transient thermoelectroelastic response of a piezoelectric material Strip with two parallel cracks in arbitrary positions. J Thermal Stresses in press.

19. Ueda S, Iogawa T (2010) Two parallel penny-shaped or annular cracks in a functionally graded piezoelectric strip under electric loading. Acta Mechanica 210: 57-70.

20. Ueda S, Ueda T (2013) Transient thermoelectroelastic response of a functionally graded piezoelectric strip with two parallel Axisymmetric Cracks. J Thermal Stresses 36: 1027-1055.

21. Sneddon IN, Lowengrub M (1969) Crack Problems in the classical theory of elasticity John Wiley and Sons Inc New York.

22. Erdogan F, Wu BH (1996) Crack problems in FGM Layers under thermal stresses. J Thermal Stresses 19: 237-265.

23. Erdogan F, Gupta GD, Cook TS (1972) Methods of analysis and solution of crack problems in GC Sih (edn) Noordhoff Leyden.

24. Miller MK, Guy WT (1966) Numerical inversion of the laplace transform by use of jacobi polynomials. SIAM J Numerical Analysis 3: 624-635.

25. Wang BL, Mai YW (2004) Impermeable crack and permeable crack assumptions which one is more realistic? transactions of the ASME J Applied Mechanics 71: 575-578.

On Nonlinear Vibration Analysis of Shallow Shells – A New Approach

Banerjee MM[1]* and Mazumdar J[2]

[1]202 Nandan Apartment, Hill view (N), Asansol-713304, West Bengal, India
[2]School of Electrical and Electronic Engineering and School of Mathematical Sciences, the University of Adelaide, Australia

Abstract

A method for the analysis of nonlinear vibration of shallow shells of arbitrary shape is presented. The method is based upon the concept of constant deflection contours on the surface of the shallow shell. The constant deflection contour method has previously been found to be a simple tool for the study of linear vibration analysis of shallow shells of arbitrary shape. A new approach has been made here to utilize this concept to study the large amplitude vibration of shallow shells in conjunction with the Galerkin method. A number of illustrative examples are included to demonstrate the accuracy of the proposed method.

Keywords: Iso-amplitude contour lines; Constant deflection contours; Shallow shells

Introduction

In the linear theory of motion of elastic plates and shallow shells, the strain of the middle surface can be neglected when the deflections are assumed to be small compared with the thickness of the surface. However, in most practical cases, this basic assumption is no longer valid; instead the deflections have the magnitude of the thickness of the surface. Hence the derivation of governing differential equations exhibiting large deflections needs special attention in such analyses.

The importance of the inclusion of nonlinear effects in problems relating to the strength and stability of modern flight structures has been made initially by von Kármán [1]. Indeed, the von Karman theory is widely used to account for the influence of large deflection in plates and shells. In fact, more than half a century ago it was Herrmann [2] who first proposed the nonlinear plate theory of motion corresponding to the dynamic analogue of the von Kármán theory.

It is well-known that the nonlinear dynamic behavior of thin shallow shell structures is of much technical importance to designers due to its wide range of applications in many fields of engineering. Containers, tanks, domes etc. are common examples of practical importance of such structures. However, the papers on nonlinear vibrations of shallow shell structures to date are limited in number.

The problems of nonlinear vibration of shallow shells have attracted the attention of relatively few investigators in the past [3-6]. Due to the very complicated nature of the basic equations governing the motion of a structure exhibiting large deflection, it has always been a difficult task for investigators to obtain even an approximate solution. Mazumdar in 1970 proposed a new approach which appeared to be quite suitable for bending analysis of elastic plates of arbitrary shapes based on the concept of iso-deflection contour lines on the bent surface of the plate [7]. This simple but efficient method is best known as Constant Deflection Contour Method or CDC-Method. Subsequently, the same method has been extended to the vibration analysis of plates and shallow shells [8,9].

The CDC method has so far been restricted to linear analysis until an attempt has been made recently to extend it to nonlinear analysis of plates [10,11]. In the present paper a similar approach as in [11] is undertaken for extension of the study to shallow shell analysis. This paper is therefore regarded as a sequel to earlier papers and deals with the nonlinear vibration of shallow shells based on the CDC Method. Some specific examples on nonlinear vibrations of shallow shells have been included to show the efficacy of the method, and that the results are in excellent agreement with known results in the literature.

Derivation of Basic Equations

Consider an elastic, isotropic shallow shell of uniform thickness h subject to a continuously distributed normal load q. Let the equation of the middle surface of the shell referred to a system of orthogonal coordinates xyz, be given by [9].

$$z = \frac{x^2}{2R_x} + \frac{xy}{R_{xy}} + \frac{y^2}{2R_y} \qquad (2.1)$$

Where $r = \sqrt{x^2 + y^2}$ is small compared to the least of the radii of curvature R_x, R_y, R_{xy} and (supposed to be constants).

If the shell is assumed to be comparatively thin and the displacements (u, v, w) are predominantly flexural, the strain components can be written as

$$\varepsilon_x = \frac{\partial u}{\partial x} + \frac{w}{R_x} + \frac{1}{2}\left(\frac{\partial w}{\partial x}\right)^2 - z\frac{\partial^2 w}{\partial x^2} = \frac{\sigma_x - \nu\sigma_y}{E}, \quad \varepsilon_y = \frac{\partial v}{\partial y} + \frac{w}{R_y} + \frac{1}{2}\left(\frac{\partial w}{\partial y}\right)^2 - z\frac{\partial^2 w}{\partial y^2} = \frac{\sigma_y - \nu\sigma_x}{E}$$

$$\varepsilon_{xy} = \frac{\partial v}{\partial x} + \frac{\partial u}{\partial y} + \frac{\partial w}{\partial x}\frac{\partial w}{\partial y} + \frac{2w}{R_{xy}} - 2z\frac{\partial^2 w}{\partial x \partial y} = \frac{2(1+\nu)}{E}(\sigma_{xy}) \qquad (2.2)$$

With usual notations, the total strain energy is given by

$$U = \frac{1}{2}\iiint (\sigma_x\varepsilon_x + \sigma_y\varepsilon_y + \sigma_{xy}\varepsilon_{xy})\, dz\,dx\,dy, \qquad (2.3)$$

Whereas the kinetic energy is

$$T_e = (\rho h/2)\iint \left(\dot{u}^2 + \dot{v}^2 + \dot{w}^2\right) dx\,dy, \qquad (2.4)$$

And the work done is

$$W_k = \iint p w\, dx\,dy \qquad (2.5)$$

***Corresponding author:** Banerjee MM, 202 Nandan Apartment, Hill view (N), Asansol-713304, West Bengal, India
E-mail: muralimohan_banerjee@yahoo.com

Formulating the Lagrangian with the help of the above expressions and applying Hamilton's principle, a straightforward application of the variational calculus yield the following equations of motion [3]

$$D\nabla^4 w = hS(F,w) - h\left(\frac{F,_{yy}}{R_x} + \frac{F,_{xx}}{R_y} - 2\frac{F,_{xy}}{R_{xy}}\right) + q - \rho h w,_{tt} \tag{2.6}$$

and

$$D\nabla^4 F = -\frac{E}{2} hS(w,w) + E\left(\frac{w,_{yy}}{R_x} + \frac{w,_{xx}}{R_y} - 2\frac{w,_{xy}}{R_{xy}}\right) \tag{2.7}$$

Where the operator $S\ (w, F)$ stands for

$$S(w,F) \equiv \frac{\partial^2 w}{\partial x^2}\frac{\partial^2 F}{\partial y^2} - 2\frac{\partial^2 w}{\partial x \partial y}\frac{\partial^2 F}{\partial x \partial y} + \frac{\partial^2 w}{\partial y^2}\frac{\partial^2 F}{\partial x^2}$$

Here 'F' denotes the Airy-Stress function, defined by

$$\int_{-h/2}^{h/2} \sigma_{xx} dz = N_x = h\frac{\partial^2 F}{\partial y^2}, \int_{-h/2}^{h/2} \sigma_{yy} dz = N_y = h\frac{\partial^2 F}{\partial x^2}, \int_{-h/2}^{h/2} \sigma_{xy} dz = N_{xy} = -h\frac{\partial^2 F}{\partial x \partial y}, \tag{2.8a}$$

Whereas

$$M_x = \int_{-h/2}^{h/2} \sigma_x z dz = -D\left[w,_{xx} + \nu w,_{yy}\right], M_y = \int_{-h/2}^{h/2} \sigma\, z dz = -D\left[w,_{yy} + \nu w,_{xx}\right], M_{xy} = \int_{-h/2}^{h/2} \sigma_{xy} z dz = -D(1-\nu) w,_{xy} \tag{2.8b}$$

And the (,) notation signifies partial derivative with respect to the suffix.

A New Approach

Mazumdar in [7] put forward a simple method, the so-called CDC-Method to solve the static and dynamic problems of elastic plates of arbitrary shapes. Mazumdar et al [8,9,12-14] applied this method for solving various problems of elastic plates and shells of arbitrary shapes, restricted to linear cases only. Following Mazumdar [7], a new idea has been put forward by Banerjee [10] to study the dynamic response of structures of arbitrary shapes based on the CDC method. While Mazumdar utilized the concept of Deflection Contour method to deduce the basic dynamical equations using elementary theory of plates and shells [15-17], the authors in [11] found it easy to arrive at the final equations by straightforward utilization of von Kármán field equations and then utilizing the required transformations to u-variables. In most practical cases, it is found that von Kármán field equations in conjunction with the CDC-Method make it easy to apply for nonlinear analyses of plates and shells.

Theory and Derivation of Governing Equations

When the plate or the shallow shell vibrates in a normal mode, then at any instant t_θ, the intersections between the deflected surface and the parallels z=constant yield contours which after projection onto the base plane z=0 are a set of level curves, u(x,y)=constant, called the "Lines of Equal Deflections" [9], which are, in fact, iso-amplitude contour lines . The boundary of the plate or the shell irrespective of any combination of support is also a simple curve belonging to the family of lines of equal deflections.

As defined by Mazumdar [7] this family of nonintersecting curves may be denoted by C_u, where $0 \le u \le u^*$, so that C_0 (u=0) is the boundary and C_u^* coincides with the point(s) at which the maximum u=u* is attained.

Let u=u(x,y)=constant be a member of the family of iso-deflection or iso-amplitude contour lines.

Using the following transformations.

$$\frac{\partial w}{\partial x} = w_x = \frac{dw}{du}u_x, \qquad w_{xx} = \frac{d^2 w}{du^2}u_x^2 + \frac{dw}{du}u_{xx}, \qquad \text{etc.} \tag{4.1}$$
$$w_y = \frac{dw}{du}u_y, \qquad w_{xy} = \frac{d^2 w}{du^2}u_x u_y + \frac{dw}{du}u_{xy}$$

Equations (2.6) and (2.7) can be written as

$$D\sum_{i=1}^{4}\lambda_i \frac{d^{5-i}w}{du^{5-i}} = h[\ \lambda_5 \frac{d}{du}\left(\frac{dw}{du}\frac{dF}{du}\right) + \lambda_6 \frac{dw}{du}\frac{dF}{du}] - $$
$$-h\left[\lambda_7 \frac{d^2 F}{du^2} + \lambda_8 \frac{dF}{du}\right] + q - \rho\ h\ w,_{tt} \tag{4.2}$$

$$\sum_{i=1}^{4}\lambda_i \frac{d^{5-i}F}{du^{5-i}} = -\frac{E}{2}[\ \lambda_9 \frac{d}{du}\left(\frac{dw}{du}\right)^2 + \lambda_{10}\left[\frac{dw}{du}\right]^2 +$$
$$+E\left[\lambda_{11}\frac{d^2 w}{du^2} + \lambda_{12}\frac{dw}{du}\right] \tag{4.3}$$

Where

$$\lambda_1 = \left(u_x^2 + u_y^2\right)^2 \quad \lambda_2 = u_x^2\left(6u_{xx} + 2u_{yy}\right) + u_y^2\left(6u_{yy} + 2u_{xx}\right) + 8u_x u_y u_{xy}$$

$$\lambda_3 = 3\left(u_{xx}^2 + u_{yy}^2\right) + 4\left(u_x u_{xxx} + u_y u_{yyy}\right) + 4\left(u_x u_{xyy} + u_y u_{xxy}\right) + 4u_{xy}^2 + 2u_{xx}u_{yy}$$

$$\lambda_4 = \left(u_{xxxx} + 2u_{xxyy} + u_{yyyy}\right),\ \lambda_5 = \left(u_x^2 u_{yy} + u_y^2 u_{xx} - 2u_x u_y u_{xy}\right) = \lambda_9$$

$$\lambda_6 = 2\left(u_{xx}u_{yy} - u_{xy}^2\right) = \lambda_{10}\ \lambda_7 = \left(\frac{u_y^2}{R_x} + \frac{u_x^2}{R_y} - 2\frac{u_x u_y}{R_{xy}}\right) = \lambda_{11}$$

$$\lambda_7 = \left(\frac{u_{yy}}{R_x} + \frac{u_{xx}}{R_y} - 2\frac{u_{xy}}{R_{xy}}\right) = \lambda_{12} \tag{4.4}$$

Since Eqns. (4.2) and (4.3) are valid for all points on the surface of the shell, we can have

$$\iint_\Omega \{\ D\sum_{i=1}^{4}\lambda_i \frac{d^{5-i}w}{du^{5-i}} - h[\ \lambda_5 \frac{d}{du}\left(\frac{dw}{du}\frac{dF}{du}\right) + \lambda_6 \frac{dw}{du}\frac{dF}{du}]$$
$$+h\left[\lambda_7 \frac{d^2 F}{du^2} + \lambda_8 \frac{dF}{du}\right] - q + \rho\ h\ w,_{tt}\ \}\ d\Omega = 0 \tag{4.5}$$

And

$$\iint_\Omega\ \{\ \sum_{i=1}^{4}\lambda_i \frac{d^{5-i}F}{du^{5-i}} + \frac{E}{2}([\ \lambda_9 \frac{d}{du}\left(\frac{dw}{du}\right)^2 + \lambda_{10}\left[\frac{dw}{du}\right]^2) +$$
$$-E\left[\lambda_{11}\frac{d^2 w}{du^2} + \lambda_{12}\frac{dw}{du}\right]\ \}\ d\Omega = 0 \tag{4.6}$$

The integration is over the region bounded by any contour C_u. While performing the above integrals it would be more convenient to utilize the formula in the modified form

$$\iint_\Omega \Psi_1\left(u, u_x, u_{xx}, \ldots\ldots, \frac{dw}{du}, \frac{d^2 w}{du^2}, \ldots \frac{d^n w}{du^n}\right) d\Omega = -\int_{u^*}^{u} \Psi_2(u)\left\{\oint\ \Psi_3(x,y)\frac{ds}{\sqrt{\lambda_1}}\right\} du \tag{4.7}$$

Which is a generalization of the formula adopted in Ref.[9]. Often it has been encountered that in the contour integral appearing in Eqn. (4.7), the integrand turns out to be dependent on *u*, and hence care should be taken to evaluate first the contour integral. Sometimes, it is useful to use the following relations for evaluation of the contour integral

$$\sqrt{\lambda_1} = u_x^2 + u_y^2 = \frac{4}{p^2}, \quad 2\frac{ds}{p} = \frac{dx}{u_y} = \frac{dy}{u_x} \tag{4.8}$$

Evaluation of the above integrals will yield two ordinary differential equations. Thus, in accordance with the present method, the basic fourth order partial differential equations reduce to ordinary differential equations which make it rather easy for further study.

Method of Solution

It should be noted here that the above analysis is valid for any shallow shell structure. It has been already stated that equations (4.5) and (4.6) will yield two ordinary differential equations. For nonlinear analysis one may have to seek an approximate solution for which the form of the deflected function w must be first assumed compatible with the boundary conditions (Figure 1). Mathematically, this may be explained in the following way. Let u(x,y)=u denote a typical member of the family of the iso-deflection curves, then for any prescribed boundary conditions the deflection function w(u,t) can be assumed to take the form

$$w=A\,W(u)\,f(t) \qquad (5.1)$$

Where f(t) is an unknown function of time to be determined. Using this expression for the deflection function in the resultant equation of (4.6), we get the stress function in the form

$$F=\Phi\{u,f(t)\} \qquad (5.2)$$

With this expression for the stress function and previously assumed form of W, the resultant equation of (4.5) will yield, after using the Galerkin procedure, an ordinary time differential equation. Let us suppose that Eqn. (4.5) in combination with (5.1) and (5.2), yields the error function in the form

$$\varepsilon=\Lambda_1\left[u,\lambda_1..\lambda_{12}f(t),f^2(t)\;\ddot{f}(t)\right] \qquad (5.3)$$

Because of the approximate nature of equation (5.1), the associated error function may be minimized using Galerkin method. The appropriate orthogonally condition applied to Eqn. (4.5) will yield the following "Time Differential Equation" with known constants in the form.

$$\ddot{F}(t)+\alpha_1F(t)+\alpha_2F^2(t)+\alpha_3f^3(t)=q* \qquad (5.4)$$

The solution of which can be obtained and from which the subsequent analysis can be performed.

Eqn. (5.4) can be studied for the following cases:

(a) Free linear vibration of Plates (when R_x *and* $R_y \rightarrow$ *infinity*) and shell (*as the case may be*)

(b) Free nonlinear vibration of plates and shells

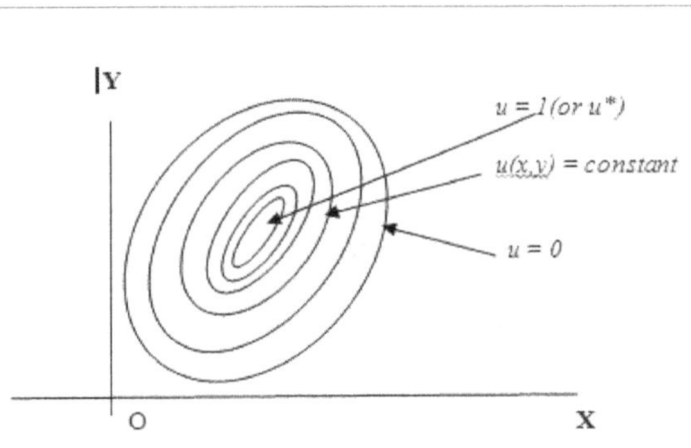

Figure 1: Iso-deflection curves.

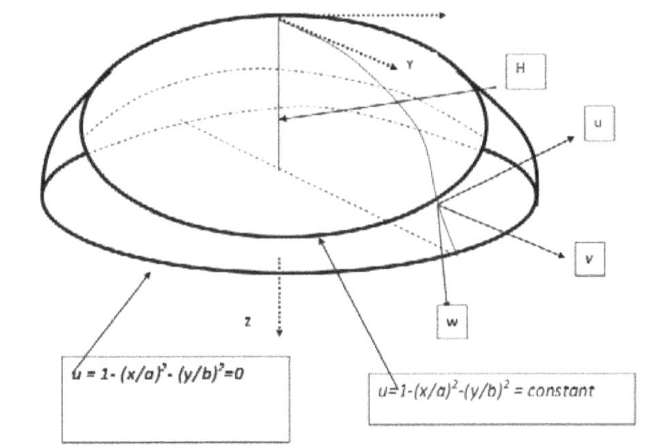

Figure 2: Shallow dome placed on an elliptical base.

(c) Static analysis of plates and shells.

Specific Illustration

Large vibration of a shallow dome upon an elliptical base

Consider the vibration of a shallow dome of nonzero Gaussian curvature upon an elliptic base. Figure 2 depicts the geometry of the shell. The edges are clamped and immovable. When the shell vibrates in a normal mode, the lines of equal deflections, as described in Sec.3, may reasonably be taken as

$$u(x,y)=1-\frac{x^2}{a^2}-\frac{y^2}{b^2} \qquad (6.1)$$

Clearly, in this case u= 0 on the boundary and u=u*=1 at the center of the shell.

The corresponding values of are given by

$$\lambda_1=16\left(\frac{x^2}{a^4}+\frac{y^2}{b^4}\right)^2,\ \lambda_2=-48\left(\frac{x^2}{a^6}+\frac{y^2}{b^6}\right),\ \lambda_3=4\frac{3a^4+2a^2b^2+3b^4}{a^2b^2}$$

$$\lambda_4=0,\lambda_5=-\frac{8(1-u)}{a^2b^2}=\lambda_9,\ \lambda_6=\frac{8}{a^2b^2}=\lambda_{10}$$

$$\lambda_7=4\left(\frac{x^2}{a^4R_y}+\frac{y^2}{b^4R_x}\right)=\lambda_{11},\ \lambda_8=-2\left(\frac{1}{a^2R_y}+\frac{1}{b^2R_x}\right)=\lambda_{12} \qquad (6.2)$$

Substituting the above values in Equations (4.5) and (4.6), and utilizing the formula given by (4.7) one gets

$$\left[(1-u)^2\frac{d^4w}{du^4}-4(1-u)\frac{d^3w}{du^3}+2\frac{d^2w}{du^2}\right]=-\alpha\,[(1-u)\frac{d}{du}((1-u)\frac{dw}{du}\frac{dF}{du}) \\ -\beta\frac{d}{du}\left[(1-u)\frac{dF}{du}\right]+\frac{q}{2DP}-\frac{\rho h}{2DP}w_{,tt} \qquad (6.3)$$

$$\left[(1-u)^2\frac{d^4F}{du^4}-4(1-u)\frac{d^3F}{du^3}+2\frac{d^2F}{du^2}\right]=\gamma\,[\frac{d}{du}\{(1-u)(\frac{dw}{du})^2\}] \\ +\delta\frac{d}{du}\left[(1-u)\frac{dw}{du}\right] \qquad (6.4)$$

Where

$$\alpha=\frac{4h}{DPa^2b^2},\quad \beta=\frac{h\,\kappa}{DP}\quad \gamma=\frac{2E}{Pa^2b^2}$$

$$\delta=\frac{E\kappa}{P}\quad P=\frac{3a^4+2a^2b^2+3b^4}{a^4b^4}\quad \kappa=\left(\frac{1}{a^2R_y}+\frac{1}{b^2R_x}\right)\tag{6.5}$$

It appears that the exact solutions of Eqns. (6.3) and (6.4) are not possible to find. So in order to obtain approximate solutions, let us assume

$$w(u,t)=W(u)\,f(t)\approx Au^2 f(t)\tag{6.6}$$

Where f (t) is an unknown function of time to be determined.

Substitution of (6.6) in (6.4) the first integral of (6.4) yields

$$\frac{d}{du}\left\{(1-u)\frac{dF}{du}\right\}=\frac{4\gamma}{3}A^2f^2(t)u^3+\delta A\,f(t)u^2+A_1\tag{6.7}$$

Which further integration reduces to

$$\left\{(1-u)\frac{dF}{du}\right\}=\frac{\gamma}{3}A^2f^2(t)u^4+\frac{1}{3}\delta A\,f(t)u^3+A_1u+A_2\tag{6.8}$$

Considering the case for a clamped immovable edge condition we set the following conditions:

$$\left.\frac{dF}{du}\right|_{u=0}=0\quad\text{and}\quad\left.\left\{(1-u)\frac{d^2F}{du^2}-2(1-\nu)\frac{dF}{du}\right\}\right|_{u=0}=0\tag{6.9}$$

Which make both A_1 and A_2 to be zero and Eqn. (6.8) reduces to

$$\left\{(1-u)\frac{dF}{du}\right\}=\frac{\gamma}{3}h^2f^2(t)u^4+\frac{1}{3}\delta h\,f(t)u^3\tag{6.10}$$

Substituting Equations (6.10) and (6.6) into Equations (6.3) and applying Galerkin procedure one gets after some mathematical operations

$$\rho h^2\ddot{f}+\alpha_1 f+\alpha_2 f^2\alpha_3 f^3=Q^*\tag{6.11}$$

Where

$$\left.\begin{aligned}&\alpha_1=\frac{DP}{}\left(\frac{40}{3}+2\beta\,\delta\right)A\qquad \alpha_2=\frac{10DP}{}\left(\frac{2}{9}\beta\gamma+\frac{4}{9}\delta\,\alpha\right)A^2\\&\alpha_3=\frac{100\,DP\gamma\alpha\,A^3}{21},\qquad Q^*=\frac{5\,q}{3}\end{aligned}\right\}\tag{6.12}$$

Where

$$\alpha=\frac{4h}{DPa^2b^2},\quad \beta=\frac{h}{DP}\left(\frac{1}{a^2R_y}+\frac{1}{b^2R_x}\right)\quad \gamma=\frac{2E}{Pa^2b^2}$$

$$\delta=\frac{E}{P}\left(\frac{1}{a^2R_y}+\frac{1}{b^2R_x}\right)\quad P=\frac{3a^4+2a^2b^2+3b^4}{a^4b^4}\tag{6.13}$$

An indirect verification of the correctness of the time differential equation may be made by considering the case for a flat plate problem. When $\beta\to 0,\ \delta\to 0,$ it implies $\alpha_2=0$ and further if a=b the problem reduces to that of a circular plate for which Eqn. (6.11) takes the form (for ν = 0.3)

$$\rho h^2\ddot{f}+\frac{Eh^4}{a^4}\left[9.756\,f+4.762\,f^3\right]=\frac{5}{3}q\qquad (\textit{present study})$$

$$\rho h^2\ddot{f}+\frac{Eh^4}{a^4}\left[9.768\,f+4.602\,f^3\right]=\frac{5}{3}q\tag{6.12}$$

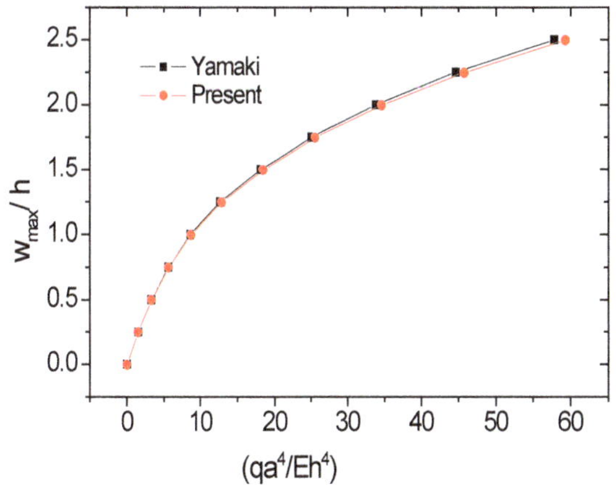

Figure 3: Comparison of result with that of Yamaki [16].

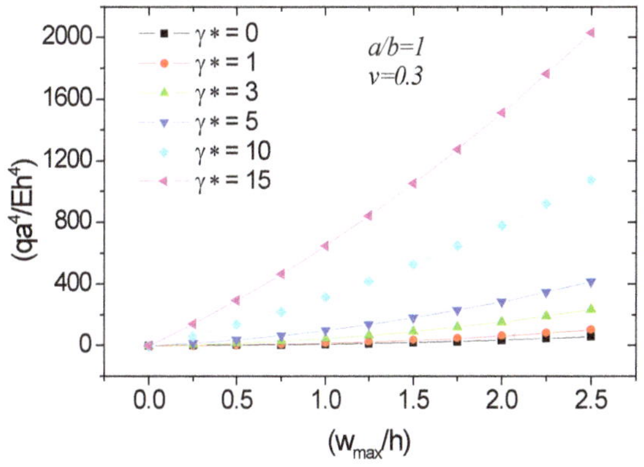

Figure 4: Load-deflection curve for a spherical shell of radius a, $\nu=0.3$ and γ^* = 0, 1, 3, 5, 10 and 15.

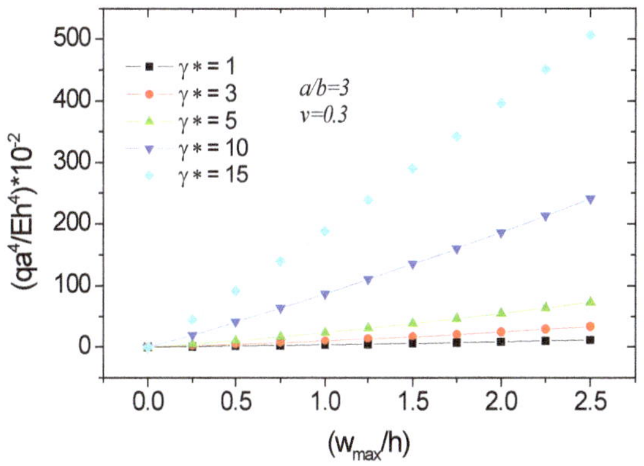

Figure 5: Load-Deflection Curve for (a/b)=3, $\nu=0.3$ and γ^* = 1, 3, 5, 10 and 15.

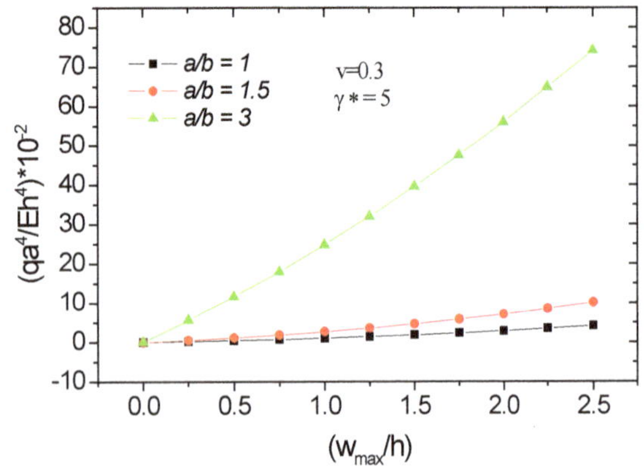

Figure 6: Load-Deflection curves for a/b=1, a/b=1.5, a/b=3 with $\gamma^*=5$.

Which is excellent agreement considering the fact that only a single term approximation for the deflection function has been made for the present study (Figures 3-6).

Free linear vibration

Set, α_2, α_3 *and* Q^* each equals to zero, when the linear frequency is given by

$$\omega_L^2 = \frac{DP}{\rho h}\left[\frac{40}{3}+2\beta\delta\right] \text{ or } \omega_L^2 = \frac{Eh^2P}{8\rho}\left[\frac{(320/3)}{12(1-\nu^2)}+4\left(\frac{2\gamma_1}{h}\right)^2\right] \tag{6.13}$$

Where $\gamma_1 = \kappa / P$ and $\frac{2\gamma_1}{h} = \gamma^*$ represents the measure of shallowness of the shell.

Eq. (6.13), on simplification and with a little rearrangement of the parameters, becomes

$$\omega_L^2 = \frac{Eh^2P}{8\rho}\left[\frac{\lambda^4}{12(1-\nu^2)}+\frac{M^4}{12(1+\nu)^2}\right] \tag{6.14}$$

Where $M^4 = \frac{192\gamma_1^2}{h^2}(1+\nu)^2$ and $\lambda^4 = (320/3) \approx (3.196)^4$ (for fundamental mode of vibration) has been introduced for comparison [9, 15].

If ω_0 be the value of ω_L corresponding to *M=0* and $\nu = 0$, that is the value of the frequency for a flat plate with vanishing Poisson's ratio, then

$$\omega_0 = 2.984\left[Eh^2(3a^4+2a^2b^2+3b^4)/(8\rho a^4 b^4\right]^{1/2} \tag{6.15}$$

and

$$\frac{\omega_L}{\omega_0} = \left[\left(\frac{1}{(1-\nu^2)}\right)+\left(\frac{M}{\lambda_0}\right)^4\frac{1}{(1+\nu)^2}\right]^{1/2} \tag{6.16}$$

Which are in exact agreement with that of [9].

If the second term in the expression for ω_L dominates the first then

$$\omega_{L2} = \left(\frac{2E}{P\rho}\kappa^2\right)^{1/2},$$

Which is exactly the same as that [9]. It may be noted here that following Reissner the first term is predominant when γ^* or *H/h < or=25* and the second term is predominant when *H/h≥25* in order that the theory of shallow shells is applicable. Table 1 shows a close agreement for the values of fundamental frequency for a flat circular plate.

Nonlinear free vibration

Substituting *Q*=0* in Equation (6.11) one obtains

$$\rho h^2 \ddot{f} + \alpha_1 f + \alpha_2 f^2 \alpha_3 f^3 = 0 \tag{6.17a}$$

Or

$$\ddot{f} + A_1 f + A_2 f^2 A_3 f^3 = 0 \tag{6.17b}$$

This is a familiar form of time differential equation and for which the frequency ratio (Nonlinear to Linear) is given by [16]

$$\frac{\omega^*}{\omega} = \left[1+\left\{\frac{3}{4}\frac{A_3}{A_1}-\frac{5}{6}\left(\frac{A_2}{A_1}\right)^2\left(\frac{A}{h}\right)^2\right\}\right]^{1/2} \tag{6.19}$$

From which one can find the nonlinear effect on the frequency. The results have been presented in the form of graphs (Figures 7-9).

Static deflection

Neglecting the inertial term, Equation (6.11) can be written as

$$\alpha_1 f + \alpha_2 f^2 \alpha_3 f^3 = \frac{5}{3}q \tag{6.20}$$

Which after simplification reduces to (here *f* stands for maximum static deflection).

$$\frac{qa^4}{Eh^4} = L\left[\left\{\frac{2}{3(1-\nu^2)}+0.3\left(\frac{2\gamma_1}{h}\right)^2\right\}f + \frac{20}{3N}\left(\frac{2\gamma_1}{h}\right)f^2 + \frac{160}{7N^2}f^3\right] \tag{6.21}$$

ν	Ref. 15	Present Study	Re. 16
0	2.948	2.948	___
0.3	3.091	3.125	3.125

Table 1: Values of coefficient of $\frac{h}{a^2}\sqrt{\frac{E}{\rho}}$ is in the expression for the fundamental frequency for a circular plate.

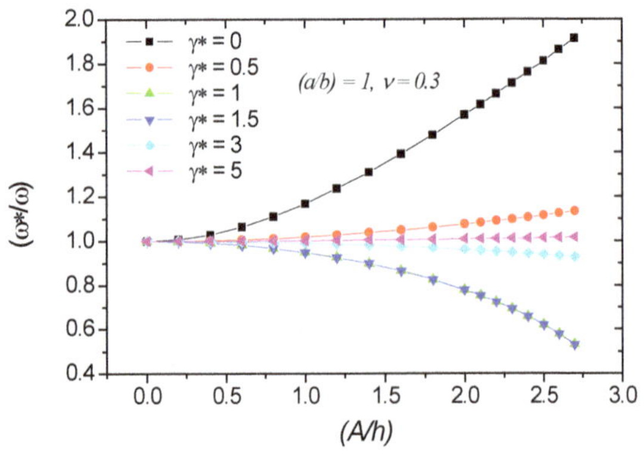

Figure 7: (ω^*/ω) *vs.* relative amplitude *(A/h)* for a spherical shell.

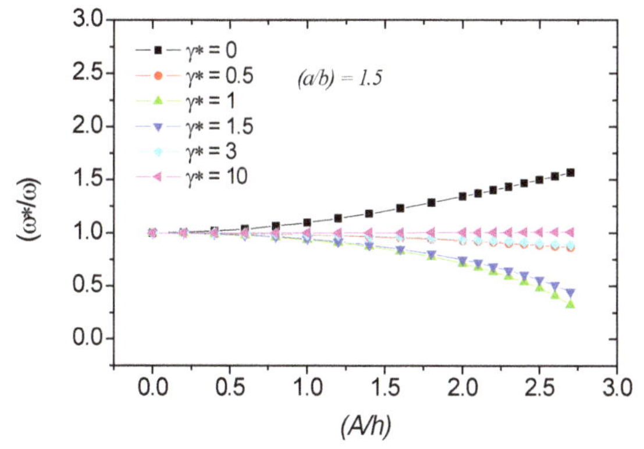

Figure 8: (ω^*/ω) vs. relative amplitude *(A/h)* for *a/b=1.5, v.*

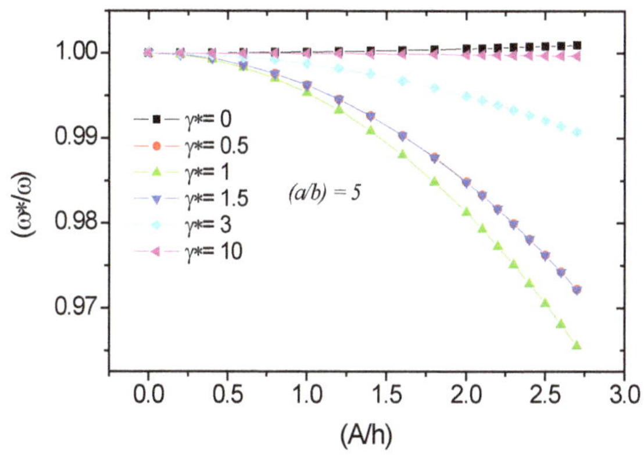

Figure 9: (ω^*/ω) vs. relative amplitude *(A/h)* for *a/b=5, v=0.3.*

Where $L=\left\{3\left(\frac{a}{b}\right)^4+2\left(\frac{a}{b}\right)^2+3\right\},\ N=\left\{3\left(\frac{a}{b}\right)^2+2+3\left(\frac{b}{a}\right)^2\right\}$ (6.22)

Since in the literature, no result on the static large deflection of the dome on an elliptic base is available, we may verify the results with that for a flat circular plate in the limiting case (Tables 2 and 3). When $\kappa \to 0,\ a=b$, equation (6.19) represents the static behaviour of a flat circular plate of radius "*a*" with clamped immovable edges. Equation (6.21) shows a comparative study for the same.

$$\frac{qa^4}{Eh^4}=\begin{cases}5.8608\ \bar{w}_m+2.857\ \bar{w}_m^3 & (present\ study)\\ 5.861\ \bar{w}_m\ \ +2.761\ \bar{w}_m^3 & (Yamaki)\\ 5.848\ \bar{w}_m\ \ +2.754\ \bar{w}_m^3 & (.Timoshenko)\end{cases} \quad (6.23)$$

Where $\bar{w}_m$ stands for

$\bar{w}_m = w_m/h=\left[f(t)\right]_{max}$ [Ref. 16] (6.24)

= (maximum deflection divided by the plate thickness)

The graphical representation of the above results has been made in Figure 3 validating the correctness of the present method.

Results and Discussion

Frequency analysis

Table 1 shows the values of linear frequency for a circular plates obtained using different approaches. It justifies the present approach (CDC method). Further discussion on the linear frequency is considered to be irrelevant as Equations (6.13-6.16) are exactly the same as those obtained in [9,15] and the authors have already made detailed discussion on it.

Static analysis

The results for a shallow shell resting on an elliptical base have been shown in Figures 3-6. Figure 3 gives comparison of results for maximum deflection for a circular plate obtained through a classical approach and through the CDC-method. Figure 4 shows the load-deflection behavior for a spherical shell for different values of V and $\gamma^*=(2\gamma/h)$. It shows that there is no significant difference for the load-deflection curve for a spherical shell for $\gamma^*=(2\gamma/h)<5$. But the measure of shallowness affects the results when $\gamma^*=(2\gamma/h)\geq 5$ and greater is the measure, lower is the deflection. Figure 4 shows the effect of on the load-deflection curve of the shell for a fixed ratio of the aspect ratio of the elliptic base. In this case it is observed that greater is the measure of shallowness lower is the deflection. Figure 4 shows that for a particular load, deflection increases with the increase of shallowness of the shell.

A comparison of results shown in Figures 5 and 6 indicate that for a certain load the deflection increases with the increase of or with the increase in the aspect ratio (a/b).

γ^*	$\nu=0$ (ω^*/ω)	$\nu=0.3$ (ω^*/ω)	$\nu=0.5$ (ω^*/ω)
0	1.077697	1.070936	1.058808
0.5	1.054078	1.048371	1.038602
1	1.495356	1.452167	1.374265
1.5	2.13267	2.04855	1.892266
2	2.817702	2.696798	2.469124
2.5	3.512205	3.357009	3.062966
3	4.207663	4.019331	3.661462
3.5	4.902489	4.681548	4.261079
5	6.983613	6.665906	6.06027
10	13.91613	13.27773	12.05965
20	27.79355	26.51485	24.07448

Table 2: Values of (ω^*/ω) for a/b=1for different values of ν and γ^*.

1	1	1
0.996857	0.996912	0.997036
0.987368	0.987589	0.988091
0.971348	0.971853	0.973
0.948465	0.949384	0.951471
0.918206	0.919689	0.923053
0.879812	0.882039	0.887084
0.832155	0.835358	0.842598
0.773527	0.778023	0.788154
0.70118	0.707448	0.721502
0.610256	0.619123	0.638831
a/b=1.5	1.5	1.5
g=5	5	5
v=0	0.3	0.5

Table 3: Values of (ω^*/ω) for a/b=1.5 and γ^* =5 for different values of ν.

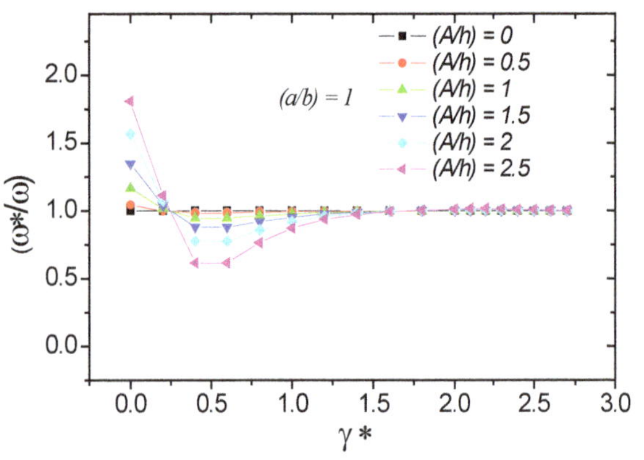

Figure 10: (ω^*/ω) vs. measure of shallowness for a spherical shell for various amplitudes.

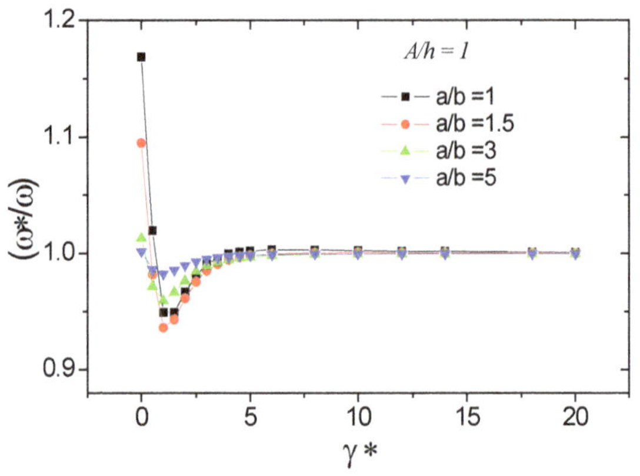

Figure 13: (ω^*/ω) vs. measure of shallowness for various aspect ratios for relative amplitude *A/h=1.*

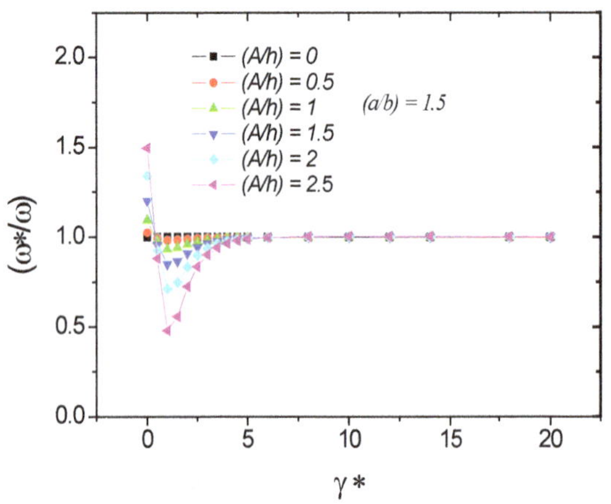

Figure 11: (ω^*/ω) *vs.* measure of shallowness for aspect ratio 1.5 and for various amplitudes.

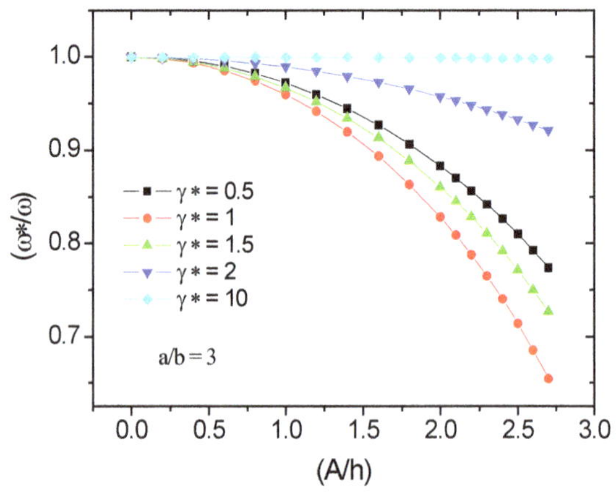

Figure 14: (ω^*/ω) vs. relative amplitude *A/h* for various values of measure of shallowness and for aspect ratio *a/b=3.*

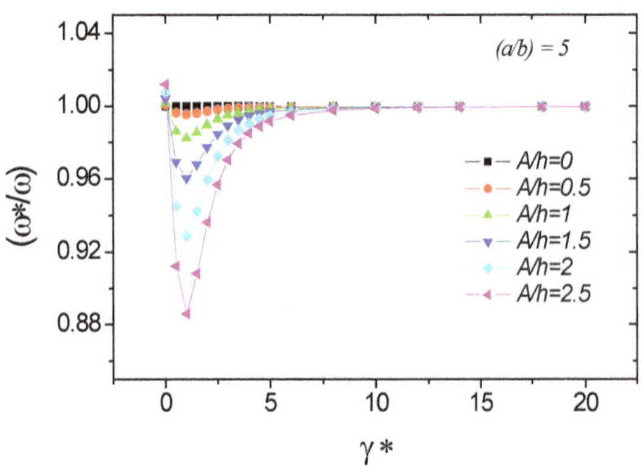

Figure 12: (ω^*/ω) vs. measure of shallowness for various amplitudes for aspect ratio *a/b=5.*

Vibration analysis

Figures 7-14 show the dependence of nonlinear to linear frequency ratio on γ^* and aspect ratio a/b. Figures 7 and 10 show the result for a spherical shell (γ^* =0). It has been observed that the dependence on the Poisson's ratio is not so much significant though the nonlinear effect is comparatively a little lower for higher values of ν. Figure 13 makes a comparative study of dependence of the relative frequency ratio γ^* on the aspect ratio (a/b) of the axes of the elliptic base of the dome. The nonlinear effect is significant when value of a/b decreases. Figure 13 confirms that the nonlinear effect is not so much dependent on aspect ratio for γ^* ≥1.5-2. Considering all aspects as relevant from the Figures 7-14, it appears that the values of γ^* in the range of 1-2 affect the nonlinear behaviour of the vibrating shell.

Conclusion

In conclusion it can be said that the method proposed in this paper offers a new approach to deal with problems involving large amplitude vibrations of plates and shallow shells. The application of polynomial expressions for the deflection and the stress functions in conjunction with the Galerkin procedure appears to produce highly accurate results. The comparison of results shows that using a moderately approximated expression for the deflection function yields results which are comparable to the previously obtained results using other approximate methods. It can therefore be concluded that the CDC method appears to be a simple tool to deal with the problems of nonlinear vibration of plates and shallow shells of arbitrary shapes.

References

1. Karman V (1910) Festigkts problems in mechanical engineering. Encyclopedia of Mathematical sciences 4: 311-385.
2. Herrmann G (1955) Influence on large amplitudes on flexural motions of elastic plates. NACA Tech 41: 501-510.
3. Leissa WA, Kadi AS (1971) Curvature effects on shallow shell Vibrations. Jl of Sound and Vivration 16: 173-187.
4. Mayers J, Wrenn BG (1967) Developments in Mechanics. Proceedibgs of the Tenth Midwestern Mechanics Conference 4, On the nonlinear free vibration of thin cylindrical shells. New York.
5. Evensen DA, Fulton RE (1965) Some studies on the nonlinear dynamic response of shell-type structures. International Conference on Dynamic Stability of Structures, Evanston.
6. El-Zaouk BR, Dym CL (1973) Nonlinear Vibrations of Orthotropic Doubly-Curved Shalow shells. Jl of Sound and Vivration 31: 89-103.
7. Mazumdar J (1970) A method for solving problems of elastic plates of arbitrary shapes. J Aust Math Soc 11: 95-112.
8. Mazumdar J (1971) Transverse vibration of elastic plates by the method of constant deflection lines. J Sound Vib 18: 147-155.
9. Jones R, Mazumdar J (1974) Transverse vibrations of shallow shells by the method of constant-deflection contours. J Acoust Soc Am 56: 1487-1492.
10. Banerjee MM (1997) A new approach to the nonlinear vibration analysis of plates and shells. Trans 14th Intl Conf On Struc Mech In Reactor Tech (SMIRT-13), France.
11. Banerjee MM, Rogerson GA (2002) On the application of the constant deflection-contour method to non-linear vibrations of elastic plates. Archive of Applied Mechanics 72: 279-292.
12. Jones R, Mazumdar J, Chiang FP (1975) Further studies in the application of the method of constant deflection lines to plate bending problems. Intl J Eng Sc 13: 423-443.
13. Mazumdar J (1971) Buckling of elastic plates by the method of constant deflection lines. J Aust Math Soc 13: 91-103.
14. Bucco D, Mazumdar J (1979) Vibration analysis of plates of arbitrary shape-A new approach.
15. Bucco D, Mazumdar J, Sved G (1979) Vibration analysis of plates of arbitrary shape–A new approach. Journal of Sound and Vibration 67: 253-262.
16. Reissner E (1955) On axi-symmetrical vibrations of shallow spherical shells. Quar Appl Math 13: 279-290.
17. Yamaki N (1961) Influence of large amplitude on flexural vibration of elastic plates. ZAMM 41: 501-510.
18. Timoshenko S, Woinowski-Krizer S (1959) Theory of Plates and Shells (2ndedn.) McGraw-Hill Book Co, New York.

8

Investigation on Last Stage High Pressure Steam Turbine Blade for Producing Electricity

Vijendra Kumar[1]* and Viswanath T[2]

[1]*Professor, Manager, Jagadguru Dattatray College of Technology, Indore, M.P, India*
[2]*Deputy General Manager, Kruthi Computer Services Pvt Ltd, Bangalore, Karnataka, India*

Abstract

The investigation on design of high pressure steam turbine blade addresses the problems of steam turbine efficiency. A precise focus on aerofoil profile for high pressure turbine blade and it gages the effectiveness of certain like Chromium and Nickel in resisting creep and fracture in the turbine blades. The thermal and chemical conditions in blades are, substrate from to prevent the corrosion when exposed to wet steam. The efficiency of the steam turbine is a key factor in both the environmental and economical collisions of any coal-fired in power stations. The increasing efficiency of a typical 500 MW turbine by 1% reduces emissions of CO_2 from the turbine location, with corresponding reductions in NOx and SOx.

In this connection an attempt is made on steam turbine blade performance is important criterion for retrofit coal fired in power plant. Based on the research presented modifications to high pressure steam turbine blades can be made to increase turbine efficiency a turbine. The results and conclusions are presented for a study concerning the durability problems experienced with steam turbine blades.

Keywords: Steam turbine; High pressure blade; Stresses in blade

Introduction

Blades are the heart of a steam turbine, as they are the elements that convert the thermal energy into kinematic energy. The efficiency and consistency of a turbine depends on the proper design of the blades. It is therefore necessary for engineers involved in the steam turbines field to have an overview of the importance and the basic design aspects of the steam turbine blades. The blade design is a multi-disciplinary task. It involves like thermodynamic, aero-dynamic, mechanical and material science restraints. The total development of a new blade is thus possible only when experts of all these fields come together as a panel. Efficiency of the turbine is depends on the parameters like, Inlet and outlet angle of blade, blade Materials, blade profile and Surface finishing of the blade and etc.

The most critical aspect of steam turbine reliability centre on design of buckets. Buckets or rotating blades are subjected to unsteady steam forces during operation, the phenomenon of vibration significance must be measured. Resonance happens when an exciting frequency coincides with a natural frequency of the system. At timbre conditions, the amplitude of vibration is related primarily to the amount of stimulus and damping present in this system. A high bucket reliability requires design with minimum resonant vibration. The design procedure starts with precise calculation of bucket natural frequencies in the tangential, axial, tensile, and complex modes, which are verified by in the data. In addition, improved aerodynamic nozzle shapes and generous stage axial clearances are used to reduce stimulus bucket. Bucket covers are used on some of stages or all stages to attenuate prompted vibration. These design practices and composed with advanced precision manufacturing techniques, ensured the necessary bucket reliabilities. Almost all of the blading used in modern mechanical drive steam turbines is either of drawn or milled type construction and drawn blades are machined from extruded airfoil shaped pieces of material stock of material. Milled blades are machined from a rectangular piece of bar stock.

The purpose of this paper is to examine the causes for these seemingly contradictory results. An attempt will be made here to review the previous studies to look into future possibilities of high pressure blade from the view point of datum and modified design.

Literature Survey

Ghosh and Bansal [1] states that the limited primary energy sources and awareness of environmental pollution has led to ever increasing end over to develop new steam turbine power plants with the highest possible efficiency. Considering their output, even small increase in efficiency can result in saving for the customers. As overall cycle efficiency is strongly dependent on steam turbine performance, Constant development are sought to increase turbine efficiency. These effects are directly primarily towards improvements are blading as the key component of the turbine. This paper presents the BHEL to meet the requirement of higher efficiency by adapting newer blading, which can substantially improve stages efficiency and hence overall performance of the turbine

Kenji Nakamura [2] response to global environmental productions, higher efficiency and improved operating reliability are increasingly being requested for steam turbines are vital role for thermal power plants, by increasing the temperature and pressure of the steam turbine working conditions, more efficient power generation is recognized, and in order to realize a turbine applied with the higher temperature conditions of 700°C for the upcoming, Fuji Electric is contributing in the METI-sponsored development of advanced ultra-supercritical

***Corresponding author:** Vijendra Kumar, Professor, Manager, Jagadguru Dattatray College of Technology, Indore, M.P, India
E-mail: vijendravk@gmail.com

power generation, and is evaluating and verifying the reliability of materials used for high temperature controllers. In addition to geothermal steam turbines, Fuji has developed surface coatings and other technology for enhancing corrosion resistance in order to develop reliability. Moreover Fuji is moving ahead with the development of geothermal binary power-generating turbines that utilize a low boiling point medium.

Zachary Stuckl [3] addresses steam turbine efficiency by discussing the overall design of steam turbine blades with a specific focus on blade aerodynamics, materials are used in the manufacture of steam turbine blades, and the factors that cause turbine blade failure and therefore the failure of the turbine itself. This paper enumerates and describes the currently available technologies that enhance the overall efficiency of the generator and prevent turbine failure due to blade erosion and blade cracking. The stresses developed in the blade as a result of steam pressure, steam temperature, and the centrifugal forces due to rotational movement are delineated; current designs calculated to counter the fatigue caused by these stresses are existing. The aerodynamic designs of impulse and reaction turbine blades are compared and contrasted, the effect of those designs have on turbine efficiency are debated. Based on the research unfilled herein, this paper presents a complete summary of what modifications to existing steam turbine generator blades can be made to increase turbine efficiency.

Mısekl [4] have developed 3000 rpm 1220 mm blade for a steam turbine was developed with application of new design structures. The last stage stirring blade is designed with an integral cover, a mid-span tie-boss connection with a fir-tree dovetail joint. With this configuration the blades are continuously coupled by the blade untwist due to the centrifugal force when the blades rotate at high speed, so that vibration control and structural damping are provided. The last stage airfoil was optimalized from view of mineralization of its centrifugal force. In order to develop a speed of 3000 rpm 1220 mm blade, the advanced analysis methods to predict dynamics behaviour of the bladed structure were applied. To validate calculated results the verification measurement such as rotational vibration tests was carried out in the high speed test rig. The relation of the friction damping of the bladed structure on amount of excitation level was also monitored and evaluated.

Tulsidas [5] have said large variety of turbo-machinery blade root geometries used in industry prompted the question if an optimum geometry could be found. An optimum blade root stood defined, as a root with useful geometry when loaded returns the minimum stress concentration factor at the fillet. The present paper outlines the design modification for fillet stresses and a special attention is made on SCF of the blade root. Finite Element Analysis is used to determine the fillet stresses and Peterson's Stress Concentration Factor chart is effectively utilized to modify root of blade. The root is modified due to difficulty in manufacturing the butting surface of the tang which grips the blade to the disk crowns having small contact area.

Yasutomo Kaneko [6] to improve the reliability and the thermal efficiency of High Pressure (HP) end blades of steam turbine, new standard series of HP blades has remained and developed. The new HP blades are categorized by the Integral Shroud Blade (ISB) structure. In the ISB structure, blades are continuously coupled by blade untwist due to centrifugal force when the blades rotate at great speed. One of the probable failure modes of the ISB structure seems to be fret fatigue, because the ISB utilizes friction damping between adjacent shrouds then stubs. Therefore, in order to design a blade with high reliability, the design technique for evaluating the fretting fatigue strength was established by the model test and the nonlinear contact analysis. This paper boons the practical design method for predicting the fretting fatigue strength of the ISB structure, and the some solicitations are explained.

Stanisaa [7] has suggest to the erosion caused by wet steam flow reduces the efficiency of the last stage rotor blades of condensing steam turbines and makes their service life tinier. Today there is insufficient data on the erosion process which the steam turbine rotor blades are subject to during the working data which could be an origin for development and verification of mathematical models to estimate the service life of eroded rotor blades. This rag reviews the results of many years monitoring and researching of the laws of the erosion process and its mechanism for rotor blades of condensing steam turbines. On the beginning of the obtained regulations of the rotor blades erosion process and a simplified model their service life is estimated.

Christoph-Hermann Richter [8] to provide an overview of the structural design of modern steam turbine blades at Siemens power generation using the FE method. The altered types of blades are described in detail regarding their geometry of loading. The segmental building block approach of modelling is shown to be importance for the different analysis, a fatigue post-processor has been applied as well as an optimization tool. Both of these are in-house codes be briefly presented.

Pavlos K. Zachos [9] investigated the effect of blade twist, caused by developed inaccuracies, on the performance of a two stage axial steam turbine. A high reliability 3D coordinate measurement machine has been employed to obtain the exact geometrical model of the blades. A Streamline Curvature solver was used to predict the overall performance of turbine. In the manufacturing process of the casts and of the blades themself several types of errors can happen which lead to a different geometry envisaged by the designer. A high fidelity measurement of the actual geometry of both stator and rotor blades has been carried out. Finally, a comparison with the performance plots of the original geometry has been accepted. A assessable change of efficiency as well as in the total power delivered by the turbine was originated. This proposes that the accumulated error caused during the manufacturing procedure plays a significant role in the overall performance of the machine by making it less efficient by more than 1%. Reverse engineering techniques can be applied to predict and alleviate these errors leading thereby to a final design of each stage with improved performance.

Ahmad [10] on a droplet size influence on low pressure steam turbine blade erosion in the last stage of steam turbines, huge droplets is generated from the flow of wet steam. These droplets collide with the following rotating blades with almost the peripheral speed of the rotor. This high speed impact is observed in the form of erosion of low pressure steam turbine blades. Among others, impacting droplet size is a key parameter contributing to the erosion of low-pressure steam turbine blades. At Institute of Thermal Turbo machinery and Machinery Laboratory Stuttgart, the effect of droplet size on the erosion of steam turbine blade has been investigated with the help of a corrosion test rig. The experiments confirmed that the erosion increases with increasing droplet sizes.

Sandeep Soni [11] wet steam flow reduces the efficiency of the last stage rotor blades of steam turbines and makes their service shorter life. Water droplet corrosion is one of the major concerns in the design of modern steam turbine because it causes serious operational problems such as performance degradation and reduction of service life. A model has been used in the present study for the prediction of water droplet erosion of rotor blades operated in wet steam conditions. It is used to analyze the erosion behaviour of nickel coated glass epoxy steam turbine blades. The major erosion parameters to find growth time is rate of mass loss under varying conditions of dryness fraction of steam (x) ,steam temperature (T), coating thickness and size of the water droplets(d) are involved in the model so that it can also be used for engineering purpose at the design stage of rotor blades and these results are showing better improvement in the erosion characteristics like incubation period and rate of mass loss due to application of Ni coating on the glass -epoxy blades. Accordingly to that suitable operational factors have been defined to obtain the best possible performance of steam turbines.

Sevidova [12] to evaluation of the protective properties of multilayer coatings for steam turbine blades protective properties of multilayer ion-plasma coatings relative to the conditions of their exploitation on steam turbines are defined. It was established that the protection properties of coatings on 20X13 steel in an aggressive NaCl environment of various concentrations increase according to the sequence [Cr + (Cr,Ti)N]10 < (Ti + TiN)10 < (Cr + CrN)10. It was also found that a breach in the coating integrity can lead to the appearance of macrogalvanic couples. Their activity considerably increases (by 4-5 times) during the mechanical passivation of the surface under the conditions of drop-collision erosive wear. The maximum values of the EMF in stationary conditions are generated between the 20X13 steel and Ti + TiN coating.

Blade Material

Composition and Microstructure of Corrossive-Resistant High-Alloy Cast Steel

Alloy: CA-6 NM (Chromium and Nickel)

Microstructure: Martensitic Tempered Overall (Table 1), it is the material properties that make a blade consistent to failure. The yield strength, tensile strength, corrosion resistance, and modulus of elasticity all play a role in determining whether or not a blade will fail under operating loads.

Experimental high pressure blade design: Theory behind static analysis

In the static analysis we calculate the Centrifugal stresses.

Centrifugal stresses: The centrifugal forces exert the tensile stresses at the blade root, which pulls the blade away from the disc or the rotor. So sufficient section must be provided to the blades at the root and the material capable of withstanding the stresses without fatigue must be selected. Blades of area A, with angular frequency ω and density ρ exert centrifugal force,

$F_c = \rho A h \omega^2 r$, It is also given by following equation, $F_c = \dfrac{\rho A A_a}{2\pi(\omega)^2}$

Where, A=annulus area= $\dfrac{2\pi r}{(t-2rn)}$

1	Alloy	CA-6NM Heat
2	Treatment	> 955OC, Air cool, Tempering
3	Tensile strength	827 MPa
4	Yield strength (0.2% offset)	689 MPa
5	Elongation in 50 mm (2 in.), %	24%
6	Reduction in area, 60% Hardness (HB)	269
7	Charpy impact energy	94.9 J

Table 1: Mechanical Properties Alloys.

The centrifugal forces at the blade root section is the centrifugal force divided by area of the blade section at the root. $\sigma_c = \dfrac{\rho A_a}{2\pi(\omega)^2}$

Theory behind thermal analysis

The design features of the turbine segment of the steam turbine have been taken from the "Preliminary design of a control turbine. It was observed that in the above design after the rotor blades begins designed they were analysed only for mechanical stresses but there was no evaluation of thermal stresses. As the temperature takes significant effect on the overall stresses in the rotor blades a detailed study is carried out on the temperature effects to have a clear understanding of the combined mechanical and thermal stresses and the Radial elongation resulting from the Axial and Centrifugal forces.

The gas forces namely Tangential and Axial were determined by constructing velocity triangles at the inlet and exit of the rotor blades for obtaining the temperature distribution. The convective heat transfer coefficients on the blade surface exposed to the gases are fed in to the software. The radial elongation in the blade is also calculated. Temperature distributions and elongations are evaluated at several sections in the rotor blade [13-15].

The blades are designed for strength on the basis of the total effects of both static and dynamic stresses since the blades are designed to these stresses at one and the same time. The centrifugal forces causes tensile and bending stresses of constant magnitude, whereas the gas pressure causes bending stresses due to centrifugal forces are known as static stresses and those due to gas pressure are known as dynamic stresses.

The most dangerous of a constant section is the one at the root since it is weakened by the presence of reverting holes etc. If a blade is acted up on by instantaneous forces free vibrations are setup. The frequency of these vibrations depends on the dimensions of the blade or blade assembly and their mounting on the disc. There is a lot of stress concentration entailed in the root portion of the blade, so care should be taken to reduce this concentration. For blades with constant blade section along its length, the stresses at the weaker section are:

$$\sigma = \frac{C_o}{F_o} = \frac{C_b + \Sigma C_s}{F_o}$$

Where, Co=Centrifugal forces of blade, shroud etc,

Fo=area of the weakest blade section (root section).

Centrifugal forces of constant section blade will be:

$$C_b = G_b \gamma_{xy} \frac{\omega^2}{g} = F_o \gamma_{xy} \omega^2 \frac{h}{g}$$

Where, G=weight of the blade

h=height of the blade,

$\gamma_{xy=}$mean diameter

ω=angular velocity.

The centrifugal forces of the shrouding are obtained as:

$$C_s = G_s r_s \frac{\omega^2}{g} = F_s \gamma r_s \omega^2 \frac{I_s}{g}$$

Where, γ is the specific weight of the material from which the blades are made.

r is radius of the strip centroids.

The centrifugal forces of binding wire: $C_w = G_w r_w \dfrac{\omega^2}{g} = F_w I_w r_w \gamma \dfrac{\omega^2}{g}$

Bending and twisting calculations: Maximum tangential stress produced in the shaft

$$T_{\max} = \frac{1}{2} W \sqrt{(M_b)^2 + (M_t)^2 W} = 32\pi d$$

Where,

d=diameter of the shaft

M_b=bending moment

M_t=twisting moment

Twisting moment at the chosen section is given by

$$M_t = 97300 \frac{N}{n}$$

Where,

N=total power developed in KW and n speed of turbine in RPM.

Maximum bending moment can be calculated graphically by shear force bending moment diagrams. For obtaining the stresses the T shaped root node degree of freedom are constrained in the U_X, U_Y and U_Z directions and tangential, axial and centrifugal forces are applied at the centroid. The axial and tangential forces results from the gas momentum change and from pressure differences through the blades, which are evaluated by creating velocity triangles at the inlet and outlet of the rotor blades (Figure 1).

Inlet velocity triangle

The tangential and axial forces result from the gas momentum changes and from pressure difference across the blades, which are evaluated by constructing velocity triangles at the inlet and outlet of the rotor blades.

From the inlet velocity triangle of a rotor blade we get,

Whirl velocity V_{w2}=422.74 m/s.

Flow velocity V_{f2}=186.89 m/s.

Relative velocity V_{r2}=265.09 m/s.

Blade angle at the inlet (θ_3)=135.17^0

The tangential and axial force results from the gas momentum changes from pressure difference through the blades, which are

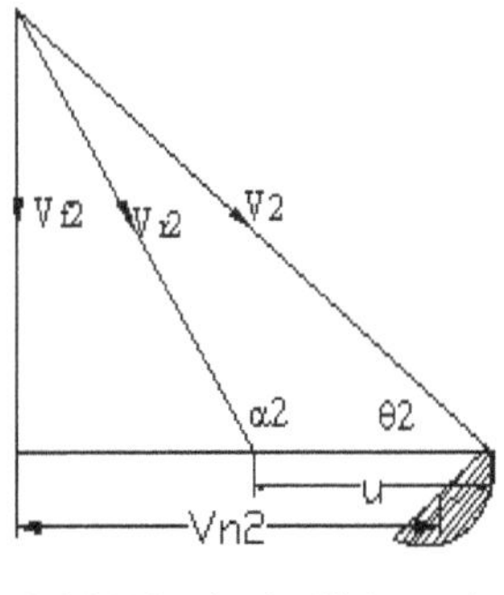

Figure 1: Inlet triangles for 1st stage rotor blade.

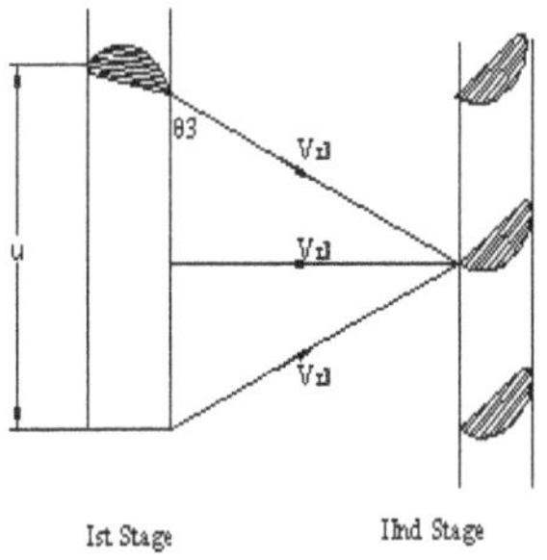

Figure 2: Exit velocity triangles for 1st stage rotor blade.

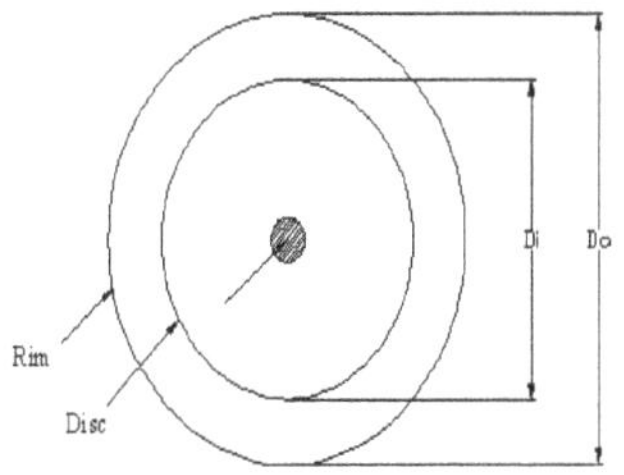

Figure 3: 1st Stage Rotors.

calculated by creating velocity tangles at the inlet and outlets of the rotor blades.

Exit velocity triangle

At the exit of first stage rotor blades,

Flow velocity V_{f3}=180.42 m/s.

Relative flow angle ψ_3=37.88°

Whirl velocity V_{w3}=2.805 m/s.

Relative velocity V_{r3}=293.83 m/s.

Please refer above Figure 2 for more understanding.

Evaluation of Tangential (Ft), Axial force (Fa) and Centrifugal force (Fc) on each rotor:

(A) Calculation of gas forces on first stage rotor:

At the inlet of first stage rotor blades,

Absolute flow angle (α_2)=22.850

Absolute velocity (V_2)=462.21 m/s.

Dia of blade mid span (γ)=1.3085 m.

Design speed of turbine (u)=πDN/60.

Tangential force Ft=m(V_{w2} + V_{w3}) Newton.

Axial force Fa=m(V_{f2} + V_{f3}) Newton.

Where, m is mass flow rate of gases through the turbine.

Referring to the above Figure 3.

$$M = \frac{\rho_2 (D_0 - D_i) V_{f2}}{4}$$

Where, ρ_2=density of gases at the entry of first stage rotor,

ρ_2=0.8900 kg/m^3,

m=70.925 kg/s.

Total axial force on first stage rotor Fa=458.88N.

Total tangential force on each rotor Ft=29783.88N.

Number of blade passages in first stage rotor=120.

Tangential forces on each rotor blade, $F_t = \frac{F_t}{No.of.blade.passages}$ =248.199N

Axial force on each rotor blade, $F_a = \frac{F_a}{No.of.blade.passages}$ =3.82N.

From Euler's Energy Equation,

Power developed in First stage rotor

P=m (V_{w2} U + V_{w3} U)

Using the above Equation

P=6.991MW

The distance $X = \frac{(m_1 \times x_1 + m_2 \times x_2 + m_3 \times x_3)}{(m_1 + m_2 + m_3)}$

Where, m_1, m_2 and m_3 are masses of volume 1,2 and 3.

x_1, x_2 and x_3 distances of the centroids of volumes, 1, 2 and 3 from the axis of revolution.

The material density ρ is graphically measured to be:

ρ=8900 kg/m^3

m_1=0.382 kg,

$m_2=\rho \times V_2$,

$m_3= \rho \times V_3$

Where, V_2 and V_3 are volumes of portions 2 and 3 of rotor blades,

The distance X is calculated and is 648.85 mm.

Total mass M=m_1+ m_2+m_3.

Centrifugal force Fc =M$(2\pi n/60)^2$

X=38038.33N.

Conclusions

The implementation of robust turbine blades are designed in accordance with the modern material technologies and able to withstand the most of circumstances, in combination with the use of clean and renewable fuel presents an efficient method of generating substantial amount of electricity. An enhanced blade design, focused on resisting the effects of stresses, corrosive agents, and creep-inducing temperatures, will raise the turbine efficiency, consequently leading to an increase in the power plant overall efficiency reduction of the quantity of fuel consumed, and ultimately a decrease in operational costs. To improve efficient blade design, will help to reduce operating costs even further and minimize the environmental impacts of steam turbines. Generally such a combination of technologies would benefit society by providing an effective, viable, and invulnerable means of generating electrical energy.

This example demonstrates that under normal operating conditions, stresses are sound and tinny acceptable limits. However, what the exemplar also shows are the large forces and stresses involved in such rotating machinery and how important factors such as design viewpoint, manufacture and maintenance strategy are to ensure safe operation.

The results and conclusions are presented for a study concerning the durability problems experienced with steam turbine blades. The maximum operational Von-Mises Stresses are within the yield strength of the material but the deformation is comparatively better for material CA-6 NM (Chromium Nickel).

Modified solutions for Steam turbine blade values to machines to maximize their reduce life cycle costs, efficiency and improve reliability.

References

1. Ghosh TK, Bansal SM (2005) Latest Trends in Large Rating Steam Turbine Blading", BHEL, Hardwar.
2. Nakamura K, Takahiro Tabei, Tetsu Takano "Recent Technologies for Steam Turbines", Energy Solution Group, Fuji Electric Systems Co., Ltd.
3. Stuck Z (2012) Steam Turbine Blade Design, Twelfth Annual Freshman Conference 2214 United state of America, Conference Session B6, 14th APR 2012.
4. Mısek T (2008) Static and Dynamic Analysis of 1220 mm Steel Last Stage Blade for Steam Turbine. Applied and Computational Mechanics 133-140.
5. Tulsidas D, Shantharaja M, Kumar K (2012) Design modification for fillet stresses in steam turbine blade. International Journal of Advances in Engineering & Technology (IJAET) 3: 343-346.
6. A book of Analysis of Fatigue Strength of Integral Shroud Blade for Steam Turbine, Proceedings of the International Conference on Power Engineering 2007.
7. Stanisaa B (2011) Erosion behavior and mechanisms for steam turbine rotor blades. International Journal of Advances in Engineering & Technology (IJAET) 2: 110-117.
8. Anestis K (2008) Turbine Blading Performance Evaluation Using Geometry Scanning and Flow field Prediction Tools. Journal of Power and Energy Systems 2: 1345-1358.
9. Ahmad M (2013) Casey Experimental investigation of droplet size influence on low pressure steam turbine blade erosion 303: 83-86.
10. Sandeep S (2011) Erosion behaviour of steam turbine blades of glass-epoxy. IJAET 2: 110-117.
11. Sevidova EK, Kazak KV, Vakulenko (in press) Evaluation of the protective properties of multilayer coatings for steam turbine blades, IJAET Volume 4 Issue 2.
12. Kiyoshi S, Yoshio S (2002) Development of a Highly Loaded Rotor Blade for Steam Turbines. Japan Society Mechanical Engineering (JSME) International Journal Series B 45: 881-890.
13. Arkan K, Husain Al-T (2008) Stress Evaluation of Low Pressure Steam Turbine Rotor Blade and Design of Reduced Stress Blade, Iraqi Academic Scientific Journal Eng Tech 28(2).
14. Daniel A, Snyder (2010) A Modeling Study of the Sensitivity of Natural Frequency of Vibration to Geometric Variations in a Turbine Blade, Rensselaer Polytechnic Institute Hartford, CT, December 2010.
15. Forces on Large Steam Turbine Blades. RWE power, Basic Engineering Mathematics, John Bird, 2007, published by Elsevier Ltd.

9

Ship Energy Efficiency Performance Estimation using Normal Daily Report Operational Data

Rajendra Prasad Sinha[1]* and Pragasen T Kunjambo[2]

[1]*Malaysian Maritime Academy, Advance Marine Engineering, Kuala Sungai Baru, Masjid Tanah, Melaka, Melaka 78300, Malaysia*
[2]*Malaysian International Shipping Corp, Malaysia*

Abstract

A ship is designed to consume ceratin amount of fuel to meet its business objective. But as operating hours build up the state of machinery and surface condition of underwater hull change resulting in increased fuel consumption and rising operating cost. This calls for close monitoring and regular energy efficiency analysis of the ship's energy systems. A true comprehensive energy analysis of the ship requires taking into consideration energy flow across each major power producing and consuming components of the energy systems including those originating from environmental and human factors such as hull fouling, wind, wave, current, ship's draft, and sea temperature. The overall impact of all these factors on ship's energy demand is extremely complex and have been rarely ever correctly assessed. The most effective approach so far to ship energy performance analysis/monitoring has been to quantify the contribution of each energy element by removing the effects of remaining.

The authors in this paper conduct heat balance analysis of the steam power plant and apply filtering technique to the data from ship's daily report to assess the effects of external factors such as hull fouling trim and wind resistance on fuel consumption to estimate the overall energy efficiency performance (EEP) of an LNG ship.

Keywords: Energy efficiency; Specific fuel rate; Shaft power; Hull fouling; Power plant; Liquefied Natural Gas (LNG)

Glossary

EEP: Energy Efficiency Performance

HPT: High Pressure Turbine

LPT: Low Pressure Turbine

T/G: Turbo Generator

LNG: Liquefied Natural Gas

FO: Fuel Oil

LOG: Equipment to measure ship speed with respect to water in nautical miles per hour

HCV: High Calorific Value

Prop SFR: Propulsion Specific Fuel Rate

Ship SFR: Ship Specific Fuel Rate

Introduction

Efficient ship operation means all energy producing and consuming systems in the ship utilize least amount of fuel for a given power output [1]. Therefore, to analyse and estimate fuel consumption of the ship all its major energy exchange processes must be identified and investigated for their energy flow patterns. In steam powered LNG ships the major energy producing/consuming systems are

(i) Main Steam Boilers

(ii) Steam Turbine Propulsion Engines

(iii) Steam condensers and auxiliary machinery

(iv) Turbo Genrators and Motors that produce and consume electrical power

(v) Boiler Feed Pumps

(vi) Ship's hull and propeller which receive power from steam turbines to overcome resistance from water, wind, current, and wave effects.

(vii) Ship's rudder and steering system

To achieve overall best fuel economy by the ship each subsystem needs to be analysed for its energy efficiency [2]. In this paper, the energy efficiency analysis of the thermal system has been carried out by using heat balance diagram of machinery operating data. The EEP of the hull and propeller system has been carried out using data recorded in daily and voyage reports. One major drawback in using operating ship data to estimate EEP of the hull is the presence of large spurious external noise in the data which if not adequately filtered out can affect accuracy and reliablity of the performance baseline. To overcome this problem, authors in this paper use conventional filtering technique to a large volume of actual ship operating data to establish reliable hull condition trends.

Data Acquisition

The ship is fitted with a high capacity data equisition system to record operating parameters automatically and store as excel data sheets. The following operational parameters, as in Table 1, are recorded and forwarded to the head office as the ship's daily/voyage reports for energy analysis by the shore staff [3]. Six months operating

***Corresponding author:** Rajendra Prasad Sinha, Malaysian Maritime Academy, Advance Marine Engineering, Kuala Sungai Baru, Masjid Tanah, Melaka 78300, Malaysia, E-mail: rajendra@alam.edu.my

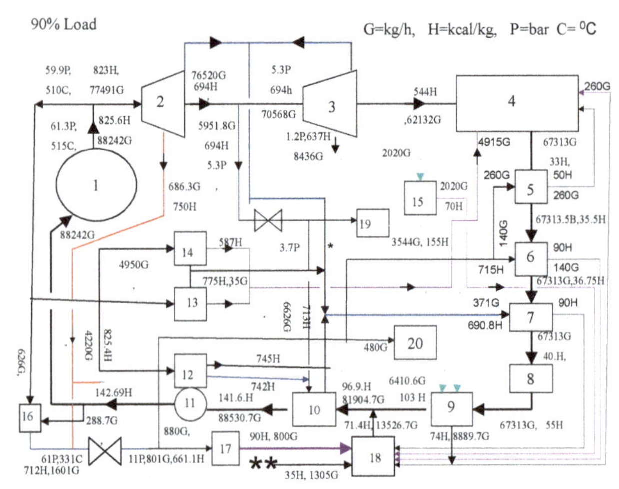

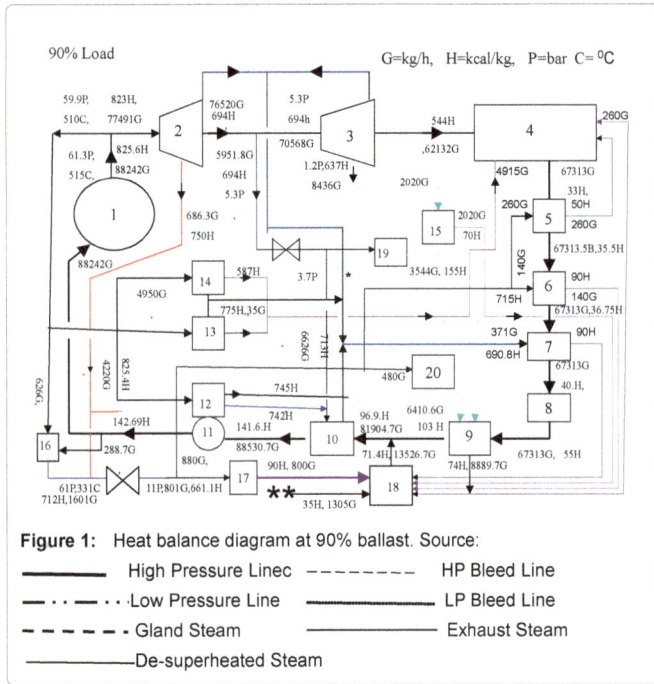

Figure 1: Heat balance diagram at 90% ballast. Source:

High Pressure Linec — HP Bleed Line — Low Pressure Line — LP Bleed Line — Gland Steam — Exhaust Steam — De-superheated Steam

S.No	Parameter	Unit
1	Duration of operationa at sea	Hours
2	Ship Velocity	Knots
3	Shaft Power	kWh
4	T/G 1 Power	kWh
5	T/G 2 Power	kWh
6	D/G Power	kWh
7	Shaft speed	rpm
8	Fuel consumed at sea	tons/day
9	F.O consumed in maneuvr/port	tons/day
10	Distance covered by LOG	nm
11	Draft forward	m
12	Draft aft	m
13	Wind speed	Knots
14	Wind direction	Degree

Table 1: Operating Parameters.

SHP, kW	F.O, t/h	SFR, g/kWh	HCV, kJ/kg	BlrEff (oil)	BlrEff (gas)	T/G1, kW	Evap, t/d	S.W °C
22693	6514.45	299.81	43052	88.5	84	1210	30	24

Table 2: Sea Trials with 90% Ballast Load.

data from January-June 2014 has been used in this investigation to estimate energy efficiency of the ship. As the main machinery operating parameters are not entered in the daily report, the power plant EEP has been estimated using the archieved sea trials data.

Machinery Performance and Heat Balance

Knowledge of how well energy exchange between various components of the ship propulsion system taking place is best illustrated by preparing a heat balance diagram [4]. The heat balance analysis is a method based on the energy conservation principle of the first law of thermodynamics and commonly used in process and power plant industry to measure energy efficiency. It essentially provides the energy flow topology of the power plant giving visual realizations of energy exchanges taking place within the system and its environment. It is of great assistance in identifying areas of plant improvements for better fuel economy. The heat balance diagram prepared from the data recorded during the full power sea trials at 90% ballast load condition is shown in Figure 1. Table 2 shows the baseline data for the sea trial.

Using energy flow information in Figure 1, the enthalpy drop and power produced in different components of the steam power plant

S.No	Equipment	Mass flow (kg/s)	h(kJ/kg)	m.x h (kW)	Remarks
1	Boiler	24.51	2860.4	70108.5	---
	Main Turbines				
2(a)	HP Turbine	21.52	542.25	**11669.3**	Net shaft power
(b)	Power from HPT Bleed	0.19	306	**58**	
3(a)	LP Turbine	17.255	628.2	**10838.9**	
(b)	Power from LPT Bleed	2.336	238.77	**557.8**	
	Power produced by HP and LP turbines			**23124**	
4	Condenser				
(a)	Heat from LPT exhaust	17.255	2140.06	36926.9	Heat lost to sea
(b)	Heat from TG exhaust	1.365	2320.15	-3167	
(c)	Heat from Air ejec drain	0.0722	71.196	-5.14	
	Heat lost to cooling sea water			40099	
	Auxiliary machinery and systems				Heat loss, kW
5(a)	Air ejector(steam)	0.0722	2774.55	-200.3	Negligible
(b)	Air ejector(FW)	18.7	10.51	196.53	
6(a)	After condenser(steam)	0.0388	2604.93	-101.07	Negligible
(b)	After condenser (FW)	18.7	5.235	98	
7(a)	Gland cond.(FW)	18.74	13.61	254.5	Negligible
(b)	Gland cond(Steam)	0.103	2516.15	-259.16	
8	Fresh W Gen (FW)	18.7	62.82	1174.73	
9(a)	LP heater (FW)	18.7	200.02	3759.15	Negligible
(b)	LP heater(bleed)	1.448	2357.84	-3414.15	
(c)	LP htr (exst steam)	0.984	339.2	-333.8	
10(a)	De-aerator (FW)	22..75	187.2	+4259..25	145.53
(b)	De-aerator(steam)	1.84	2393.8	-4404.6	
11	Feed pump(FW)	24.51	0.444	133.44	---
12	Feed PP (steam)	1.171	349.28	**409**	---
13(a)	Turbo Gen2	0	0	0	---
(b)	Turbo Gen1(drain)	0.0097	211.91	-2.05	---
14	Turbo Gen1	1.375	998.42	**1373.4**	---
15	Evaporator(inlet steam)	0.561	2374.59	-1332.41	---
16(a)	De-super heater(steam)	0.2277	474.08	-107.94	17.94
(b)	De-super heater(FW)	0.0377	2387.16	90	
17(a)	Aux de-super heater(FW)	--	--	--	
(b)	Aux de-super heater (steam)	--	--	--	
18	Auxiliary De-sup services	0.222	2391.5	-530.91	---
19(a)	Atmospheric drain tank	0.362	152.44	55.2	Negligible
(b)	Atmospheric drain tank	2.47	36.01	-26.89	
(c)	Atmospheric drain tank	0.222	77.9	-17.32	
(d)	Atmospheric drain tank	0.103	77.9	-8.02	
(e)	Atmospheric drain tank	0.0388	77.9	-3.03	
20	Heat to combustion air	0.984	2394.3	2355.97	
21	Atomizing and soot blow	0.133	2558	341	
	Remarks: +ve temperature rise , -ve Temperature drop.				

Table 3: Heat balance analysis data [2].

1	Mean Draft	10, m
2	Trim	<± 0.5, m
3	Speed from LOG	>14 knot
4	Speed difference (LOG-GPS)	<± 0.3 knot
5	Beaufort Sea State	≤3
6	Wind Speed	<15 knot

Table 4: Baseline Operating Conditions [3].

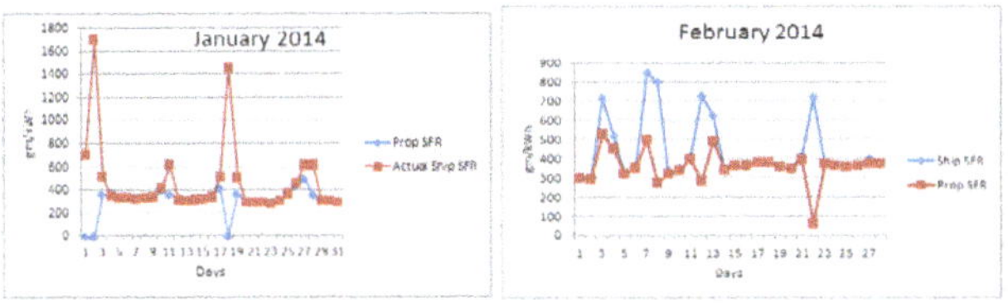

(a) (b)

Figure 2: Ship SFR- Prop. SFR.

have been calculated and placed in Table 3. The High Pressure Turbine (HPT), Low Pressure Turbine (LPT), Turbo Generators (T/G) and feed pumps are the units of the power plant which actually convert heat into useful mechanical shaft power. The remaining components referred to as auxiliary units of the plant essentially exchange heat to raise temperature of water condensate, combustion air, liquid/gas fuel and sea cooling water of the evaporator to improve overall plant efficiency.

A comparison of shaft power together with T/G power recorded during the sea trial (Table 2) is in close agreement within error margin of 2% with the ship power calculated from Heat Balance analysis in Table 3. The overall thermal efficiency of 35.52% calculated from heat balance for steam power plant is consistent with Industry standards for marine steam turbines.

Hull and Propeller Performance Analysis

Major hurdle in hull performance analysis arises from the interference of environmental and operational factors such as wind, wave, current, draft/trim, sea water temperature and rudder transients [3,5]. Because of that the actual power needed to move the ship through water has very complex functional dependence on these variables and requires conducting special experiments with application of advanced mathematical tools to investigate their individual impact on the overall fuel consumption. To estimate true impact of hull and propeller fouling on fuel consumption will require complete elimination of interference from those factors. In this paper, the authors use conventional data filtering technique as in [3,5] to eliminate effects of undesirable external disturbances by establishing a baseline operating condition as filter (Table 4).

Remarks

(a) As draft/trim has influence on skin friction through the wetted surface area of the hull it must be maintained constant for comparative analysis. The standard mean draft is selected based on the historical archived data from previous voyages.

(b) Although skin friction due to hull fouling is highly sensitive to speed through water its effect below 10 knots is not so significant. Therefore, ship speeds by LOG below 10 knots have not been considered to ensure that the impact of hull/propeller fouling on fuel consumption is effectively captured.

(c) The effect of ocean currents and rudder oscillations is reflected in the difference between speed through water and speed over ground. Heavy weather condition results in increased difference between LOG speed and GPS speed signalling greater impact on increased hull resistance and vice versa. For this reason, any data with speed difference exceeding 3% has not been considered for analysis.

(d) Effects from wind and waves have been filtered out by considering data only where wind speed is below 15 knots. Similarly to eliminate effects of bad weather conditions the data has been considered only when sea state was below Beaufort scale of 3.

Data Analysis

Three months operating data from archived daily reports have been taken and filtered to baseline conditions for estimating energy efficiency performance indicators. The propulsion specific fuel rate (Prop SFR) and ship specific fuel rate (ship SFR) are defined as fuel consumed in grams to produce one kWh of power and has been taken as the ship's energy efficiency performance indicator [3,6].

$$Prop\,SFR = \frac{Fuel\,Consumed\,at\,sea\,per\,Day}{Total(SHP + T/G\;power)\;kWh}$$

$$Ship\,SFR = \frac{Fuel\,Consumed\,at\,(sea + port)\,per\,Day}{Total(SHP + T/G\;power)\;kWh}$$

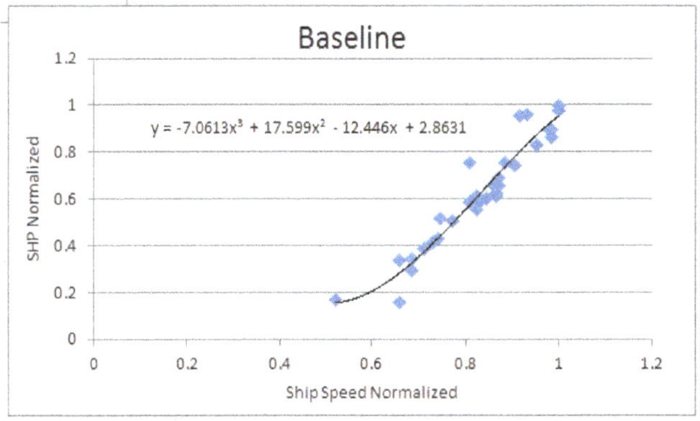

Figure 3: Ship speed–Shaft power.

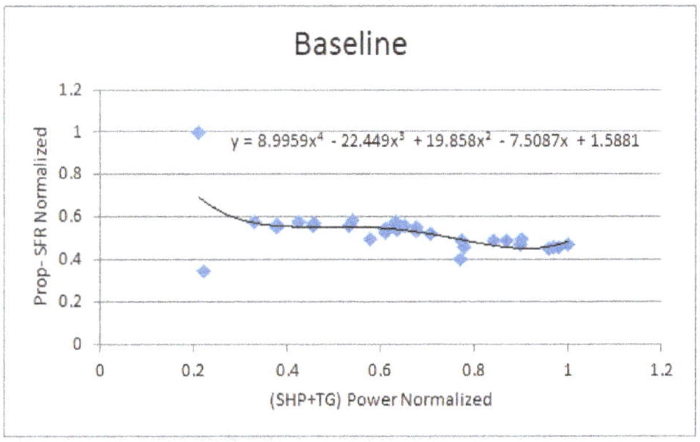

Figure 4: Shaft power – Prop SFR.

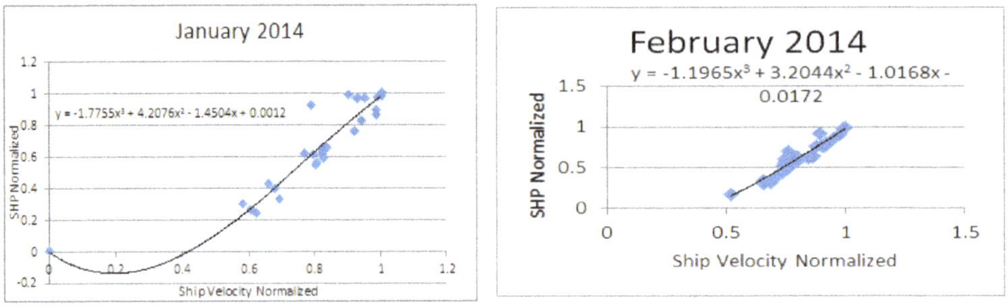

(a) (b)

Figure 5: Ship speed – Shaft power.

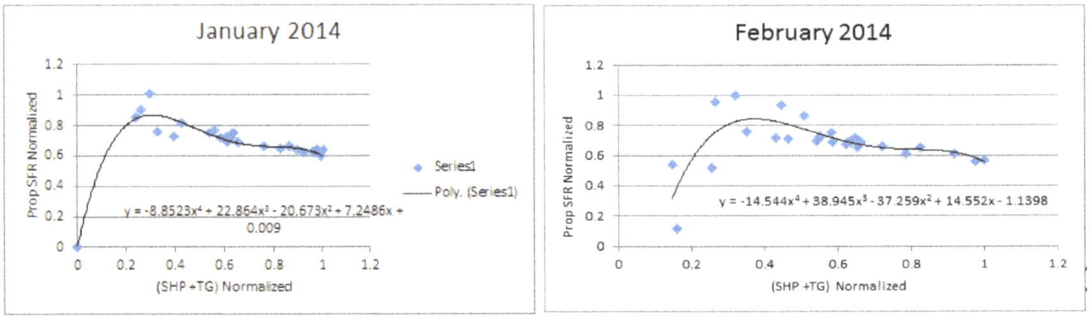

Figure 6: Shaft power- Prop. SFR.

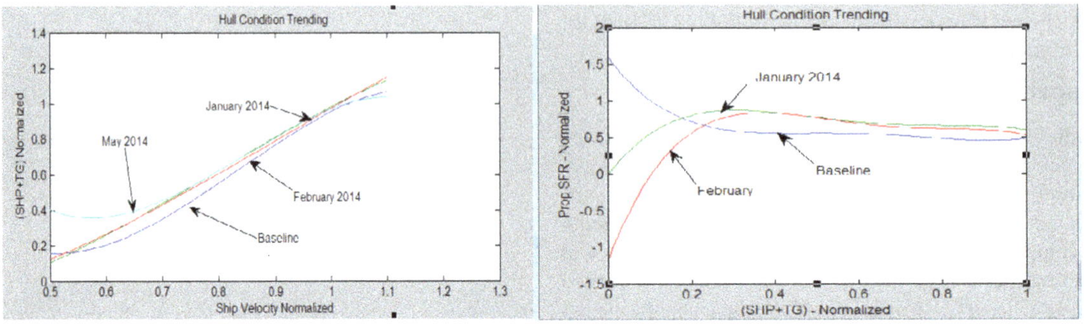

(a) (b)

Figure 7: Hull fouling – Ship speed, Prop. SFR.

From definitions, the ship and propulsion SFRs are expected to be the same during sea voyage and will differ only when ship is in port, as in Figure 2a and 2b. The separation of two energy efficiency terms has been made to keep track on fuel consumption during cargo operations in the port.

Figures 3 and 4 show ship's baseline EEP plots obtained from the filtered data, As expected, Figure 3 shows a 3rd order functional relationship between ship speed and shaft power. A decrease in propulsion SFR (Figure 4) with rising shaft power indicates improvement in fuel consumption at higher powers. But in this study because of the narrow data base of this investigation the minimum fuel estimate of 300 gms/kWh (0.45N) as shown in the graph may not be the true conclusive baseline EEP. A larger data base is expected to provide more accurate and convincing result. Besides that the actual data in daily report is referenced to Beaufort sea state 7 as against 3 originally intended to be used for data filtering which is also expected to introduce some error in the baseline performance estimate. But since no other better data is currently available until the ship begins her next post docking operating cycle these estimates are treated as benchmark for ship energy EEP comparison [7].

The minimum ship SFR estimate of 0.45 N (300 gms/kWh) (Figure 4) is about 10% higher compared to 272.53 gms/kWh obtained in power plant heat balance analysis. This 10% variance in ship SFR estimate is considered little high which may be due to error in data filtering and also the fact that the ship is already in operation for over half its next dry docking cycle. After setting the baseline for EEP, daily report data from January-May 2014 has been analysed to estimate impact of hull fouling on ship's energy efficiency. The results of analysis is discussed and presented in the following section [8].

Result of Investigation

Figures 5a,5b and 6a,6b show plots of ship velocity versus SHP and SHP versus prop SFR respectively. These plots have been obtained after normalizing the original daily report data to remove error in curve fitting arising from poorly conditioned data points. Although as expected, the ship Velocity–SHP plots in Figures 5a and 5b retain 3rd order functional dependence but model coefficients show some variations from baseline (Figure 3), due to effect of hull fouling. Similar, result can be observed in Figure 6a and 6b with respect to the specific fuel rate consumption of the ship during sea passage [9].

As the impact of hull fouling on ship speed is less detectable at lower velocities, ship speeds of 15 knots and above only have been considered for estimating its impact on fuel consumption. Also the hull fouling being a slow process progressing with time its impact on energy efficiency performance is visible only by way of gradual loss in ship's speed and increased fuel consumption. Figure 7a shows deviation in ship's speed from the baseline for the January and February 2014 data. Although one month time interval to measure effect of hull fouling on ship speed is too short to detect any significant variation but still at speed above 15 knots the trend indicates slight drop in ship speed from January-May 2014. But as expected, the drop in speed due to hull fouling with reference to baseline for the same period is clearly noticeable. Similar conclusion is also drawn from the plots in Figure 7b which shows small increase in propulsion SFR from baseline in February 2014.

Conclusions

A practical and operator friendly approach, free from complex mathematical computations to estimate ship energy efficiency performance, using operational data from daily/voyage report has been presented. The energy efficiency performance of major power plant components such as boiler, turbines, condenser and Turbo Generators has been estimated using heat flow diagram. The effect of hull fouling on vessel speed and increased SFR has been estimated by eliminating the external disturbing factors through use of data filters. Result of the investigation may be summarized as follows.

(i) The method proposed in this paper is simple and economical to implement.

(ii) Reliability of result will depend on the accuracy of data filter used.

(iii) The R2 values and corresponding residues of data fits in Figures 3-7 show satisfactory result.

(iv) In present investigation the data base was very small hence result is more qualitative than quantitative. Further investigation with wider data base is continuing.

(v)The estimated baseline Prop SFR of 300 gms/kWh is only approximate and needs further refinement by using post dry dock operational data.

Acknowledgement

The authors express sincere appreciation to Malaysian International Shipping Corporation Berhad for continuous encouragement and providing noon report data from their LNG fleet to carry out this research. The work presented here is part of ongoing Ship Energy Efficiency Research project at the Malaysian Maritime Academy.

References

1. Bazari Z, Longva T (2011) Assessment of IMO mandated energy efficiency measures for International Shipping.
2. Sinha RP, Wan M, Wan Nik N (2012) Energy analysis of steam power plant of a medium size LNG Tanker, International Conference on Maritime Technology, Kuala Terengganu.
3. Logan KP (2011) Using ship propeller for hull condition monitoring, presented at ASNE Intelligent Ship Symposium IX, Philadelphia, PA,USA.
4. Rashad A, Maihy AEL (2009) Energy and exergy analysis of steam power plant in Egypt, International Conference on Aerospace Sciences and Aviation Technology, ASAT-13.
5. Benjamin P, Larsen J (2009) Prediction of full scale propulsion power using artificial neural networks, Department of Informatics and Mathematics, Technical University of Denmark.
6. Sinha RP (2013) Fuel efficient ship operation-An optimization approach, 6th International Engineering Conference (ENCON 2013), Kuching, Malaysia.
7. Petursson S (2009) Predicting optimal trim configuration of marine vessels with respect to fuel usage, Master Thesis, School of engineering and natural Sciences, University of Iceland, Reykjavik.
8. Kallman J (2012) Ship power estimation for marine vessels based on system identification, Thesis, Department of electrical engineering, LIKOPINGS University, SF-581 83 Linkoping, Sweden.
9. Hellstrom T (2004) Optimum pitch, speed and fuel control at Sea. Journal of Marine Sciences and Technology 12: 71-77.

10

Effect of Axial Groove on Steady State and Stability Characteristics of Finite Two-Lobe Hybrid Journal Bearing

Lintu Roy*

Department of Mechanical Engineering, National Institute ofTechnology, Silchar-788010, India

Abstract

The study of the steady state and stability characteristics including whirl instability of finite two-lobe hybrid journal bearing with an axial groove, located at the top from which oil is supplied at a constant pressure is obtained theoretically. The two lobes of 1600 arc each are separated by two axial oil groove of 20° circumferential extensions in the horizontal direction. A two-dimensional finite difference solution was used to predict the performance of finite length externally pressurized two-lobe hydrodynamic hybrid journal bearings.The Reynolds equation is solved numerically using finite difference method satisfying the appropriate boundary conditions to obtain the effect of speed parameter on bearing performance.The stability characteristics were found out using first-order perturbation method. With the change of speed Stiffness and damping coefficients behaviour was determined at various eccentricity ratios. The bearing load-carrying capacity, stiffness, lubricant flow rate, attitude angle and frictional torque due to bearing rotation increase with increase in eccentricity ratio and speed. A comparison is also made with plain cylindrical axial grooved journal bearing. It is found that externally pressurized two-lobe hybrid bearing is more superior to plain cylindrical axial-grooved oil journal bearing in terms of load capacity, improved end flow, friction losses and stability. The bearing is generally stable at high values of the eccentricity ratio and speed parameter.

Keywords: Axial groove; Two-lobe hydrodynamic journal bearing; Steady state and stability characteristics

Introduction

The quantity of oil flow in a journal bearing plays an important role in maintaining an uninterrupted oil film and removing most of the frictional heat to cool the bearing. The oil flow rate depends on several factors, such as the viscosity of the lubricant, the geometry (length, diameter and radial clearance) of the bearing, operating eccentricity, the inlet oil pressure and the arrangement of feeding sources. The pressure developed in the film due to journal motion also contributes to the flow. One of the simplest ways of feeding oil is a single hole through the bearing which is usually a stationary member at the unloaded region. This ensures higher pressure development in the larger land area in the clearance space. The external radial load on such a bearing should be unidirectional and constant in magnitude. However, the direction of applied load may vary only within relatively narrow limits so that the oil hole remains always in the unloaded region.

In an internal combustion engine bearing both the magnitude and direction of load changes. In such a situation the location of an oil hole in the unloaded region is not possible. This is overcome by feeding oil through a circumferential groove at the mid plane. This will naturally reduce the load capacity because of reduced land area. This bearing can cope up easily with the condition when both the direction and magnitude of load vary. However an elaborate arrangement of the feeding system is to be designed. Oil is also fed by providing an axial groove in the unloaded region. Ordinary circular bearings were not found to be very stable at such high speeds. This gave rise to some new designs of bearings by changing their geometries, such as multilobe bearings and pressure dam bearings, which were found to possess better stability. Hydrodynamic bearings operating at high speeds are often confronted with problems of instability, known as whirl and whip. Instability may ruin not only the bearings but the machine itself. Satisfactory dynamic characteristics are an essential requirement of a good bearing design and bearings of non-circular cross-section hold good promise for applications where bearing stiffness and stability are major considerations. Non-circular bearing geometry enhances shaft stability and under proper conditions, this will also reduce power losses and increase oil flow (as compared to circular bearings), thus reducing bearing temperatures. Among the non-circular sleeve bearings, elliptical and three lobe bearings are most commonly used. Extensive literature is available on circular bearings but the data available for the design of non-circular bearings is comparatively scarce. The steady state load capacity and power losses for elliptical bearings have been calculated by Pinkus [1,2] using finite difference method. The computation procedure for the stiffness of externally pressurized bearings relies on an analytical description given in some previous papers [3,4]. Lund [5] developed the stability criterion for a multilobe bearing based on linearization of Reynolds equation by small perturbation theory. Falkenhagen and Gunter [4] investigated the stability of a vertical rotor and evaluated the hydrodynamic forces by finite difference analysis and an approximate method. In an internal combustion engine both magnitude and direction of load changes. A novel method to cope up with this situation is the use of submerged oil bearing proposed by Floberg [6]. The friction characteristics of externally pressurised bearings are investigated in [7] over a complete range of operating bearing conditions. Geometrical characterizations of externally pressurized journal bearing have been defined [8]. Falkehagen et al. [9] investigated the stability characteristics and transient motion of a finite width three-lobe bearing for a wide range of ellipticity ratio and offset factor. Lund and Thomson [10] gave some design data which included both static and dynamic characteristics for laminar as well as turbulent flow regimes. A comparison of non-dimensional values of steady state and dynamic characteristics for two-lobe bearing has been made with

***Corresponding author:** Lintu Roy, National Institute of Technology, Silchar-788010, India, E-mail: lintu2003@gmail.com

the published results of Lund et al. [10] for L/D=1 with two 20° axial groove.

Incompressible fluid externally pressurized bearing has proved over the first few years the ability to satisfy the most demanding lubrication requirements of machine tools. An excellent result has been obtained in regard to stiffness both with flat bearings and with journal bearings. Externally pressurized journal bearing have proved, with special purpose design, equal to stiff roller bearings and distinctly superior to ordinary hydrodynamic bearings.

This arrangement of oil supply in the present case is somewhat similar to an externally pressurized bearing having a single recess without any restrictor located between the supply manifold and the bearing .On the other hand a multi-recess externally pressurized oil journal bearing must have a compensating element (restrictor) to be located before the each recess, this elaborate arrangement is quite expensive and often a standby system is to be provided to take care of the possible failure of the supply system. In the present cases the radial load is applied from the top. Thus when a bearing is to be operated with an axial groove its location is important from the operational point of view. One such arrangement is feeding from the top. In this case the radial load is applied from the top. The arrangement with feeding from the top will ensure higher load capacity, as more pressure will be developed in the bottom region due to hydrodynamic action. The arrangement with feeding from the top will ensure higher load capacity as more pressure will be developed in the bottom region due to hydrodynamic action. So our interest here is to study the characteristics of this type of bearing where is a constant pressure of oil is supplied from outside source.

The purpose of this study here is to determine the steady state characteristics of an axial grooved two lobe hydrodynamic oil journal bearing since a bearing having single oil hole for lubrication has somewhat greater load carrying capacity than bearings with circumferential grooves having the same characteristics and operating under the same conditions, so our interest here is to study the characteristics of this type of bearing where we add a constant pressure of oil. If the lubricant is supplied to the axial feed groove under pressure an additional flow through the bearing film will occur. Multilobe hydrodynamic journal bearings have been investigated for their antiwhirl characteristics by many researcher [1,2,11-13]. Malik [11] theoretically studied an elliptical hydrodynamic journal bearing and compared its performance over a wide range of load conditions and provided the comprehensive design data including the static and dynamic characteristics for the two-lobed journal bearing for different aspect ratios. The study dealing with the effects of surface ellipticity on the dynamically loaded cylindrical bearing was carried out by Goenka and Booker [12] for an optimum bearing shape on the basis of maximizing the minimum film thickness. The notable observation about most of these studies is that they are all concerned with hydrodynamic journal bearing systems. Few studies dealing with the noncircular multirecess hydrostatic/hybrid journal bearing systems have also been reported in literature recently [14-16]. The recessed journal bearings are unable to generate a substantial hydrodynamic action because recess constitutes a large bearing area, thus leaving very less area for lands. Thus, recessed bearings when operating at higher speeds are not suitable for heavily loaded applications. Hence, non-recessed journal bearings, is used to generate substantial hydrodynamic action. Such bearings give better performance than the recessed bearings. It is worth pointing out that this description covers not only the conventional hydrostatic bearings but also systems where the hydrodynamic effects induced by the journal rotation are dealt with. Therefore to harness the maximum advantages of both hydrostatic and hydrodynamic actions in a more efficient way non-recessed (i.e. hole-entry) journal bearings were developed and are used frequently. Non-recessed multilobe journal bearings give superior performance to recessed or pocketed bearings in addition to their relative ease in manufacturing. Since no work on this topic could be found in the literature, the author studied the type of configuration where the lubricant is supplied at a constant pressure with an attempt to find out a better new configuration.

Theory

The governing equation is the Reynolds equation is a partial differential equation governing the pressure distribution of an incompressible and isoviscous fluid was first derived by Osborne Reynolds [17] in two dimensions for an incompressible fluid (Figure 1). It can be written in dimensionless form as [18]

$$\frac{\partial}{\partial\theta}\left(\bar{h}^3\frac{\partial\bar{p}}{\partial\theta}\right)+(D/L)^2\,\bar{h}^3\frac{\partial^2\bar{p}}{\partial\bar{z}^2}=\Lambda\frac{\partial\bar{h}}{\partial\theta}+2\Lambda\lambda\frac{\partial\bar{h}}{\partial\tau} \qquad (1)$$

Steady state characteristics

Under steady state condition equation (1) can be reduce to

$$\bar{h}_0^3\frac{\partial^2\bar{p}_0}{\partial\theta^2}+3\bar{h}_0^2\frac{\partial\bar{h}_0}{\partial\theta}\frac{\partial\bar{p}_0}{\partial\theta}+(D/L)^2\,\bar{h}_0^3\frac{\partial^2\bar{p}_0}{\partial\bar{z}^2}-\Lambda\frac{\partial\bar{h}_0}{\partial\theta}=0 \qquad (2)$$

For axially grooved journal bearings, the boundary conditions are $\bar{p}_0=1$ in the groove, $\bar{p}_0=0$ at the bearing ends and the pressure is set equal to 0 when the pressure falls below zero. (3)

Swift -Strieber boundary condition was applied at the cavitation boundary

The equation (2) is solved using Gausss-Siedel method with successive over-relaxation technique .The grid size used is 88 x 24. The non-dimensional steady pressure distribution on each bearing lobe is calculated. The convergence criterion adopted for pressure calculation is $\left|1-\frac{\sum\bar{p}_{old}}{\sum\bar{p}_{new}}\right|\leq 10^{-4}$

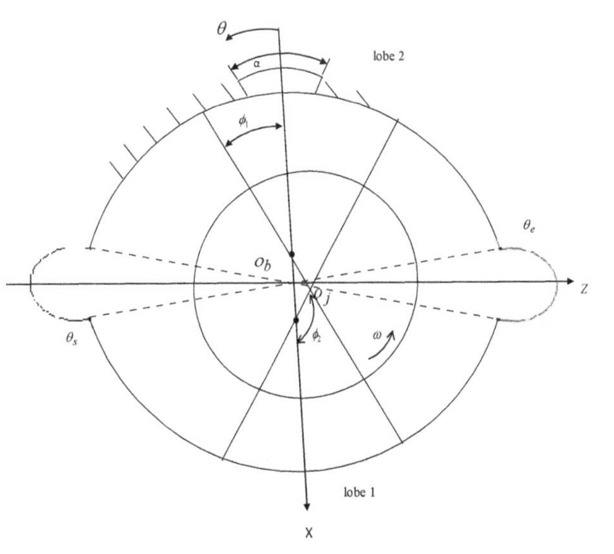

Figure 1: Co-ordinate system of the bearing configuration.

The non-dimensional steady state load components are given by

$$\overline{W}_X = \frac{4W_{Xs}}{LDp_s} = \int_{\theta s}^{\theta e}\int_{0}^{1} \overline{p}_0 \cos\theta d\theta d\overline{z} \quad (4a)$$

$$\overline{W}_Z = \frac{4W_{Zs}}{LDp_s} = \int_{\theta s}^{\theta e}\int_{0}^{1} \overline{p}_0 \sin\theta d\theta d\overline{z} \quad (4b)$$

The two-lobe bearings are suitable for a vertical load support, for calculating the vertical load an eccentricity ratio and attitude angle picked at random which results in magnitude of forces generated due to pressure wedge in the bearing can be calculated. The horizontal force ($\overline{W}_Z$) in the pressure wedge must be zero. If it is not this case a different value of attitude angle is chosen, where the sum of all the forces in the horizontal direction is again calculated. This will eventually locate the shaft at correct attitude angle and where the force in the horizontal direction is zero. Then for this equilibrium position, the vertical force ($\overline{W}_X$) gives the load carrying capacity $\overline{W}_0$

The Sommerfeld number can be given as S= $\frac{\Lambda}{3\pi \overline{W}_0}$

The end flow in each lobe in the dimensionless form can be written as

$$\overline{Q} = \frac{4Q\eta L}{C^3 D p_s} = -\frac{1}{3}\int_0^{2\pi} \overline{h}_0^3 \frac{d\overline{p}_0}{d\overline{z}} \Big|_{\overline{z}=1} d\theta \quad (5)$$

The friction variable is given by $\overline{\mu} = (R/C)\mu = \frac{\overline{F}}{\overline{W}_0}$ where

$$\overline{F} = \left(\frac{F}{2LCp_s}\right) = \iint 2\left(\frac{h_0}{4}\frac{dp_0}{d\theta} + \frac{\Lambda}{12}\frac{1}{h_0}\right) d\theta dz \quad (6)$$

Dynamic characteristics

The Reynolds equation under dynamic condition is the equation (1). The pressure and film thickness can be expressed for small amplitude of vibration as:

$$\overline{p} = \overline{p}_0 + \varepsilon_1 e^{i\tau}\overline{p}_1 + \varepsilon_0\phi_1 e^{i\tau}\overline{p}_2 \quad (7)$$

$$\overline{h} = \overline{h}_0 + \varepsilon_1 e^{i\tau}\cos\theta + \varepsilon_0\phi_1 e^{i\tau}\sin\theta \quad (8)$$

$\overline{h}_0$ = the steady state dimensionless film thickness.

$$\varepsilon = \varepsilon_0 + \varepsilon_1 e^{i\tau}$$
$$\phi = \phi_0 + \phi_1 e^{i\tau} \quad (9)$$

And $|\varepsilon_1| << \varepsilon_0$ and $|\phi_1| <<$

Substitution of equations (7) and (8) into the equation (1) and retaining the first linear terms, gives the three differential equations in $\overline{p}_0, \overline{p}_1$ and $\overline{p}_2$. The equations for $\overline{p}_1$ and $\overline{p}_2$ are solved satisfying the modified boundary conditions of equation (3) and known values of $\overline{p}_0$.

Dynamic loads due to $\overline{p}_1$ and $\overline{p}_2$ are given by

$$\overline{W}_{X1} = \int_{\theta_s}^{\theta_e}\int_0^1 \overline{p}_1 \cos\theta d\theta d\overline{z} \quad \overline{W}_{Z1} = \int_{\theta_s}^{\theta_e}\int_0^1 \overline{p}_1 \sin\theta d\theta d\overline{z} \text{ and}$$

$$\overline{W}_{X2} = \int_{\theta_s}^{\theta_e}\int_0^1 \overline{p}_2 \cos\theta d\theta d\overline{z} \quad \overline{W}_{Z2} = \int_{\theta_s}^{\theta_e}\int_0^1 \overline{p}_2 \sin\theta d\theta d\overline{z}$$

Dynamic forces of each lobe are added and total horizontal and vertical components are determined.

Stiffness and damping coefficients

It is found that the fluid film, which supports the bearing, is equivalent to a spring mass damping system. Since the journal executes small harmonic oscillations about its steady state position; the dynamic load carrying capacity can be expressed as a spring and a viscous damping force. The stiffness and damping coefficients are given by

$$\overline{K}_{XX} = -\mathrm{Re}(\overline{W}_{Xt1});\quad \overline{K}_{ZX} = -\mathrm{Re}(\overline{W}_{Zt1});\quad \overline{K}_{XZ} = -\mathrm{Re}(\overline{W}_{Xt2});$$
$$\overline{K}_{ZZ} = -\mathrm{Re}(\overline{W}_{Zt2})$$

$$\overline{C}_{XX} = -\mathrm{Im}(\overline{W}_{Xt1});\quad \overline{C}_{ZX} = -\mathrm{Im}(\overline{W}_{Zt1});\quad \overline{C}_{XZ} = -\mathrm{Im}(\overline{W}_{Xt2});$$
$$\overline{C}_{ZZ} = -\mathrm{Im}(\overline{W}_{Zt2})$$

Mass parameter and whirl ratio

The mass parameter ($\overline{M}$) and whirl ratio (λ) according to [19] are related as

$$\lambda^2\overline{M} = \frac{\overline{K}_{XX}\overline{C}_{ZZ} + \overline{K}_{ZZ}\overline{C}_{XX} - (\overline{K}_{XZ}\overline{C}_{ZX} + \overline{K}_{ZX}\overline{C}_{XZ})}{\overline{C}_{XX} + \overline{C}_{ZZ}} = k_0$$

So, $\lambda^2 = \frac{(\overline{K}_{XX} - k_0)(\overline{K}_{ZZ} - k_0) - \overline{K}_{XZ}\overline{K}_{ZX}}{\overline{C}_{XX}\overline{C}_{ZZ} - \overline{C}_{XZ}\overline{C}_{ZX}}$ and $\overline{M} = \frac{k_0}{\lambda^2}$

Results and Discussion

A computer program was developed, based on the present theory, to analyze the performance of a finite two-lobe hydrodynamic journal bearing with an axial groove. Before going to the present analysis the values of Sommerfeld number and stiffness and damping co-efficient obtained from the computer simulation of the ordinary two lobe bearing is compared with [10] and the obtained values are found to be matching. When the bearing operates at a small speed, the hydrodynamic effect is not predominant. The hydrodynamic pressure developed due to hydrodynamic action is insufficient to balance the applied load when fed from top. Thus it is difficult to run the bearing at low speeds. Therefore, there is a speed below which the bearing cannot be operated. In this present analysis, it has been found that the limiting value of non-dimensional speed parameter is Λ=6.5. To be on the safe side, we have considered the speed parameter Λ is above 10. The variation of load carrying capacity ,friction variable, end flow , attitude angle, stiffness co-efficient, mass parameter and whirl ratio for a bearing having groove length=¼ of the total length of the bearing and 10° groove angle for 20° lobe angle is shown in Figures 2-17. Load capacity and friction variable increases with an increase in eccentricity ratio and speed. The load capacity increases with bearing number, which is a function of journal speed (Figure 2). This increase is sharp at higher eccentricity ratio. The rise in friction is particularly high at higher eccentricity ratios, as shown in Figure 3. The end flow increases with eccentricity and speed parameter (Figure 4). The attitude angle decreases with the increases in eccentricity ratio but it increases with the increase in speed parameter (Figure 5). A comparison has been done with plain cylindrical axial grooved oil journal bearing having groove geometry 18° and groove length ½ and ¼ of the total bearing length (Table 1). It is observed that in comparison to the plain axial grooved bearing two lobe hydrodynamic journal bearing having improved performance in terms of load carrying capacity, end flow, friction characteristics and stability. The stability also improves for smaller groove angle and groove length. From the comparison with

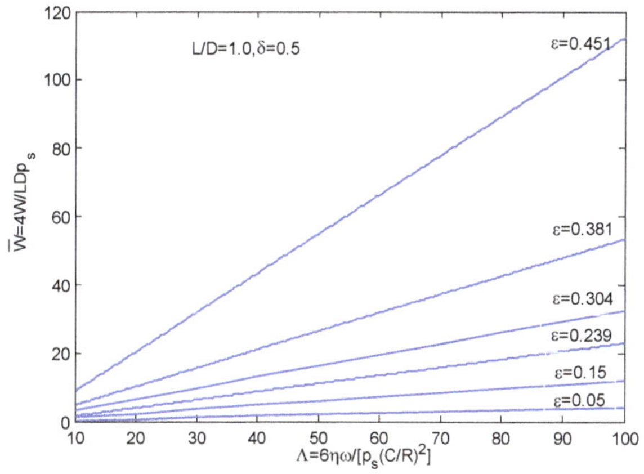

Figure 2: Variation of load carrying capacity with speed parameter.

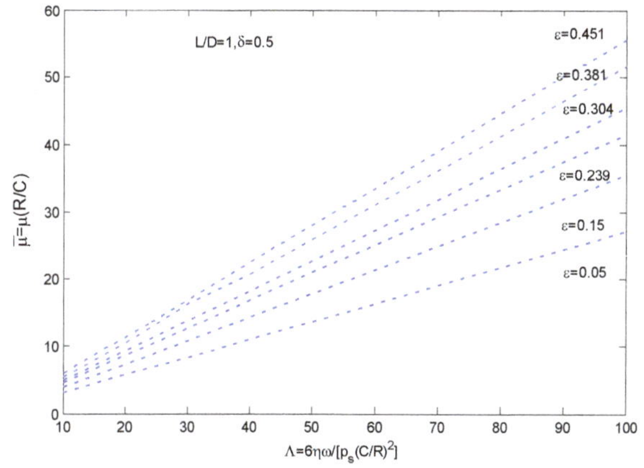

Figure 3: Variation of friction variable with speed parameter.

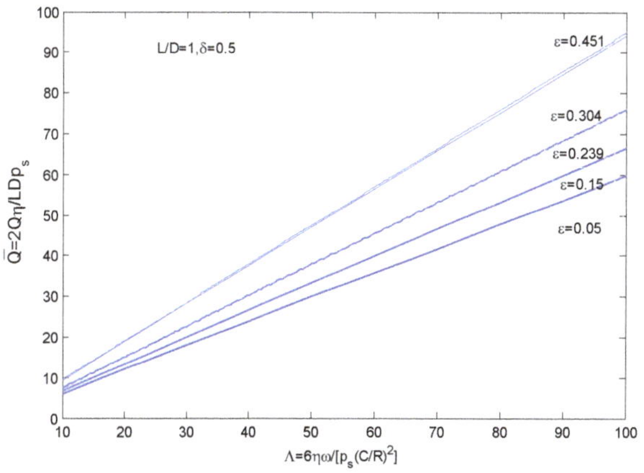

Figure 4: Variation of end flow with speed parameter.

ordinary plain two lobe journal bearing having L/D=1.0,δ=0.5, Λ =10.0, lobe angle=20° with present configuration of bearing with feeding groove angle=18° and groove length ½ of the total bearing length, it is found that at higher bearing number the load capacity , mass parameter value tends to increase, hence an increase in critical mass parameter value (Tables 2 and 3). Direct stiffness co-efficient $\overline{K}_{XX}$ is found to be

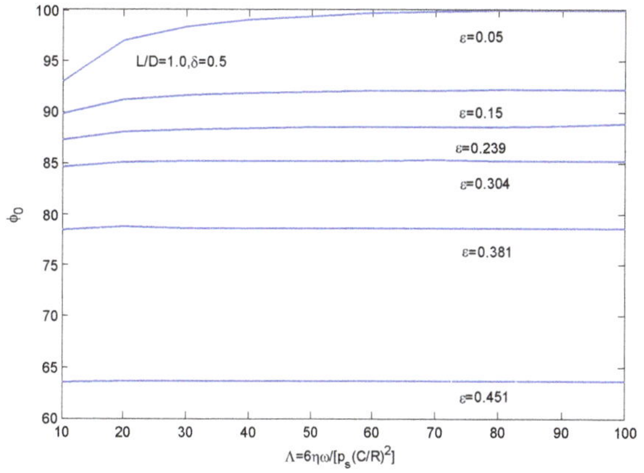

Figure 5: Variation of attitude angle with speed parameter.

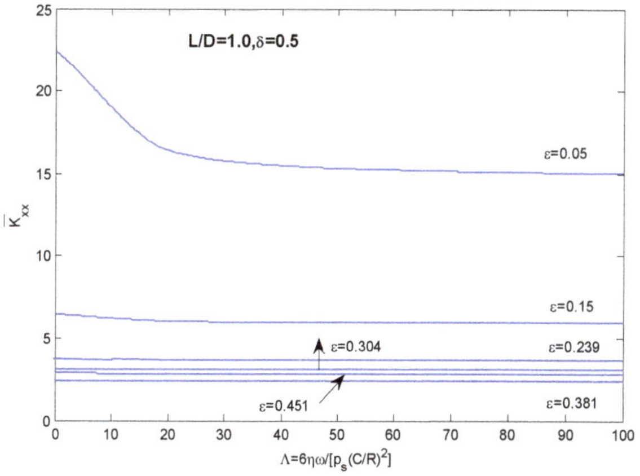

Figure 6: Variation of $\overline{K}_{XX}$ with speed parameter.

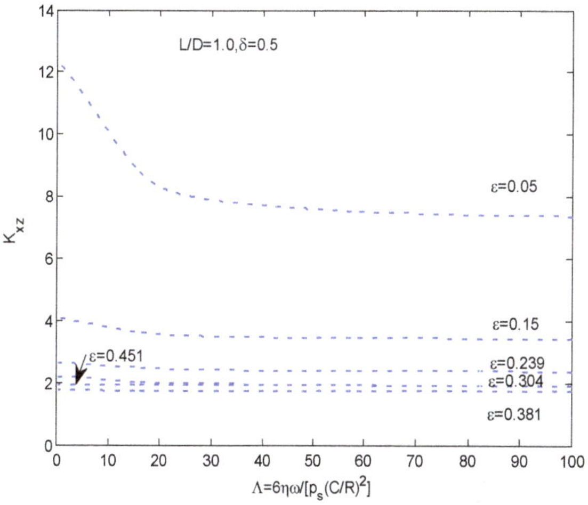

Figure 7: Variation of $\overline{K}_{XZ}$ with speed parameter.

decreased with speed at lower eccentricity but at high eccentricity ratio the changes in stiffness magnitude less with speed (Figure 6). A similar pattern is found in case of $\overline{K}_{ZZ}$ (Figure 9). The cross coupling stiffness $\overline{K}_{ZX}$ and $\overline{K}_{XZ}$ is found to increase in magnitude with eccentricity but the change is very little with the change of speed (Figures 7 and

ε	Groove length	$\overline{W}$	ϕ_0	$\overline{Q}$	$\overline{\mu}$	$\overline{M}$	λ
0.2	½ ¼	1.6069 (0.1145) 1.6551 (0.1865)	87.14 00 (83.0441) 88.4100 (81.8925)	7.2384 (0.6979) 7.2001 (1.2346)	4.1807 (49.0074) 4.3296 (30.1540)	3.2546 (2.9987) 5.9365 (2.9270)	0.4642 (0.5326) 0.4660 (0.5543)
0.4	½ ¼	6.2487 (0.6831) 6.2937 (0.7897)	75.7600 (68.7400) 75.9000 (67.6610)	9.5725 (2.9266) 9.5807 (2.4645)	5.4969 (8.8515) 5.6831 (7.6900)	10.7621 (6.8051) 10.9166 (6.5390)	0.2804 (0.5740) 0.2778 (0.5826)
0.451	½ ¼	11.6817 (0.81414) 11.7020 (0.93633)	63.3500 (65.4927) 63.5100 (64.5818)	9.4575 (1.8555) 9.4699 (1.5448)	5.9374 (7.6416) 6.1549 (6.6804)	16.7095 (10.154) 18.2425 (8.6813)	0.5550 (0.5737) 0.5570 (0.5819)

Table 1: Comparison of results with plain cylindrical groove angle bearing having L/D=1.0, δ=0.5, Λ =10.0, groove angle=18°and groove length=½ and ¼ of total length of the bearing.

For the above table the numbers in the bracket's indicate the data obtained for axial grooved plain cylindrical journal bearing

ε	$\overline{W}$	ϕ_0	$\overline{Q}$	$\overline{\mu}$	$\overline{M}$	λ
0.2	1.6069 (1.7078)	87.14 00 (90.35)	7.2384 (7.1704)	4.1807 (3.4042)	3.2546 (6.2912)	0.4642 (0.4526)
0.4	6.2487 (6.3348)	75.7600 (76.0)	9.5725 (9.586)	5.4969 (4.8363)	10.7621 (11.1789)	0.2804 (0.2742)
0.451	11.6817 (11.7221)	63.3500(63.65)	9.4575 (9.4819)	5.9374 (5.0725)	16.7095 (21.4991)	0.5550 (0.2743)

Table 2: Comparison of results with conventional two lobe bearing having δ=0.5,L/D=1.0, Λ =10.0, lobe angle=20°.

For this table the numbers in the bracket's indicate the data for conventional two lobe bearing with feeding groove angle=18°

ε	$\overline{W}$	ϕ_0	$\overline{Q}$	$\overline{\mu}$	$\overline{M}$	λ
0.2	17.0204 (17.0887)	90.06 (90.33)	71.756 (71.7058)	4.1807 (38.5205)	4.2098 (6.2876)	0.4642 (0.4526)
0.4	63.3480 (63.348)	75.9965 (63.348)	95.8578 (95.8597)	5.4969 (52.8413)	12.6402 (11.1789)	0.2804 (0.2742)
0.451	117.275 (63.635)	63.6260 (63.635)	94.8004 (94.8074)	5.9374 (55.1996)	22.085 (21.7493)	0.47665 (0.2741)

Table 3: Comparison of results with conventional two lobe bearing having δ=0.5,L/D=1.0, Λ =100.0, lobe angle=20°

For this table the numbers in the bracket's indicate the data for conventional two lobe bearing with feeding groove angle=18°

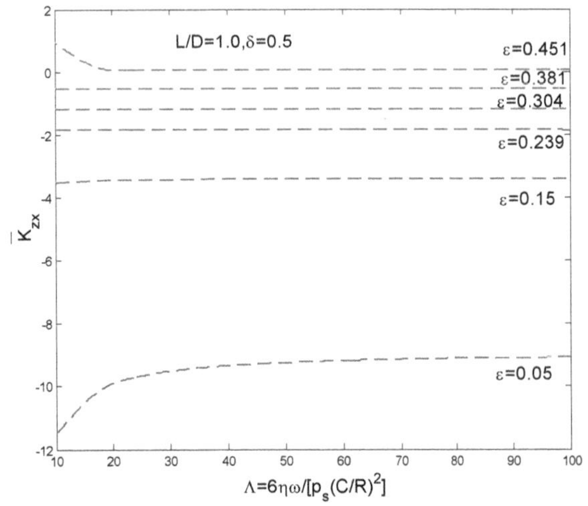

Figure 8: Variation of $\overline{K}_{ZX}$ with speed parameter.

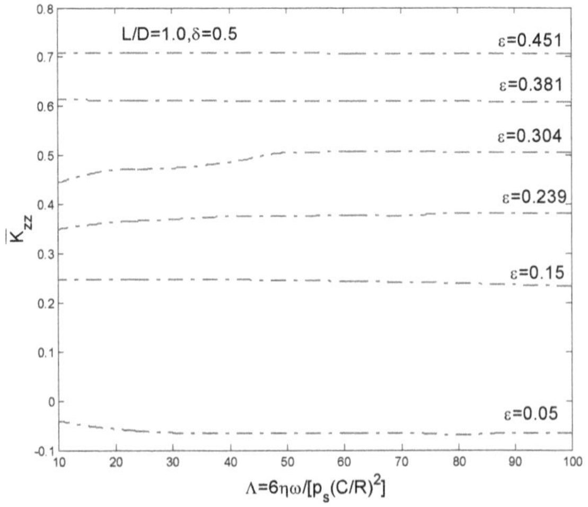

Figure 9: Variation $\overline{K}_{ZZ}$ with speed parameter.

8). Direct stiffness is not dependant on the speed generally, the cross stiffness affects the stability of the rotor and therefore its increase with speed generally indicates reduced stability of the rotor supported by the bearing [15]. Direct damping co-efficient $\overline{C}_{XX}$ and $\overline{C}_{ZZ}$ decreases at a low value of eccentricity ratio but at higher eccentricity ratio the change is very little with speed increase (Figures 10 and 13). Both the cross damping coefficient $\overline{C}_{XZ}$ and $\overline{C}_{ZX}$ increases gradually with the increase of speed (Figures 11 and 12). Direct damping is helpful in stabilizing the rotor supported by the bearing [15]. The mass parameter $\overline{M}$ and whirl ratio λ are used as a measure of stability. These are plotted in Figures 14 and 15. The upper portion of the curve is unstable and the lower portion of the curve is stable. The stability is found to increase with the increase of speed and eccentricity. It is observed that load capacity and stability also improves when smaller groove dimensions (Table 1) (i.e. smaller groove length and smaller groove angles) are used at higher speeds. As the mass parameter of the bearing increases and whirl ratio decreases as shown in Figures 14 and 15. This signifies that the bearing is more stable as the load carrying capacity increases

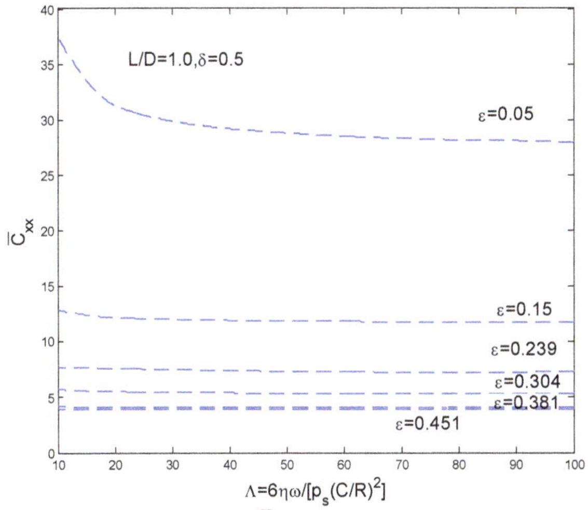

Figure 10: Variation $\bar{C}_{XX}$ with speed parameter.

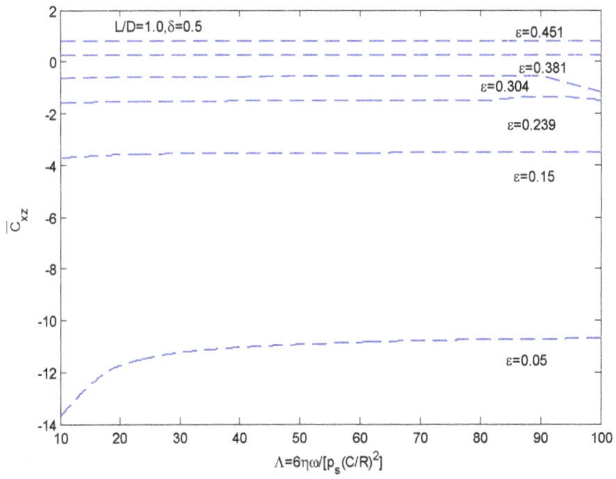

Figure 11: Variation of $\bar{C}_{XZ}$ with speed parameter.

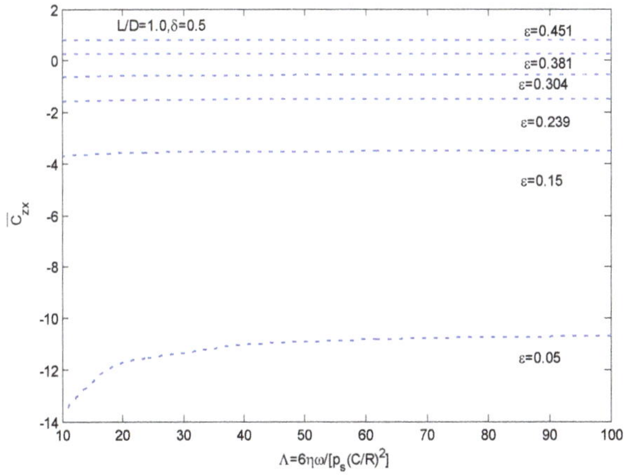

Figure 12: Variation of $\bar{C}_{ZX}$ with speed parameter.

with eccentricity ratio. The mass parameter and whirl ratio variation with speed at a constant eccentricity ratio is shown in Figures 16 and 17. The mass parameter and whirl ratio decreases when groove length changes from ½ to ½ Tables 1-3.

Conclusions

1. The bearing load capacity, the lubricant flow rate increases with increases in eccentricity ratio and speed parameter. This is due to the increase in journal speed.

2. The frictional torque due to journal rotation increases with

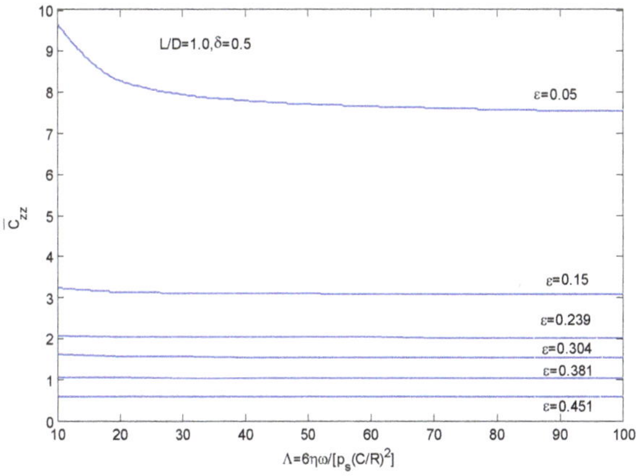

Figure 13: Variation of $\bar{C}_{ZZ}$ with speed parameter.

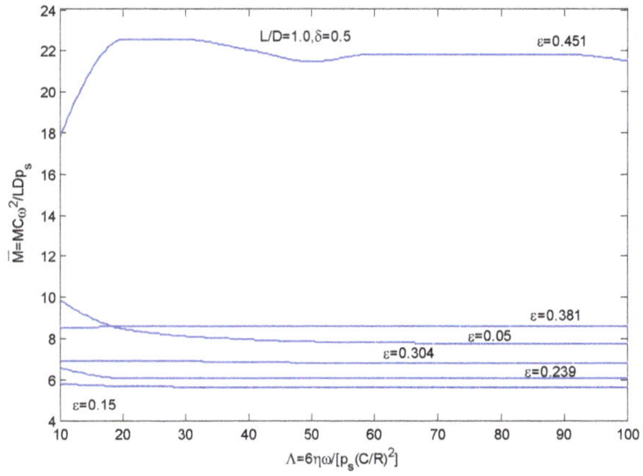

Figure 14: Variation of mass parameter with speed parameter.

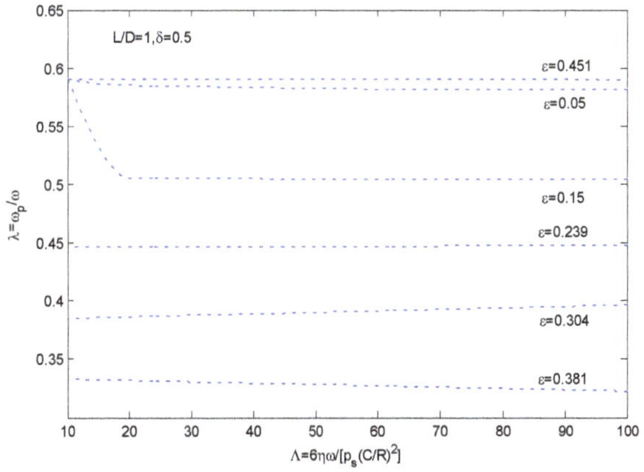

Figure 15: Variation of whirl ratio with speed parameter.

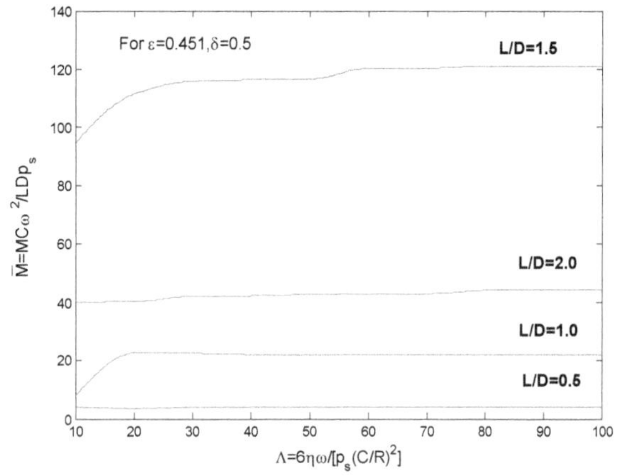

Figure 16: Variation of mass parameter with speed parameter for different L/D ratios.

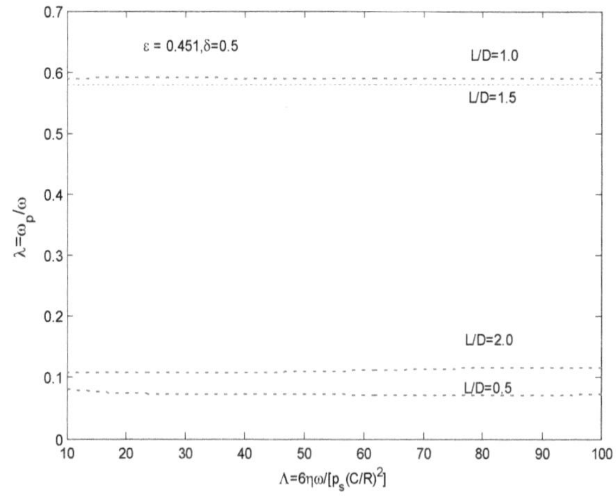

Figure 17: Variation of whirl ratio with speed parameter for different L/D ratios.

increases in the eccentricity ratio and speed of rotation.

3. At lower value of eccentricity ratio when the speed of the bearing increases the critical mass parameter remains almost the same and at higher value of eccentricity ratio it increases with speed and the whirl ratio decreases. At higher value of eccentricity ratio when the speed of the bearing increases the bearing becomes more stable.

At a lower value of eccentricity ratio both direct stiffness ($\overline{K}_{XX}$ and $\overline{K}_{ZZ}$) and damping ($\overline{C}_{XX}$ and $\overline{C}_{ZZ}$) co-efficient found to be decrease while at higher eccentricity ratio it remains more or less same as the speed increases.

At a higher value of eccentricity ratio the cross stiffness ($\overline{K}_{XZ}$, $\overline{K}_{ZX}$) and cross damping coefficients ($\overline{C}_{XZ}$, $\overline{C}_{ZX}$) is found to be increases as the speed increases.

4. The stiffness and damping coefficient magnitude is higher for the bearing fed from a smaller groove angle.

5. A bearing having smaller groove angle gives higher load capacity. This is due to high pressure in the land region.

6. The attitude angle increases with increases in the speed parameter, and it is generally stable for higher values of the eccentricity ratio and speed parameter.

The data from the analysis these are presented in dimensionless form may be used in the design of such bearings.

References

1. Pinkus O (1956) Analysis of elliptical bearings Trans. ASME, 78: 965.
2. Pinkus O, Mass WL (1956) Power losses in elliptical and three lobe bearings Trans. ASME, 78: 899.
3. Skinkle JM, Hornung RG (1965) Frictional characteristics of liquid hydrostatic bearings. J Basic Eng 87: 63-169.
4. Wilcock DF (1967) Externally pressurized bearings as servomechanisms. I— The Simple Thrust Bearing. J Tribol 89: 418-424.
5. Lund J (1968) Rotor-bearing dynamics design technology, Part VII: The three-lobe bearing and floating ring bearing Technical Report AFAPL-TR: 64-45.
6. Floberg L (1968) On hydrodynamic lubrication with special reference to sub-cavity pressures and number of streamers in cavitations region. Acta Poly Scand.
7. Ghigliazza R, Michelin RC (1968) Comparative investigation of friction in externally pressurized journal bearing, Wear 12: 241-256.
8. Michelini KC, Ghiliazza R (1968) Optimum geometrical design of multipad externally pressurized journal bearings. Meccanica 3: 231-241.
9. Falkenhagen GL, Gunter EJ and Shuller FT (1972) Stability and transient motion of a vertical three-lobe bearing system. Trans J Engg Ind 94: 665-677.
10. Lund W, Thomson KK (1978) A Calculation Method and Data for the Dynamic Coefficients of Oil Lubricated Journal Bearings. Proceedings of the ASME Design and Engineering Conference, Minneapolis.
11. Malik M (1983) A comparative study of some two-lobed journal bearing configurations. Tribology Transactions 26: 118-124.
12. Goenka PK, Booker JF (1983) Effect of surface ellipticity on dynamically loaded cylindrical bearing. Journal of Lubrication Technology, 105:1-12
13. Pai R, Majumdar BC (1992) Stability of submerged four-lobe oil journal bearing under dynamic load. Wear 154: 95-108.
14. Ghosh MK, Satish MR (2003) Stability of multilobe hybrid bearing with short sills—part II. Tribology International 36: 633–636.
15. Ghosh MK and Satish MR (2003) Rotor dynamic characteristics of multilobe hybrid bearings. Tribology International 36: 625-632.
16. Ghosh MK, Nagraj A (2004) Rotor dynamic characteristics of a multilobe hybrid journal bearing in turbulent lubrication. Proceedings of the Institution of Mechanical Engineers, Journal of Engineering Tribology, 218: 61–67.
17. Reynolds O (1889) On the theory of lubrication and its application to Mr. Beauchamps Tower's Experiments, including an experimental determination of the viscosity of olive oil. Phil. Trans. Roy. Soc. 177:157–234.
18. Majumdar BC (2011-2012) Introduction to Tribology of Bearings. S.Chand, New Delhi, India.
19. Hamrock BJ (1994) Fundamentals of film lubrication. McGraw-Hill: 244-252.

Theoretical Results for Utilizing Nozzle between the Wind-Way and Wind Turbine in Roof of the Buildings - Wind Speed Increase for Wind Turbine to Produce Electricity

Bazgir AS*

Department of Aeromechanics and Flight Engineering, Moscow Institute of Physics and Technology, 140180, Gagarina Street, 16, Zhukovsky, Russia

Abstract

In order to increase the electrical power of the wind turbines, the velocity of the wind blowing on the wind turbine, is the most important factor that has to increase. In this paper it has been recommended that Contraction Nozzles could be applied between Wind Turbines and wind-way to provide the wind through themselves with more velocity. For all cases analysed in this paper, a three dimensional contraction nozzle with the same length (3 meters) but different input (in which wind blows through it) and output segment areas is considered. In the presented calculations, the inlet mean velocity is considered a constant value of $3\frac{m}{s}$ in a windward area. Numerical solutions and CFD results have been the same for increased wind velocity in each Nozzle outlet or wind turbine's inlet. Furthermore, the electrical power of related wind turbines has been numerically calculated in the presence and absence of the nozzles between wind turbines and wind flow and it shows a dramatic increase in wind turbines power.

Keywords: Wind energy; Wind turbines; Contraction nozzle; Electrical power; Increasing wind velocity

Introduction

It has been inventing different machinery devices which can convert different kind of energies to each other. Using all energy resources make humanity's life more comfortable. The electricity is a secondary energy source, which we can obtain it from the conversion of different kind of primary sources like oil, nuclear power, natural gas, coal and other natural sources.

As we know, the flows, which without any limitation are produced in the nature and on the buildings (External Flows), blow abundantly especially in the windward areas. It is also common for homeowner to install Wind Turbine near their home to make use of wind energy. In this case, the velocity of wind is so important because every wind turbine needs an appropriate and specific amount of velocity to be turned by wind.

Almost in all over the world, most of the electrical power of the buildings is made in power stations. Turbines, that are turned very quickly, are large machines used in power stations to produce electricity by conversion of mechanical energy into electrical power (electrical current is made by the relative movement between magnetic field and conductor) so that a large amount of energy is needed to turn the turbines for producing electrical energy in power stations for household uses.

In other side, according to overt researches, nowadays the consumption of all sorts of energy has been increased. As a matter of fact, with daily development of science and technology and decreasing of natural and non-renewable resources, the optimal use of all sorts of energy is so important to human. Due to huge amount of energy for power stations to start electric energy, the governments tend to optimally use renewable energy sources. There is no doubt that Wind Energy, which is renewable energy, has obvious role in different energy supply such as Electrical power. Due to the fact that wind does not emit greenhouse gases and utilizes no fossil fuel, it has been known as one of the greenest forms of energy. Most homes that uses wind turbine's electrical power are still tied into the power grid when the wind blows slowly [1]. Recently with the development of making of the wind turbines, it has been easier to install wind turbines in a residential area, on the ground or on the roof of the buildings [2].

As it has been said Wind Turbines are one of the most significant electrical energy resources, it may be practical that, in the windward areas, each building to provide its required electrical energy by installing wind turbines on the buildings or close the building where wind blows (especially in rural regions).

Providing the electricity energy of buildings by wind turbine is practical but the main problem of this kind of source energy is that it is only useable for wind-ward areas and therefore the main problem for areas with low speed-blowing wind is the velocity of the wind to spin a wind turbine. According to aerodynamic principles, there must be methods to increase the wind speed. In the windward areas, even low-wind-speed areas, some contraction nozzle can be horizontally installed in a roof of a building, located between wind turbines and wind-way to increase the velocity of incoming wind and then increase the overall electrical power of the wind turbines. It is the aim of this project to bring increased wind energy, caused by contraction nozzle, to household us even in low-speed wind-ward areas.

Calculations and Results

As we know the external flows, which without any limitation are produced in the nature and on the buildings, blow repeatedly in the

***Corresponding author:** Bazgir AS, Department of Aeromechanics and Flight Engineering, Moscow Institute of Physics and Technology, 140180, Gagarina Street, 16, Zhukovsky, Russia, E-mail: alibazgir71@yahoo.com

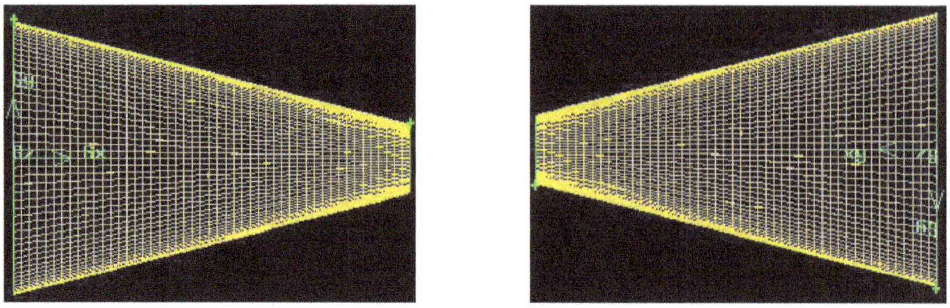

Figure 1: (A) Converging nozzle, (B) diverging nozzle.

windward areas. Low-velocity wind flow commonly blows on some wind-ward regions in the world. Although the wind blows non-stopped in these wind-ward areas but the produced wind in such a wind-ward areas is known to be exceedingly small, and it is not the most effective way to generate energy for our purposes. According to the power equation of wind turbines, the velocity is the most effective factor for spinning a wind turbine and then producing electric energy. There must be another effective ways to increase the velocity of the wind but here we numerically demonstrate that a contraction nozzle or converging duct can easily increase the flow velocity so that if we install such a nozzle between wind stream and wind turbine, the wind turbine will spin faster and as a consequence the power of wind turbine can be increased.

According to the following assumptions, we derive the inlet velocity for starting the wind turbine (in the real design, the numerical amounts can be easily changed):

1. Temperature is constant (= 25°C). Otherwise, we would add heat transfer equations in radial systems to obtain temperature in each section.

2. The material, which is used for creating of contraction nozzle, is steel.

3. The wind flow throughout the contraction nozzle should be maintained as incompressible and low-speed flow to be prevented from noises. Moreover, the compressibility affects on flow characteristics and the contraction nozzle only increase the velocity of the flow for velocities less than velocity of the sound.

4. Regardless of the calculations of boundary layer in the converging nozzle, we have calculated the velocity of the air at the output of the nozzle. It is because of the fact that calculation of boundary layer is not important in converging nozzles. For example, the ratio of boundary layer to the length of converging nozzle can be any amount.

5. In this study, we only propose that such a method for increasing the wind velocity in wind-ward areas can be practical in future applications to install wind turbines in low-velocity areas. Considering different factors for optimal design of a wind turbine with more efficiency is outside the scope of this study.

6. This method can be practical in both small wind turbines in roof of the buildings or near the buildings and big wind turbines in deserts.

7. For all the following calculations, the density of the incoming air and power efficiency of the wind turbine is considered to be constant amounts of = 1.169/3 and = 0.4 respectively.

In aerodynamic applications, converging and diverging nozzles are used to, corresponding to velocity needs, increase or decrease the velocity of incoming flows.

Figure 1 shows two kinds of nozzle that are used to increase or decrease the velocity of incoming flow. First of all, we need the Mach number, equation (1), to determine the kind of flow which approaches to the entrance of the nozzles. Therefore the Mach number would be given as:

$$M = \frac{v}{c} \tag{1}$$

Where c and v and are respectively the velocity of sound and the velocity of incoming air [3]. The velocity of sound can be given by:

$$C = F(\mathrm{T}) \rightarrow \mathrm{C} = \sqrt{KRT} = \sqrt{1.4 \times 287 \times (273 + 25)} = 346.02$$

Where K, R, T are ratio of special heat, gas constant and temperature respectively [4]. For subsonic wind velocities (M< 1), an increase in area of the nozzle causes wind velocity to reduce but for supersonic wind velocities (M>1), an increase in area causes wind velocity to increase. We can use the equation (2) to understand better why a supersonic wind flow accelerates through the divergent nozzle while a subsonic wind flow decelerates through a divergent nozzle [5].

$$\frac{dA}{A}\left(\frac{1}{M^2 - 1}\right) = -\frac{d\rho}{\rho V^2} \tag{2}$$

According to equation (2), for subsonic wind (incompressible) flows, the density of the wind flow remains constant, so when area of the nozzle increases, the velocity of the wind flow tends to reduce but for supersonic wind (compressible) flows, ρ and v are changing as we change the area.

Now we can suppose that we are applying this project in a windward area that the velocity of incoming wind blowing on wind turbine is less than the velocity of sound, therefore, according to aerodynamics principles for subsonic wind velocities, we have to use a converging nozzle, which increase the wind velocity by reducing the segment area of the nozzle, to increase the velocity of the incoming wind blowing on wind turbine. For all cases analyzed below, a three dimensional contraction nozzle (which converts the compressive energy to kinetic energy) with the same length (3 meters) but different input and output segment areas is considered. In these presented calculations, the inlet mean velocity is considered a constant value of $3\frac{m}{s}$ in a windward area.

The simulations of (Figure 2) are carried out by FLUENT software to show the changes of wind speed caused by nozzle. From the Figure 2 it has been found out that, for all cases, by decreasing the segment area of the nozzle, the wind speed obviously gets increase and it reaches peak at nozzle output or wind turbine inlet.

So it is showed that the velocity of the incoming wind blowing on the wind turbine blades can be easily increased by utilizing a well-designed contraction nozzle in the path of the wind used in the roof of

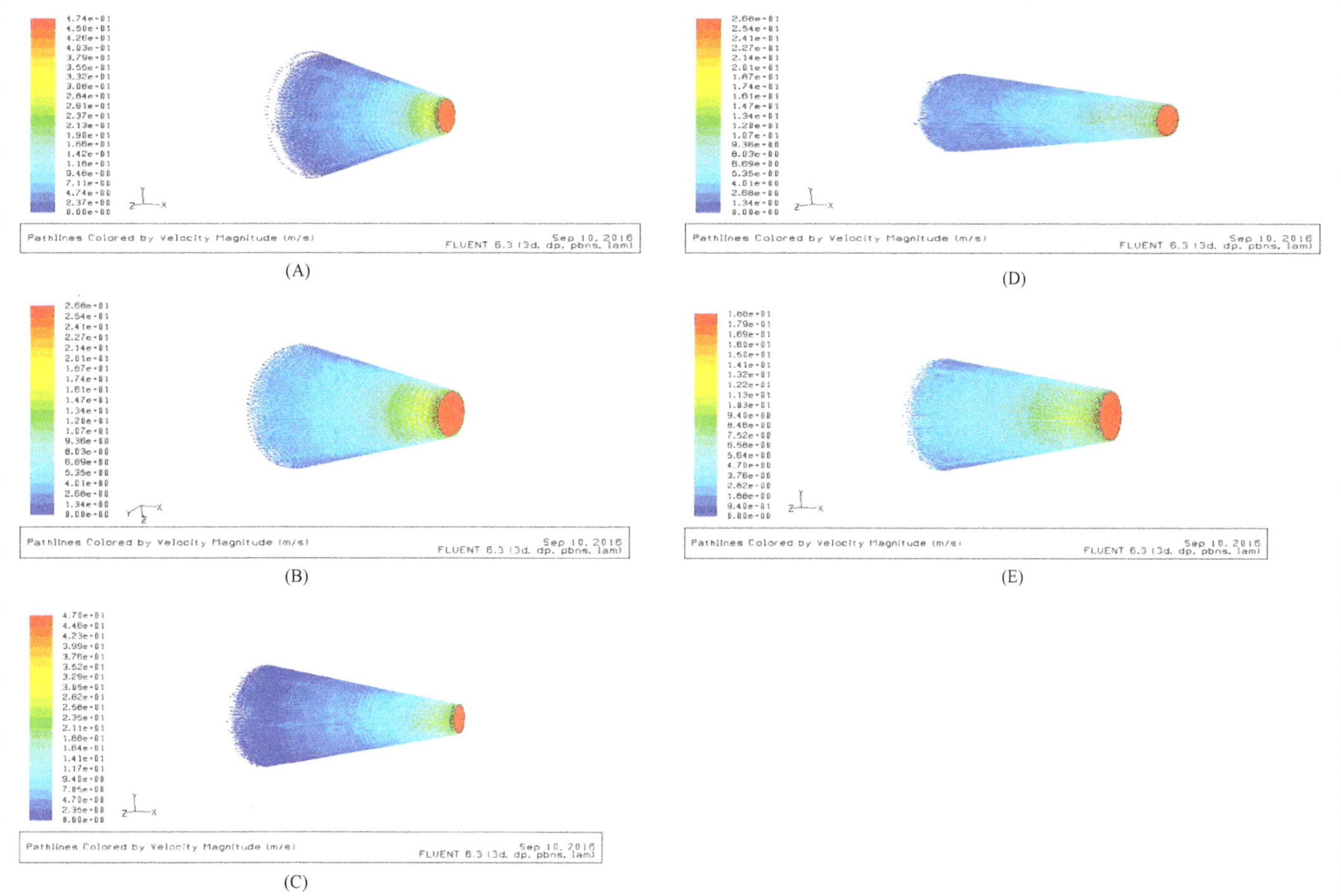

Figure 2: Comparison of increasing of inlet velocity for a wind turbine after passing through different nozzles with different segment areas.

a building or anywhere else that there is turbine. Now it is of critical economic importance to know the power and hence energy produced by different types of Wind Turbine that are located after the small area segment of contraction nozzle. Each wind turbine produces a specific power depending on the speed, mass and amount of the incoming wind hence the power of the wind turbine, which has been converted from mechanical energy to the electrical power, can be calculated by using the following equation [6]:

$$P = \frac{1}{2}\rho A V^3 C_p \tag{3}$$

Where P, V, ρ, A, and CP are total produced power, the speed of the wind blowing on wind turbine, the incoming air density, the swept area of the wind turbine that can be calculated by $A = \pi r^2$ (where is the blade length of the wind turbine), and power coefficient. The ratio of power extracted by the wind turbine to the total contained in the wind resource is called the power coefficient. Albert Betz calculated that it is impossible that a wind turbine convert more than 59.3 % of the kinetic energy of the blowing wind into mechanical energy turning a rotor. This coefficient, which is known as the Betz Limit [6,7], is used for the theoretical maximum coefficient of electrical power for any wind turbine.

According to velocity calculations for incompressible flows, the continuity equation has been utilized [4] to obtain the velocity of the flow in the small diameter of the contraction nozzle (small segment area of the nozzle) or inlet velocity of wind turbine, which is given by:

$$\dot{m}_{in} = \dot{m}_{out} \rightarrow \rho V_{in} A_{in} = \rho V_{out} A_{out} \rightarrow V_{out} = \frac{V_{in} A_{in}}{A_{out}} \tag{4}$$

The equation (3) and the continuity equation (mass flow rate) $\dot{m} = \rho V A$= constant (where ρ, V and A are density of the air, velocity of the incoming wind and segment area respectively) have been utilized to get some information of Table 1. Moreover, the FLUENT Software calculations has been taken to account to obtain the wind velocity at outlet of the contraction nozzles or wind velocity that blows on wind turbines and as we see, there is no important difference between numerical calculations and FLUENT results for wind amount.

Some results of (Table 1) performed by FLUENT software. For example, the wind turbine inlet velocity calculated by continuity equation is approximately the same as CFD results. The power converted from the wind into rotational energy (electrical power of wind turbines) with different inlets wind speed and different lengths of wind turbine's blade have been numerically calculated by applying equation (3). This is the main result of this paper that shows we can install some nozzle in a building's roof between the wind turbines and wind-way and then produce the electrical power for household consumption. Figure 3 shows the main above results in a graph. It has been found that the power converted from the wind into rotational energy clearly depends on the wind velocity and blade length of the wind turbine (short diameter of the nozzle) (Figure 4).

The graph shows that a well-designed nozzle, for increasing the wind speed, and along with suitable blade length of the wind turbine can be practically designed and performed in a roof of a building to produce the electrical power of a house, of course more efficiently rather than installing wind turbine without nozzle, independent from

Nozzle Number	Height(L)	Large Diameter (M)	Short Diameter (M)	Inlet Velocity (M/S)	Outlet Velocity (Wind Turbine Inlet) (M/S) (Numerical)	Outlet Velocity (M/S)(Cfd)	Maximum Pressure(Pascal) (Cfd)	Electrical Power (Kw)
A	3	4	1	3	48	47.38	1490.84	20.297
B	3	3	1	3	27	26.75	458.27	3.618
C	3	2.5	1	3	18.75	18.7	218.64	1.209
D	3	2	0.5	3	48	46.99	1419.17	5.074
E	3	1.5	0.5	3	27	26.75	445.69	0.903

Table 1: Comparison of the growth of wind speed caused by different nozzles and electrical power produced by different wind turbines.

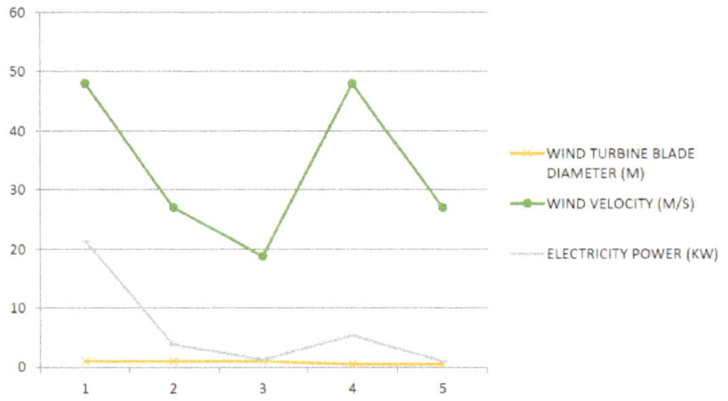

Figure 3: Comparison of effect of different wind speeds and blade length of wind turbines on electrical power.

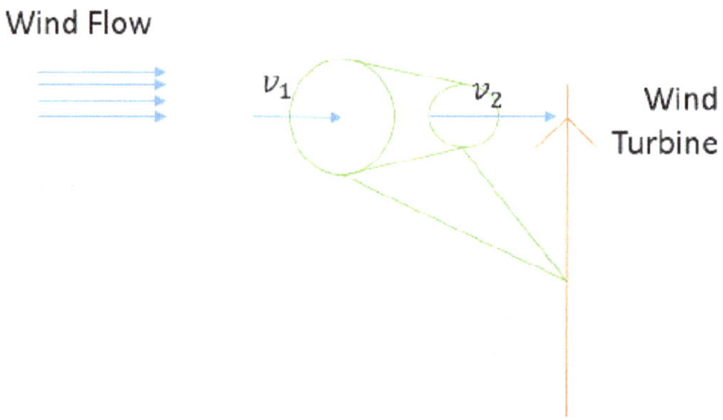

Figure 4: Schematic of the proposed method for increasing the velocity of the wind, where V1 < V2.

electrical power plants. That is, it is only needed to design suitable blade length for wind turbine corresponds to design of nozzle for increasing the wind speed, afterward, to put the nozzle between the wind-way and wind turbine and then it could be used of the electrical power of the wind turbines more efficiently which can be economically critical for many urban and rural householders.

Table 2 shows the electrical power for both common wind turbines and nozzle-attached wind turbines. By relying on the results obtained numerically for utilizing such a method for increasing of the incoming wind velocity, it could be found out that there will be more efficiency to apply such a method for using wind turbines electrical power. As it has been shown, nozzles with different diameters alter the amount of velocity of the wind and according to equation (3), the velocity is the most effective factor for increasing the electrical power of the wind turbines. The optimal design of wind turbines and nozzles for getting the most efficiency design is not in the scope of this study but according to the increased velocities caused by different nozzles, it could certainly be expected to experimentally apply such a method therefore wind turbine's producer factories can optimally design wind turbines with more efficiency even for low-speed wind-ward areas.

Conclusion

In this study it has been recommended that contraction nozzle can be installed between wind turbine and wind-way to increase the velocity of the wind blowing on wind turbine blades. From the CFD and Numerical analyses, it was found out that by decreasing the segment area of the nozzle, the wind speed gets increase through the nozzle and it reaches peak at nozzle output or wind turbine inlet. The power calculations for typical wind turbines in the presence and absence of the contraction nozzle demonstrated that power of the wind turbines will dramatically increase if a well-designed nozzle was used for wind speed increase. Since the power of the wind turbine can be increased by such a method, it would provide a basis for making the buildings self-sufficient from the electrical power plants even in low speed wind-ward areas. Although a single wind turbine is not enough for electrical power.

Wind Turbine Numbers	Electrical Power without Nozzle(KW)	Electrical Powerwith Nozzle(KW)
A	0.005	20.297
B	0.005	3.618
C	0.005	1.209
D	0.0012	5.074
E	0.0012	0.903

Table 2: Comparison the electrical power of wind turbines in the same conditions of common wind and increased wind velocity by nozzle.

References

1. Wollenhaupt G (2010) 3 ways to generate electricity at home. Proud Green Home.
2. Wind turbines (2016) Alternative energy, AE newsletter, USA.
3. Streeter VL (1951) Fluid mechanics. McGraw-Hill, USA.
4. Borgnakke C, Richard ES (2005) Fundamentals of thermodynamics. (6th edn), Wiley, New Jersey, USA.
5. Shames IH (1992) Mechanics of fluids. (4th edn), McGraw-Hill, USA.
6. Kalmikov A, Dykes K (2010) Wind power fundamentals. MIT Wind Energy Group and Renewable Energy Projects in Action, USA.
7. John B (2007) Basic engineering mathematics. Elsevier LTD, Amsterdam, The Netherlands.

12

The Finite Element Analysis and Optimization of an Elliptical Vibration Assisted Cutting Device

Guilin Shi[1], Chen Zhang[1*], Yingguang Li[1], Kornel F Ehmann[2], Yun Song[1] and Ming Lu[1]

[1]College of Mechanical and Electrical Engineering, Nanjing University ofAeronautics andAstronautics, China

[2]Department of Mechanical Engineering, Northwestern University, USA

Abstract

The 2D elliptical vibration assisted cutting(EVC), which has significant advantages in the tool wearing restriction, cutting heat reduction, the quality of the finished surface improvement, is widely supposed to be the most promising machine method to difficult-to-cut materials and micro-texturing formation. Based on the certain 2D EVC structure, three structures of different angles (30°, 60°and 90°) for optimization and the corresponding finite element models of the three structures are created. The simulation analyses of static structure, modal and harmonic response of these structures have been conducted. The 60° topological structure was selected for the final design model after the comprehensive comparison about the resonance frequency, vibration modes and vibration amplitudes of the elliptical trajectories in the tool tip. The performance test system has been established for the testing of the actual 2D EVC device, and the comparison analyses between the results of finite element analysis and the experiment results have been detailed. The experimental data showed that the optimized structure could generate the required elliptical locus for the elliptical vibration cutting. Comparing with the finite element analysis results, the guidance effect of the finite element analysis method to the structure optimization and trajectory prediction has been successfully verified.

Keywords: Elliptical vibration; Cutting; Finite element analysis; Optimization

Introduction

As the fast development of the micro machine in recent years, micro-texturing has attracted more and more attention in research and application. Comparing with laser ablation, LIGA and the other means of micro-texturing formation, the EVC (elliptical vibration assisted cutting) technology has many advantages in terms of costing, efficiency and time saving, which is becoming more and more popular in surface micro topography fields [1]. The EVC technology was first introduced by Hong and Ehmann [2], it adds the vibration to the tool tip in depth of cut direction and cutting direction, which would bring better cutting effect to the finished surface [3,4]. The current research showed the performance of the micro-texturing is determined by its shapes and distribution style [5], but the EVC was mainly focus on keeping the vibration of the tool tip stable to obtain a single shape, which restricted the EVC application in the micro fields [6]. Ping and Kornel have developed an EVC device [7], which has been verified successfully in the ultrasonic micro-texturing machine, but the shapes of texturing are single micro-grooves. For all the above reasons, in this paper the authors have designed three structures of different angles based on the structure and kinematic analysis of Ping's EVC device, then the simulations of vibration and deformation have been conducted.

Finite element method (FEM) analysis has been proved significantly useful in prediction by simulation in the field of vibration assisted cutting. Amini studied the machining forces and stresses acting on the work piece during the ultrasonic assisted turning by FEM analysis [8]. Lu has developed finite element models to analyze the effect of vibration frequency on cutting forces and cutting temperature during elliptical vibration turning [9]. Vivekananda has used FEM in the calculation of the natural frequency and amplitude of EVC device in his work [10], and later he has proved the optimization effect of FEM in process parameters selection [11]. Huang enhanced the cutlery life and drilling process quality by investigating the micro drill's natural properties by FEM [12].

In this paper, FEM is firstly used for the performance optimization (including resonance frequency, vibration amplitude, and etc.) of the 2D EVC structure, and then the actual EVC device based on the optimization is manufactured for the following experiment [13]. The corresponding theoretical model of the tool tip trajectory is established and the elliptical locus is proved. The comparisons between the experimental data and the simulation data are detailed in the aspects of excitation amplitude and phase difference. The predictability and the reliability of the created finite element model are verified by the comprehensive comparisons.

Principle and Dynamic Analysis of the EVC Structure

Principle and of EVC and the EVC structure

The work piece surface is selected to be the reference for the illustration of the EVC principle. As shown in Figure 1, the cutting trajectory (dash-line) left on the work piece surface is contributed by the synthesis of the tool's vibration and the work piece's rotation. During the machine, different from the tool constant contacting with the work piece in conventional cutting, EVC creates the separation intermittent between the tool and the work piece, which is advantageous for the heat dissipation, the wearing suppression and longer tool life. The elliptical trajectories of the tool edge left on the work piece forms the required micro-texturing finally.

***Corresponding author:** Chen Zhang, College of Mechanical and Electrical Engineering, Nanjing University of Aeronautics and Astronautics, China
E-mail: meeczhang@nuaa.edu.cn

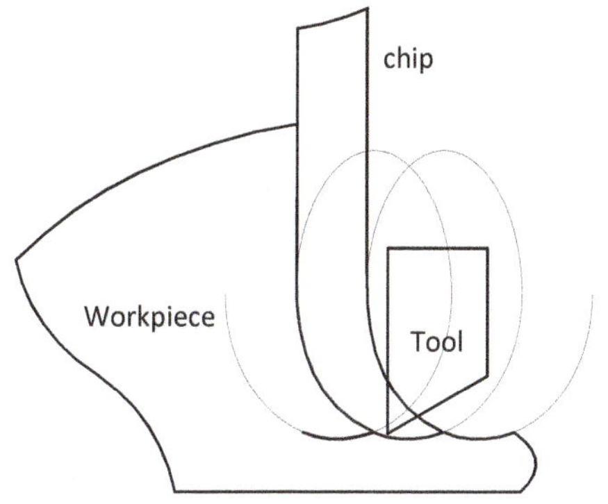

Figure 1: Principle of the EVC.

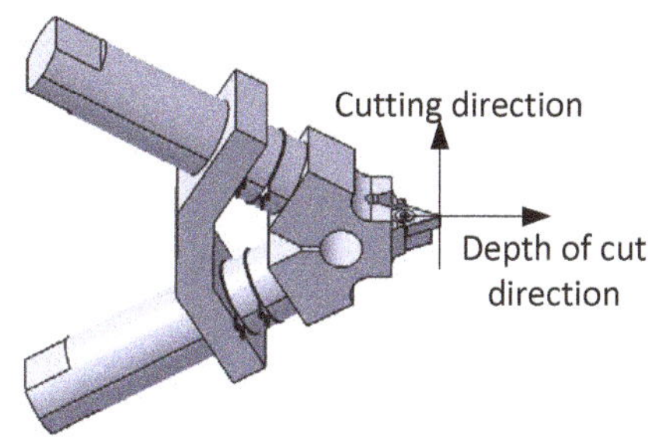

Figure 2: CAD model of the EVC structure.

The CAD model of the EVC structure studied in this paper is showed in Figure 2. This structure is composed by two Langevin transducers which are fixed on the base by bolts. The header works as the tool holder and synthetizes the two transducer's displacement to the tool. When the excitations applied on the two transducers are in the same phase, the transducers' movements in the cutting direction are cancelled out by each other for the symmetry of the structure, the tool tip moves along the depth of cut direction; when the excitations are in the opposite phase, the transducers' movements in the depth of cut direction are cancelled out by each other and the tool tip moves along the cutting direction. In the recent research, EVC has two work modes, which are resonant mode and non-resonance mode. By the common sense, the amplitude of the EVC in the resonant mode is several times or even more than which in the non-resonance mode. In the fields of micro-texturing, the big amplitude provides the EVC with a broad range of elliptical locus, which is advantageous to the diversification of the micro-texturing. As the consequence, in this paper, the resonant work mode is selected for the FEM simulation and the actual experiments.

Kinematic analysis of the 2D EVC equipment

The principle of the elliptical vibration formation of the EVC equipment used in this paper is shown in Figure 3. For the convenience of the specification, some simplification is been made with the main structure unchanged. The excitations applied on the two transducers have the same amplitude and frequency, the phase difference is 90°, as shown in Figures 3 and 4. Figure 5 is the tool's positions correspond to the different phases in Figure 3. The transducers are separately named as rod A and rod B as shown in Figure 4. The position of the tool is contributed by the two Langevin transducers' displacements. Assuming one vibration cycle begins at the point whose phase value is a (in the next paper, phase a, is short for this point) in Figures 2 and 3, at this moment, the displacements of rod A and rod B have the same displacement. To the tool tip, the horizontal components (the cutting direction in Figure 4) are cancelled out by the structures' symmetry, and the vertical components (the depth of cutting direction) superimpose each other. In the whole vibration locus, the position of the tool tip in phase a is on the peak in depth of cut direction. In phase b, rod B displacement reaches the small position with the rod A position unchanged, the tool tip reaches the left limit of the locus. The positions in phase c and d are the same as in phase a and b, when the tool run to the phase a' in next round cycle, the whole elliptical vibration is completed. Adjust the excitation frequency to the resonance mode, the bigger amplitude achieved and the device can be used for the actual experimental cutting.

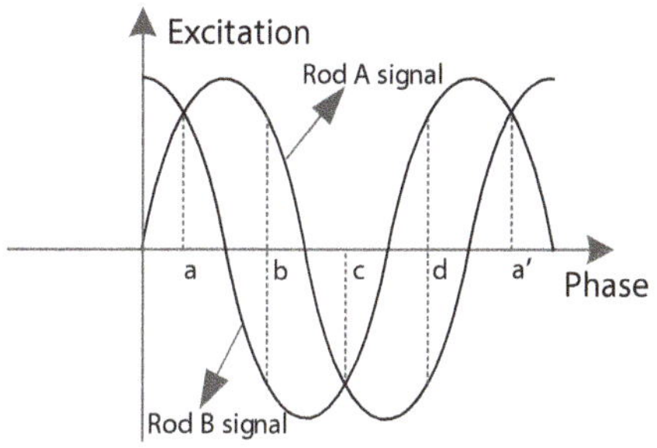

Figure 3: Excitation signals applied on the transducers.

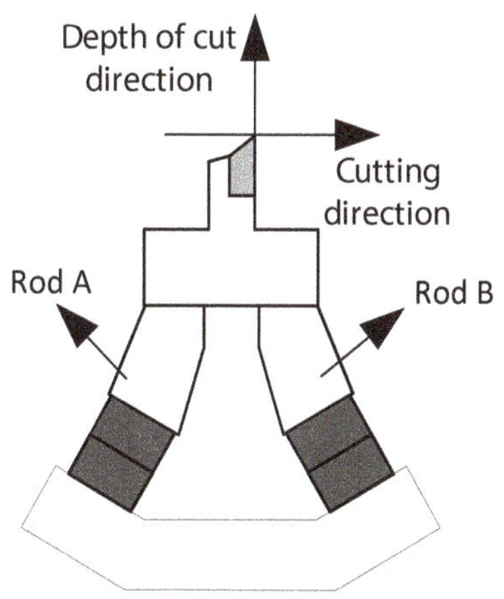

Figure 4: Illustration of rods and directions.

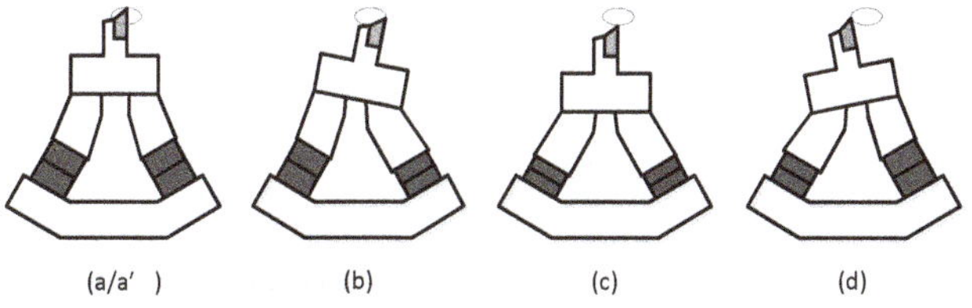

Figure 5: Generation of the elliptical trajectory in the tool tip.

Based on the conclusion above, the EVC equipment in this paper is mainly influenced by the two transducers' amplitudes and phase differences, the elliptical locus can be modified by controlling the two factors. The following content analysis the relations between the amplitudes and phase differences and the tool tip elliptical locus, the results is verified in experiments.

Mathematical model of the trajectory at the tool tip

Based on the analyses above, the mathematical model of the trajectory of the tool tip is established as shown in Figure 6a. The transducers are named as rod A and rod B, and the amplitude of each other is *A* and *B*, the angle between the rods is θ, the frequencies is the same and represent as *F*, the phase difference is φ, *t* is time. In the initial condition (without any excitation input), PD_0 is the point of the tool tip, PM_0 is the middle point of the simplified connection rod, PA_0 and PB_0 is the endpoint of rod A and rod B separately. θ_D is the angle between the line from PM_0 to PD_0 and the horizontal line, can be calculated by $\theta_D = \tan^{-1}(\frac{T}{L})$.

For the kinematic model created above, at time t, the position of tool tip point $PD_t(pd_x, pd_y)$ in two directions (*ox* and *oy*) can be formulated into Equation (1).

$$\begin{cases} pd_x = pm_x - \sqrt{L^2+T^2}\cos(\theta_t - \theta_D) \\ pd_y = pm_y - \sqrt{L^2+T^2}\sin(\theta_t - \theta_D) \end{cases} \tag{1}$$

In the Equation (1), as shown in Figure 6b where *T* is the thickness of the tool, *L* is the distance between *PD* and the connection rod. *D*is the length from PB_0 (or PA_0) to, $PM_0 \theta_t$ is the angle between the tool holder and the horizontal line at time t, PM_t is the middle point of the connection line at time t, pm_x and pm_y are the corresponding coordinates. Considering the symmetric of the structure, pm_x and pm_y can be formulated in to Equations (2) (Figure 7).

$$\begin{cases} pm_x = \dfrac{pa_x + pb_x}{2} \\ pm_y = \dfrac{pa_y + pb_y}{2} \end{cases} \tag{2}$$

In the Equation (2), where pa_x and pa_y are the coordinates of, PA_t. pb_x and pb_y are the coordinates of PB_t.PA_t and PB_t. are the endpoints of rod A and rod B at time *t* separately.

The rods are excited by the sinusoidal signals in Figure 3, pa_x and pa_y can be formulated in to Equations. (3).

$$\begin{cases} pa_x = pa_{0x} - A\sin(2\pi ft)\cos(\frac{\theta}{2}) = -A\sin(2\pi ft)\cos(\frac{\theta}{2}) \\ pa_y = pa_{0y} - A\sin(2\pi ft)\sin(\frac{\theta}{2}) = D - A\sin(2\pi ft)\sin(\frac{\theta}{2}) \end{cases} \tag{3}$$

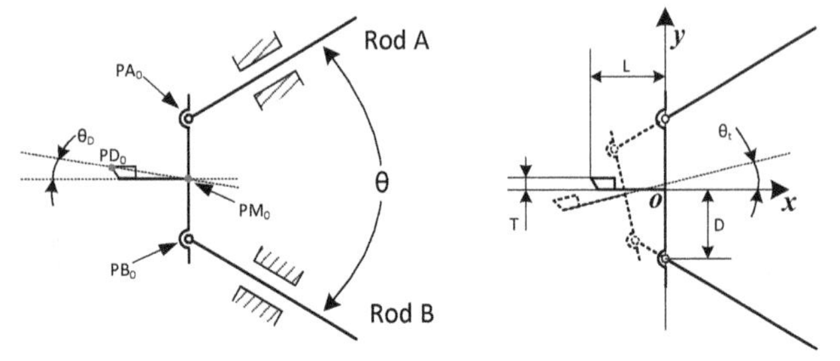

(a)Simplifiedstructure (b) Established coordinate system

Figure 6: Kinematic model of the tool tip vibration.

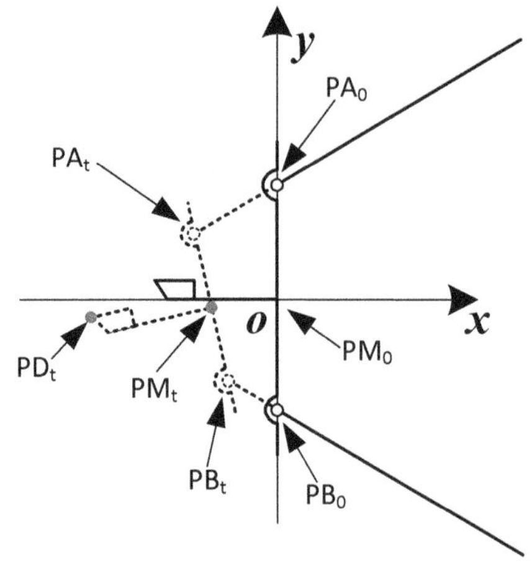

Figure 7: Illustration of representation.

In the Equation (3), wherepa_{0x} and $pa_{0y}$are the coordinates of PA_0, as the given parameters, $pa_{0x} = 0, pa_{0y} = D$. Equation (4) shows the coordinates of $PB_t(pb_x, pb_y)$, the formula reasoning procedure will not be repeated for its similarity with PA_t.

$$\begin{cases} pb_x = pb_{0x} - B\sin(2\pi ft + \varphi)\cos(\frac{\theta}{2}) = -B\sin(2\pi ft)\cos(\frac{\theta}{2}) \\ pb_y = pb_{0y} + B\sin(2\pi ft + \varphi)\sin(\frac{\theta}{2}) = D + B\sin(2\pi ft)\sin(\frac{\theta}{2}) \end{cases} \tag{4}$$

The coordinates of PD_t and θ_t can be obtained by substituting the intermediate variables as shown in Equation (5) and (6).

$$\begin{cases} pd_x = \dfrac{-A\sin(2\pi ft)\cos(\frac{\theta}{2}) - B\sin(2\pi ft + \varphi)\cos(\frac{\theta}{2})}{2} - \sqrt{L^2+T^2}\cos(\theta_t - \theta_D) \\ pd_y = \dfrac{-A\sin(2\pi ft)\sin(\frac{\theta}{2}) + B\sin(2\pi ft + \varphi)\sin(\frac{\theta}{2})}{2} - \sqrt{L^2+T^2}\sin(\theta_t - \theta_D) \end{cases} \tag{5}$$

$$\theta_t = \frac{\pi}{2} + \tan^{-1}\left(\frac{2D - A\sin(2\pi ft)\sin(\frac{\theta}{2}) - B\sin(2\pi ft + \varphi)\sin(\frac{\theta}{2})}{-A\sin(2\pi ft)\cos(\frac{\theta}{2}) + B\sin(2\pi ft + \varphi)\cos(\frac{\theta}{2})}\right) \tag{6}$$

In this paper, the amplitudes*A* and *B* are equal, the phase difference is 90°. To a certain structure, L, T, θ_D are constant. The position of the tool tip can be simplified as follows:

$$\begin{cases} pd_x = \dfrac{-A\sin(2\pi ft)\cos(\frac{\theta}{2}) - A\cos(2\pi ft)\cos(\frac{\theta}{2})}{2} - \sqrt{L^2+T^2}\cos(\theta_t - \theta_D) \\ pd_y = \dfrac{-A\sin(2\pi ft)\sin(\frac{\theta}{2}) + A\cos(2\pi ft)\sin(\frac{\theta}{2})}{2} - \sqrt{L^2+T^2}\sin(\theta_t - \theta_D) \end{cases} \tag{7}$$

During the simplification of the formulas, $-\theta_D$ substitutes $\theta_t - \theta_D$ for $\theta_t \ll \theta_D$, then the position equations can express as:

$$\begin{cases} \dfrac{pd_x + \sqrt{L^2+T^2}\cos(-\theta_D)}{-\frac{A}{2}\cos(\frac{\theta}{2})} = \sin(2\pi ft) + \cos(2\pi ft) \\ \dfrac{pd_y + \sqrt{L^2+T^2}\sin(-\theta_D)}{-\frac{A}{2}\sin(\frac{\theta}{2})} = \sin(2\pi ft) - \cos(2\pi ft) \end{cases} \tag{8}$$

The sum (9) of the squaring of Equation (8) is a standard ellipse equation with the center coordinate being $(-\sqrt{L^2+T^2}\cos(-\theta_D), -\sqrt{L^2+T^2}\sin(-\theta_D))$, the length of the semi-major axis being $(\frac{A}{\sqrt{2}}\cos(\frac{\theta}{2}))$ and the length of the semi-minor axis being $(\frac{A}{\sqrt{2}}\sin(\frac{\theta}{2}))$.

$$\left(\frac{pd_x+\sqrt{L^2+T^2}\cos(-\theta_D)}{\frac{A}{\sqrt{2}}\cos(\frac{\theta}{2})}\right)^2+\left(\frac{pd_y+\sqrt{L^2+T^2}\sin(-\theta_D)}{\frac{A}{\sqrt{2}}\sin(\frac{\theta}{2})}\right)^2=1 \quad (9)$$

Finite Element Analyses of the 2D EVC Devices

The piezoelectric ceramics convert the electrical energy to the mechanical energy when the EVC works in physical condition, namely simple harmonic motion can be achieved by the transduces when the excitations are sine waves. The preload of the device implied on the PZT rings makes the vibration transfer to the whole device, as well as the tool tip. As the consequence, the elliptical vibration can be obtained on the tool tip. Considering the efficiency and necessaries, the simulation won't directly perform on the PZT's output performance, the amplitude is getting by the experimental parameters and functions, and the excitation used in the finite element analysis is imported as the displacement. In another word, during the finite element analysis to each device, the input parameters are the displacement amplitudes, frequency and the phase difference, the parameters and calculation functions are given in the following section.

Preparing for the simulation

The displacement amplitudes of PZT under inverse piezoelectric effect can be calculated by the following formula [10]:

$$\delta = n \bullet d_{33} \bullet U_3 \quad (10)$$

In the formulas, where n is the quantity, d_{33} is the coefficient in direction, U_3 is the voltage amplitude. The$d_{33}$of the PZT used in this paper is 220×10^{-12} m/V. The voltage amplifier has the bipolar range of 0 to ± 350 V, and the slew rate is 500 V/us which are enough to the experiment. The single transducer is composed of two mechanical series PZT rings which are parallel connected in circuit, so the value of n is 2. The displacement amplitude of the PZT in the above condition can be calculated as follows:

$$\delta = 2\times 350\times 220\times 10^{-12} = 15.4\times 10^{-8} m \quad (11)$$

The devices of three different common angles are shown in Figure 8 (a is θ=30° , b is θ=60° , c is θ=90°), whose main structures remain unchanged. The following section studies the influence of the angles to the tool tip's trajectory.

The simulations in this paper are performed in ANSYS Workbench (in the next, WB is short for it). There are three main steps for common simulations analysis in WB, which include the foundation of the model, the meshing grid and the post processing. Considering the complexity of the model creating in WB and the compatibility to IGES (the general 3D type data), the models of the structures are created in CATIA, saved as igs files and then imported to WB.

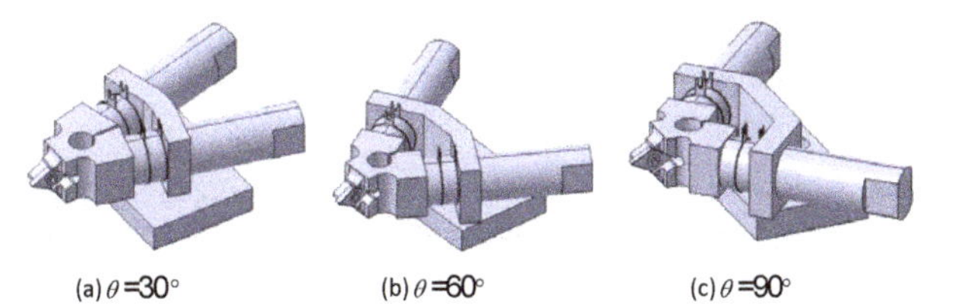

Figure 8: CAD models of different angles.

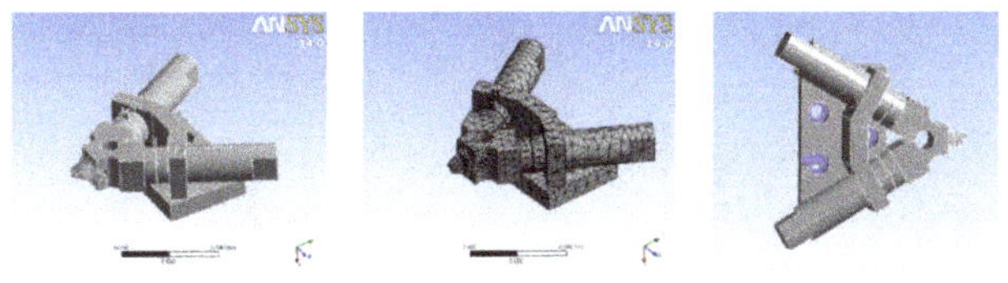

Figure 9: Processes of the simulation in WB.

Modal analysis is adopted to confirm the nature frequency and vibration modes of the structure in common kinematic simulation, and the results can also provide the other kinematic analysis (the harmonic response analysis, random vibration analysis, etc.) parameters with reference. Harmonic response confirms the structures' steady state response in the loads whose amplitudes and frequency are already known. Harmonic response includes two kinds of transient dynamic analysis, linear and non-linear. The former is suitable for small strain and stress of the model structure and the linear analysis also has two algorithms which are direct way and mode suppression. Considering the property of the device in this paper, the linear analysis and mode suppression algorithm are selected for the FEA simulation.

When to the simulation in WB of ANSYS, after the running of the soft and establishing the analysis program, import the model of IGS to the program by the interface provided by WB, as shown in Figure 9a, the model is imported successfully. For the fully understand of the kinematic property of the structure, set the grid mesh accuracy to medium by the smart mesh Figure 9b. The fixed constrains are applied on the torus of counter bores surfaces in the bottom of the model Figure 9c. The theoretical results of the PZTS's output are applied on the structure as the amplitudes of excitations.

Modal and harmonic response analyses

The harmonic responses are separately performed to the three structures by the method of mode suppression in the same condition (the material, the constrains and loads and the grid meshing). For the convenience of data comparison, the amplitudes (in cutting direction and depth of cutting direction) of simulation results of the three structures are shown in one figure. The comparison of the amplitudes in depth of cutting direction of different structures is shown in Figure 10. The range of the frequency is 0 to 50 kHz and the step size is 500 Hz.

As shown in Figure 10a, in terms of the depth of cut direction, there are obvious differences in resonance frequency and amplitude between different structures. The 30° structure has a bigger amplitude in high frequency, and resonances in 13 kHz, 35 kHz, 35 kHz, and reaches the peak amplitude of 2.72 μm in 44.4 kHz; the 30° structure has the superiority in the range of low frequency comparing with the other structures except some certain frequency point and reaches the amplitude peak 3.43 μm in 6 kHz; the 90° structure has a relative minor amplitude in the whole simulation range.

For clearly comparing of the amplitudes of three structures in different vibration modes, Figures 10b and 11 are made for comparison. In these two figures, the value in horizontal axes is the vibration mode order.

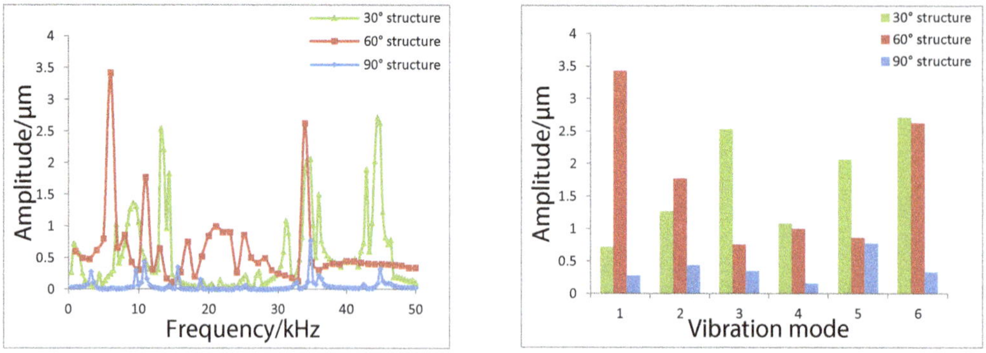

(a) Comparisonof frequency (b) Comparisonof vibration mode

Figure 10: Influence of frequency on the amplitude in depth of cut direction.

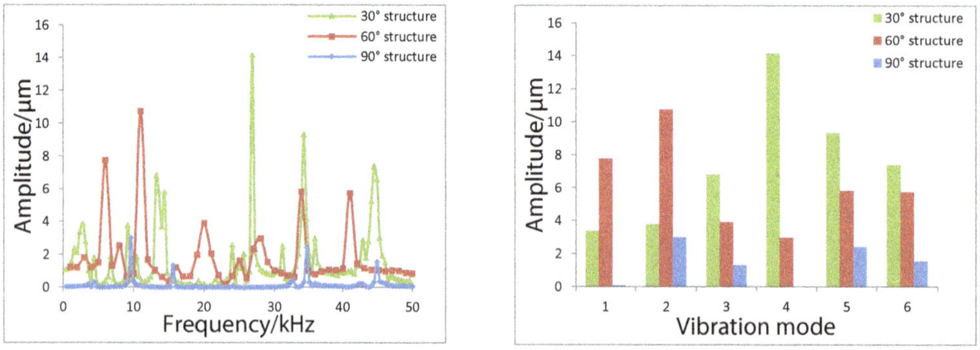

(a) Comparisonof frequency (b) Comparisonof vibration mode

Figure 11: Influence of frequency on the amplitude in cutting direction.

In terms of the amplitudes in cutting direction of the tool tip, 30° structure reaches the amplitude peak 14.165 μm in 26.8 kHz and resonances at 13 kHz, 34.4 kHz, 44.4 kHz; 60° structure reaches the amplitude peak 10.753 μm in 11 kHz and resonances at 6 kHz, 20 kHz, 34 kHz, 41 kHz; 60° structure has the obvious shortage in amplitude except some certain frequency when compare to the other two structures (Table 1).

As seen from the comparisons above, in both depth of cut direction and cutting direction, the 90° structure has a clear defection in amplitude comparing to the other two structures, which is disadvantage to the following operation. When to the 30° structure, there are some superiorities in amplitudes in some certain frequency (34.4 kHz, 44.4 kHz), but in a certain frequency range, there are some vibration modes combine together, which can be seen in figure as the extreme points in a small frequency range. These gather may lead to the instability of the trajectory of the tool tip, which is harmful to the machine accuracy, and the manufacture error of the structure may even deteriorate this phenomenon. The vibration modes distribution of 60° structure are discrete in the simulation range which can avoid the instability when physical machine theoretically.

θ/°		30	60	90
Vibration stability[1]	Depth direction[2]	0.26	0.14	0.2
	Cutting direction	0.34	0.2	0.12
Average amplitude[3]/μm	Depth direction	1.737	1.742	0.392
	Cutting direction	7.418	6.17	1.411
Average resonance frequency[4]/kHz	Depth direction	22.13	19	21.33
	Cutting direction	21.8	23.33	23.66

[1]Vibration stability is ratio of the number of extreme points and the number of the whole simulation points, the higher the ratio value is, the lower stability the structure performs.

[2]Depth direction is short for depth of cut direction.

[3]Average amplitude is the average value of amplitudes of the former six vibration modes.

[4]Average resonance frequency is the average value of the former six vibration mode frequencies.

Table 1: Comprehensive comparison of three structures.

Stress and strain analysis

The stress finite element analyses are simulated to verify if the intensity of the vibration structure meet the stiffness requirements. Figure 12a is the structure's strain cloud chart and Figure 12b is the stress cloud chart. As shown in the figure, the strain of the structure reaches the maximum in the tool tip, which is agreed with the design objective.

The stress concentration occurs in the flexible hinge (Figure 12b), which may exceed the fatigue limit of the material. So the detailed discussion of the structure stress in flexible hinge is given in the following. The stress in flexible hinge in different excitation frequency is shown in Figure 13. The red line in the figure represents the allowable

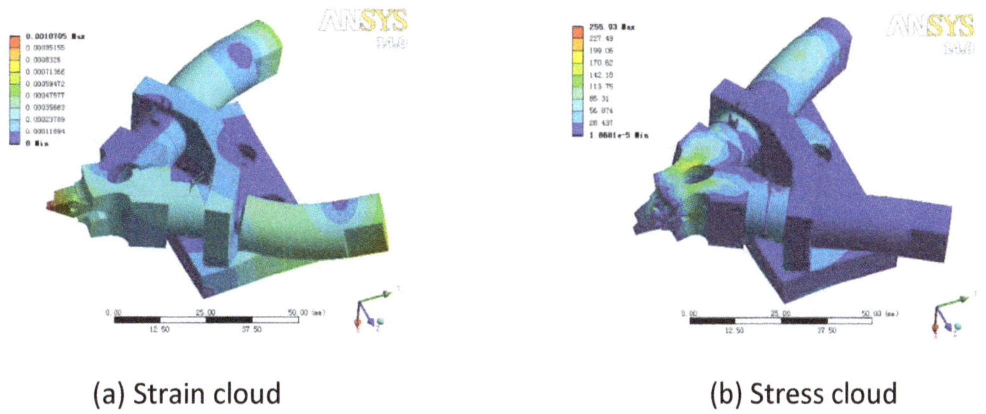

Figure 12: Strain and stress cloud of the structure.

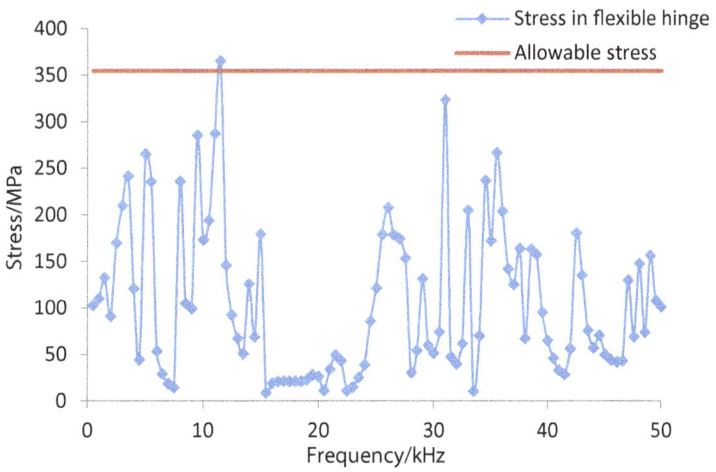

Figure 13: Stress in flexible hinge in different frequency.

stress (350 MPa) of the material; the blue line is the stress in flexible hinge by different frequency. The figure clearly shows that, except the point of 11.5 kHz, the stress is under the allowable stress, which means the structure meets the requirements in the stress field in most frequency.

As the consequence, considering the consequence performance (amplitudes, stability and stress) of three structures, the 60°structure is finally adopted for the following manufacturing and physical experimental testing.

Experiment and Comparison

Experiment condition

Figure 14 shows the experiment condition, the material is 45# steel. The signals are generated by Data Acquisition Card which is programed by LABVIEW on PC, then amplified by the power amplifier which has a 0 ± 350 V bipolar output. The two Langevin transducers are excited by the outputs of the amplifier. The displacements of the tool tip are measured by the Micro Sense capacitance sensors and processed by the Data Acquisition Card. The displacement data are shown in real time by the LABVIEW and recorded as LVM files in PC for post processing.

Frequency response tests

The frequency response data of the tool tip in experimental is also recorded and analyzed. The voltage amplitude applied on the PZT in this experiment is 350 V. Comparing with the simulation results (Figures 15 and 16), there is a certain difference between the theoretical and experimental results. Figures 15a and 16a show the comparison of amplitude in depth of cut direction and cutting direction separately. The analyses of the differences in resonance frequency and amplitude are given in the following section.

The comparison of the simulation and experimental results are shown in Figure 15b, the resonance frequencies of the physical device basically consistent with the simulation results but a little deviation; the amplitudes differences are much more apparent in the figure and the simulation results are less than the experimental results, this is may possibly cause by the PZT's calibration coefficient error. The amplitudes in 23 kHz and 29 kHz are much bigger than common, this may cause by the manufacture and assembly error.

In terms of amplitudes in cutting direction of the tool tip, the trends of two results are basically in accordance with each other. As shown in Figure 16b, the experimental results verify most of the simulation

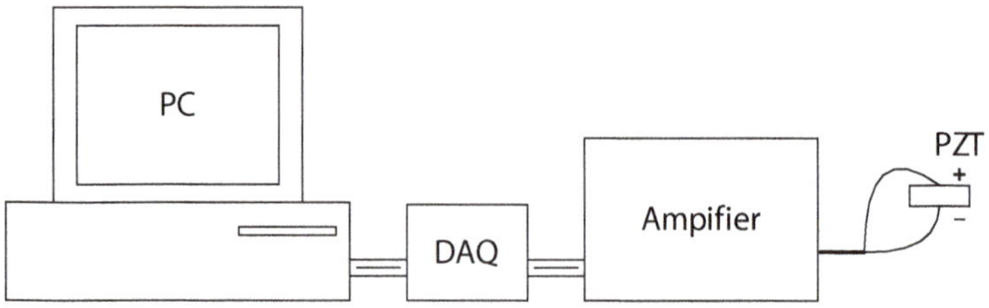

Figure 14: Experiment condition.

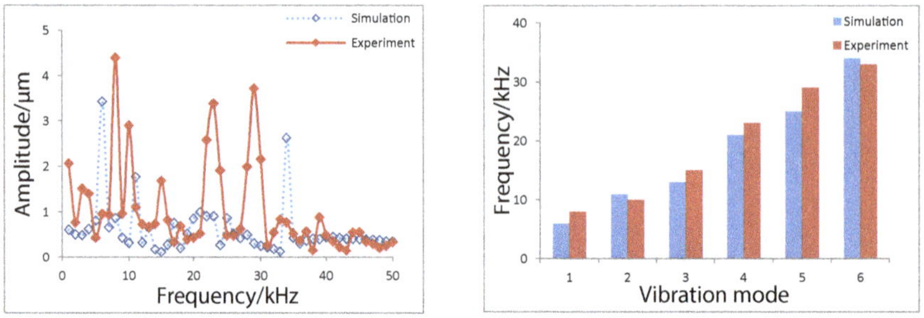

(a) Comparisonof frequency (b) Comparisonof vibration mode

Figure 15: Simulation and experiment comparison in depth of cut direction.

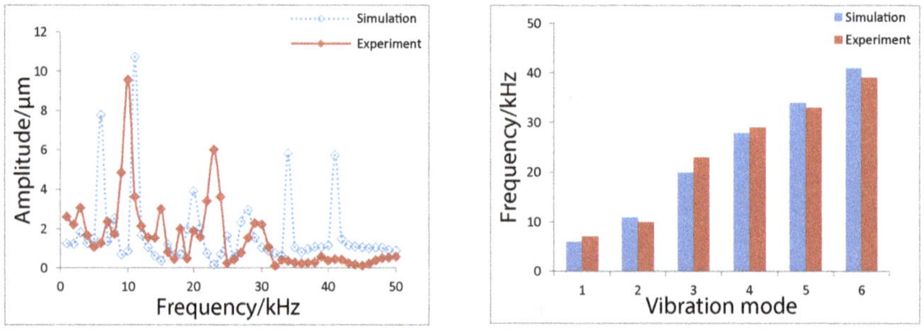

(a) Comparisonof frequency (b) Comparisonof vibration mode

Figure 16: Simulation and experiment comparison in cutting direction.

prediction with a little deviation. At the same time, the great decrease in amplitude of the experimental vibration mode in 34 kHz and 41 kHz cannot be neglected.

From the analyses above, it can be obviously seen that in the low frequency range (0-20 kHz), the simulations make a good agreement with the experiments, when to the high frequency range (above 33 kHz), the deviations arise to the extent which cannot be ignored, which provides little reference for the experiments.

Excitation phase difference tests

In Figure 3, the phase of the excitation applied on rod B is 1/4 cycle time earlier than which applied on rod B, the direction of the elliptical locus at the tool tip is clockwise which can be used for EVC machine. The simulations show that the excitation phase differences have an effect on the elliptical motion of the tool tip in both amplitude and direction, which means in some phase difference range, the device is not suitable for EVC. Simulations and corresponding experimental tests of this effect are conducted and the results are discussed in the following.

The experimental results under the effect of the phase differences are solid line in blue color in next two figures (Figures 17 and 18), and the simulation results are dash line in red color. The trends of the amplitude changes of the tool tip in depth of cut direction and cutting direction are in consistent, when the phase difference is 160°, the maximum amplitude in both directions achieved, and 320° the minim amplitude achieved. The cycle of the tool tip vibration is the same as the simple harmonic excitations' cycle.

Figure 17 shows the comparison of the amplitudes in the depth of cut direction of the tool tip trajectory between simulations and experimental results influenced by the phase difference. Except a certain difference in the amplitudes in some certain points, the simulation results match the experimental results well in both varying trends and specific values.

In terms of the amplitudes in cutting direction in the elliptical trajectory of the tool tip, as shown in Figure 18, the trend of the variation of amplitude in the simulation matches the experimental data well, the corresponding values can also consistent with which measured in the

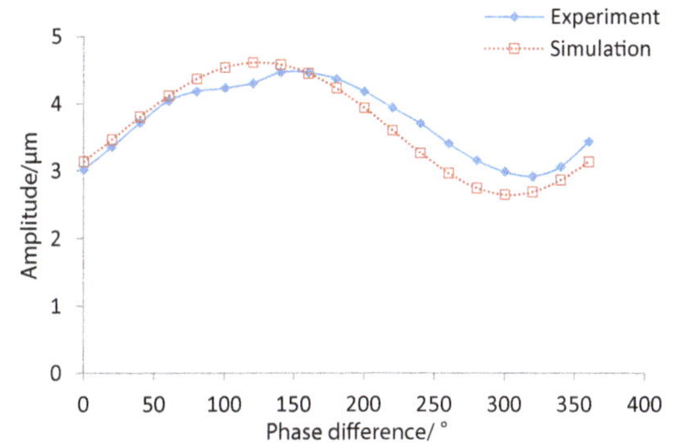

Figure 17: Simulation and experiment comparison in depth of cut direction amplitude.

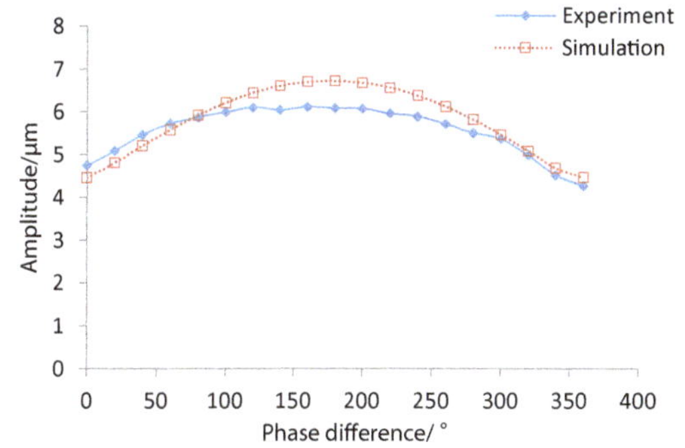

Figure 18: Simulation and experiment comparison in cutting direction amplitude.

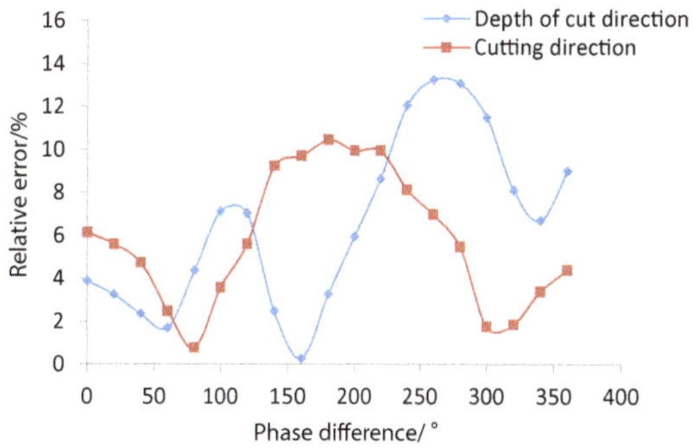

Figure 19: Relative error between simulation and experiment.

experiments considering a certain error.

As shown in the two figures (Figures 17 and 18), no matter in the depth of cut direction or in the cutting direction, the variation trends of the amplitudes by the phase difference are both apparent in either simulations or experiments, and two trends exactly consistent with each other. To the amplitude values in a certain phase difference, for the clear comparing, the figure about the error between the simulation and experimenting two directions is shown in Figure 19. The value of the relative error in Figure 19 is calculated as the following formula (12):

$$\text{Relative error} = \frac{\text{simulation data-experiment data}}{\text{experiment data}} \qquad (12)$$

As shown in this the figure, the error is under 10% in most phase difference, which is acceptable considering the factors of the environment effect, simplification and device manufacturing. The

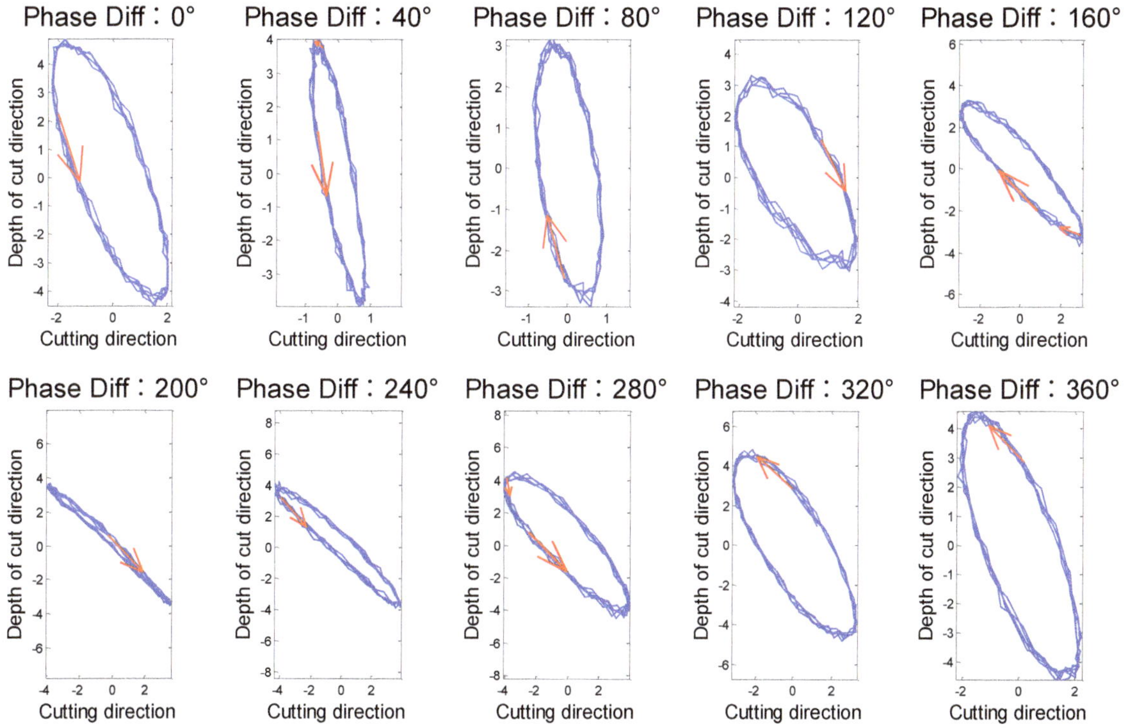

Figure 20: Influence of phase difference on the trajectory.

error in both directions is basically in the same trend, but in the phase difference range 120°-280°, the error in depth of cut direction goes through with a peak while the other with a valley. During this range, the error in depth of cut direction reaches the maximum 10.46% when the phase difference is 180°, and the error in cutting direction reaches minimum 0.29% when the phase difference is 160°. In the whole phase difference cycle, the maximum error in depth is 13.25% at 260°, and the minimum error in cutting direction is 0.78% at 80°.

To the structure placed by the means in this paper (Figure 4), only when the tool tip moves along the clockwise direction, the tool works in EVC condition. Figure 20 shows in effect of the phase difference to the direction of the tool's trajectory. The amplitude of the excitation voltage is 350 V.

As shown in the figure, the tip moves in the anticlockwise direction when the phase difference in the range of 0° to 60°and 240° to 360°, which cannot be used for EVC machine; when the phase difference in the range of 60° to 240°, the tool tip moves in clockwise direction, which can be used for EVC machine. When the phase difference close to the boundary (60° and 240°), the minor axis of the elliptical locus decreases to zero, the locus nearly becomes a straight line in a certain phase difference, then the move direction of the tool tip changes as the phase difference increases furtherly.

From the Figure 20, when the phase difference changes, the slope of the elliptical trajectory of the tool varies. When the phase difference is 90°, the major axis of elliptical trajectory approximately along the depth of cut direction. The slope increases when the phase difference away from 90°. When the phase difference reaches 180°approximately, the slope reaches the maximum, after then the slope decrease with the phase difference increasement until the next cycle.

Conclusion

Based on the structure analysis and kinematic analysis of an EVC device, FEM is used for a serious simulations includes locus amplitudes of the tool tip, the resonance frequency and the vibration modes in different angles of device. After comprehensive comparison of the simulation results, the 60° structure is selected for the final optimization. The actual device is manufactured by the optimization and tested in a series of experiments, the simulation results are verified by the experimental data. The comparison shows that, ignoring a certain manufacture and measurement errors, the simulation models established in this paper can optimize the vibration structure and predict the locus of the tool tip successfully, which make a significantly meaning to the wide application of the EVC device in the following research.

References

1. Zhang J, Suzuki N, Wang Y, Shamoto E (2014) Fundamental investigation of ultra-precision ductile machining of tungsten carbide by applying elliptical vibration cutting with single crystal diamond. Journal of Materials Processing Technology 214: 2644-2659.
2. Hong MS, Ehmann KF (1995) Generation of engineered surfaces by the surface-shaping system. International Journal of Machine Tools and Manufacture 35:1269-1290.
3. Zhang JG, Suzuki N, Kato T, Hino R, Shamoto E (2012) Influence of material composition on ductile machining of tungsten carbide in elliptical vibration cutting. In Key Engineering Materials 523: 113-118.
4. Arif M, Rahman M, San WY (2011) Analytical model to determine the critical feed per edge for ductile–brittle transition in milling process of brittle materials. International Journal of Machine Tools and Manufacture 51: 170-181.
5. Ziki JDA, Didar TF, Wüthrich R (2012) Micro-texturing channel surfaces on glass with spark assisted chemical engraving. International Journal of Machine Tools and Manufacture 57: 66-72.
6. Kim GD, Loh BG (2007) An ultrasonic elliptical vibration cutting device for micro V-groove machining Kinematical analysis and micro V-groove machining characteristics. Journal of materials processing technology 190:181-188.
7. Guo P, Ehmann KF (2013) Development of a tertiary motion generator for elliptical vibration texturing. Precision Engineering 37: 364-371.
8. Amini S, Soleimanimehr H, Nategh MJ, Abudollah A, Sadeghi MH (2008) FEM analysis of ultrasonic-vibration-assisted turning and the vibratory tool.Journal of materials processing technology 201: 43-47.
9. Lu D, Cai LG, Cheng Q (2014) Finite Element Study of Ultrasonic Elliptical Vibration Turning of Ti6Al4V. Applied Mechanics and Materials 494: 383-386.
10. Vivekananda K, Arka GN, Sahoo SK (2014) Finite element analysis and process parameters optimization of ultrasonic vibration assisted turning (UVT). Procedia Materials Science 6: 1906-1914.
11. Vivekananda K, Arka GN, Sahoo SK (2014) Design and Analysis of Ultrasonic Vibratory Tool (UVT) Using FEM and Experimental Study on Ultrasonic Vibration-assisted Turning (UAT). Procedia Engineering 97: 1178-1186.
12. Huang BW, Chen WT, Tseng JG (2015) Finite Element Analysis on Natural Properties of a Micro Drill with Ultrasonic Vibration. Applied Mechanics and Materials 764: 285-288.
13. Jaffe B (2012) Piezoelectric ceramics. Elsevier.

13

An Experimental Investigation into the Thermal Properties of Nano Fluid

Tushar A Sinha[1]*, Amit Kumar[1], Nikhilesh Bhargava[1] and Soumya S Mallick[2]

[1]*Department of Mechanical Engineering, Thapar University, Patiala 147004, India*
[2]*Faculty, Department of Mechanical Engineering, Thapar University, Patiala 147004, India*

Abstract

In this paper, the results of the experimental investigation on the thermal properties of Nano fluid are presented. The effect of sonication time, settling time and temperature on the thermal conductivity, viscosity and specific heat of zinc oxide (ZnO, 14 nm and 25 nm size) and single walled carbon nanotube (SWCNT, 10nm size) based Nano fluid are investigated and the results of ZnO with DI water and ethylene glycol (EG) as base fluids are compared. The experimental results indicate that the studied parameters have a remarkable effect on the thermal properties of Nano fluid. The rate of enhancement in thermal conductivity of EG based Nano fluid is found to be less than that of water based Nano fluid. The SWCNT based DI water Nano fluid found to be very unstable i.e. the nanoparticles settle down very rapidly. The 0.02% volume fraction of SWCNT nanoparticles suspension results in 10% increase in the specific heat of DI water. A decrement of 24% and 13% in the specific heat of 14 nm size ZnO based Nano fluid were obtained at a volume fraction of 0.001% and 0.002% respectively.

Keywords: Nano fluid; Heat transfer; Nanoparticles; Viscometer; Torque meter

Introduction

Nano fluid came into picture in the field of heat transfer in systems since it was introduced by Choi [1]. The heat transfer coefficient of a fluid depends on thermal properties like conductivity, viscosity and specific heat. So far the effect of particle size, volume fraction and temperature was studied by many researchers [2-5] but the effect of sonication and settling time on Nano fluid is studied by few researchers [6-11]. Calvin [12] investigated effect of volume fraction and temperature on the CuO and Al_2O_3 nanoparticles based water Nano fluid and the results showed an increase of 52% in the thermal conductivity of DI water, when CuO nanoparticles were dispersed at a volume fraction of 6% [13]. Also, an increase of 30% in the conductivity of Al_2O_3 based Nano fluid was reported at volume fraction 10% in a temperature range of 27.5 to 34.7°C. Jang [14] investigated the effect of temperature and volume fraction on the viscosity Al_2O_3 nanoparticles dispersed in water, the results reported an increase of 2.9% in the viscosity of base fluid at a volume fraction of 0.3% and with the increase in temperature the viscosity of Nano fluid decreases continuously [15,16]. The results of Zhou [17] showed that the specific heat of water decreases by 50%, when Al_2O_3 nanoparticles were dispersed in a volume fraction range of 0 to 21.7% [18]. In this paper, the effect of volume fraction, sonication time, settling time, diameter of particles and temperature on the thermal properties of zinc oxide and single walled carbon nanotube based Nano fluid s is presented.

Experimental Procedure

In order to study the effect of various parameters on thermal properties (Thermal conductivity, viscosity and specific heat) of Nano fluid, zinc oxide (ZnO) nanoparticles and single walled carbon nanotube (SWCNT) has been purchased from Rainste Nano Ventures Pvt. Ltd., Noida. The average diameter of ZnO powders are 14 nm and 25 nm with 1nm surfactant coating of oleic acid and of SWCNT is 10 nm. The density of ZnO nanoparticles and SWCNT is 5600 kg/m^3 and 0.05 kg/m^3 respectively. The as received powders are sealed, dried and loosely agglomerated. The two step method was used to prepare ZnO based ethylene glycol (EG) Nano fluid s of two different volume fractions; 0.01% and 0.05%. The relative thermal conductivity and viscosity of EG based ZnO Nano fluid s are compared with DI water based ZnO Nano fluid s. The experiments were also carried out with SWCNT as nanoparticles and water as base fluid for a volume fraction of 1%. The thermal conductivity was measured by Decagon devices KD2Pro Thermal Properties Analyzer (Decagon Devices Inc., Pullman, WA, USA). The Viscosity of the sample is measured by the instrument named Brookfield DV III Rheometer (Brookfield DV III Ultra Manual). This rheometer is a cone plate viscometer which is a precise torque meter.

Results and Discussion

Several measurements were carried out for DI water and ethylene glycol as basefluids, to investigate the effect of sonication time, settling time and temperature on the thermal properties of the Nano fluid. In Figure 1, the relative thermal conductivity of ZnO Nano fluid s increases for both basefluids ethylene glycol and DI water. The relative thermal conductivity of ZnO ethylene glycol based Nano fluid s shows low enhancement as compare to DI water based Nano fluid s, because the ethylene glycol has more viscosity due to which ZnO nanoparticles takes more time to disperse in ethylene glycol basefluids. As a result, surface to volume ratio decreases in ethylene glycol based Nano fluid s. In Figure 2, the samples were prepared by dispersing nanoparticles in ultra-bath sonicator up to 8 hours with basefluids ethylene glycol and DI water. The relative thermal conductivity of ZnO ethylene glycol and DI water based Nano fluid s decreases as the settling time increases because, as time increases the nanoparticles which were dispersed in basefluids, starts agglomerated due to which cluster formed and settle down as the time increases. The decreases rate of relative thermal conductivity in ethylene glycol based Nano fluid s is less as compare to DI water based Nano fluid s because, the viscosity of ethylene glycol is

***Corresponding author:** Tushar A Sinha, Department of Mechanical Engineering, Thapar University, Patiala 147004, India
E-mail: honeysinha497@gmail.com

more due to which nanoparticles will not settle down, as in DI water based Nano fluid s (Figures 1 and 2).

Figure 3 shows the relative thermal conductivity of ZnO-ethylene glycol and ZnO-DI water Nano fluid s increases as the temperature increases because with the increase in temperature, Brownian motion of nanoparticles increase which excites the particles due to which random motion of nanoparticles increases and the particles starts strikes with each other and transfer the heat energy. Figure 4 shows that the relative viscosity of EG-ZnO Nano fluid decreases with the increase in sonication time, but increases in the case of water-ZnO Nano fluid. The viscosity of water based Nano fluid is more than ethylene glycol based Nano fluid. Also with the increase in size and volume fraction of

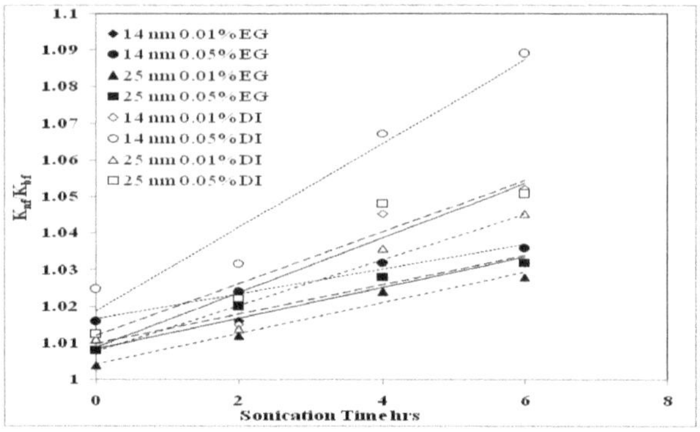

Figure 1: Comparison of relative thermal conductivity with basefluidsEG and DI water versus sonication time.

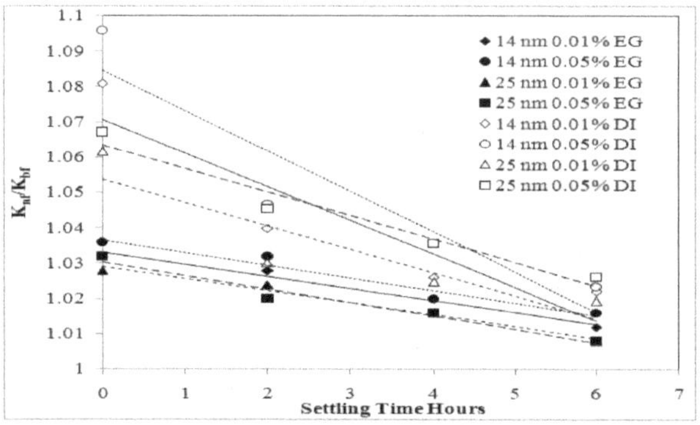

Figure 2: Comparison of relative thermal conductivity with basefluids EG and DI water versus settling time.

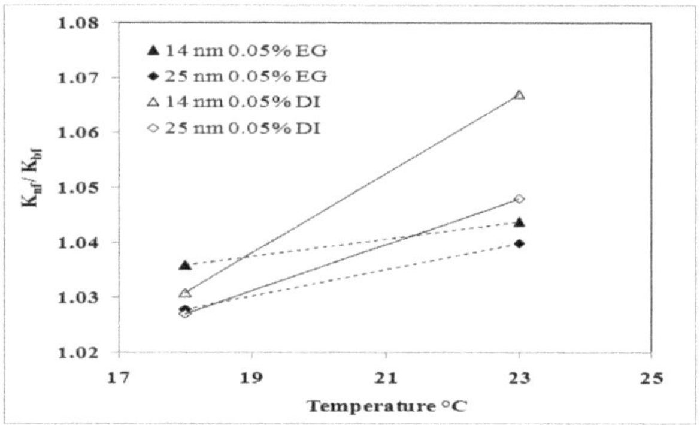

Figure 3: Comparison of relative thermal conductivity with basefluids EG and DI water versus temperature.

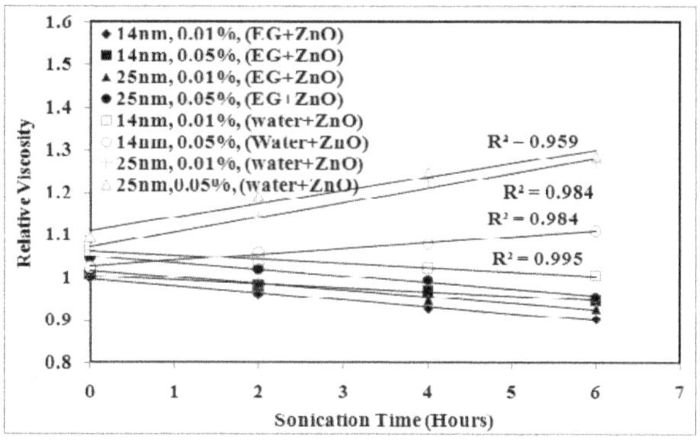

Figure 4: Relative viscosity versus sonication time for EG-ZnO and water-ZnO nanofluid.

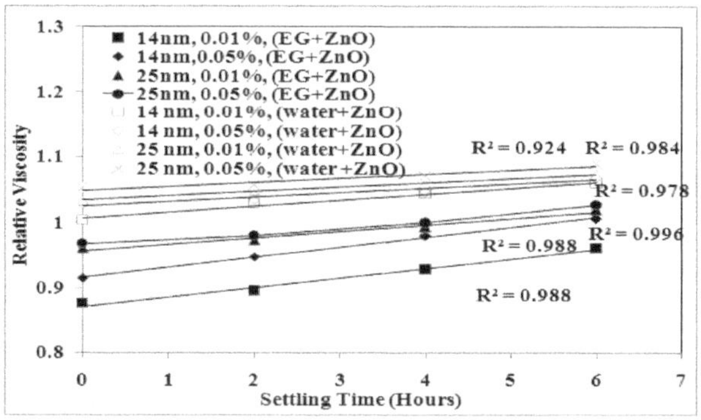

Figure 5: Relative viscosity versus settling time for EG-ZnO and water-ZnO nanofluid.

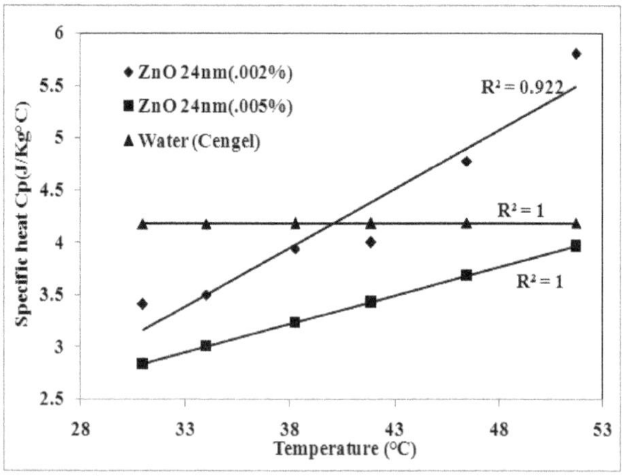

Figure 6: Variation of specific heat of ZnO-water nanofluid with temperature.

nanoparticles in base fluid, the relative viscosity of Nano fluid increases continuously. The relative viscosity of ethylene glycol based Nano fluid comes very close to 1 at zero hours of sonication. In Figure 5, the rate of enhancement in viscosity of EG-ZnO Nano fluid is less as compared to water-ZnO Nano fluid. This is because the particles in Nano fluid start to settle down and agglomerate as the sonication of Nano fluid is stopped. As the viscosity of ethylene glycol is comparatively higher than that of water, which makes the particles suspended for a longer time due to which, there is less increase in the viscosity of ethylene glycol based Nano fluid (Figures 3-5).

In Figures 6 and 7, the variation of specific heat of 14 nm and 25 nm sized ZnO-water Nano fluid with temperature is shown. The specific heat of Nano fluid is found to be increasing with the increase in temperature for a range of 30° to 50°C. But the specific heat of water remains unchanged [11] for the specified range (Figures 6 and 7).

In Figures 8 and 9, the relative thermal conductivity of SWCNT - water Nano fluid increases with the increase in sonication time and decreases with the increase in settling time. In Figures 10 and 11, the viscosity of Nano fluid is decreasing with the increase in sonication time and increases with the increase in settling time. Also with the increase in power of sonication, the conductivity of Nano fluid is increasing and viscosity of Nano fluid is decreasing (Figures 8-11).

Conclusion

The results of the current investigation clearly indicate that the thermal conductivity of Nano fluid increases with the increase in the sonication time, but the viscosity of Nano fluid decreases with it. Also with the increase in settling time, the thermal conductivity decreases and viscosity increases. With the increase in temperature, the thermal conductivity and specific heat of Nano fluid increases and viscosity decreases.

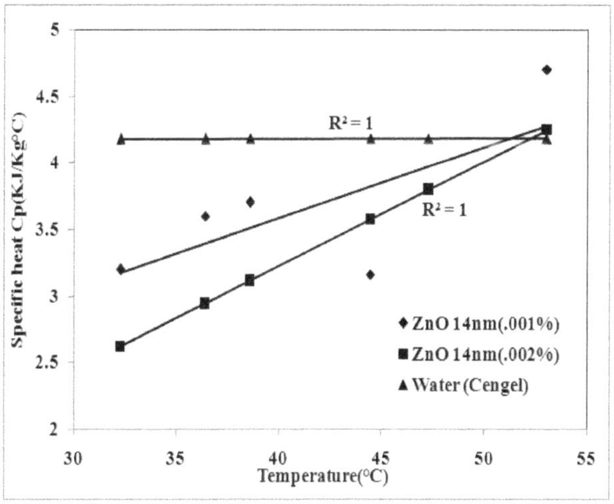

Figure 7: Variation of specific heat of ZnO-water (14 nm) with temperature.

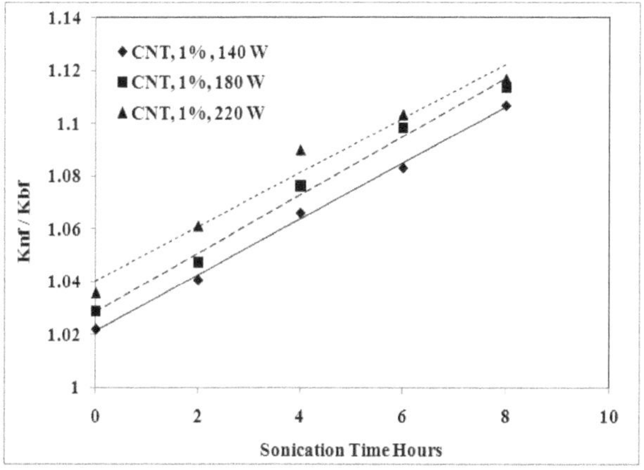

Figure 8: Thermal conductivity ratio variation with sonication time at different sonication power.

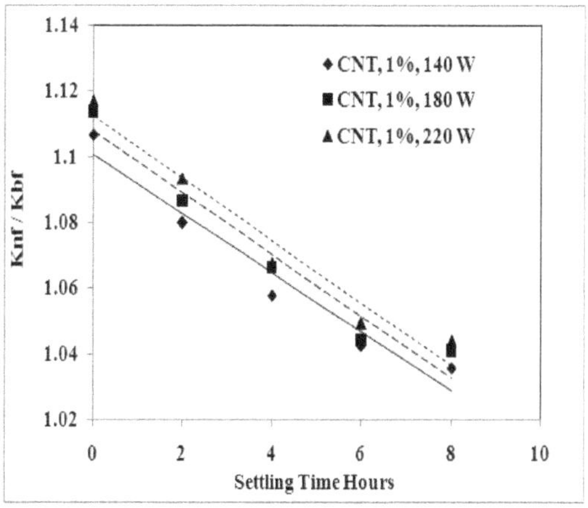

Figure 9: Thermal conductivity ratio variation with settling time at different sonication power.

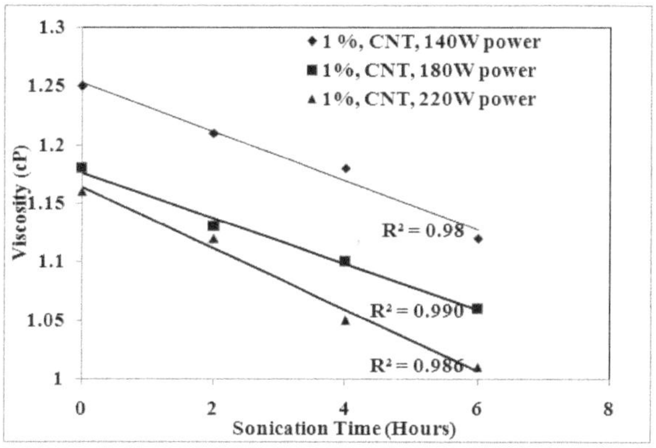

Figure 10: Variation of viscosity of water based single walled carbon nanotube nanofluids with sonication time.

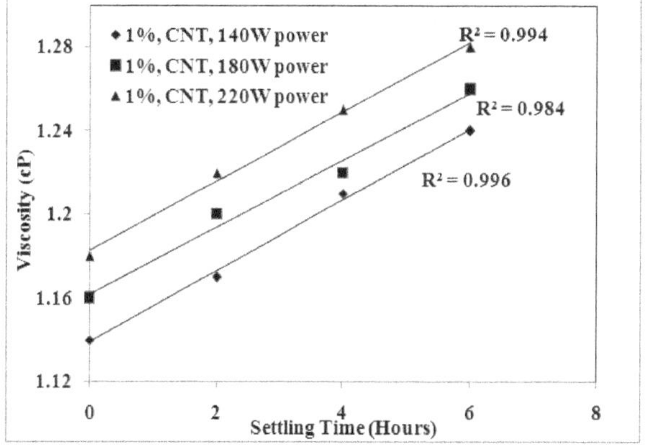

Figure 11: Variation of viscosity of water based single walled carbon nanotube nanofluid with settling time.

Acknowledgement

Thapar University-Seed Money Grant (Financial Assistance).

Nomenclature

K=Thermal conductivity

DI=Deionized

nf=nano fluid

bf=base fluid

References

1. Choi SUS, JA Eastman (1995) Enhancing thermal conductivity of fluids with nanoparticles, in Developments and Applications of Non-Newtonian Flows, D. A. Singer and H. P. Wang, Eds., American Society of Mechanical Engineers, New York, 99-105.
2. Mintsa HA, Roy G, Nguyen CT, Doucet D (2009) New temperature dependent thermal conductivity data for water-based nanofluids. International Journal of Thermal Sciences 48: 363-371.
3. Jang SP, Hwang KS, Lee JH, Kim JH, Lee BH, et al. (2007) Effective Thermal Conductivities and Viscosities of Water-based Nano fluids Containing Al_2O_3 with Low Concentration, International Conference on Nanotechnology, August 2-5, 2007, Hong Kong.
4. Nguyen CT, Desgranges F, Roy G, Galanis N, Mare T, et al. (2007) Temperature and particle-size dependent viscosity data for water-based nanofluids–Hysteresis phenomenon. International Journal of Heat and Fluid Flow 28: 1492-1506.
5. Shin D, Banerjee D (2011) Enhanced Specific Heat of Silica Nanofluid. Journal of Heat Transfer 133.
6. Kwak K, Kim C (2005) Viscosity and thermal conductivity of copper oxide nanofluid dispersed in ethylene glycol. Korea-Australia Rheology Journal 17: 35-40.
7. Yu W, Xie H (2012) A Review on Nanofluids: Preparation, Stability Mechanisms and Applications. Journal of Nanomaterials 2012.
8. Buongiorno J (2006) Convective transport in nanofluids. ASME J Heat Transfer 128: 240-250.
9. Das SK, Choi SUS, Yu W, Pradeep T (2007) Nanofluids: Science and Technology. Wiley, New Jersey.
10. Wong KV, Leon OD (2010) Applications of nanofluids: current and future. Adv Mech Eng.
11. Wen D, Lin G, Vafaei S, Zhang K (2011) Review of nanofluids for heat transfer applications. Particuology. 7: 141-150.
12. Calvin H Li, Peterson GP (2006) Experimental investigation of temperature and volume fraction variations on the effective thermal conductivity of nanoparticle suspensions (nanofluids). J Appl Phys 99.
13. Kakaç S, Pramuanjaroenkij A (2009) Review of convective heat transfer enhancement with nanofluids. Int J Heat Mass Transfer 52: 3187-3196.
14. Jang SP, Hwang KS, Lee JH, Kim JH, Lee BH, et al. (2007) Effective Thermal Conductivities and Viscosities of Water-based Nanofluids Containing Al_2O_3 with Low Concentration, International Conference on Nanotechnology, August 2-5, 2007, Hong Kong.
15. Saidur R, Leong KY, Mohammad HA (2011) A review on applications and challenges of nanofluids. Renewable and Sustainable Energy Reviews 15: 1646-1668.
16. Mahian O, Kianifar A, Kalogirou SA, Pop I, Wongwises S (2013) A review of the applications of nanofluids in solar energy. Int J Heat Mass Transfer 57: 582-594.
17. Zhou SQ, Ni R (2008) Measurement of the Specific Heat Capacity of Water-Based Al_2O_3 Nanofluid. Appl Phys Lett 92.
18. Cengal YA (2007) Heat and Mass Transfer, A Practical Approach; Tata Mcgraw Hill Publication.

14

Fourier Analysis of Milling Force for General Helical Cutters via Space-Time Convolution, Applications for Common Cutters and Model Validation

Zheng CM[1]* and Junz Wang JJ[2]

[1]*Department of Mechanical and Electrical Engineering, Quanzhou Institute of Information Engineering, Quanzhou 36200, PR China*
[2]*Department of Mechanical Engineering, National Cheng Kung University Tainan, 701, Taiwan*

Abstract

Part 2 of this study illustrates the applications of the frequency domain force model put forth in Part 1 for three common helical cutters: the square, taper, and ball end mills. The respective geometric and boundary functions required for the evaluation of the force spectra are derived by applying differential geometry to these three types of cutters including cutters of constant helix angle and constant helix lead. By virtue of the explict expression of Fourier coefficients of milling force, the differences between cutting forces generated by two cutters with constant helix angle and constant helix lead can be described quantitatively. In slot (or half slot) milling for the taper end mills with a constant helix angle and constant helix lead, the strategy for selecting axial depths of cut to reduce force pulsation is presented respectively. Also derived are the specific expressions for the average forces of these three helical cutters in common cutting configurations. Moreover, as an inverse application, a linear equation is formulated for the identification of six shearing and ploughing cutting constants from the measured average cutting forces for a general helical cutter. The frequency domain force model and the identification of the cutting constants are finally demonstrated and validated through experiments with all three types of milling cutters.

Keywords: Ball-end milling; Taper-end milling; General helical mill; Differential geometry

Introduction

The convolution force model and the force spectra characteristics of a general helical milling cutter has been presented and analyzed in Part 1 of this paper. It has been shown that the composing structures of the milling force in the angular and the frequency domain are the same for different types of helical cutters. Also the frequency spectra of the milling forces are shown to be characterized by the geometric functions of the helical cutting edge and by the cutting boundary functions which are determined by the axial and radial cutting depths. Therefore, the task of establishing the specific milling force models for any type of helical cutter is reduced to finding the analytic expressions for the line geometry of its helical cutting edge, as well as the expressions for the cutting boundaries including the entry and exit angle and the limits of the radial angle immersion. Although this generalized analytical model is applicable for any cutter with an analytically definable cutting edge, the second part of this paper will only illustrate its application for three common industrial helical end mills: the square, taper, and ball end mills.

Kinematics and cutting forces for the square end mill dates back to Martellotti [1,2]. However, force models for cutters other than the face milling cutter and square end mill only began in the 1990's. Yang and Park [3] could have been the first to present a force model for the ball-end mill. In their work, the force model is based on the fundamental mechanics of orthogonal cutting and explicitly considers the effects of the shear angle, shear stress and friction angle. Other researchers [4-7] have studied the force model for the ball end mill with a mechanistic local force model, in which cutting constants are assumed to be proportional to the uncut chip area and are obtained through milling experiments. Unlike the mechanistic local force model, Sonawane and Joshi [8] presented a analytical force model for the ball-end milling of superalloy Inconel 718 considering strain, strain rate, and temperature dependence of work material shear strength by applying Johnson–Cook material mode. The model analysis shows that there is a significant compression in the chip along its length. The taper end mill is not as common as the square or ball end mills in its industrial uses and has received less attention in the study of its force model. Ramaraj and Eleftheriou [9] used oblique cutting theory in establishing the force model for a taper end mill. Based on the mechanistic local force model, Huang and Whitehouse [10] obtained the total forces through numerical integration for a taper end mill. Instead of dealing with a helical cutter of specific geometry, Altintas et al. [11,12] presented force and process models for a general helical cutter through numerical integration and simulation. With emphasis on the composing structure of the milling forces, the present frequency domain force model is established using a systematic bottom-up approach through the principles of differential geometry, and can be de-generalized to accommodate any helical cutter of definite geometry. In the following, the required geometric and cutting boundary functions will be derived for the cylindrical, taper and ball end mill to complete the force model for each respective cutter. Expressions for their average forces are also derived and subsequently used for the identification of shearing and ploughing constants for each cutter. The dual-mechanism frequency domain force model and the cutting constant identification formula are finally validated through milling experiments with all three types of cutters.

***Corresponding author:** Zheng CM, Department of Mechanical and Electrical Engineering, Quanzhou Institute of Information Engineering, Quanzhou 36200, PR China, E-mail: cmzheng1206@gmail.com

Geometric Functions and Cutting Boundary Functions for Common Helical Cutters

The evaluation of the Fourier coefficients for a specific cutter requires its geometric functions in $\psi(\beta)$ and $h'(\beta)$, as well as the cutting boundary functions in $\beta 1$, $\beta 2$, $\theta 1(\beta)$ and $\theta 2(\beta)$. Although the geometric functions can be completely determined for any mathematically definable cutter, the boundary functions depend not only on the cutter geometry but also on the axial and radial depths of cut and the relative position of the work with respect to the cutter. As illustrative examples, the boundary conditions are derived based on the typical cutting configurations shown in Figure 1.

Square end mills

For the square end mill shown in Figure 1a, it is clear that the cutter radius is a constant R and the axial elevation angle is also a constant with ψ=90°. It is assumed that the helix angle is constant and so is the helix lead,

$$h'(\beta)=\frac{R}{\tan\alpha} \text{ and } h(\beta)=\frac{R}{\tan\alpha}\beta \tag{1}$$

From (1), the end points of the radial cutting range are found to be $\beta 1=0$ and $\beta 2=da\tan\alpha/R$. With a constant radial depth of cut and radius, the entry/exit angles are also constants, which are

$\theta_1=0$ and $\theta_2=\cos^{-1}(1-\frac{dr}{R})$ in up milling and

$\theta_1=\cos^{-1}(1-\frac{dr}{R})$ and $\theta_2=180$ in down milling

Thus, for square end mill, the geometric and boundary functions all have constant values. Substituting these constants, the expression for the Fourier coefficients in Part 1 of this paper is de- generalized to a closed form expression similar to that presented in ref. [13]:

$\theta_1=0$ and $\theta_2=\cos^{-1}(1-\frac{dr}{R})$ in up milling and

$\theta_1=\cos^{-1}(1-\frac{dr}{R})$ and $\theta_2=180$ in down milling

Thus, for square end mill, the geometric and boundary functions all have constant values. Substituting these constants, the expression for the Fourier coefficients in Part 1 of this paper is de- generalized to a closed form expression similar to that presented in ref. [13]:

$$A[Nk]=\frac{N}{2\pi}F_1(Nk)=\frac{N}{2\pi}\int_{\beta_1}^{\beta_2}h'(\beta)e^{-jNk\beta}\,d\beta\sum_{i=1}^{2}\mathbf{q}_i\mathbf{P}_{iw}(Nk)=\frac{N.H'(Nk)}{2\pi}\sum_{i=1}^{2}\mathbf{q}_i\mathbf{P}_{iw}(Nk)$$

$$=\frac{N.H'(Nk)}{2\pi}\left(\begin{bmatrix}1 & k_{rs} & 0\\ -k_{rs} & 1 & 0\\ 0 & 0 & -k_{as}\end{bmatrix}\mathbf{P}_{1w}(Nk)+\begin{bmatrix}1 & k_{rp} & 0\\ -k_{rp} & 1 & 0\\ 0 & 0 & -lk_{ap}\end{bmatrix}\mathbf{P}_{2w}(Nk)\right) \tag{2}$$

where $H(Nk)$ is the Fourier transform of the windowed chip width density function, $H'(\omega)$, evaluated at $\omega=Nk$. The frequency spectra are explicitly determined by the product of $\mathbf{P}_{iw}(Nk)$ and $H'(Nk)$. The values of $\mathbf{P}_{iw}(Nk)$'s are related to the radial depth of cut and have been discussed in Part 1 of this paper. Spectra characteristics of $H'(\omega)$ are similar to those of the radial cutting window function, $W_r(\omega)$, discussed in Part 1 with periodic zeros at $\omega=2k\pi/\beta_2$. β_2 is in turn determined by the axial depth of cut through Eq. (1); therefore, the value of $H'(Nk)$ can be shown to vanish at axial depths of cut with

$$da=\frac{2m\pi R}{N.\tan\alpha} \quad m=1,2... \tag{3}$$

Under these axial depths of cut, the dynamic forces will vanish completely regardless of the values of $\mathbf{P}_{iw}(Nk)$'s and only the average forces remain in the X, Y and Z directions.

Taper end mills

For a taper end mill shown in Figure 1b, the axial elevation angle ψ_0 is a constant and the radius of the cutter is a linear function of h:

$$R(h)=R_o+h\cot\psi_o \tag{4}$$

The radial depth of cut for the configuration shown becomes

$$dr(h)=dr_o+h\cot\psi_o \tag{5}$$

The entry and exit angles are therefore no longer constant as in the case of the cylindrical end mill and can be determined from Eq. (2) as a function of h. The curvilinear geometry of the cutting edge is generally defined by two types of helical functions: constant helix angle or a constant helical lead.

i) For a taper end mill with a constant helix angle, α_0, the helix lead is

$$h'(\beta)=\frac{dh}{d\beta}=\frac{R(\beta)}{\tan\alpha_o} \tag{6}$$

Substituting Eq. (4) into (6) and considering the boundary condition of $\beta=0$ at $h=0$, $h(\beta)$ can be shown to be:

$$h(\beta)=\frac{Ro}{\cot\psi_O}(\exp(\frac{\cot\psi_O}{\tan\alpha_O}\beta)-1) \tag{7}$$

Substituting Eq. (6) into Eqs. (3), (4) and (5) results in

$$R(\beta)=R_o\exp(\frac{\cot\psi_o}{\tan\alpha_o}\beta) \tag{8}$$

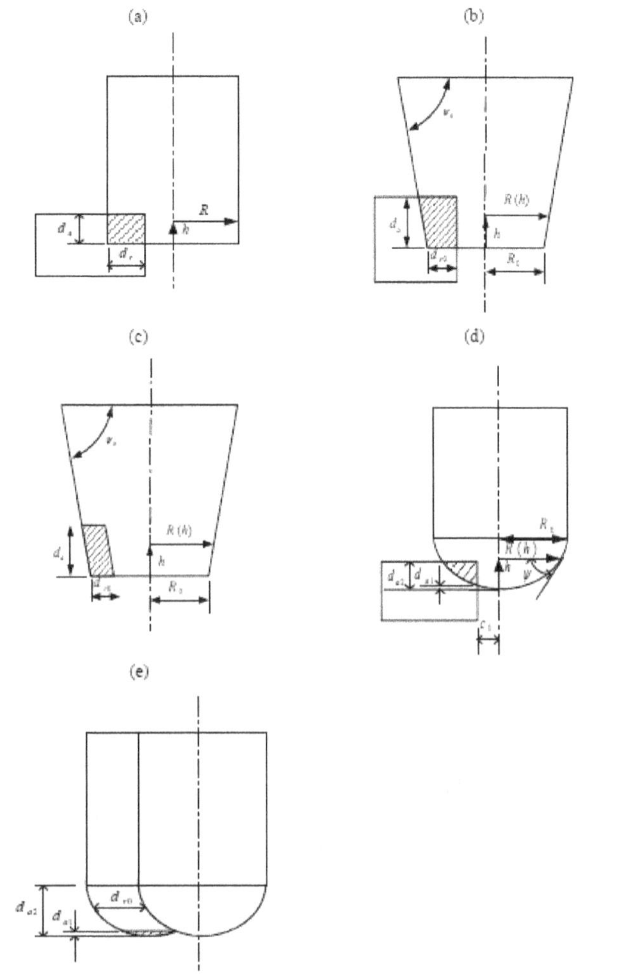

Figure 1: Cutting geometry for (a): a cylindrical end mill, (b) & (c): a taper end mill, and (d) & (e): a ball end mill.

$$dr(\beta)=dr_0+R_0(\exp(\frac{\cot\psi_o}{\tan\alpha_o}\beta)-1) \text{ and } h'(\beta)=\frac{R_0}{\tan\alpha_o}\exp(\frac{\cot\psi_o}{\tan\alpha_o}\beta) \quad (9)$$

From Eqs. (1) and (8), the entry/exit angles along the cutting edge, $\theta_1(\beta)$ and $\theta_2(\beta)$, can be obtained. The radial range of immersion starts at β_1=0 with h_1=0 and ends at β_2, which can be found from Eq. (7) to be

$$\beta_2=\frac{\tan\alpha}{\cos\psi_o}\ln(1+\frac{d_a\cos\psi_o}{R_0}) \quad (10)$$

ii) For a taper end mill with a constant helix lead with

$$h'(\beta)=\frac{R_o}{\tan\alpha_0} \quad (11)$$

where α_o is the nominal helix angle at the bottom of the cutter at h=0 with β=0, its cutting edge is defined by the following constraint equation,

$$h=\frac{R_o}{\tan\alpha_0}\beta \quad (12)$$

The cutter radius as a function of β can be found from Eq. (4) to be

$$R(\beta)=R_o+h\cot\psi_o=R_o+\frac{R_o\cot\psi_o}{\tan\alpha_o}\beta \quad (13)$$

For the cutting configuration shown in Figure 2b, the radial depth of cut as a function of β becomes

$$dr(\beta)=(R_o-C_o)+\frac{R_o\cot\psi_o}{\tan\alpha_o}\beta \quad (14)$$

For a milling process with consecutive passes as shown in Figure 2c, the radial side step dr_0 will be the radial depth of cut for each cutting point so that $dr(\beta)=dr_0$. The entry and exit angles as a function of β and can be found from Eq. (2) together with Eqs. (12) and (13). The two end points of the radial engagement are

$$\beta_1=0 \text{ and } \beta_2=\frac{d_a\tan\alpha_0}{R_o} \quad (15)$$

The required boundary functions and the helix lead for two types of taper end mills has been derived in the above so that Fourier coefficients of the milling forces can be calculated. Although these two types of taper end mills have different mathematical representations, β as function of h for both cutters are found to be practically the same for a typical α_0=30° cutter with h/R_0<1 as shown in Figure 2 for cutters of four different taper angles. Therefore, the helix lead and the boundary functions can be treated equally for both types of taper end mills and their milling forces can be inferred to be practically the same.

It should be noted that the Fourier coefficients of cutting forces for taper end mills in slot milling or half slot milling can be expressed in a closed form like Eq. (3):

$$\mathbf{A}[Nk]=\frac{N.H'(Nk)}{2\pi}(\mathbf{q}_1(\psi)\mathbf{P}_{1w}(Nk)+\mathbf{q}_2(\psi)\mathbf{P}_{2w}(Nk)) \quad (16)$$

Since $H'(\omega)$have periodic zeros at $\omega=2k\pi/\beta_2$, $H'(Nk)$ can be shown to vanish at axial depths of cut with

$$da_{t1}=\frac{2m\pi R_0}{N.\tan\alpha_0} \quad m=1,2\ldots \text{ (for mill with a constant helix lead)} \quad (17a)$$

$$da_{t2}=\frac{R_o}{\cot\psi_o}(\exp(\frac{2m\cot\psi_o}{N.\tan\alpha_o})-1) \quad m=1,2\ldots \text{ (for mill with a constant helix angle)} \quad (17b)$$

which indicates that selecting the axial depths of cut as in (17a) or (17b) in slot milling and half slot milling can only get average, or DC, force component, in which dynamic force components due to shearing and ploughing mechanism will vanish completely regardless of the flute number of taper end mill. However, when N < 8 (i.e., $\beta_2>^{\pi/4}$), da_{t2} is larger than da_{t2} under the same R_0 for ψ > 40° as shown in Figure 2.

Ball end mills

The axial cross section profile of the ball-end cutter shown in Figure 1d is defined by

$$dh=R_0\sin\psi d\psi \quad (18)$$

As for the taper end mill, two types of curvilinear edges will be used.

i) A ball-end cutter with constant helix lead will have

$$h'(\beta)=\frac{R_o}{\tan\alpha_0} \quad (19)$$

where R_0 is the ball radius and α_0 the nominal helix angle at $h=R_0$.

Combining Eqs. (15) and (18) with the boundary condition of ψ=0 at β=0, ψ can be shown to be

$$\psi=\cos^{-1}(1-\frac{\beta}{\tan\alpha_0}) \quad (20)$$

The radius and radial depth of cut for a point at β along the helical edge thus can be expressed as

$$R(\beta)=R_o\sqrt{\frac{\beta}{\tan\alpha_0}(2-\frac{\beta}{\tan\alpha_0})} \text{ and } dr(\beta)=R_0\sqrt{\frac{\beta}{\tan\alpha_0}(2-\frac{\beta}{\tan\alpha_0})-c_0} \quad (21)$$

from which the entry and exit angles as function of β can be found. For the configuration shown in Figure 2d for shoulder milling, the end points of the radial engagement are determined by

$$\beta_1=\tan\alpha_0(1-\sqrt{1-(\frac{c_0}{R_0})^2}),\ \beta_2=\frac{\tan\alpha_0 d_a}{R_0} \quad (22)$$

ii) For a ball-end cutter with constant helix angle, $\alpha 1$, its helix lead becomes a variable defined by

$$h'(\beta)=\frac{R(\psi)}{\tan\alpha_1}=\frac{R_o\sin\psi}{\tan\alpha_1} \quad (23)$$

Combining Eqs. (15) and (22) will have

$$\frac{d\psi}{d\beta}=\frac{1}{\tan\alpha_1} \quad (24)$$

Given the boundary condition of ψ=0 at β=0, ψ can be shown to be

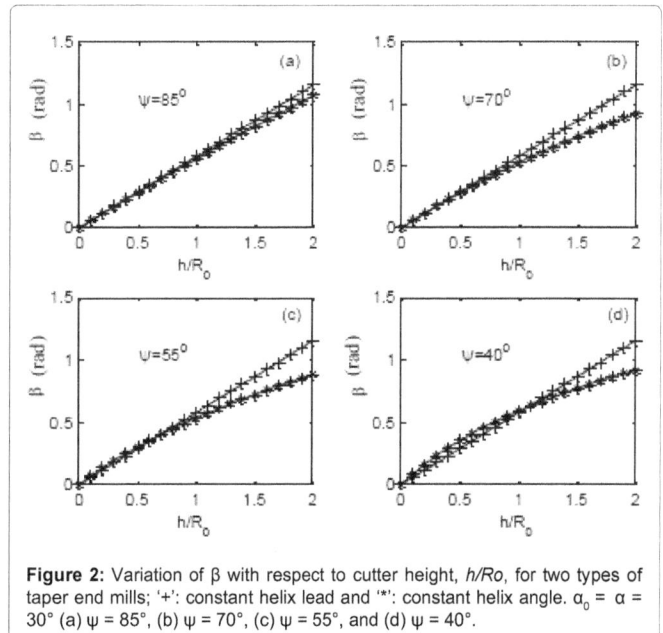

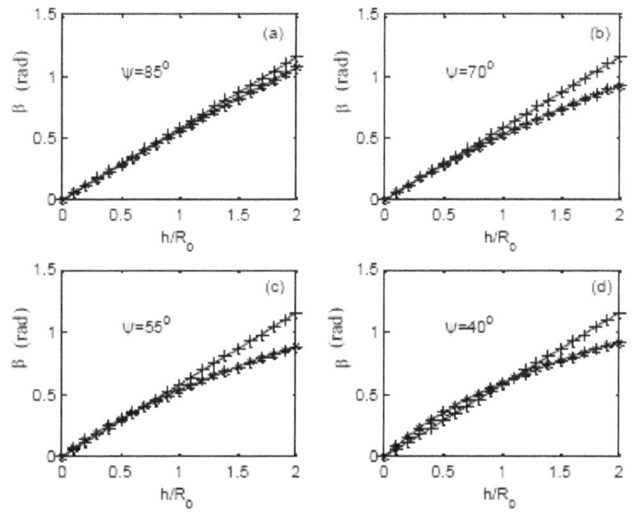

Figure 2: Variation of β with respect to cutter height, *h/Ro*, for two types of taper end mills; '+': constant helix lead and '*': constant helix angle. α_0 = α = 30° (a) ψ = 85°, (b) ψ = 70°, (c) ψ = 55°, and (d) ψ = 40°.

$$\psi = \frac{\beta}{\tan\alpha_1} \tag{25}$$

The radius and radial depth of cut for a point at β along the helical edge thus becomes

$$R(\beta) = R_o \sin(\frac{\beta}{\tan\alpha_1}) \text{ and } dr(\beta) = R_o \sin(\frac{\beta}{\tan\alpha_1}) - c_0 \tag{26}$$

The integration limits β_1 and β_2 are determined by the upper and lower limits of the axial depth, da_1 and da_2 by

$$\beta_1 = \tan\alpha_1 \cos^{-1}(1-\frac{da_1}{R_0}) \text{ and } \beta_2 = \tan\alpha_1 \cos^{-1}(1-\frac{da_2}{R_0}) \tag{27}$$

The major difference between cutters of constant helix lead and constant helix angle is reflected in their $\beta(\psi)$ functions, which can be obtained from Eqs. (20) and (25) as

$$\beta_{cl} = \tan\alpha_0(1-\cos\psi) \text{ and } \beta_{ca} = \psi\tan\alpha_1 \tag{28}$$

where β_{cl} and β_{ca} represent the radial angle of the cutter with constant lead and constant helix angle respectively. β_{cl} and β_{ca} as a function of ψ are plotted in Figure 3 with the assumption of $\alpha_0 = \alpha_1 = 30°$. It is shown that β_{cl} and β_{ca} are almost in parallel to each other at $\psi>0.57$ or equivalently $h/R_0>0.15$. Within this region, β_{cl} and β_{ca} can be related to each other by the following expression,

$$\beta_{cl} \approx \beta_{ca} - (\frac{\pi}{2}-1)\tan\alpha_1 \tag{29}$$

at a given ψ or h position. Furthermore, the helix lead $dh/d\beta$ can be shown to also be the same.

Assuming the cutting region is not confined to the bottom center, as in the case of Figure 2d, the phase difference in Eq. (29) will be reflected in the phase shift of the total milling forces based on the modified convolution theorem as presented in this study. Given the same cutting conditions and cutting constants for both types of cutters, the Fourier coefficients of the milling forces for the two types of ball-end cutters can be shown to have the following relationship:

$$A_{cl}[Nk] = EXP(-jNk(\frac{\pi}{2}-1)\tan\alpha_1 A_{ch}[Nk] \tag{30}$$

There exists a phase difference of $Nk(\pi/2-1)\tan\alpha_1$ for each kth harmonic coefficient while the phase difference is a constant in the angle domain. Figure 4 shows the results of numerical simulation for the two

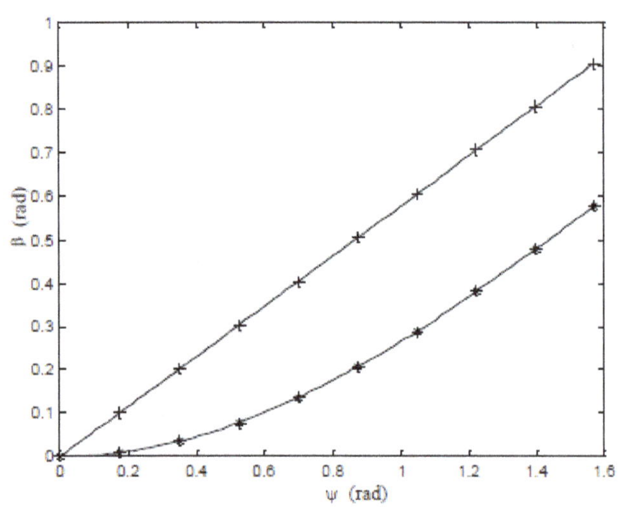

Figure 3: Variation of β with respect to the axial elevation angle ψ for two types of ball-end mills; '+': constant helix angle '*': constant helix lead. $\alpha_0 = \alpha = 30°$.

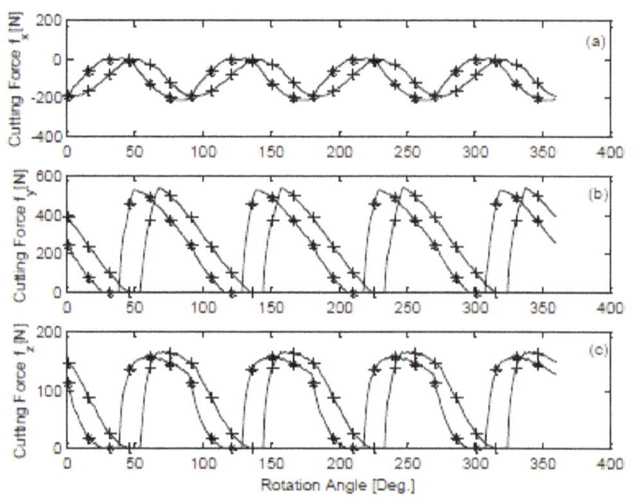

Figure 4: Simulated cutting forces for two types of ball-end mills; '+': constant helix angle and '*': constant helix lead. $\alpha_0 = \alpha = 30°$, $R_o = 8$ mm, $da = 7$ mm, $c_o = 3$ mm, $tx = 0.07$ mm/tooth, $N = 4$, down milling.

types of cutters with $\alpha_0=\alpha_1=30°$ in the shoulder milling configuration as shown in Figure 1d. The phase difference is shown to be about 18°. Figure 1e shows the common step-over milling process for a ball-end mill. For the purpose of finding the milling forces, this milling process can be treated as the combination of a slot milling process and a regular milling process with a constant radial depth of cut. The total milling forces are therefore the sum of forces from these two milling processes:

(a) For $0 \le h \le da_1$, this is a slot milling process with the axial depth, da_1, determined by the ball radius R_0 and the side step dr_0 with

$$da_1 = R_0 - \sqrt{R_0^2 - (\frac{dr_o}{2})^2} \tag{31}$$

Within this cutting region, the entry and exit angles are $\theta1=0$ and $\theta2=\pi$, and the axial cutting range starts from 0 and ends at β_2, which is determined by $da1$ through Eq. (22) or Eq. (27).

(b) For $da_1 \le h \le da_2$, the entry and exit angles are determined from Eq. (2) by the step size, dr_0, and $R(\beta)$, which is derived from Eq. (21) or (26). The end points of the radial engagement range are β_2 and β_3, which correspond to the axial depths da_1 and da_2 according to Eq. (22) or Eq. (27). Geometric functions thus have been derived for cutters of three external profiles and of five different helical geometries. The entry and exit angles of the boundary functions are obtained mostly indirectly by showing the cutter radius and radial depth of cut as a function of β. Fourier coefficients can be readily evaluated for these cutters and the selected cutting configurations.

Except for the case of the cylindrical end mill, a closed form expression for the Fourier coefficients of the frequency domain forces model cannot be derived owing to the presence of transcendental functions, $\psi(\beta)h'(\beta)\mathbf{P}_{iw}(\omega, \beta)$, in the integrand, and numerical integration will be required for their evaluation.

Average Forces for the General and Common Helical Cutters

Among the Fourier coefficients for the milling forces, the average forces are the easiest to obtain. They have also been known to have a strong influence on the stability and dimensional error problems of the milling processes. Analytical expressions for the average forces should facilitate the analysis and understanding of these problems. In this section, general expressions for the average forces for the helical end mills and specific expressions for three types of cutters will be derived and their characteristics discussed.

A general expression for the average forces of a general helical end mill can be written by setting $\omega=0$ in the Fourier coefficient expression of the milling forces such that:

$$\mathrm{A}[0]=\frac{N}{2\pi}\sum_{i=1}^{2}\int_{\beta_1}^{\beta_2}\mathbf{q}_i(\beta)h'(\beta)\mathbf{P}_{i\mathrm{w}}(0,\beta)d\beta \tag{32}$$

Where

$$\mathbf{P}_{1\mathrm{w}}(0,\beta)=\begin{pmatrix}-\frac{1}{4}\cos 2\theta\\ \frac{1}{2}(\theta-\frac{1}{2}\sin 2\theta)\\ -\cos\theta\end{pmatrix}_{\theta=\theta_1(\beta)}^{\theta_2(\beta)} \quad \text{and } \mathbf{P}_{2\mathrm{w}}(0,\beta)=k_{tp}\begin{pmatrix}\sin\theta\\ -\cos\theta\\ \theta\end{pmatrix}_{\theta=\theta_1(\beta)}^{\theta_2(\beta)} \tag{33}$$

Eq. (32) shows that the average forces are proportional to the flute number and are physically the sum of the local average forces contributed by all cutting points along the cutting edge. Eqs. (32) and (33) are applicable for all types of helical cutters. Eq. (33) can be simplified for the common up and down cut configurations. For up milling, $\theta_1=0$ and Eq. (33) can be reduced to

$$\mathbf{P}_{1\mathrm{w}}(0,\beta)=k_{ts}t_x\begin{pmatrix}\frac{1}{4}(1-\cos 2\theta_2)\\ \frac{1}{2}(\theta_2-\frac{1}{2}\sin 2\theta_2)\\ 1-\cos\theta_2\end{pmatrix} \quad \text{and } \mathbf{P}_{2\mathrm{w}}(0,\beta)=k_{ts}t_x\begin{pmatrix}\sin\theta_2\\ 1-\cos\theta_2\\ \theta_2\end{pmatrix} \tag{34}$$

And for down milling, $\theta_2=\pi$, Eq. (33) becomes

$$\mathbf{P}_{1\mathrm{w}}(0,\beta)=k_{ts}t_x\begin{pmatrix}\frac{1}{4}(\cos 2\theta_1-1)\\ \frac{1}{2}(\pi-\theta_1)+\frac{1}{4}\sin\theta_1\\ 1+\cos\theta_1\end{pmatrix} \quad \text{and } \mathbf{P}_{2\mathrm{w}}(0,\beta)=k_{ts}t_x\begin{pmatrix}-\sin\theta_1\\ 1+\cos\theta_1\\ \pi-\theta_1\end{pmatrix} \tag{35}$$

With the geometric functions and the boundary functions derived in the previous sections, the average forces in Eq. (32) can always be obtained by the numerical integration presented in part 1. However, simpler forms of expressions for the average forces can be derived under some special cutting conditions and their evaluation can be simplified.

For a cylindrical end mill as shown in Figure 1a, Eq. (32) can be reduced to the following closed form expression:

$$\mathrm{A}[0]=\frac{Nd_a}{2\pi}\left(\begin{bmatrix}1 & k_{rs} & 0\\ -k_{rs} & 1 & 0\\ 0 & 0 & -k_{rs}\end{bmatrix}\mathbf{P}_{1\mathrm{w}}(0)+\begin{bmatrix}1 & k_{rp} & 0\\ -k_{rp} & 1 & 0\\ 0 & 0 & -k_{rp}\end{bmatrix}\mathbf{P}_{2\mathrm{w}}(0)\right) \tag{36}$$

For the taper end mill with constant helix angle, Eq. (32) is reduced to

$$\mathrm{A}[0]=\frac{N}{2\pi}\frac{R_o}{\tan\alpha_0}\sum_{i=1}^{2}\mathbf{q}_i(\psi_o)\int_0^{\beta_a}\exp(\frac{\cot\psi_o}{N.\tan\alpha_0})\mathbf{P}_{i\mathrm{w}}(0,\beta)d\beta \tag{37}$$

and for the taper end mill with constant helix lead, Eq. (32) becomes

$$\mathrm{A}[0]=\frac{N}{2\pi}\frac{R_o}{\tan\alpha_0}\sum_{i=1}^{2}\mathbf{q}_i(\psi_o)\int_0^{\beta_a}\mathbf{P}_{i\mathrm{w}}(0,\beta)d\beta \tag{38}$$

The evaluation of average forces still requires numerical integration for most cutting configurations except for the following special cutting conditions.

In the special cases of half slot and slot milling conditions a single closed form expression for the average forces can be obtained for all types of helical cutters. For the slot milling process, $\theta_1=0$ and $\theta_2=\pi$, the two vector functions in Eq. (33) are reduced to constant values with

For half slot milling, these two vectors also have constant values. For up cut configuration, they are

$$\mathbf{P}_{1\mathrm{w}}(0)=k_{ts}t_x\begin{pmatrix}P_1(0)\\ P_2(0)\\ P_3(0)\end{pmatrix}=k_{ts}t_x\begin{pmatrix}0.5\\ 0.25\pi\\ 1\end{pmatrix} \tag{39}$$

and

$$\mathbf{P}_{2\mathrm{w}}(0)=k_{tp}\begin{pmatrix}P_4(0)\\ P_5(0)\\ P_6(0)\end{pmatrix}=k_{tp}\begin{pmatrix}1\\ 1\\ 0.5\pi\end{pmatrix} \tag{40}$$

and for down cut,

$$\mathbf{P}_{1\mathrm{w}}(0)=k_{ts}t_x\begin{pmatrix}P_1(0)\\ P_2(0)\\ P_3(0)\end{pmatrix}=k_{ts}t_x\begin{pmatrix}-0.5\\ 0.25\pi\\ 1\end{pmatrix}$$

and

$$\mathbf{P}_{2\mathrm{w}}(0)=k_{tp}\begin{pmatrix}P_4(0)\\ P_5(0)\\ P_6(0)\end{pmatrix}=k_{tp}\begin{pmatrix}-1\\ 1\\ 0.5\pi\end{pmatrix} \tag{41}$$

Since these $\mathbf{P}_i(0)$'s are constant, Eq. (32) becomes an analytically integrable form for the slot and half slot milling conditions:

$$\mathrm{A}[0]=\frac{N}{2\pi}\sum_{i=1}^{2}\left(\int_0^{\beta_a}\mathbf{q}_i(\beta)h'(\beta)d\beta\right)\mathbf{P}_i(0)=\frac{N}{2\pi}\sum_{i=1}^{2}\left(\int_0^{d_a}\mathbf{q}_i(\psi)dh\right)\mathbf{P}_i(0)=\frac{N}{2\pi}\sum_{i=1}^{2}\mathbf{Q}_i\mathbf{P}_i(0) \tag{42}$$

where $\mathbf{Q}_i$ can be shown to be

$$\mathbf{Q}_1=R_o\begin{bmatrix}Q_1 & k_{rs}Q_3+k_{as}Q_2 & 0\\ -(k_{rs}Q_3+k_{as}Q_2) & Q_1 & 0\\ 0 & 0 & k_{rs}Q_2-k_{as}Q_3\end{bmatrix} \tag{43}$$

$$\mathbf{Q}_2=R_o\begin{bmatrix}Q_5 & k_{rp}Q_6+k_{ap}Q_4 & 0\\ -(k_{rp}Q_6+k_{ap}Q_4) & Q_5 & 0\\ 0 & 0 & k_{rp}Q_4-k_{ap}Q_6\end{bmatrix} \tag{44}$$

In which

$$\begin{aligned}
&Q_1=\frac{1}{R_o}\int_0^{d_a}dh=\frac{d_a}{R_o},\ Q_2=\frac{1}{R_o}\int_0^{d_a}\cos\psi\, dh=\frac{1}{R_o}=\frac{1}{R_o}\int_0^{\psi_a}\cos\psi\frac{dh}{d\psi}d\psi\\
&Q_3=\frac{1}{R_o}\int_0^{d_a}\sin\psi\, dh=\frac{1}{R_o}\int_0^{\psi_a}\sin\psi\frac{dh}{d\psi}d\psi,\ Q_4=\frac{1}{R_o}\int_0^{d_a}\cot\psi\, dh=\frac{1}{R_o}\int_0^{\psi_a}\cot\psi\frac{dh}{d\psi}d\psi\\
&Q_5=\frac{1}{R_o}\int_0^{d_a}\csc\psi\, dh=\frac{1}{R_o}\int_0^{\psi_a}\csc\psi\frac{dh}{d\psi}d\psi,\ Q_6=\frac{1}{R_o}\int_0^{d_a}dh=\frac{d_a}{R_o}
\end{aligned} \tag{45}$$

The axial immersion angle, ψ_a, in Eq. (45) for the ball-end cutter can be found directly from Eq. (20) or (25). Values of Q_i's for the square, taper, and ball end mills are listed in Table 1.

Eqs. (42-45) for the average forces in the slot or half slot milling conditions are applicable not only for the three types of cutters discussed here, but also for all helical cutters of different external profile and curvilinear geometry. Assuming the cutting constants are the same, it is therefore shown through these equation that the average forces in slot or half slot milling are independent of their helical lead, $h'(\beta)$, and are only dependent on the external profile of the cutter through function $h(\psi)$ in Eq. (45).

Identification of the Shearing and Ploughing Cutting Constants

For all types of milling force models, both in the numerical or analytical form and in the angle or frequency domain, the accuracy of the force prediction bears on the trueness of the cutting coefficients. However, cutting constants are difficult to predict and are mostly taken or calculated from a pre-established database through elaborative cutting tests. The analytical nature of the presented frequency domain milling force model allows the direct identification of the six unknown cutting constants from the measured milling forces. With the least signal processing and mathematical complexity required as well as considering the simplicity for the test set up, the cutting constants can be identified using the closed-form expression of Eq. (42) for the average cutting forces in the slot or half slot milling operation. Eq. (42) can be rearranged as the following linear equation in the unknown shearing and ploughing cutting constants:

$$\begin{bmatrix} A_x[0] \\ A_y[0] \\ A_z[0] \end{bmatrix} = \begin{bmatrix} A_x[0] \\ A_y[0] \\ A_z[0] \end{bmatrix}_{shearing} + \begin{bmatrix} A_x[0] \\ A_y[0] \\ A_z[0] \end{bmatrix}_{ploughing} = \frac{NR_o t_x}{2\pi}\mathbf{T}_s\mathbf{k}_s + \frac{NR_o}{2\pi}\mathbf{T_p}\mathbf{k_p} \quad (46)$$

$$\text{where } \mathbf{T}_s = \begin{bmatrix} Q_1P_1(0) & Q_3P_2(0) & Q_2P_2(0) \\ Q_1P_2(0) & -Q_3P_1(0) & -Q_2P_1(0) \\ 0 & Q_2P_3(0) & -Q_3P_3(0) \end{bmatrix};\ \mathbf{k}_s = \begin{bmatrix} k_{ts} \\ k_{rs}k_{ts} \\ k_{as}k_{ts} \end{bmatrix} \quad (47)$$

$$\text{and } \mathbf{T_p} = \begin{bmatrix} Q_5P_4(0) & Q_6P_5(0) & Q_4P_5(0) \\ Q_5P_5(0) & -Q_6P_4(0) & -Q_4P_4(0) \\ 0 & Q_4P_6(0) & -Q_6P_6(0) \end{bmatrix};\ \mathbf{k_p} = \begin{bmatrix} k_{ts} \\ k_{rp}k_{tp} \\ k_{ap}k_{tp} \end{bmatrix} \quad (48)$$

Three equations for the three average force components can be obtained from each cutting test. Therefore, two sets of measured average cutting forces data with different cutting conditions are required to make up the system of equations in solving the six cutting constants. By regrouping the matrix equation in (46), a closed form formula for the identification of the six cutting constants can be written for a general helical end mill as follows:

$$\begin{bmatrix} k_{ts} \\ k_{rs}k_{ts} \\ k_{as}k_{ts} \\ k_{tp} \end{bmatrix} = \frac{2\pi}{NR_o}\mathbf{T}^{-1} \begin{bmatrix} A_{x1}[0] \\ A_{y1}[0] \\ A_{z1}[0] \\ A_{x2}[0] \\ A_{y2}[0] \\ A_{z2}[0] \end{bmatrix} \text{ where } \mathbf{T} = \begin{bmatrix} t_{x1}\mathbf{T}_{s1} & \mathbf{T_{p1}} \\ t_{x2}\mathbf{T}_{s2} & \mathbf{T_{p1}} \end{bmatrix} \quad (49)$$

and subscripts '1', '2' indicate the two different cutting tests.

Both the cutting tests and the computations can be further simplified if the same cutting conditions, except the feed speed are chosen for these two slot or half slot milling processes. In that case, the **T** matrix in Eq. (49) becomes

$$\mathbf{T} = \begin{bmatrix} t_{x1}\mathbf{T}_s & \mathbf{T_p} \\ t_{x2}\mathbf{T}_s & \mathbf{T_p} \end{bmatrix} \quad (50)$$

Eq. (49) or Eq. (50) has provided a convenient formula for the identification of six shearing and ploughing cutting constants for all types of helical cutters through two cutting tests with the slot or half slot milling condition.

Experimental Validation

Milling experiments were carried out to verify the frequency domain force model and the identification formula for the cutting constants. The cutting forces were measured with the Kistler 9255B dynamometer. Three different work/cutter pairs are used: a tapered end mill with AL7075-T6, a ball-end mill with AL2024-T4 and a cylindrical end mill with AL2024-T4. Typical yield and tensile strengths are 503 MPa and 572 MPa respectively for 7075-T6, and 325 MPa and 470 MPa for AL2024-T4. With average uncut chip thickness ranging from 0.02 to 0.1 mm, five sets of cutting constants are identified from 10 slot milling tests for each type of cutter and are shown with respect to the

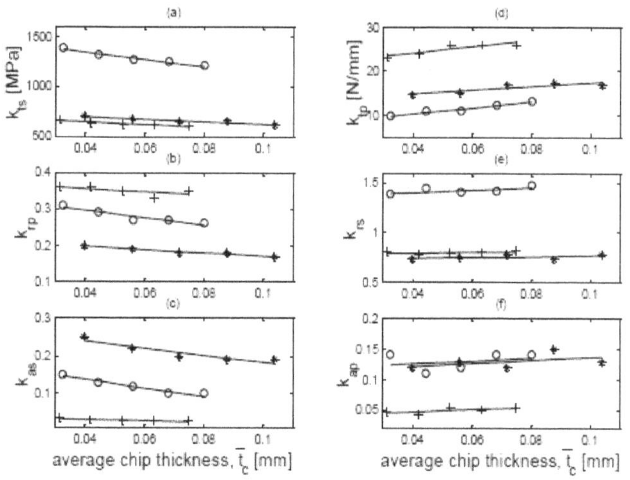

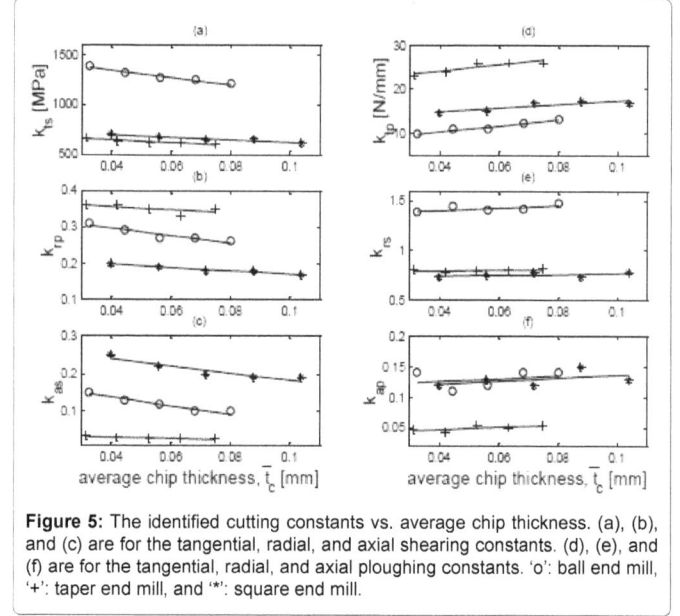

Figure 5: The identified cutting constants vs. average chip thickness. (a), (b), and (c) are for the tangential, radial, and axial shearing constants. (d), (e), and (f) are for the tangential, radial, and axial ploughing constants. 'o': ball end mill, '+': taper end mill, and '*': square end mill.

Cutter type	Coefficients due to shearing			Coefficients due to ploughing		
	$Q_1(0)$	$Q_2(0)$	$Q_3(0)$	$Q_4(0)$	$Q_5(0)$	$Q_6(0)$
Cylindrical end mill	$\bar{d}_a$	0	$\bar{d}_a$	0	$\bar{d}_a$	$\bar{d}_a$
Taper end mill	$\bar{d}_a$	$\bar{d}_a \cos\psi_0$	$\bar{d}_a \sin\psi_0$	$\bar{d}_a \cot\psi_0$	$\bar{d}_a \csc\psi_0$	$\bar{d}_a$
Ball end mill	$\bar{d}_a$	$\frac{\sin^2\psi_a}{2}$	$\frac{1}{2}(\psi_a - \frac{\sin 2\psi_a}{2})$	$\sin\psi_a$	$\bar{d}_a$	$\bar{d}_a$

Table 1: Directional coefficients for the average cutting forces in slot milling $\bar{d}_a = d_a / R_0$.

No.	da(mm)	tx1(mm/tooth)	tx2 (mm/tooth)	t (mm)
1	4	0.0563	0.0688	0.0398
2	4	0.0812	0.0928	0.0557
3	4	0.106	0.119	0.0716
4	4	0.131	0.144	0.0875
5	4	0.156	0.169	0.1035
6	6	0.0469	0.0563	0.0327
7	6	0.0656	0.075	0.0446
8	6	0.0844	0.0938	0.0565
9	6	0.103	0.113	0.0684
10	6	0.122	0.131	0.0803
11	3	0.0668	0.0843	0.0315
12	3	0.0938	0.106	0.0421
13	3	0.119	0.131	0.0526
14	3	0.144	0.156	0.0631
15	3	0.169	0.188	0.0749

Table 2: Cutting conditions for the identification of cutting constants. Spindle speed = 400 rpm, dry slot cut. No.1-5 are the cylindrical end mill with N = 2, R = 5 mm, α = 30°; No.6-10 are for the taper end mil with N = 4, R_o = 5 mm, α = 30°, ψ_o = 85°; No.10-15 are for the ball end mill with N = 2, R_o = 5 mm, α_o = 30°.

Cutter / work material	Shearing constants			Ploughing constants		
	kts (MPa)	*krs*	*kas*	*ktp* (N/mm)	*krp*	*kap*
Taper end mill/ AL7075-T6	1285	0.28	0.12	11.3	1.42	0.13
Cylindrical end mill/ AL2024-T4	660	0.18	0.21	16.1	0.76	0.13
Ball end mill/ AL2024-T4	631	0.35	0.03	25	0.8	0.05

Table 3: Averages of identified cutting constants.

average chip thickness in Figure 5. The three shearing related constants in Figure 5a-c for each work/material pair are shown to increase slightly with decreasing chip thickness, which could be possibly explained by the size effect. The power related tangential shearing constant *kts* is the most significant contributor of all to the milling forces. The tangential shearing constant of the AL7075-T6 with the taper end mill is almost twice that of the AL2024-T4 with the ball and cylindrical end mills. The greater difference in the shearing cutting constant seems to reflect more than the mechanical strength of the work material and could be

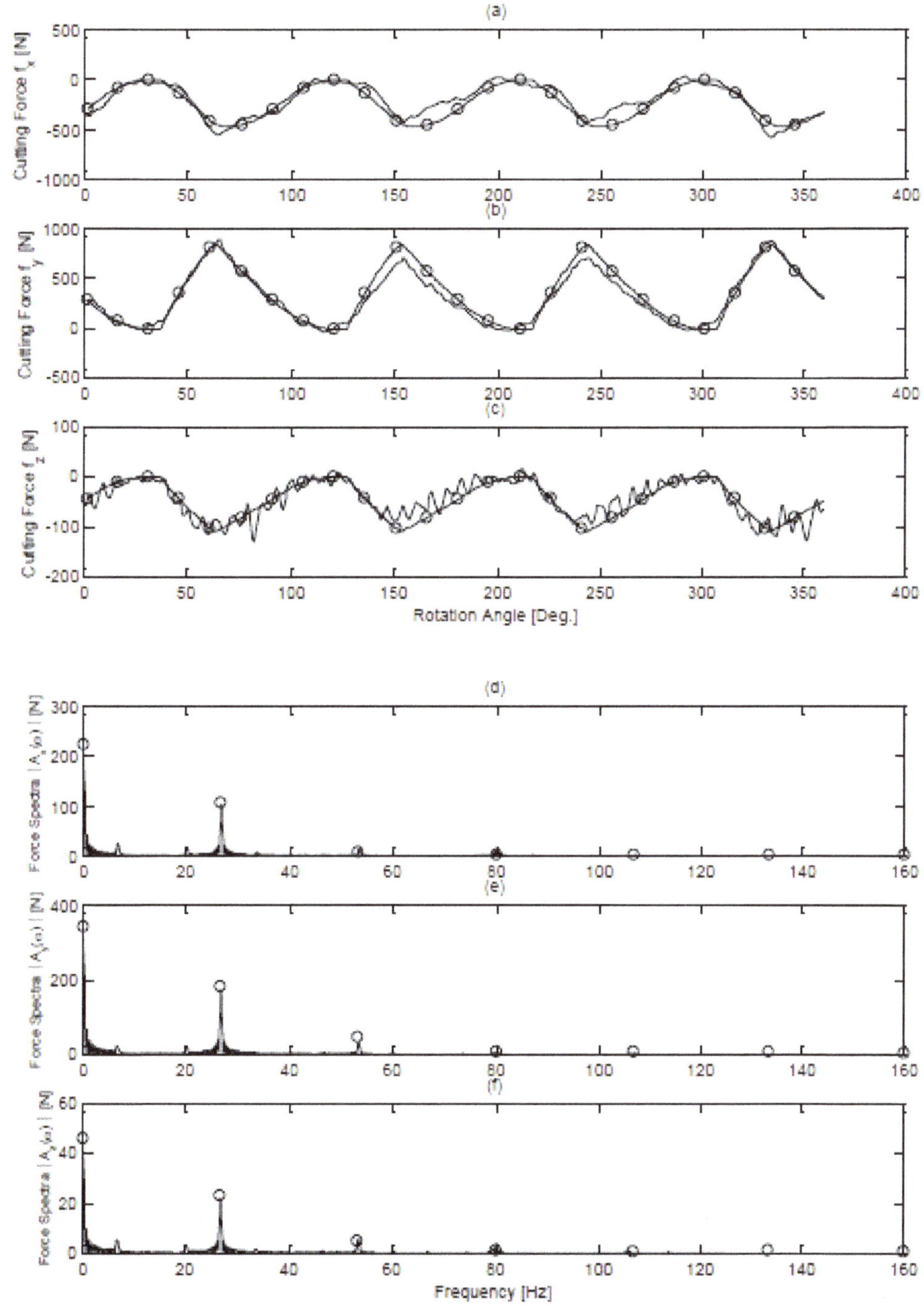

Figure 6: Predicted and measured forces for a taper end mill. Work material: AL7075-T6. $N = 4$, $R_0 = 5$ mm, $\psi_o = 85°$, $\alpha = 30°$, $da = 8$ mm, $dr_0 = 4$ mm, $tx = 0.125$ mm/tooth, 400 rpm, dry cut. '-': measured, 'o': predicted.

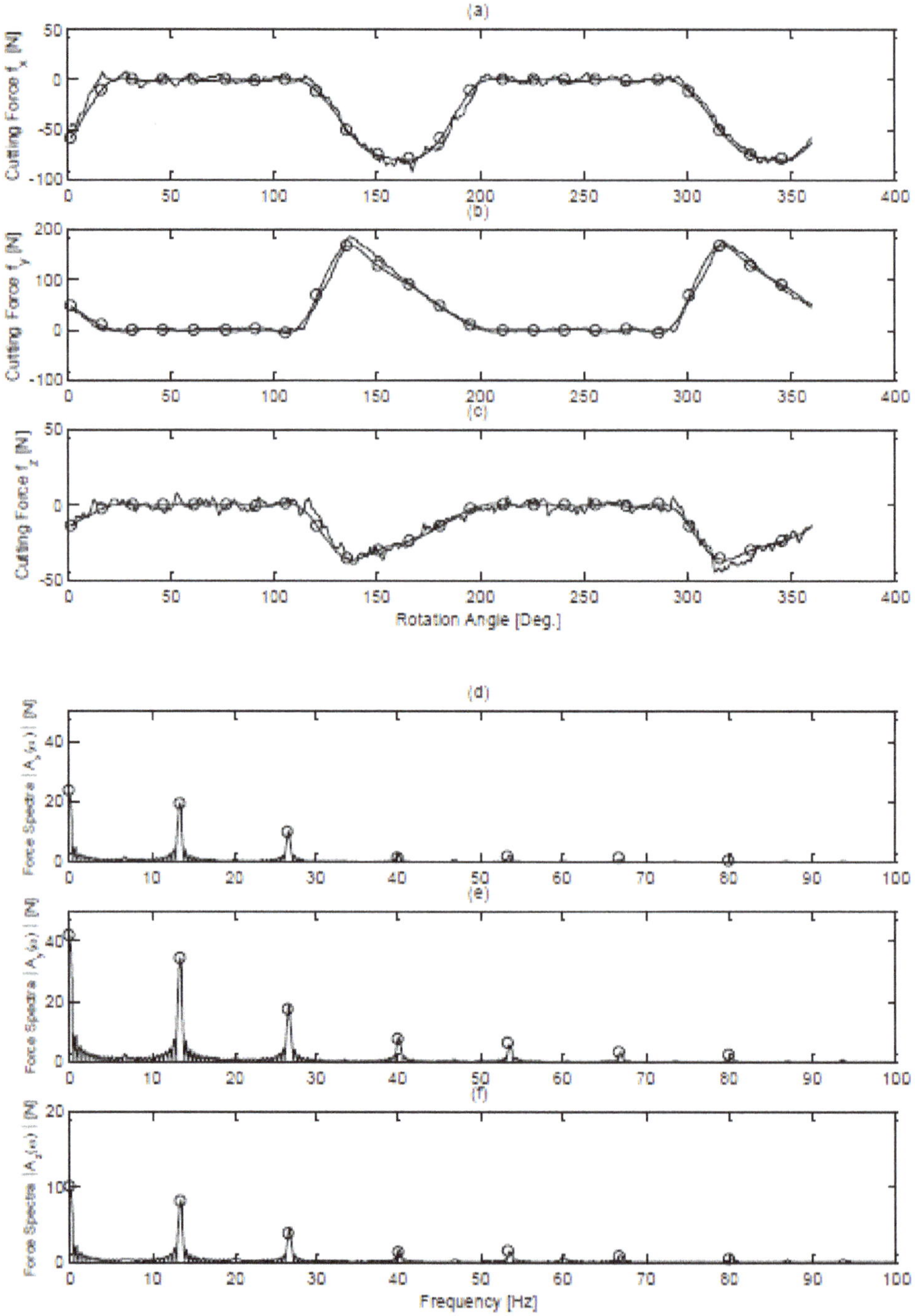

Figure 7: Predicted and measured forces for a cylindrical end mill. Work material: AL2024-T4. N = 2, Ro = 5 mm, da = 3 mm, dr = 3 mm, tx = 0.075 mm/tooth, N = 4 flutes, α = 30°, 400 rpm, dry cut. '-': measured, 'o': predicted.

partly attributed to the worn tool edge of the taper end mill. The ball and square end mills are both new cutters while the taper end mill is in a used condition. All three ploughing constants for three cutters also remain relatively flat, even showing some reverse trend, Figure 5d-f. This slight decrease of ploughing constants with decreasing chip thickness might be explained by the smaller tool edge deformation thus less flank contact area due to the smaller chip load. Compared with the shearing constant, the tangential ploughing constant has a relative small value. However, the ploughing force will be the same as the shearing force for chip thickness in the range of 0.01 to 0.03mm, which is well within the range encountered in finish milling. As the feed per tooth and the radial depth of cut get smaller such as in high speed milling, the ploughing force will become more significant and its stronger presence should warrant closer examination of its effect on the milling process.

From these identified cutting constants, it is reasonable to use the average cutting constants for cutting force predictions without

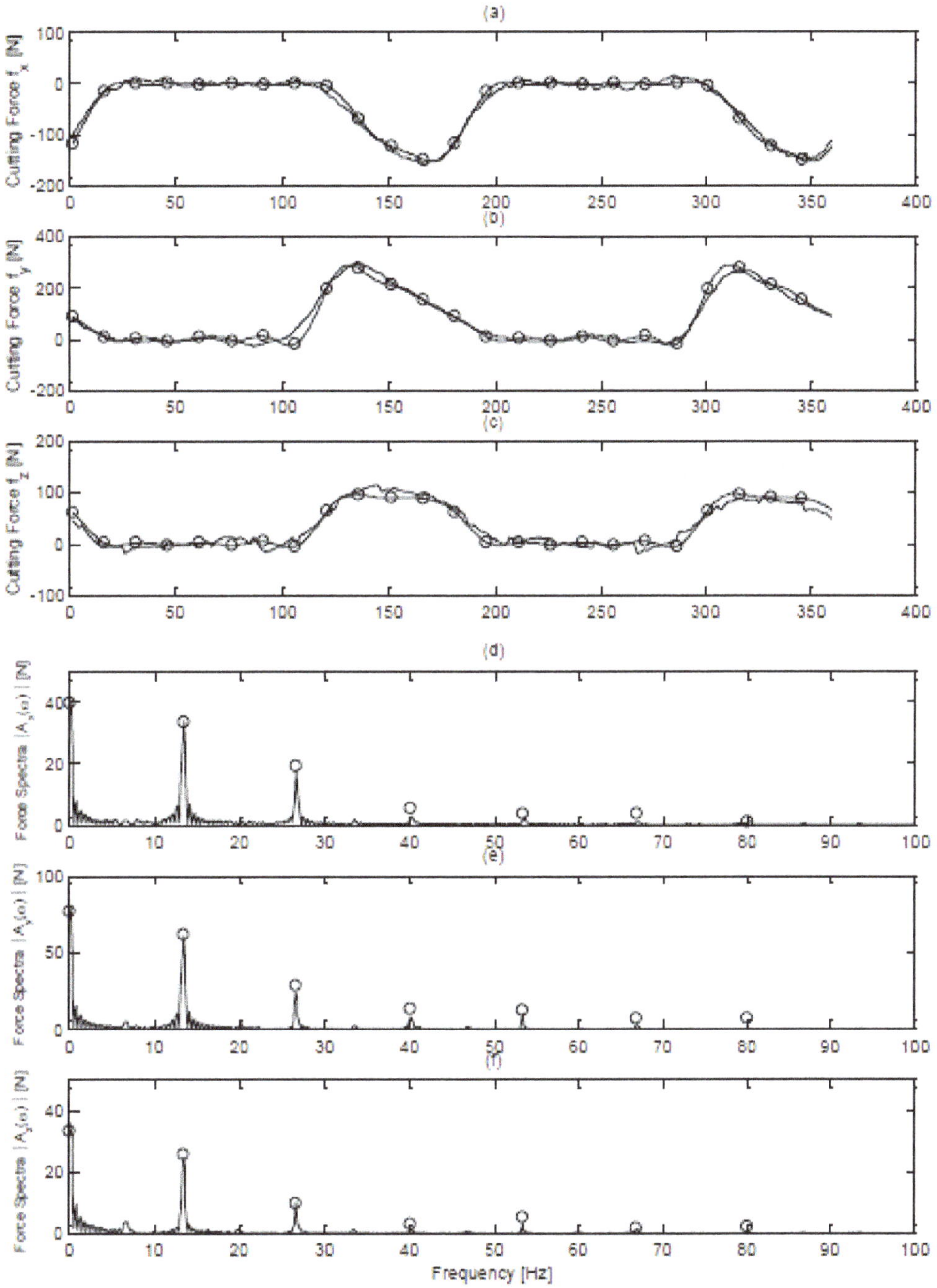

Figure 8: Predicted and measured forces for a ball end mill. Work material: AL2024-T4. N = 2, R_0 = 5 mm, da = 3 mm, c_0 = 1 mm, tx = 0.1 mm/tooth, α_0 = 30°, 400 rpm, dry cut. '-': measured, 'o': predicted.

incurring significant error. The average cutting constants for different cutters are listed in Tables 2 and 3 and are used to predict the milling forces for additional cutting tests with different cutting conditions and configurations. The predicted cutting forces are calculated by first finding the Fourier coefficients up to the sixth harmonics by numerical integration with QUAD8 function in Matlab. The angle domain forces are subsequently obtained through Fourier series expansion using inverse Fourier transform function in Matlab. The predicted and experimentally measured milling forces are shown to coincide very well in both the angular and frequency domain as illustrated in Figure 6-8. The Frequency spectra of these milling forces are shown to decrease rapidly as predicted and assume significant magnitude only up to the third harmonics. Sine the Fourier coefficients $\mathbf{A}[Nk]$ and $\mathbf{A}[-Nk]$ are complex conjugates, only four numerical values for k from 0 to 3 for each force component would be adequate in representing the periodic milling forces.

The forward application of the force model in predicting forces and the inverse application for the identification of cutting constants thus proves the validity of the presented frequency domain force model for a generalized helical end mill. Not only are the composing structures and the derived force expressions verified, the basic assumption of constant cutting coefficients with the dual cutting mechanisms for the local forces are also found to be acceptable for the milling process. The fact that these cutting coefficients can be treated as constants within a reasonable range of chip thickness is of great significance in validating the presented model as a linear force model, thus facilitating its further application in the analysis of milling processes.

Conclusions

Specific applications of the frequency domain force model are illustrated for three common cutters. Geometric functions for the cutter profile, helix lead and entry/exit angles are derived for the square, taper and ball end mills so that numerical values of the force spectra can be calculated for these cutters. For both taper and ball-end cutters, geometric functions for the cutters with a constant helix lead and constant helix angle are separately derived. Assuming the cutting constants are the same, the forces for these two types of taper end mills are shown to have little difference when the axial depth of cut is smaller than the bottom radius. The axial depths of cut can be selected to reduce the force pulsation for the square end mill in peripheral milling and taper end mill in slot and half slot milling. The chosen depth of cut for taper end mill with a constant helix angle is larger than that with a constant helix lead under the same bottom radius when $N < 8$ and $\psi > 40^o$.

For the ball-end mills, different helical geometry is shown to only result in a predictable phase difference in the milling forces when the range of axial depth of the cut is not constrained within the bottom two tenths of the ball radius.

A general expression for the average forces of common cutting configurations has been derived for all types of helical cutters. The average forces for slot and half slot milling conditions are presented in a simple closed-form expression and are shown to only be dependent on the external profile of the helical cutter. From this expression for the average forces, an inverse application of the analytical force model is demonstrated to identify the unknown cutting constants from the measured average forces. Milling experiments for each type of cutter were carried out to verify the predictive accuracy of the frequency domain force model and the effectiveness of the identification formula for the six cutting constants.

Based on its analytical nature, this convolution force model can be further explored for other potential applications including the monitoring and identification of other process parameters and using the model as a tool in the design of special helical cutter for the desired force characteristics.

Acknowledgement

The authors gratefully acknowledge the financial support from Quanzhou Institute of Information Engineering.

References

1. Martellotti ME (1941) An Analysis of the Milling Process. Transaction of ASME 63: 677-700.
2. Martellotti ME (1945) An Analysis of the Milling Process Part 2: Down Milling. Transaction of ASME 67: 233-251.
3. Yang M, Park H (1991) The Prediction of Cutting Force in Ball End Milling. International Journal of Machine Tools & Manufacture 31: 45-54.
4. Feng H, Menq C (1994) The prediction of cutting force in ball-end milling process-Part I: Model formulation and model building procedure. Int J of Machine Tool Design and Manufacture 34: 697-710.
5. Feng H, Menq C (1994) The prediction of cutting force in ball-end milling process-Part II: Cut Geometry Analysis and Model Verification. International Journal of Machine Tools & Manufacture 34: 711-719.
6. Yucesan G, Altintas Y (1996) Prediction of ball end milling forces. ASME Journal of Engineering for Industry 118: 95-103.
7. Imani BM, Sadeghi MH, Elbestawi MA (1998) An improved process simulation system for ball-end milling of sculptured surfaces. International Journal of Machine Tools & Manufacture 38: 1089-1107.
8. Sonawane HA, Joshi SS (2010) Analytical modeling of chip geometry and cutting forces in helical ball end milling of superalloy Inconel 718. CIRP Journal of Manufacturing Science and Technology 3: 204-217.
9. Ramaraj TC, Eleftheriou E (1994) Analysis of the mechanics of machining with tapered end-milling cutters. ASME Journal of Engineering for Industry 116: 398-404.
10. Huang T, Whitehouse DJ (1999) Cutting force formulation of taper end-mills using differential geometry. Precision Engineering 23: 196-203.
11. Altintas Y, Lee P (1996)A General Mechanics and Dynamics Model for Helical End Mills. Annals of CIRP 45: 59-64.
12. Engin S, Altintas Y (2001) Mechanics and dynamics of general milling cutters Part I: helical end mills. International Journal of Machine Tools & Manufacture 41: 2195-2212.
13. Wang JJ, Zheng CM (2002) An analytical force model with shearing and ploughing mechanisms for end milling. International Journal of Machine Tools & Manufacture 42: 761-771.

15

Three-Dimensional Analysis of Mixed Convection in a Differentially Heated Lid-Driven Cubic Enclosure

Nasreddine Benkacem*, Nader Ben Cheikh and Brahim Ben Beya

Department of Physics, Faculty of sciences, Tunis ELMANAR University, Campus Universities, 2092 El-Manar II, Tunis, Tunisia

Abstract

To study the intlicate three-dimensional flow structures and the companion heat transfer rates in a differentially heated lid-driven cubic cavity, a numerical methodology based on the finite volume method and a full multigrid acceleration is utilized in this note. The four remaining walls fowling the cubic cavity are adiabatic. The working fluid is air so the Prandtl number equates to 031. Numerical solutions are generated for representative combinations of the controlling Reynolds number inside 100<Re<1000 and the Richardson number inside 0.001<Ri<10. Typical sets of streamlines and isotherms are presented to analyze the tortuous circulatory flow patterns set up by the competition between the forced flow created by the moving wall and the buoyancy force of the fluid. Correlations between the average Nusselt number through the cold wall and the Richardson number were established for the mentioned Reynolds numbers.

Keywords: Three-dimensional analysis; Lid-driven cubic cavity; Mixed convection; Numerical simulation; Multigrid method

Introduction

The problem on laminar mixed convection in cavities has multiple applications in the field of thermal engineering. Such problems are of great interest, for example in electronic device cooling, high-performance building insulation, multi shield structures used for nuclear reactors, food processing, glass production, solar power collector, etc. Numerous studies on lid-driven cavity flow and heat transfer involving different cavity configurations, various fluids and imposed temperature gradients have been continually published in the literature.

The numerical simulations of Moallemi and Jang [1] focused on two-dimensional laminar flow induced by Reynolds number $100 \le Re \le 1000$, and small-to-moderate Prandtl number $0.01 \le Pr \le 50$ on the flow and heat transfer features in a cavity for different levels of the Richardson numbers. These authors found that the influence of buoyancy on the flow and heat transfer are to be more pronounced for higher values of Pr, if Re and Gr are kept constant.

Sharif [2] performed a numerical investigation with flow visualization of laminar mixed convective heat transfer in two-dimensional shallow rectangular driven cavities of aspect ratio 10. The top moving plate of the cavity is set at a higher temperature than the bottom stationary plate. The fluid Prandtl number is taken as 6, representative of water. The effects of inclination of such a cavity on the flow and thermal fields were also investigated for inclination angles ranging from 0° to 30°. It was concluded that the average or overall Nusselt number increases mildly with cavity inclination for the dominant forced convection case dictated by Ri=0.1. In contrast, it increases much more rapidly with inclination for the other dominant natural convection case dictated by Ri=100.

Prasad et al. [3] numerically studied mixed convection inside a rectangular cavity where the two vertical walls are maintained at cold temperature. In one case, the top-moving wall is maintained at hot temperature and the bottom is at a cold temperature and in the other case, the top is at a cold temperature and the bottom is at a hot temperature. They concluded that when the negative is increased, a strong convection is manifested for aspect ratios equal to 0.5 and 1.0. Even more, a Hopf bifurcation occurs at for the aspect ratio 2.

Mohammad and Viskanta [4] numerically examined two and three-dimensional laminar flow and heat transfer in a Rayleigh-Bénard container. They established that the lid motion annihilates all forms of convective cells due to heating from below for finite size cavities. Aydin et al. [5] conducted a numerical investigation to analyze the transport mechanism of mixed convection in a shear and buoyancy-driven cavity having a locally heated lower wall and moving cooled sidewalls. In addition, other numerical studies such as Han and Kuehn [6] and Oztop and Dagtekin [7] were carried out on this topic.

Iwatsu et al. [8] performed a numerical investigation on the effect of external excitation on the flow structure in a square cavity. The results have shown similar flow structure to steady driven-cavity flows when utilizing small frequency values. Such a similarity, however, vanished when large frequency values were implemented. A subsequent work by Iwatsu et al. [9] carried out a numerical study of the viscous flow in a heated driven-cavity under thermal stratification, where the oscillating lid was maintained at a temperature higher than the lower wall. Their collection of results had revealed significant augmentation in heat transfer rate at particular lid frequency values, which convincingly indicates the existence of the resonance phenomena.

A detailed literature survey reveals that the majority of existing numerical investigations are restricted to two dimensional configurations. In this vein, 2D models are deficient because they do not always realistically capture the intricacies inherent to the flow behaviour. Because of these shortcomings, 3D models have to be undertaken to guarantee accuracy. A limited number of articles falls into this general category and has been reported in the literature.

***Corresponding author:** Nasreddine Benkacem, Department of Physics, Faculty of sciences, Tunis ELMANAR University, Campus Universities, 2092 El-Manar II, Tunis, Tunisia, E-mail: nasreddine.benkacem@gmail.com

Among others, Iwatsu [10] numerically studied three dimensional mixed convective flows in a cubical container with a steady vertical temperature stratification. He observed that the three dimensional effects are intensified as Re increases. Mohammad and Viskanta [11] conducted three-dimensional numerical simulation of mixed convection in a shallow driven cavity heated from the top moving wall and cooled from below. The cavity was filled with a stably stratified fluid encompassing a relative large range of Rayleigh and Richardson numbers. In a consecutive number of papers, Freitas et al. [12] and Freitas and Street [13] carried out a numerical study of the viscous flow in a rectangular cavity of depth-to-spanwise aspect ratio 3 at. They discovered the existence of meridional vortices and considerable flow unsteadiness.

In view of the foregoing statements, it seems that the problem of three dimensional laminar mixed convection heat transfers in a differentially heated lid-driven cubic cavity has not been addressed yet. In this paper, we undertake this task varying the Reynolds number in the Re-interval and the Richardson number in the Ri-interval for air (Pr=0.71) as the working fluid. The transport processes will be investigated with the finite volume method and the discussion will revolve around the precise determination of steady velocity and temperature fields. In addition, the average Nusselt number will be documented for all cases studied.

The paper is organized as follows: in the second section the physical system is formulated; the numerical methodology is briefly described in the third section and subsequently validated. The computed results are presented and discussed in the fourth section. In the final section, the most important findings of this study are summarized.

Physical System

The physical system under study is sketched in Figure 1. It basically consists of a cubic cavity with side filled with air. The applicable flow and temperature boundary conditions are described next. The top lid imparts a steady sliding motion with a uniform velocity, while the other walls are stationary. The cavity is differentially heated over the vertical sides. The left hot wall has a temperature and the right cold wall has a temperature where in. In addition, the remaining walls are considered adiabatic.

Numerical Methodology and Algorithm Validation

The governing equations for unsteady, incompressible laminar flow consist of the continuity equation, the Navier-Stokes equations accounting for the Boussinesq approximation and the energy equation. The non-dimensional equations are collectively written in tensor notation as follows:

Continuity equation:

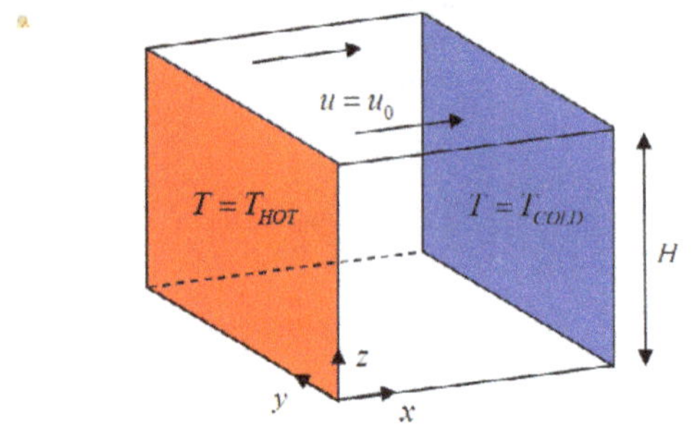

Figure 1: Physical system and Cartesian coordinates.

$$\frac{\partial u_i}{\partial x_i} = 0, \tag{1}$$

Three Momentum equations:

$$\frac{\partial u_i}{\partial t} + \frac{\partial (u_i u_j)}{\partial x_j} = -\frac{\partial p}{\partial x_i} + \frac{1}{\mathrm{Re}}\left(\frac{\partial^2 u_i}{\partial x_i \partial x_i}\right) + Ri\theta\delta_{i3}, \tag{2}$$

Energy equation:

$$\frac{\partial \theta}{\partial t} + \frac{\partial (u_i \theta)}{\partial x_i} = \frac{1}{Ra\Pr}\left(\frac{\partial^2 \theta}{\partial x_i \partial x_i}\right) \tag{3}$$

Where u_i = (u; v; w) the velocity components, pare is the kinematic pressure, and θ is the temperature, ρ is the mass density, and g is the gravitational acceleration. In Eq. (2), the symbol δ_{ij} stands for the Krönecker delta. The chosen scales in Equations (1)-(3) are the side H, the velocity $u_0 = \sqrt{g\beta H \Delta T}$, the time t_0=H/u_0 and the pressure $p_0 = \rho u_0^2$. Further, the non-dimensional temperature is defined by β $\theta = \frac{T - T_R}{T_{HOT} - T_{COLD}}$, where the reference temperature $T_R = \frac{T_{HOT} + T_{COLD}}{2}$ and the temperature scale is the lid-to-lid temperature difference $T_{HOT} - T_{COLD}$.

As presented above, the forced-natural convection problem is characterized by three non-dimensional parameters:

1) the Reynolds number $\mathrm{Re} = \frac{u_0 H}{\nu}$ where u_o is the impressed lid velocity;

2) The Prandtl number $\Pr = \frac{\nu}{\alpha}$, where ν is the kinematic viscosity, α the thermal diffusivity of the fluid;

3) The Grashof number $Gr = \frac{\beta g \Delta T H^3}{\nu^2}$ in which β is the coefficient of thermal expansion of the fluid, g the gravity and $\Delta T = T_{HOT} - T_{COLD}$ the temperature difference between the hot and cold horizontal walls. Alternatively, *Gr* and *Re* are adequately blended in the mixed-convection parameter called the Richardson number $Ri = \frac{Gr}{\mathrm{Re}^2}$.

The unsteady Navier–Stokes and energy equations are discretized by a second-order time stepping finite difference procedure. The procedure adopted here deserves a detailed explanation. First, the non-linear terms in Eqs. (2) are treated explicitly with a second-order Adams–Bashforth scheme. Second, the convective terms in Eq. (3) are treated semi-implicitly. Third, the diffusion terms in Eqs. (2) and (3) are treated implicitly. In order to avoid the difficulty that the strong velocity-pressure coupling brings forward, we selected a projection method described in Peyret and Taylor [14] and Achdou and Guermond [15].

A finite-volume method is implemented to discretize the Navier–Stokes and energy equations (Patankar [16], F. Moukhalled and M. Darwish [17], Kobayachi and Pereira and Pereira [18]). The advective terms in Eqs. (2) are discretized using a QUICK third-order scheme whereas a second-order central differencing (Hayase, Humphrey and Greif [19]) is applied in Eq. (3). The discretized momentum and energy equations are solved employing the red and black successive over relaxation method (RBSOR) in Press et al. [20], while the Poisson pressure correction equation is solved utilizing a full multi-grid method (Hortmann et al. [21], Mesquita and de Lemos [22], Nobile [23]). If specific details about the computational methodology are needed, the reader is directed to Ben Cheikh et al. [24]. Finally, the convergence of the numerical 3D velocity field and the 3D temperature field is established at each time step when all residuals are forced to stay below 10^{-6}. To secure steady state conditions the following criterion has to be satisfied:

$$\sum_{i,j,k}\left|\Phi_{i,j,k}^{m+1}-\Phi_{i,j,k}^{m}\right|\leq 10^{-5}$$

Where the generic variable Φ represents the set of four variables (u, v, w) or θ. In the above inequality, the superscript *m* indicates the iteration number and the subscript sequence (i, j, k) represents the space coordinates *x*, *y*, *z*.

For enhanced accuracy, the present numerical model was checked against the published numerical solution of Tric [25]. The outcomes of the one-to-one comparisons are documented in Table 1 for the average Nusselt number predictions and maximum velocities. It is observed here that the present numerical computations match very closely those of [25].

A second comparison to those of Iwatsu [10] relatively to a 3D mixed convection was undertaken. As shown in Table 2, good agreements are evident with respect to the result reported by [10].

Results and Discussion

The computed mixed convection flow and temperature fields in the lid-driven cubic cavity are examined in this section. The numerical results are presented in terms of streamlines and isotherms. The Reynolds number Re is varied two orders of magnitude between 100 and 1000. In addition, the Richardson number Ri is varied four orders of magnitude between 0.001 and 10. The Prandtl number is set at Pr=0.71. We ran computations for nine different pairs of Ri and Re; that is: (Ri, Re)=(10, 100), (10, 400), (10, 1000), (1, 100), (1, 400), (1, 1000), (0.001, 100), (0.001, 400) and (0.001, 1000). In harmony with this, the implications of varying Ri and Re will be adequately highlighted.

A series of trial calculation were conducted with two different variable grid distributions, i.e., 48×48×48 and 64×64×64. For the moderate case dealing with Re=400 and Ri=1.0, minor differences of less than 0.25% were detected between the flow and temperature results produced by the grid 48×48×48 and those by the grid 64×64×64. Consequently, to optimize the grid distribution appropriately, the grid 48×48×48 was deemed adequate to perform all numerical computations. For completeness, the two grids were built using a tangent hyperbolic formulation. The smallest space intervals chosen in the three coordinate directions are $\Delta x_{\min}=\Delta y_{\min}=\Delta z_{\min}=2.25\times10^{-3}$, and are localised near the moving and stationary walls to capture the growth of the flow and thermal boundary layers adjacent to them. The time step was set to Δt=0.01 for all computations.

		$RRRR$=1111^{44}			$RRRR$= 1111^{55}	
	Tric [25]	Pres. Work	Err %	Tric [25]	Pres. Work	Err %
Grille	813	48^3		813	483	
$uu_{mm}RR_{mm}$	16.719	16.634	-0.51	43.90	44.06	0.36
$vv_{mm}RR_{mm}$	2.156	2.136	-0.93	9.69	9.55	-0.15
ww_{mRRmm}	18.983	18.942	-0.22	71.06	70.85	-0.30
$NNuum_{mmm}$	2.250	2.247	-0.13	4.612	4.605	-0.15
$NNuu_{3333}$	2.054	2.054	0	4.337	4.332	-0.12

Table 1: Comparison of the computed average Nusselt number predictions and maximum velocities.

Re	*Ri*=0.001		*Ri*=1.0		*Ri*=10.0	
	Ref [10]	Pres. Work	Ref [10]	Pres. Work	Ref [10]	Pres. Work
100	1.82	1.836	1.33	1.348	1.08	1.092
400	3.99	3.964	1.50	1.528	1.17	1.130
1000	7.03	7.284	1.80	1.856	1.37	1.143

Table 2: Comparison of our results with [10].

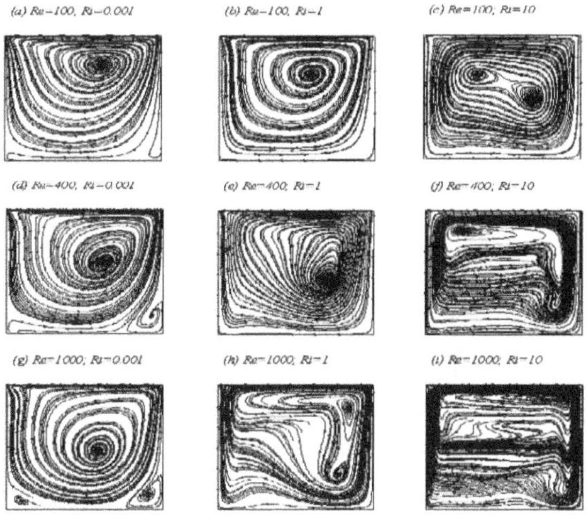

Figure 2: Trajectory particles at the *x-z* mid-plane for different combinations of *Re* and *Ri*.

The mid-plane streamlines distributions for designated values of Re and Ri are displayed in Figure 2. We note that for the lowest Richardson number employed (Ri=0.001), the trajectory of fluid particles is very similar to that corresponding to the classical lid-driven cavity [26] (Figure 2a-2d-2g). Indeed, Figure 2a shows the flow structure in the cavity at Re=100 with a primary vortex occupying the main part of the cavity. Two small recirculation cells are also emerging at the bottom corners as the Reynolds number goes through Re=400 to Re=1000.

When Ri is large (Ri=10), it is noticeable in (Re=100) the presence of two eddies localised in the proximity of the core region. With increments in Re, the right cell becomes feeble and amalgamates with the left one to provide only one stretched vortex. Interestingly, it is also noticed when Re=1000, that the direction of the lid velocity causes the centre of the vortex to move from the left side to the right side as confirmed by Figure 2i.

The case (Ri=1; Re=100) is very similar to (Ri=0.001; Re=100). In fact a primary cell is observed in the cavity with a little difference that its center is slightly moved downward. It is conspicuous in Figure 2e the effect of increasing the Reynolds number (Re=400) on the flow structure. The main vortex moves down and is somewhat dragged to the right side of the cold wall. For Re=1000, the high lid velocity causes the division of the main vortex in two cells (see Figure 2h).

The qualitative features of the temperature field are demonstrated by plotting the perspective views of isotherms, as reflected in Figure 3. In fact, it is clearly discernible from the patterns of isotherms that, for the feeble value of Richardson number (Ri=0.001), the mechanically driven forced convection dominates the buoyancy-driven convection (Figure 3a, 3d and 3g), implying that the forced convection is essentially due to the lid-movement. In contrast, as Ri increases to Ri=10 the buoyant

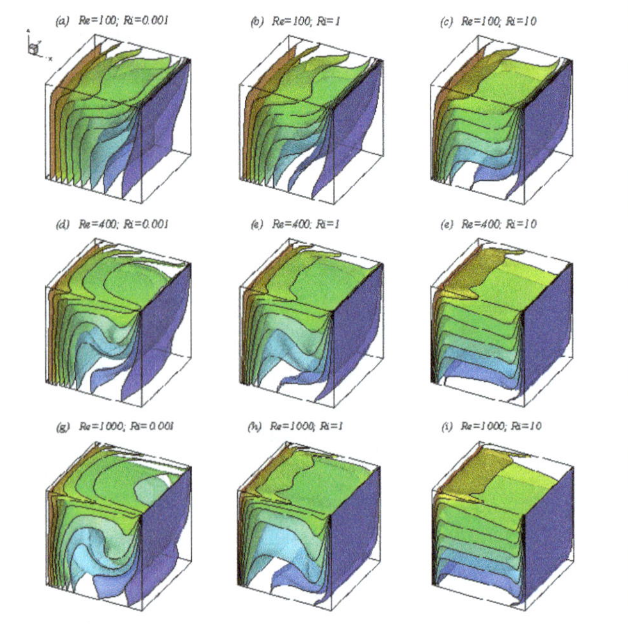

Figure 3: Isotherm plots for different combinations of *Re* and *Ri*.

Re	*Ri*=0.001	*Ri*=1.0	*Ri*=10.0
	(*NNNNh oooo*; *NNNNccoocccc*)	(*NNNNh oooo*; *NNNNccoocccc*)	(*NNNNh oooo*; *NNNNccoocccc*)
100	(2.1714; 2.1714)	(2.6876; 2.6876)	(4.3186; 4.3187)
400	(4.3276; 4.3285)	(5.7232; 5.7241)	(9.6408; 9.6420)
1000	(6.6252; 6.6294)	(9.3182; 9.3236)	(15.8457; 15.852

Table 3: The average Nusselt number Nu at the cold and hot walls.

Re	a	b	*αα*
100	0.523	2.165	0.615
400	1.421	4.303	0.575
1000	2.768	6.555	0.526

Table 4: Values of coefficients *aa*, *bbaabbcc αα*.

convection distorts the isotherm fields and three-dimensional patterns become more pronounced when Re increases (Figure 3c, 3f and 3i). The distortion of the isotherm field increases with Richardson number. In other words, the flow is principally dominated by buoyancy and the heat transfer is controlled mainly by natural convection, signifying that the forced convection due to the lid-movement is almost absent.

For Ri=1, a compromise between the two phenomena, evoked previously, is clearly seen in Figures 2b, 2e, 2h, 3b, 3e and 3h.

In order to assess the average heat transfer distribution along the vertical walls, the Nusselt number is introduced and is defined by:

$$Nu=\int_0^1\int_0^1 -\frac{\partial\theta}{\partial x}\bigg|_{x=0\ or\ x=1} dydz\ .$$

Table 3 lists the average Nusselt number Nu at the cold and hot walls for the computations obtained for the nine combinations (Re,Ri) studied. The results convincingly indicate that when Re is small (Re=100), the heat transfer through the cold and the hot walls exhibit similar trends for each value of Ri. For this same Reynolds number, the average Nusselt number increases with the Richardson number. By increasing the Reynolds number, values of Nusselt number increase and small differences between Nu_{hot} and Nu_{cold} are observed. When Ri is small at high values of Re, the difference between Nu_{hot} and Nu_{cold} augment.

Relatively to the heat transfer through the cold wall, a correlation between Nu and Ri was established. In fact, several computations (for each Reynolds number) demonstrate clearly the existence of a relation expressed as:. $Nu = a\times Ri^{\alpha} + b$. Table 4 lists the values of coefficients, b and α.

Conclusions

The current investigation addressed three-dimensional laminar mixed convection in a lid-driven cubic cavity filled with air (Pr=0.71) for suitable combinations of three different Reynolds numbers and three different Richardson numbers. The effects of varying both Reynolds and Richardson numbers on the resulting convection are investigated. Interesting behaviours of the flow and thermal fields with varying Reynolds and Richardson numbers are observed.

When small *Ri* is united with low *Re*, a primary vortex is observed occupying the main part of the cavity and its intensity is slightly modified when *Re* increased. In addition, two minor secondary recirculating vortices are observed at the bottom corners as the Reynolds number goes through *Re*=400 to *Re*=1000. Furthermore, three dimensionalities of the isotherm patterns are manifested. In this case, the mechanically driven forced convection dominates the buoyancy-driven, implying that forced convection is essential due to the lid-movement.

When large *Ri* is paired with low *Re*, two primary vortex are observed in the proximity of the core region and their intensity is considerably modified to provide only one stretched vortex when *Re* increase. It is also seen that the buoyancy-driven dominates the mechanically driven forced convection.

The heat transfer characteristics inside the cubic cavity are improved significantly for low values of *Ri* due to the dominant effect of the mechanical effect provoked by the moving lid. The effects of both *Re* and *Ri* are also apparent in the values of the average Nusselt number. For high *Ri* united with large *Re*, the overall heat transfer and the convection mode dominates the picture. Finally for Reynolds number ranging from 100 to 1000, a correlation between the averaged heat transfer (*Nu*) and *Ri* has been reported.

References

1. Moallemi MK, Jang KS (1992) Prandtl number effects on laminar mixed convection heat transfer in a lid-driven cavity. Int J Heat Mass Transfer 35: 1881-1892.
2. Sharif MAR (2007) Laminar mixed convection in shallow inclined driven cavities with hot moving lid on top and cooled from bottom. Applied Thermal Engineering 27: 1036-1042.
3. Prasad YS, Das MK (2007) Hopf bifurcation in mixed flow inside a rectangular cavity. Int J Heat Mass Transfer 50: 3583-3598.
4. Mohammad AA, Viskanta R (1992) Laminar flow and heat transfer in Rayleigh-Benard convection with shear Phys. Fluids A 4: 2131-2140.
5. Aydin O, Yang WJ (2000) Mixed convection in cavities with a locally heated lower wall and moving sidewalls, Numer. Heat Transfer, Part A: Applications 37: 695-710.
6. Han H, Kuehn TH (1989) A numerical simulation of double diffusive convection in a vertical rectangular enclosure. ASME HTD 107: 149-154.
7. Oztop HF, Dagtekin I (2004) Mixed convection in two sided lid driven

differentially heated square cavity. Int J Heat Mass Transfer 47: 1761-11769.

8. Iwatsu R, Hyun JM, Kuwahara K (1992) Numerical simulation of flows driven by a torsionally oscillating lid in a square cavity. J Fluids Engineering 114: 143-151.

9. Iwatsu R, Hyun JM, Kuwahara K (1992) Convection in a differentially-heated square cavity with a torsionally-oscillating lid. Int. J. Heat Mass Transfer 35: 1069-1076.

10. Iwatsu R, Hyun JM (1995) Three-dimensional driven-cavity flows with a vertical temperature gradient. Int J Heat Mass Transfer 38: 3319-3328.

11. Mohammad AA, Viskanta R (1995) Flow and heat transfer in a lid-driven cavity filled with a stably stratified fluid. Appl. Math. Modelling 19: 465-472.

12. Freitas CJ, Street RL, Findikakis AN, Koseff JR (1985) Numerical simulation of three-dimensional flow in a cavity. Int. J. Numer. Meth. Fluids 5: 561-575.

13. Freitas CJ, Street RL (1988) Non-linear transport phenomena in a complex recirculating flow: a numerical investigation. Int J Numer Meth Fluids 8: 769-802.

14. Peyret R, Taylor TD (1983) Methods for Fluid Flow. Springer-Verlag, Berlin, Germany.

15. Achdou Y, Guermond JL (2000) Convergence analysis of a finite element projection/Lagrange–Galerkin method for the incompressible Navier–Stokes equations. SIAM J Numer Anal 37: 799-826.

16. Patankar SV (1981) A calculation procedure for two-dimensional elliptic situations. Numer Heat Transfer 34: 409-425.

17. Moukhalled F, Darwish M (2000) A unified formulation of the segregated class of algorithm for fluid flow at all speeds. Numer. Heat Transfer, Part B: Fundamentals 37: 103-139.

18. Kobayachi MH, Pereira JMC, Pereira JCF (1999) A conservative finite-volume second-order-accurate projection method on hybrid unstructured grids. J Comput Phys 150: 40-75.

19. Hayase T, Humphrey JAC, Greif R (1992) A consistently formulated QUICK scheme for fast and stable convergence using finite-volume iterative calculation procedures. J Comput Phys 98: 108-118.

20. Press WH, Teukolsky SA, Vetterling WT, Flannery BP (1997) Numerical Recipes in Fortran 77: The Art of Scientific Computing. (second edition), Cambridge Press, London, UK.

21. Hortmann M, Peric M, Scheuerer G (1990) Finite volume multigrid prediction of laminar natural convection: benchmark solutions. Int J Numer Meth Fluids 11: 189-207.

22. Mesquita MS, De Lemos MGS (2004) Optimal multigrid solutions of dimensional convection-conduction problems. Appl Math Comput 152: 725-742.

23. Nobile E (1996) Simulation of time-dependent flow in cavities with the correction multigrid method, Part I: mathematical formulation. Numer Heat Transf Part B: Fundamentals 30: 341-350.

24. Cheikh NB, Beya BB, Lili T (2007) Benchmark solution for time-dependent natural convection flows with an accelerated full-multigrid method, Numerical Heat Transfer. Part B: Fundamentals 52: 131-151.

25. Tric E, Labrosse G, Betrouni M (2000) A first incursion into the 3D structure of natural convection of air in a differentially heated cubic cavity, from accurate numerical solutions. Int J Heat and Mass Transfer 43: 4043-4056.

26. Wong KL, Baker AJ (2002) A 3D incompressible Navier–Stokes velocity–vorticity weakform 2nite element algorithm. Int J Numer Meth Fluids 38: 99-123.

Dam Length Optimization of Two Lobe Pressure Dam Bearing using Genetic Algorithm

Lintu Roy* and Arunabh Choudhury

Department of Mechanical Engineering, National Institute ofTechnology, Silchar-788010, India

Abstract

This paper attempts to find out the optimum length of pressure dams taking into account the various steady state characteristics of two lobe bearing. The length of the dam selected varies with eccentricity ratios of operation. Determination of Optimum performance is based on maximization of non-dimensional load, maximization of flow coefficient and minimization of friction variable using Genetic Algorithm. The result obtained gives an insight on how the performance of two lobe bearing can be enhanced using pressure dam. The data obtained from the above can be used conveniently in the optimum design of such bearings, as these are presented in dimensionless form.

Keywords: Steady state characteristics; Two lobe; Optimum dam length

Introduction

The performance of a two lobe bearing is enhanced by the use of pressure dams. The analysis of multi lobe bearings was first carried out by Pinkus [1]. A comparison of non-dimensional values of steady state and dynamic characteristics has been made with the published results of Lund et al. [2] for L/D=1 with two lobe bearing with 20° axial groove. Sihasan et al. [3] worked on various configurations of multi lobe bearings. Mehta et al. [4] analyzed three-lobe bearings with pressure dams and concluded that the stability of a three-lobe bearing can be increased by simply cutting pressure dams and a relief track. Mehta et al. [5] analyzed the static and dynamic characteristics of four lobe pressure dam bearing and concluded that the performance of four lobe bearing with pressure dams is superior to that of an ordinary four lobe bearing. Bhushan et al. [6] analysed the behaviour of four lobe pressure dam bearing operating under turbulence condition and concluded that values of eccentricity ratio, friction coefficient and oil flow coefficient increases with increase in turbulence. The attitude angle increases with increase in values of Sommerfeld number less than 1.8, it decreases for values greater than 1.8. Mehta et al. [7] carried out the stability analysis of plain circular hydrodynamic pressure dam bearing operating with couple stress fluid and concluded that the use of dams in plain circular bearings increases its stability as evident by the increase in the critical mass. Mehta and Rattan [8] studied the inverted three-lobe pressure dam bearing. The stability of inverted three-lobe pressure dam bearing is found to increase with the incorporation of a pressure dam and relief tracks. Batra et al. [9] conducted a study on the effect of L/D ratio on the performance of three-lobe pressure dam bearing and observed that the stability of an inverted three-lobe pressure dam bearing increased with the decrease in L/D ratio. Batra et al. [10] carried out the static and dynamic analysis of inverted three-lobe pressure dam bearing and compared its performance to that of three lobe journal bearing. Batra et al. [11] also carried out a study on the effect of various ellipticity ratio on the performance of an inverted three lobe pressure dam bearing. It was found that for a particular Sommerfeld number, as the ellipticity ratio increased, the value of minimum film thickness, oil flow coefficient, attitude angle and eccentricity ratio decreased whereas the value of friction coefficient slightly increased. The stability of inverted three lobe pressure dam bearing increased with the increase in ellipticity ratio. As the use of pressure dams have proved to be useful to enhance the performance of multi lobe bearings, a study has been carried out to determine the optimum dam length for two lobe bearing considering the steady state characteristics of two lobe bearings. It is found that the performance of ordinary circular bearings and multi-lobe bearings is not very satisfactory when it comes to application in these high speed turbo machine components. To improve the stability of these bearings, pressure dams are incorporated in these bearings. It has been found from analytical dynamic analysis that cylindrical pressure dam bearings are very stable. Also, using analytical and experimental stability analysis, it has been found that the stability of multi-lobe pressure dam bearing also increases. So it can be said that incorporating pressure dams in bearings help in improving the performance of the bearings.

With the advancement of technology, the requirements for various engineering elements are increasing. It is not only required by the machines to handle higher speeds and loads than before, but to do so without any drastic increase in cost of manufacturing and maintenance. This requires each element to have better performance and stability than before and bearings are no exception. Incorporating pressure dams in conventional hydrodynamic bearing could provide a solution to this increase in demand for better performance. So, a study has been carried out on the effect of pressure dam bearing on parameters like load capacity, friction variable and flow of lubricants.

Geometry

The geometry of a two-lobe pressure dam bearing is usually the same as two-lobe bearing with the exception of the slot cut in one of the bearing pads and a long groove cut in the other bearing pad. Figure 1 gives the geometry of a pressure dam bearing. The upper pad of the bearing has a slot cut into it. A rectangular dam of step depth S_d and width L_d is cut circumferentially in the upper lobe. The dam starts after the oil hole and subtends an arc of θ_S degrees at the centre. A circumferential relief track or groove is of depth and width L_t is also cut centrally in the other pad of the bearing. The relief track is assumed to be so deep that its hydrodynamic effects are neglected. The pocket

***Corresponding author:** Lintu Roy, National Institute of Technology, Silchar-788010, India, E-mail: lintu2003@gmail.com

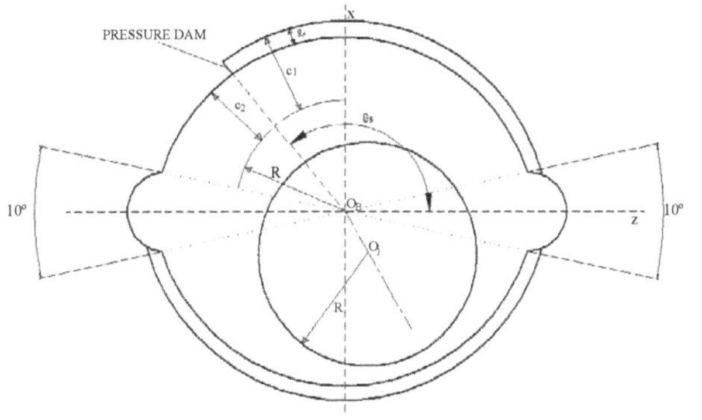

Figure 1: Geometry of a pressure dam bearing.

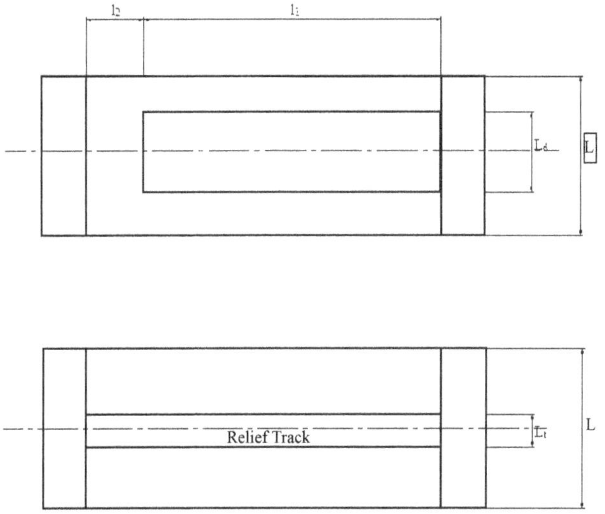

Figure 2: Top and bottom pad of pressure dam bearing.

clearance ratio K for the pressure dam is given by c_1/c_2 and the dam location is given by l_1/l_2 as shown in the Figure 2.

The pressure dam bearing is taken have two oil grooves at both the end of the two lobes. The groove angles being 10^0 each. The pressure dam is incorporated in the first lobe and a relief track is incorporated in the second lobe. The size of the pressure dam which is given in degrees is converted to be expressed in terms of the length of the arc subtended by the pressure dam. The geometry of the pressure dam and relief track can be expressed in terms of length instead of angle subtended at the centre. This helps in defining the geometry of the pressure dam and relief track in terms of discretized elements of ease of programming.

Eccentricity Ratio is given by

For lobe 1 $\varepsilon_1 = \sqrt{\varepsilon^2 + \delta^2 + 2\varepsilon\delta \cos\phi}$ (1)

For lobe 2 $\varepsilon_2 = \sqrt{\varepsilon^2 + \delta^2 2\varepsilon\delta \cos\varphi}$ (2)

Attitude angle is given by

For lobe 1 $\varphi_1 = \tan^{-1}\dfrac{e\sin\varphi}{\delta + e\cos\varphi}$ (3)

For lobe 2 $\phi_2 = \pi - \tan^{-1}\dfrac{\varepsilon\sin\phi}{\delta - \varepsilon\cos\phi}$ (4)

The circumferential length of the pressure dam is given by

$$l \quad \pi R\theta \ /180 \qquad (5)$$

Where θ_s is the angle subtended by pressure dam in degrees.

The fluid film thickness for the pressure dam bearing in the region where the pressure dam is present given as:

$$\bar{h} = 1 + S_d + \text{€}\cos\theta \qquad (6)$$

Where S_d is the step depth €

And where the pressure dam is not present

$$\bar{h} = 1 + \text{€}\cos\theta \qquad (7)$$

Theory

The Reynolds Equation has been derived from the Navier-Stokes equation and the continuity equation. The generalized Reynolds Equation is the differential equation originally developed by Reynolds restricted to incompressible flow. However, the equation can be formulated to include effects of compressibility. The simplified form of Reynolds Equation can be written as:

$$\frac{\partial}{\partial x}\left(\frac{\rho h^3}{\eta}\frac{\partial p}{\partial x}\right) + \frac{\partial}{\partial z}\left(\frac{\rho h^3}{\eta}\frac{\partial p}{\partial z}\right) = 6U\frac{\partial}{\partial x}(\rho h) + 12\frac{\partial}{\partial t}(\rho h) \qquad (8)$$

Throughout this work, a consistent and meaningful set of parameters have been used. They can be termed as the geometric conditions. One of the important non-dimensional parameter used in this work is the slenderness ratio and is defined as the ratio of bearing's length to its diameter i.e. L/D ratio.

The "unwrapping" of the circumferential coordinate such as the fluid passage is expressed in a Cartesian frame implies another common assumption in bearing analyses: that the effect of curvature is negligible. This assumption is slightly more restrictive because the curvature terms are of order C/R that is also known as the clearance ratio. For most bearings, however, C/R is of the order of 1/1000.

Where, the various non-dimensional terms that has been used,

$$\theta = \frac{x}{R}, \bar{z} = \frac{z}{L/2}, \bar{h} = \frac{h}{C}, \bar{p} = \frac{pC^2}{6\eta UR}$$

By using the above substitutions, the non-dimensionalised form of the Reynolds Equation is obtained as:

$$\frac{\partial}{\partial\theta}(\bar{h}_0^{\,3}\frac{\partial\bar{p}_0}{\partial\theta}) + \left(\frac{D}{L}\right)^2 \frac{\partial}{\partial\bar{z}}(\bar{h}_0^{\,3}\frac{\partial\bar{p}_0}{\partial\bar{z}}) = \frac{\partial\bar{h}_0}{\partial\theta} \qquad (9)$$

where, $\bar{h} = 1 + \text{€}\cos\theta$

The non-dimensional form of the Reynolds Equation is then can be used to determine the pressure at each point of a developed mesh of each side of the bearing thus giving us the pressure distribution on the bearing. The Reynolds Equation is solved using finite difference method with Gauss-Seidel method of iteration with successive over relaxation.

The Sommerfeld number is a dimensionless quantity that gives the characteristics of the bearing as it contains all the variables required to design a bearing.

It is given by

$$S = \frac{\eta N}{P}\left(\frac{R}{C}\right)^2$$

The steady state characteristics for two lobe pressure dam bearing viz. non dimensional load, friction variable and flow coefficient are defined as follows

The non-dimensional load for journal bearing is given by

$$\overline{W}_{X_0} = \int_{\theta_1}^{\theta_2}\int_0^1 \overline{p}_0 \cos\theta \, d\theta \, d\overline{z}$$

$$\overline{W}_{Z_0} = \int_{\theta_1}^{\theta_2}\int_0^1 \overline{p}_0 \sin\theta \, d\theta \, d\overline{z}$$

Therefore the load carrying capacity and attitude angle are:

$$\overline{w} = \sqrt{(\overline{w}_{x_0}{}^2 + \overline{w}_{z_0}{}^2)} \qquad (10)$$

$$\varphi = \tan^{-1}\left(\frac{\overline{w}_z}{\overline{w}_x}\right)$$

Flow coefficient which is given by [1]

$$\overline{\mu} = \mu(R/C) = \frac{\int_0^{2\pi}\left(3\overline{h}\frac{\partial \overline{p}}{\partial\theta} + \frac{1}{\overline{h}}\right)\partial\theta}{6\overline{W}} \qquad (11)$$

And friction variable which is given by

$$\overline{q}_z = \frac{1}{2}\left(\frac{D}{L}\right)^2 \int_0^{2\pi} \overline{h}_L^{\,3} \frac{\partial \overline{p}_L}{\partial \overline{z}_L} d\theta_L \qquad (12)$$

To determine the optimum dam length of two lobe pressure dam bearing by taking into consideration all the steady state characteristics of two lobe pressure dam bearing, single and multi-objective functions are defined. The single and multi-objective function is then optimized to find out the optimum dam length. The multi objective function is defined as

$$f = w_1\left(\frac{\overline{\mu}}{\overline{\mu}_{\max} - \overline{\mu}_{\min}}\right) + w_2\left(1 - \frac{\overline{q}_z}{\overline{q}_{\max} - \overline{q}_{\min}}\right) + w_3\left(1 - \frac{\overline{w}}{\overline{w}_{\max} - \overline{w}_{\min}}\right) \qquad (13)$$

Where, w_1, w_2 & w_3 are the equal weight of the functions and is taken to be 0.33

The objective functions are minimized using the Genetic Algorithm toolbox in MATLAB. The angle subtended by the pressure dam at the centre of the bearing which is θ_s is the input parameter which is finally obtained using the GA toolbox. The optimization calls the Reynolds equation solver (Gauss-Seidel Method with over-relaxation in a Finite Difference Grid then numerical integration by Simpson's rule) to estimate the required parameter. Minimization of the multi objective function using the values generated for θ_s which give us the best value for the pressure dam.

Results and Discussions

Before the analysis of results of two-lobe pressure dam bearings, the steady state result of two-lobe bearing is compared with [2]. From the Table 1 it has been observed that present results agrees well with previously published results [2]

The results for two lobe pressure dam bearing has been obtained for single objective functions for steady state characteristics of pressure dam bearing viz friction variable ($\overline{\mu}$), flow coefficient ($\overline{q}_z$) and non-dimensional load carrying capacity ($\overline{W}$). Codes were developed for single and multi-objective function for these three steady state characteristics and the optimum dam length for two lobe pressure dam bearing was determined forming single (minimization of friction variable($\overline{\mu}$), maximization of flow coefficient ($\overline{q}_z$) and maximization of non-dimensional load carrying capacity ($\overline{W}$) and multi-objective function (formed by minimization of objective function formed by combination of all the objectives as stated above). The optimum dam length for two lobe pressure dam bearing was determined for different eccentricity ratios. The optimization of dam length for the two lobe pressure dam bearing for both single and multi objective function has been done using genetic algorithm. The optimum values obtained for friction variable, flow coefficient and non dimensional load are all in non dimensional form so that they can be used for practical purposes. Manufacturers and designers will be immensely benefitted if such dam locations can be determined by some method (Figures 3-11).

From Figure 3 it can be said that the optimum dam length of two lobe pressure dam bearing considering only friction variable as objective function increases with increasing eccentricity ratios. The angle subtended by dam varies between 104° to 341° for the variation of eccentricity ratio from 0.05 to 0.451. A convergence plot considering friction variable as objective function is shown in Figures 4 and 5 it can be said that the angle subtended by dam of two lobe pressure

Eccentricity ratio	Attitude angle (ϕ)	Sommerfeld Number (S)
(ε)	Present work [2]	Present work [2]
0.050	93.812 [93.75]	1.445 [1.463]
0.100	93.125 [93.15]	0.699 [0.709]
0.150	91.980 [92.00]	0.442 [0.447]
0.200	90.387 [90.40]	0.308 [0.312]
0.239	88.831 [88.85]	0.239 [0.243]
0.250	88.284 [88.35]	0.224 [0.227]
0.260	87.896 [87.85]	0.219 [0.213]
0.304	85.561 [85.25]	0.165 [0.163]
0.350	81.853 [82.00]	0.121 [0.122]
0.381	79.029 [78.77]	0.099 [0.098]
0.451	64.194 [63.80]	0.046 [0.045]

Table 1: Validation of results for two lobe bearing.

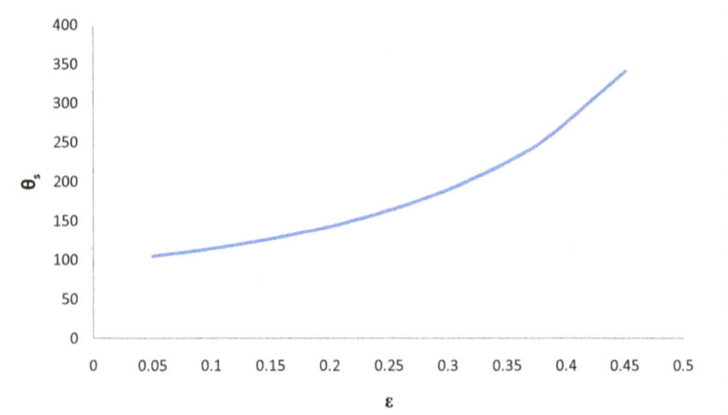

Figure 3: Variation of angle subtended by pressure dam (θ_s) with eccentricity ratio (ε) considering friction variable as objective function.

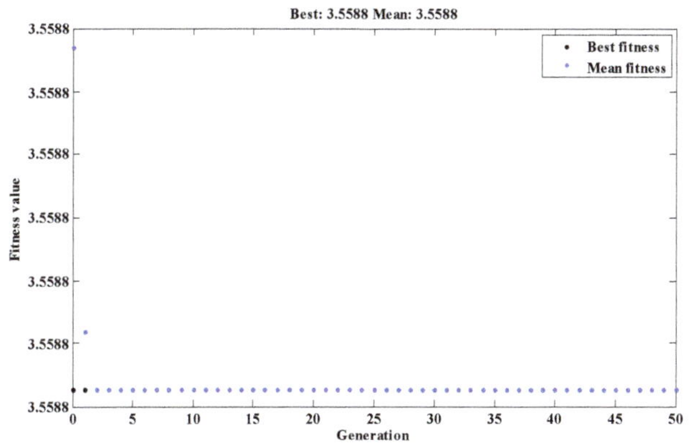

Figure 4: Convergence of best fitness value for friction variable considering eccentricity ratio 0.381.

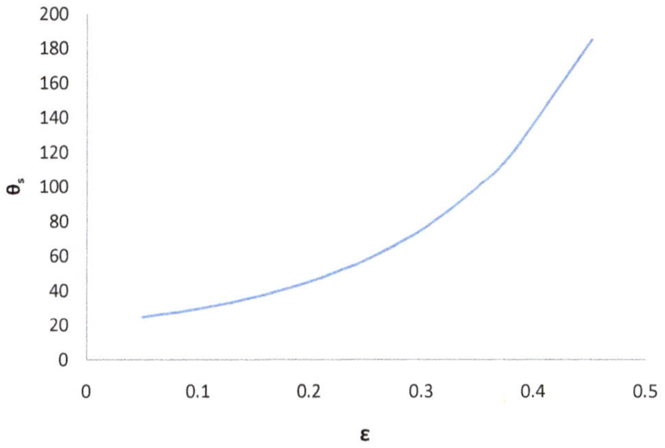

Figure 5: Variation of angle subtended by pressure dam θ_s with eccentricity ratio (ε) considering flow coefficient as objective function.

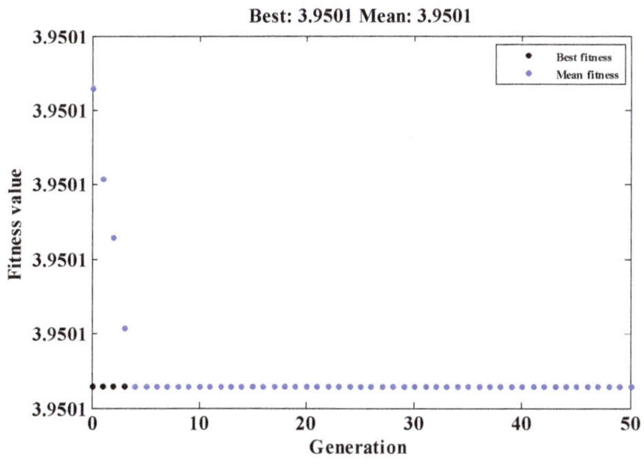

Figure 6: Convergence of best fitness value for flow coefficient considering eccentricity ratio 0.381.

dam bearing considering only flow coefficient also increases with increase in eccentricity ratio. The subtended angle varies between 24° to 185° for the variation of eccentricity ratio from 0.05 to 0.451. A convergence plot considering flow coefficient as objective function is shown in Figure 6. From Figure 7 it can be said that the optimum angle subtended by dam of two lobe pressure dam bearing considering only non-dimensional load also increases with eccentricity ratio. The dam angle varies between 19° to 37° for the variation of eccentricity ratio from 0.05 to 0.451. A convergence plot considering non dimensional load as objective function at a particular value of eccentricity ratio is shown in Figure 8. From Figure 9 it can be said that the optimum

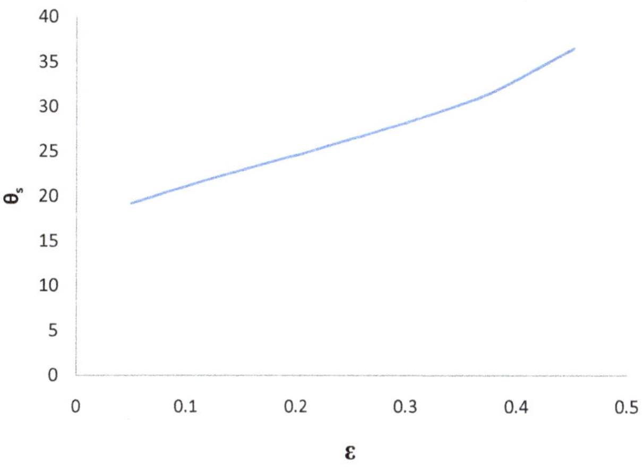

Figure 7: Variation of angle subtended by pressure dam θ_s with eccentricity ratio (ε) considering non dimensional load as objective function.

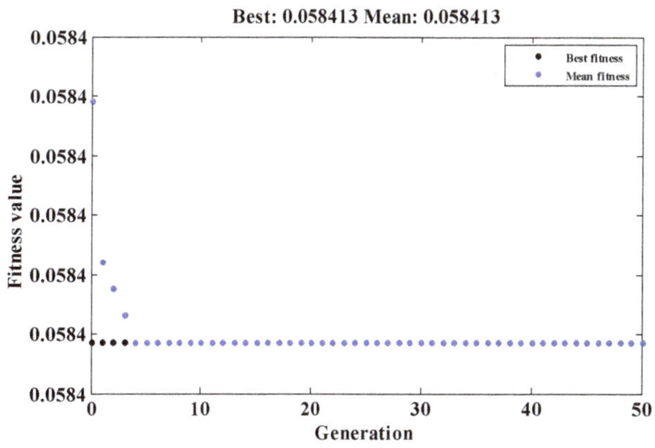

Figure 8: Convergence of best fitness value for non-dimensional load considering eccentricity ratio 0.381.

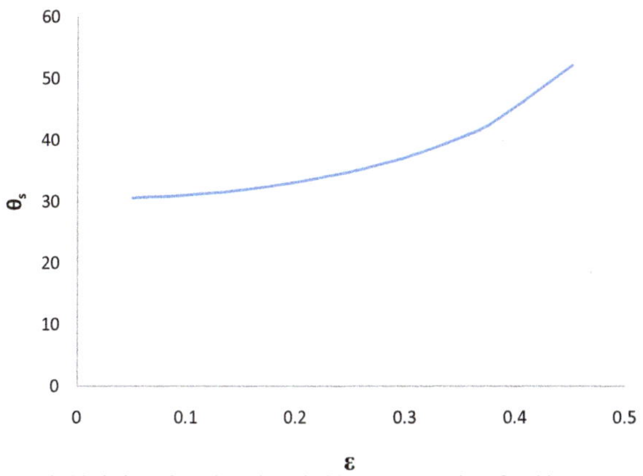

Figure 9: Variation of angle subtended by pressure dam θ_s with eccentricity ratio (ε)formed by minimization of objective function formed by combination of all the objectives.

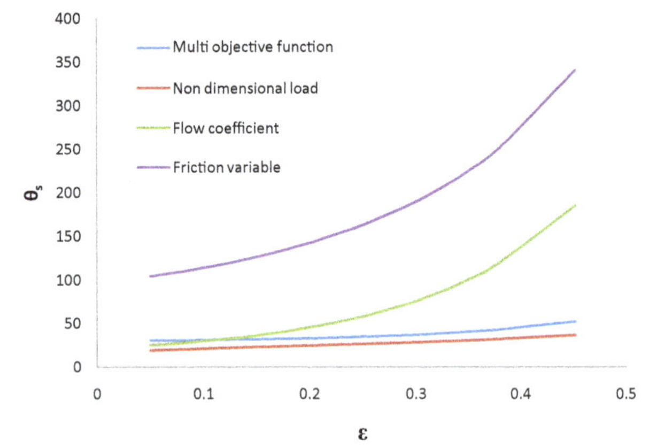

Figure10: Comparison of angle subtended by pressure dam θ_s for various objective functions.

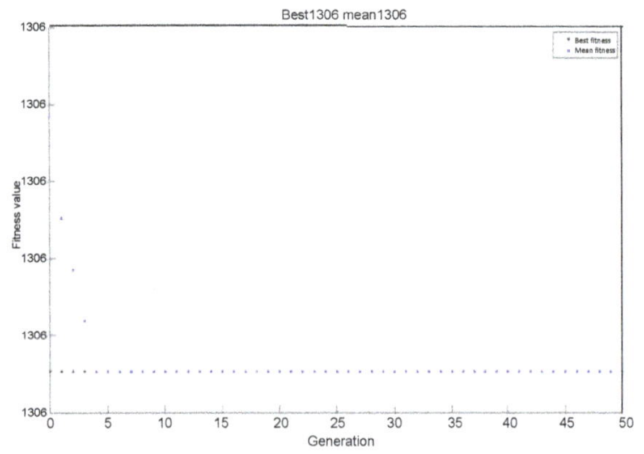

Figure 11: Convergence of best fitness value for multi objective function considering eccentricity ratio 0.304.

dam length of two lobe pressure dam bearing considering the multi objective function increases with increasing eccentricity ratio. The dam angle varies between 30° and 53° for eccentricity ratio from 0.05 to 0.451. From Figure 10 it can be said that the variation of dam length for two lobe pressure dam bearing is the highest for friction variable. The variation is less for flow coefficient and it is the least for non-dimensional load. The variation of dam length for two lobe pressure dam bearing considering multi objective function is very less and is only slightly more than the variation for non-dimensional load. A convergence plot considering multi objective function at a particular value of eccentricity ratios is shown in Figure 11.

Conclusions

In this work, optimization of dam length of two lobe pressure dam bearing is carried out using GA. Considering the different objective functions, genetic algorithm is able obtain a global solution of optimum design parameters such as friction variable, flow coefficient and non-dimensional load carrying capacity and multi objective function of the two lobe pressure dam bearing. From the discussion the following conclusions can be drawn.

In case of two lobe pressure dam bearing considering only friction variable as an objective function, the angle subtended by pressure dam comes out to be in the range of 104° to 341° for eccentricity ratio ranging from 0.05 to 0.451 respectively. The angle subtended by pressure dam for two lobe pressure dam bearing is found to increase with eccentricity ratio.

1. In case of two lobe pressure dam bearing considering only flow coefficient as an objective function, the angle subtended by pressure dam comes out to be in the range of 24° to 185° for eccentricity ratio ranging from 0.05 to 0.451 respectively. In this case also, the angle subtended by pressure dam for two lobe pressure dam bearing also increases with increase in eccentricity ratio.
2. In case of two lobe pressure dam bearing considering only non-dimensional load, the angle subtended by pressure dam is in the range of 19° to 37° for eccentricity ratio ranging from 0.05 to 0.451 respectively. The optimum dam length for two lobe pressure dam bearing increases with increase in eccentricity ratio.
3. In case of two lobe pressure dam bearing considering the multi objective function the optimized dam length comes out to be in the range of 30° to 53° for eccentricity ratio ranging from 0.05 to 0.451 respectively. In this case also, the angle subtended by pressure dam for two lobe pressure dam bearing increases with increase in eccentricity ratio.
4. The rate of increase in the angle subtended by pressure dam for two lobe pressure dam bearing is much higher while considering only friction variable as an objective function than the other cases. The rate of increase is the least while considering only non-dimensional load as an objective function. The rate of increase while considering flow coefficient is lower than that for friction variable but higher than the rest. The rate of increase for dam length for two lobe pressure dam bearing while considering the multi objective function is higher than that of non-dimensional load but lower than the rest. Dam at the top of the bearing it adds more to friction and flow than the load capacity.
5. The data obtained from the above analysis can be used conveniently in the design of such bearings, as these are presented in dimensionless form.

Identification of the optimum length of the dam is done so that the performance characteristics are near to the optimum for any loading condition (eccentricity ratio) and any objective function and will be much beneficial to the manufacturers and designers. Experimental verification of the present result may lead to a new approach of production of bearings with optimum groove locations, however, it is beyond the scope of the present work and hopefully experimentalists have a problem in hand.

References

1. Pinkus O (1959) Analysis and characteristics of three-lobe bearings. ASME J Basic Eng 81: 19.
2. Lund W, Thomson KK (1978) A Calculation Method and Data for the Dynamic Coefficients of Oil Lubricated Journal Bearings. Proceedings of the ASME Design and Engineering Conference, Minneapolis.
3. Sinhasan R, Malik M, Chandra M (1981) A Comparative study of some three-lobe bearing configurations. Elsevier Sequoia, Lausanne 72: 277-286.
4. Mehta NP, Rattan SS (1993) Stability analysis of three-lobe bearings with pressure dams. Elsevier, Sequoia, 167: 181-185.
5. Mehta NP, Rattan SS, Bhushan G (2003) Static and dynamic characteristics of a four-lobe pressure-dam bearings. Tribology Letters 15: 415-421.

6. Bhushan G, Rattan SS, Mehta NP (2007) Effect of turbulence on performance of four-lobe pressure dam bearing. IE(I) Journal-MC 87: 18-23.

7. Mehta NP, Rattan SS, Verma R (2010) Stability analysis of circular pressure dam hydrodynamic journal bearing with couple stress lubricant. Aprn J Eng Appl Sci 5: 26-33.

8. Mehta NP, Rattan SS, Bhusan G, Batra NK (2012) Stability analysis of an inverted three-lobe pressure dam bearing. J Mech Sci Technol, Springer 26: 2425-2430.

9. Batra NK, Bhushan G, Mehta NP (2012) Effect of L/D ratio on the performance of an inverted three-lobe pressure dam bearing. J Eng Technol 1: 94-99.

10. Batra NK, Bhushan G, Mehta NP (2012) Static and dynamic characteristics of an inverted three-lobe pressure dam bearing. Akademeia 2: 1523-1530.

11. Batra NK, Bhushan G, Mehta NP (2012) Effect of ellipticity ratio on the performance of an inverted three-lobe pressure dam bearing. Tribology Transactions 55: 798-804.

Design and Optimization of Screwed Fasteners to Reduce Stress Concentration Factor

Govindu N[1], Jayanand Kumar T[1] and Venkatesh S[2]*

[1]*Department of Mechanical Engineering, Godavari Institute of Engineering andTechnology, Affiliation with JNTU-K, Rajahmundry, Andhra Pradesh 530026, India*

[2]*Department of Mechanical Engineering, Raghu Engineering College,Affiliation with JNTU-K, Visakhapatnam, Andhra Pradesh 530026, India*

Abstract

The bolt and nuts are used for temporary fastening in the machines and structural assembly; they play an important role to restrict the movement of individual parts in the assembly. Around 10 to 20% of fasteners are used in various designs, sizes, strength levels and materials in complex mechanical systems for diverse applications. The unequal load distribution in the threaded fasteners causes the fatigue failure due to stress concentration

In this paper stress distribution in Buttress and ACME threads, for various symmetrical models such as groove added to bolt and nut, a step is added to nut, with or without groove on the bolt, and to the reduced diameter, a taper is added to the nut and a groove on the bolt are designed in 3D and analyzed and validated from 2D models developed to reduce stress concentration.

The objective of this study is to show the quantitative comparison between the Buttress and ACME bolt and nut design modifications to improve the design methods efficiently to reduce the stress concentration. In this study various geometrical designs are proposing to reduce the stress concentration factor.

Keywords: Fasteners; Threads; Design modification; Stress concentration

Introduction

Fasteners

Machines are assembled from individual parts. The parts may need to be joined sometimes in permanent or temporary connections. The temporary connections or joints are normally made by using screws bolts or nuts and bolts. The temporary joints render the convenience of disconnection as frequently as required for such purposes as inspection, repairs, adjustment or replacement. The most important feature of the screws or bolt is that they are standardized and are available as standard part ready to use. The threaded fasteners are used to hold two parts. The fastener consists of two parts. An externally threaded member whose length is normally greater than the diameter carries threads over entire or part length from end. The other end of this member is larger in area (you can assume that the end is upset). This part is inserted in coaxial holes in two parts taking care that the "upset" or larger end is greater in dimension than the holes so that it does not pass through the hole. The end that comes out of the holes carries thread upon which another internally threaded part is wound. Fasteners can also be used to close a container such as a bag, a box, or an envelope; or they may involve keeping together the sides of an opening of flexible material, attaching a lid to a container, etc. Fasteners used in these manners are often temporary, in that they may be fastened and unfastened repeatedly. Bolt and nut are one among the different fasteners.

Bolted joints are most common elements of structures, machines in pressure vessels, automobiles, machine tools, home appliances etc. Hundreds of different bolt designs with various sizes, strengths and materials are used in the assembly of aircrafts. On an Averages of 2.4 million fasteners are used to assemble a Boeing 747 aircraft. In bolt nut fasteners the load distribution is very unequal with high stress concentration at the thread roots. This stress concentration can cause fatigue failure in the bolt-nut fasteners. For complex structures the reliability (trust or faith in) of the bolt-nut fasteners is the most important. So, study of bolt-nut fasteners is important.

Stress

The concept of stress, introduced by Cauchy around 1822, is a measure of the average amount of force exerted per unit area of the surface on which internal forces act within a deformable body. In other words, it is a measure of the intensity, or internal distribution of the total internal forces acting within a deformable body across imaginary surfaces. These internal forces are produced between the particles in the body as a reaction to external forces applied on the body. External forces are either surface forces or body forces.

The SI unit for stress is the Pascal (symbol Pa), which is equivalent to one newton (force) per square meter (unit area). The unit for stress is the same as that of pressure, which is also a measure of force per unit area. Engineering quantities are usually measured in Mega Pascal's (MPa) or Giga Pascal's (GPa). In imperial units, stress is expressed in pounds-force per square inch (psi).

For the simple case of a body axially loaded, e.g. a prismatic bar subjected to tension or compression by a force passing through its centroid, the stress σ, or intensity of the distribution of internal forces, can be obtained by dividing the total tensile or compressive force by the cross-sectional area where it is acting upon (Figure 1). Thus, we have

$$\sigma = (F/A)$$

Where σ = Nominal Stress; F = Total tensile or compressive force; A = Cross-sectional area.

***Corresponding author:** Venkatesh S, Department of Mechanical Engineering, Raghu Engineering College, Affiliation with JNTU-K, Visakhapatnam, Andhra Pradesh 530026, India, E-mail: venkatesh.sitty@gmail.com

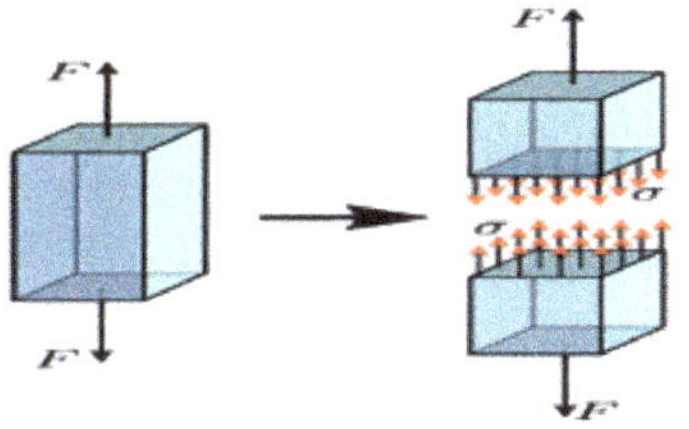

Figure 1: Tensilestress.

Experimental determination of stresses can be carried out using the photo-elastic method or FEM.

Stress concentration

A stress concentration (often called stress raisers or stress risers) is a location in an object where stress is concentrated. An object is strongest when force is evenly distributed over its area, so a reduction in area, e.g. caused by a crack, results in a localized increase in stress. A material can fail, via a propagating crack, when a concentrated stress exceeds the material's theoretical cohesive strength. The real fracture strength of a material is always lower than the theoretical value because most materials contain small cracks that concentrate stress. Fatigue cracks always start at stress raisers, so removing such defects increases the fatigue strength.

Causes: Geometric discontinuities cause an object to experience a local increase in the intensity of a stress field. The examples of shapes that cause these concentrations are: Cracks, sharp corners, holes and, changes in the cross-sectional area of the object. High local stresses can cause the object to fail more quickly than if it wasn't there. Good design of the geometry minimizes stress concentration.

Stress concentration factor

The ratio of the maximum stress and the nominal applied tensile stress is denoted as the stress concentration factor, '**K_t**'. $K_t = \sigma_m/\sigma_o$

Acme threads form: There is a special type of trapezoidal thread called ACME thread. These two threads are identical in all respects except the thread angle. In ACME thread, the thread angle is 29^0 instead of 30^0 these threads has more thickness at core diameter than that of square threads. Therefore screw with ACME threads stronger than equivalent screw with square threads. Such a screw has a large load carrying capacity. From the Figure 2, If p = pitch, d = depth of thread, and n = number of threads per inch.

Buttress threads form: Buttress threads are used where heavy axial force acts along the screw axis in one direction only. It has higher efficiency compared with ACME threads. It can economically manufactured on thread milling machine. The axial wear at the thread surface can be compensated by means of split-nut. It can transmit power and motion in one direction only. The basic profile is the theoretical profile of the thread (Figure 3). An essential principle is that the actual profiles of both the nut and bolt threads must never cross or transgress the theoretical profile. So bolt threads will always be equal to, or smaller than, the dimensions of the basic profile. Nut threads will always be equal to, or greater than, the basic profile. To ensure this in practice, tolerances and allowances are applied to the basic profile.

Applications of the screwed fasteners

1. Connecting threaded pipes and hoses to each other and to caps of fixtures.

2. Gear reduction via worm drives.

3. Moving objects connecting rotary to leaner motion (lead screw of a jack).

4. Measuring by correlation linear motion to rotary motion in Micrometer.

5. Lead screw of a lathe.

Advantages of the screwed fasteners

1. Screwed joints are highly reliable in service

2. Screwed joints are easy to assemble and disassemble.

3. A wide range of screwed joints is available to adopt under various working conditions.

4. Screws are relatively cheap.

5. Due to Standardization, cheap manufacturing processes can be adopted.

6. Screws can be used to transmit power such as lead screws.

Limitations of the screwed fasteners

1. The stress concentration in the threaded portions is variable under variable conditions.

2. Screwed joints become loose due to machine vibrations.

3. The strength of the screwed joints is not comparable with Welded or riveted joints.

Literature Review

Bolt-nut connectors are one of the basic types of fasteners used in machines and structures. They play an important role in the safety and reliability of structural systems. The load distribution in typical bolt-nut connectors is very unequal, with a high stress concentration at the thread roots. This stress concentration can cause fatigue failure

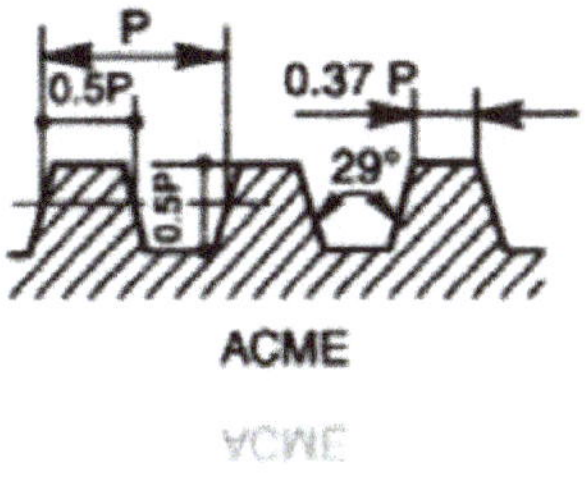

Figure 2: ACME thread form.

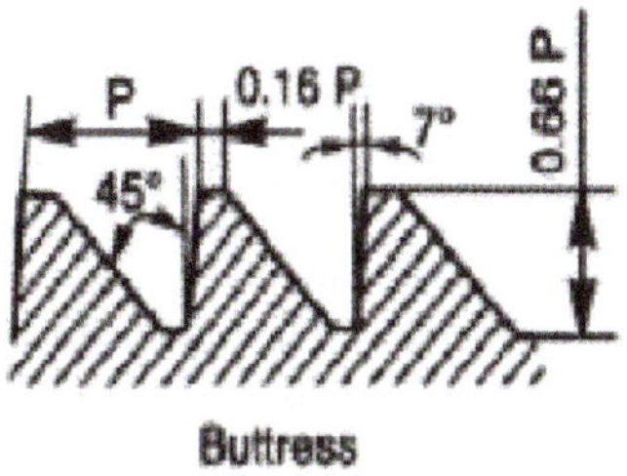

Figure 3: Buttress thread form.

in the bolt-nut connectors. For example, hundreds of different bolt designs with various sizes, strength levels, and materials are used in the assembly of an aircraft. On the average, 2.4 million fasteners are used to assemble a Boeing 747 aircraft. Of this total, 22% are structural bolts [1]. The importance of the reliability of bolt-nut connectors cannot be overemphasized in such applications. Hence, it is of considerable interest to study the stress concentration in bolt-nut connectors. In the past, several researchers have studied the stress distribution in bolt-nut connectors using computational and experimental methods. Hetenyi [2], Seika et al. [3], Patterson and Kenny [4,5] and Kenny and Patterson [6] have used experimental methods like photo elasticity to study the stress distribution in bolt-nut joints.

In present study an attempt made to understand the behavior of the bolted joints and the stress concentration factor when loaded statically with uni-axial external loads uses FEA [1]. Some journals like An innovative polariscope for photo elastic stress analysis [2], which uses photo-elasticity and Venkatesan and Kinzel [7], has studied the stress concentration in threads of bolt-nut joint using ax symmetry finite element model.

In this paper, the stress distribution in bolt-nut connectors is studied using an ax symmetric finite element model. Various geometric designs proposed in the literature were studied to determine the extent to which they reduce stress concentrations. Some well-known modifications do significantly reduce the stress concentration factor (up to 85%) while other changes produce much more modest changes. The design modifications include things such as grooves and steps on the bolt and nut, and reducing the shank diameter of the bolt. All of the changes also result in a reduction in weight.

In present study an attempt made to understand the behavior of the bolted joints and the stress concentration factor when loaded statically with uni-axial external loads [1]. Also an attempt has been made to minimize the error between experimental results and FEM simulated results by selecting the exact number of element and selecting suitable yield criteria. Linear finite element analysis method is used to determine the stress concentration factor of the threads in bolted connection.

In the current study the load transfer (tension and bending) as well as stress intensity factor solutions for various crack locations and sizes have been derived by advanced 3D FE analyses including threaded bolt and nut [8]. In addition, a large number of fatigue crack tests with Titanium and Steel fasteners have been performed to Determine realistic defect shapes for rolled- as well as machined threads, and the evolution of the shape as the defect propagate, Verify stress intensity factor and to determine the amount of bending inherent in the fastener due to the load introduction through the nut and due to the presence of a crack.

In this work the analysis of stresses and deformations arising in the bolt shank on various laws of the load distributions on the threads is considered [9]. The finite element method (FEM) [10-13] and ANSYS application on solution of this problem are used as well. Having solved the problem of finding the most loaded cross section of the bolt one can more precise to estimate the design reliability.

In the present work, joint materials are assumed to be isotropic and homogeneous, and linear elastic ax symmetric finite element analysis was performed to evaluate the member stiffness [14]. Uniform displacement and uniform pressure assumptions are employed in idealizing the boundary conditions. Wide ranges of bolt sizes, joint thicknesses, and material properties are considered in the analysis to evaluate characteristic behavior of member stiffness. Empirical formulas for the member stiffness evaluation are proposed using dimensionless parameters. The results obtained are compared with the results available in the literature.

In this paper Finite Element Analysis (FEA) with extremely fine mesh in the vicinity of the blades of Steam Turbine Rotor is applied to determine stress concentration factors [15]. A model of Steam Turbine Rotor is shown.

In this paper, the details of an experimental method to measure the clamping force value at bolted connections due to application of wrenching torque to tighten the nut have been presented. A simplified bolted joint including a holed plate with a single bolt was considered to carry out the experiments. This method was designed based on Hooke's law by measuring compressive axial strain of a steel bush placed between the nut and the plate. In the experimental procedure, the values of clamping force were calculated for seven different levels of applied torque, and this process was repeated three times for each level of the torque. Moreover, the effect of lubrication of threads on the clamping value was studied using the same method. In both conditions (dry and lubricated threads), relation between the torque and the clamping force have been displayed in graph. The current thesis extends the work of Venkatesan and Kinzel [7], in which design modifications for 1-8 UNC thread forms were considered with more of emphasis on finding the stress concentration for different thread forms. The objective of this study is to show quantitative comparison of effect of various proposed design modifications in the literature [16] on the stress concentration factor. In this thesis Buttress thread and ACME thread forms with the same design modifications proposed in the literature were considered and analyzed using ANSYS, a popular finite element analysis package.

Experimental Procedure

Bolt Design:	**Nut Design:**
Dminor=0.8797 in	W = 0.3125 in = 007.9248 mm
Dmajor =1 in	H_n = 0.690 in = 017.526 mm
Pitch (p) = 3.175 mm	R_n = H/1 = 0.2291355 mm
Dmin/2 = 11.17219 mm	Dmax/2=12.7 mm

H_b = 0.866 P

= 2.74963 mm

R_b = H/6

= .458271 mm

Where, H_b = Height of the thread, W = Width of Nut

R_b = Root radius of the bolt, H_n = Height of the nut

R_n = Radius of the nut

ACME threaded model results

Model 1: The basic model in this study used is a 23.4 mm diameter plan. Bolt with corresponding hexagonal nut with a pitch of 3.175 mm ACME threads .The bolt and nut are made of high carbon steel and it has a poisons ratio of 0.29 with an angle of 29^0. The contact between the bolt and nut is analyzed by finite element methods. The load is equal to stress 10 psi is applied on the bolt shank to produce the stress concentration (Table 1 and Figure 4).

Stress concentration factor (**K$_t$**) = $\sigma_{m/}\sigma_o$

K$_t$ =25.132/5.7489

K$_t$=4.3716

Models	1-8 UNC Thread Published values of *Kt (2D)	Whitworth thread form of *Kt (3D)	ISO-Metric of *Kt	ACME Thread Form of *Kt(3D)	Buttress thread form*Kt (3D)
MODEL 1	7.63	7.015	6.641	4.371619	3.334031
MODEL 2	5.81	5.773	5.98	4.568464	2.700146
MODEL 3	5.71	5.67	5.68	2.516703	2.241218
MODEL 4	5.4	5.363	5.499	2.575923	2.015132
MODEL 5	4.61	4.481	4.704	2.182126	2.058545
MODEL 6	5.32	5.413	5.728	2.23223	2.303686
MODEL 7	3.54	3.558	3.927	2.067838	2.7955

Table 1: Buttress threaded model results.

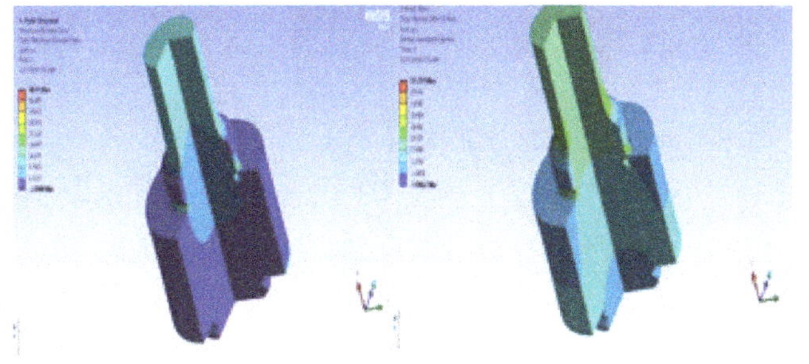

Figure 4: Model 1.

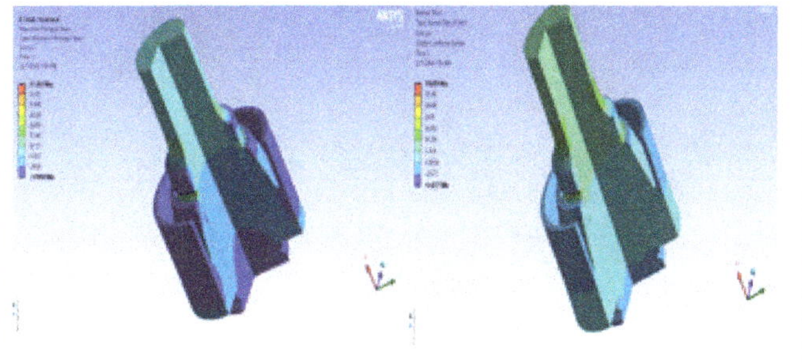

Figure 5: Model 2.

Model 2: The plane basic bolt and nut model is modified to find the stress concentration factor by adding groove to the face of the nut closes to the head of the bolt (Figure 5).

Stress concentration factor $(\mathbf{K}_t)=\sigma_{m/}\sigma_o$

$$K_t = 24.042/5.2636$$

$$K_t = 4.5684$$

Model 3: The groove is added to the lower end of the bolt in addition to the groove of the bolt (Figure 6).

Stress concentration factor $(\mathbf{K}_t)=\sigma_{m/}\sigma_o$

$$K_t = 24.334/9.669$$

$$K_t = 2.516$$

Model 4: A step is added to the nut with no grooves on the bolt and nut (Figure 7).

Stress concentration factor $(\mathbf{K}_t)=\sigma_{m/}\sigma_o$

$$=18.83/7.31$$

$$Kt = 2.5759$$

Model 5: A step is added to the nut and a groove is added to the lower end of the bolt (Figure 8).

Stress concentration factor $(\mathbf{K}_t) = \sigma_{m/}\sigma_o$

$$K_t = 19.743/9.0476$$

$$K_t = 2.1821$$

Model 6: A taper is added to the nut and a groove is added to the lower end of the bolt (Figure 9).

Stress concentration factor $(\mathbf{K}_t) = \sigma_{m/}\sigma_o$

$$K_t = 20.272/9.0815$$

$$K_t = 2.232$$

Model 7: A step is added to the nut and a groove is added to the lower end of the bolt. The shank diameter of the bolt is reduced (Figure 10).

Stress concentration factor $(\mathbf{K}_t) = \sigma_{m/}\sigma_o$

$$K_t = 15.997/7.7361$$

$$K_t = 2.0678$$

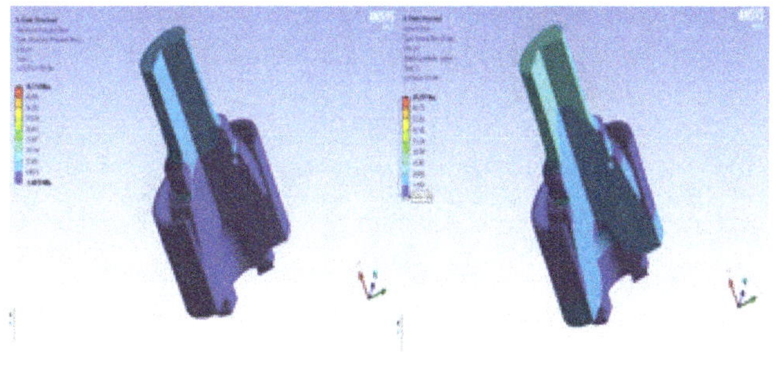

Figure 6: Model 3.

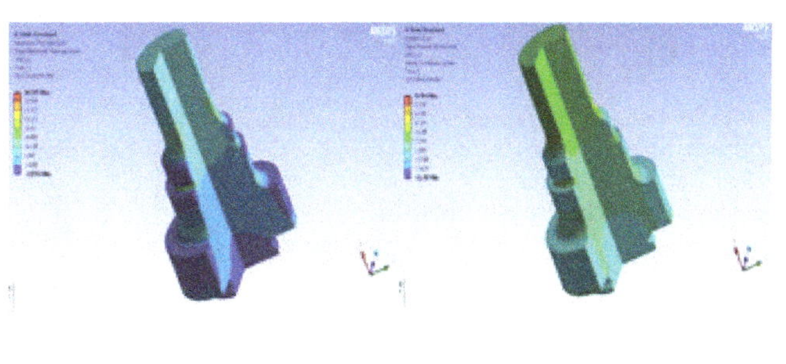

Figure 7: Model 4.

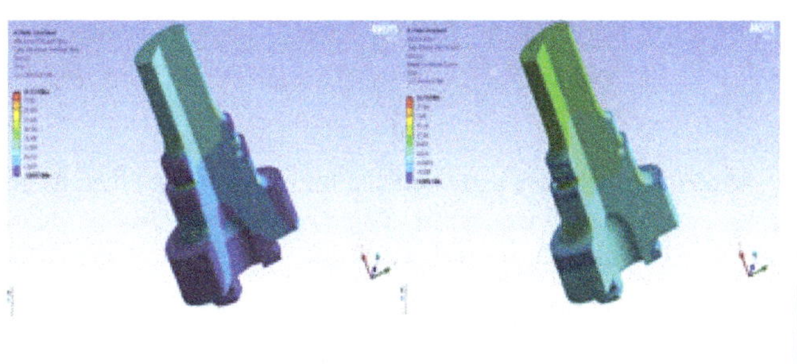

Figure 8: Model 5.

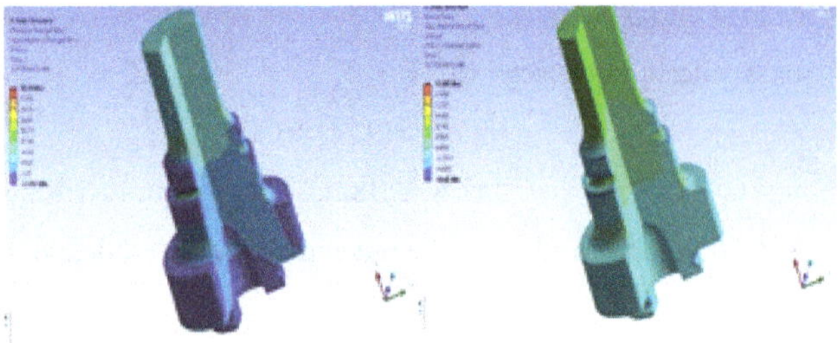

Figure 9: Model 6.

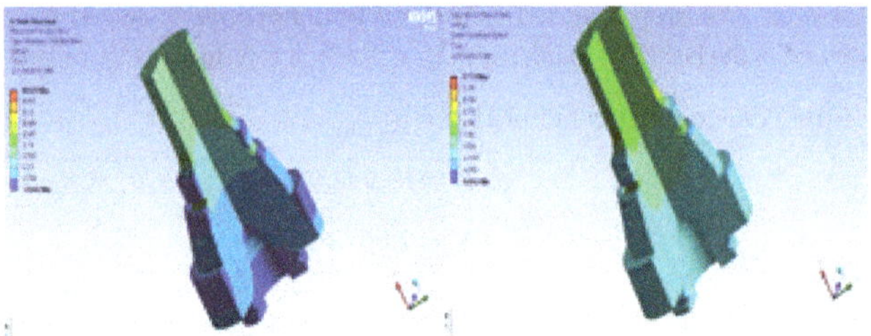

Figure 10: Model 7.

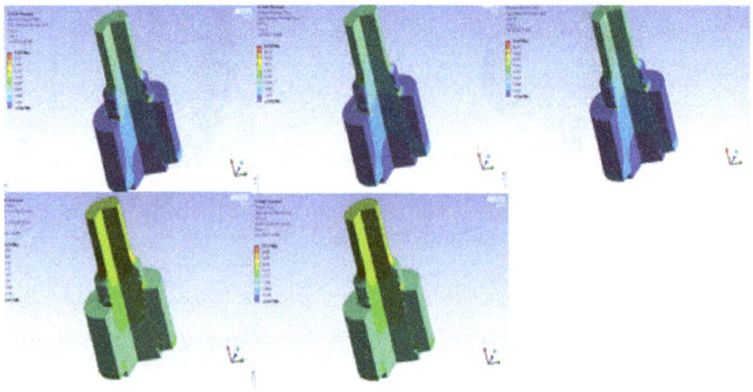

Figure 11: Buttress threaded model results (Model 1).

Buttress threaded model results

Model 1: The basic model in this study used is a 23.4 mm diameter plan. Bolt with corresponding hexagonal nut with a pitch of 3.175 mm buttress threads . The bolt and nut are made of high carbon steel and it has a poisons ratio of 0.29 with an angle of 45°. The load is equal to stress 10 psi is applied on the bolt shank to produce the stress concentration (Figure 11).

Stress concentration factor (**K_t**) = $\sigma_{m/}\sigma_o$

$$K_t = 20.721/6.215$$

$$Kt=3.334$$

Model 2: The plane basic bolt and nut model is modified to find the stress concentration factor by adding groove to the face of the nut closes to the head of the bolt (Figure 12).

Stress concentration factor (**K_t**) = $\sigma_{m/}\sigma_o$

$$K_t = 18.55/6.87$$

$$K_t = 2.70$$

Model 3: The groove is added to the lower end of the bolt in addition to the groove of the bolt (Figure 13).

Stress concentration factor (**K_t**) = $\sigma_{m/}\sigma_o$

$$K_t = 16.269/7.259$$

$$K_t = 2.241$$

Model 4: A step is added to the nut with no grooves on the bolt and nut (Figure 14).

Stress concentration factor (**K_t**) = $\sigma_{m/}\sigma_o$

$$K_t = 17.126/8.4987$$

$$K_t = 2.015$$

Model 5: A step is added to the nut and a groove is added to the lower end of the bolt (Figure 15).

Stress concentration factor (**K_t**) = $\sigma_{m/}\sigma_o$

$$K_t = 17.412/8.4584$$

$$K_t = 2.05$$

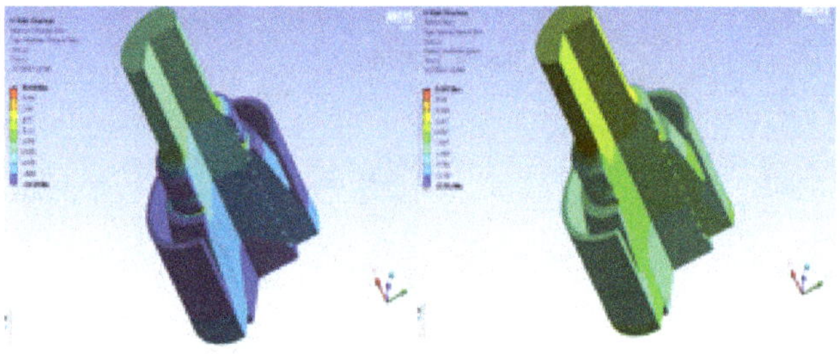

Figure 12: Buttress threaded model results (Model 2).

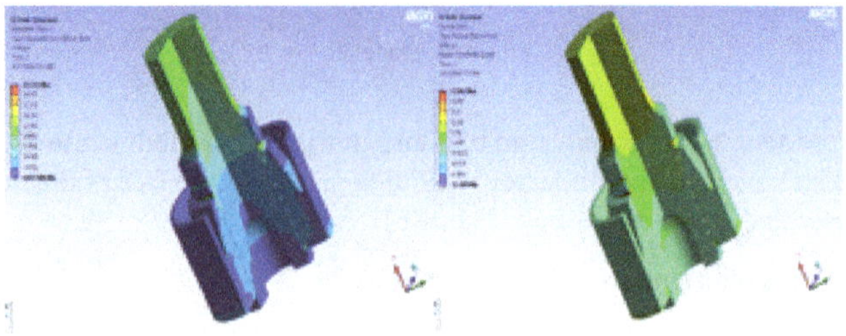

Figure 13: Buttress threaded model results (Model 3).

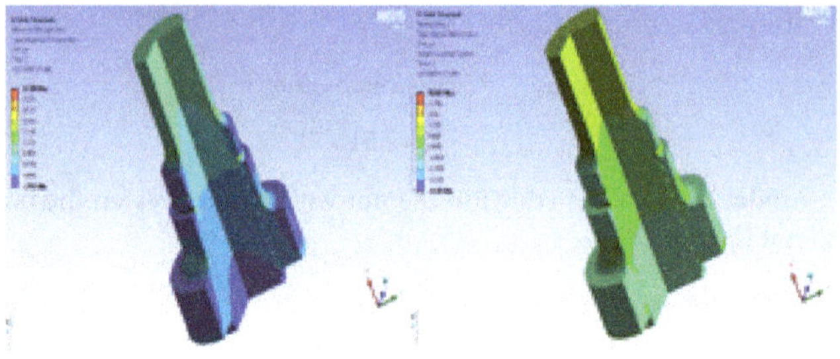

Figure 14: Buttress threaded model results (Model 4).

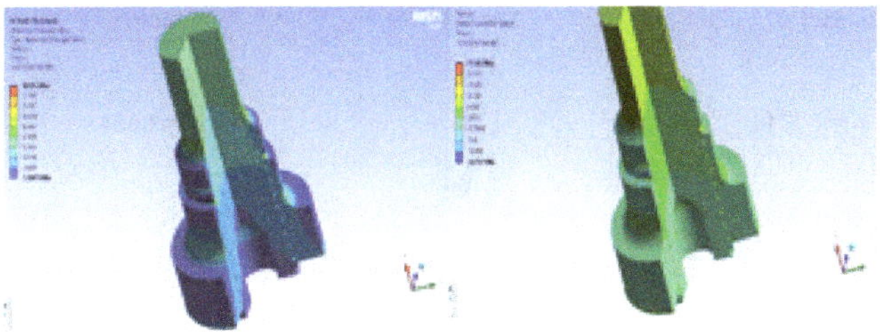

Figure 15: Buttress threaded model results (Model 5).

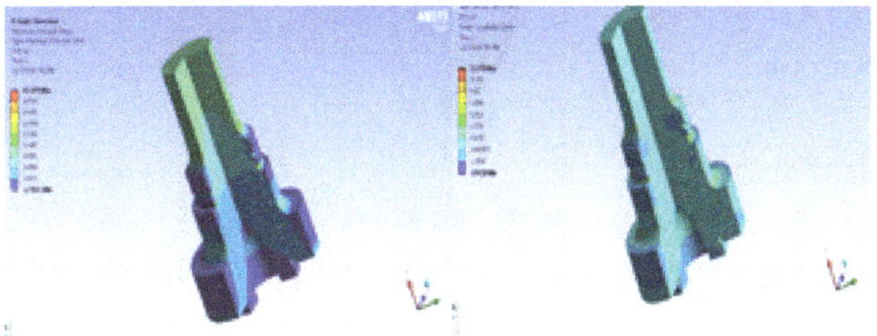

Figure 16: Buttress threaded model results (Model 6).

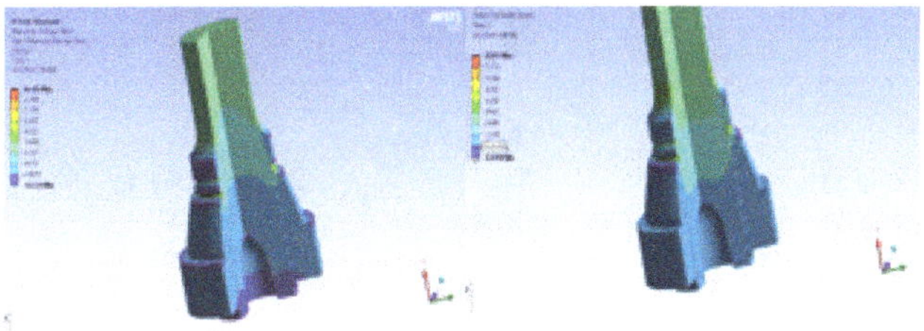

Figure 17: Buttress threaded model results (Model 7).

Model 6: A taper is added to the nut and a groove is added to the lower end of the bolt (Figure 16).

Stress concentration factor (K_t) = $\sigma_{m/}\sigma_o$

$$K_t = 17.385/7.5466$$

$$K_t = 2.30$$

Model 7: A step is added to the nut and a groove is added to the lower end of the bolt. The shank diameter of the bolt is reduced (Figure 17).

Stress concentration factor(K_t) = $\sigma_{m/}\sigma_o$

$$K_t = 18.3636.5686$$

$$K_t = 2.7955$$

Results and Discussion

The main goal of this work was to investigate the effect of relative changes in the geometries of the nut and bolt on the stress concentration for the assembly [11]. Any changes in the geometries will increase the cost of the fastener system; the specific values for the stress concentration factors will change as the size of the nut and bolt changes [12-15].

Simulation work has been done to analyze the stress concentration on various models using 3-D analysis. For this "ANSYS" has been used as simulation tool which works on finite element method [14]. The main aim of this work is to find the stress concentration factor values (K_t) for nut and bolt connectors of 1-8 UNC thread form by using 3-D modeling and compare the relative $\mathbf{K_t}$ values with the standard values from 2-D modeling and to find the optimum nut and bolt connector model among the various models of 1-8 UNC thread form. This material is a low-alloy, high-carbon steel commonly used in aerospace fasteners and has an elastic modulus of 205 GPa and Poisson's ratio of 0.29. The load applied is sufficient to produce a nominal stress of 10 psi in the shank of the bolt [15-18].

This study is to show quantitative comparison of effect of various proposed design modifications in the literature [3] on the stress concentration factor published previously to the work which has been done with ANSYS. Validation was also done for the same thread forms proposed in the literature (Figure 18).

Model graph for different thread forms

Machines are assembled from individual parts. The parts may need to be joined sometimes in permanent or temporary connections. The temporary connections or joints are normally made by using screws bolts or nuts and bolts (Table 2). The temporary joints render the convenience of disconnection as frequently as required for such purposes as inspection, repairs, adjustment or replacement. The most important feature of the screws or bolt is that they are standardized and are available as standard part ready to use. The threaded fasteners are used to hold two parts. The fastener consists of two parts. An externally threaded member whose length is normally greater than the diameter carries threads over entire or part length from end. The other end of this member is larger in area (you can assume that the end is upset). This part is inserted in coaxial holes in two parts taking care that the "upset" or larger end is greater in dimension than the holes so that it does not pass through the hole. The end that comes out of the holes carries thread upon which another internally threaded part is wound (Figure 19).

The main goal of this work was:

- To find the effect of relative changes in the geometries of the nut and bolt on the stress concentration for the assembly on different thread (Buttress Thread Form) after validating the results of published journal where he used 1-8 UNC Thread Form.
- To find the optimum design for bolt and nut connector.

With this in mind, the following conclusions can be drawn from the study

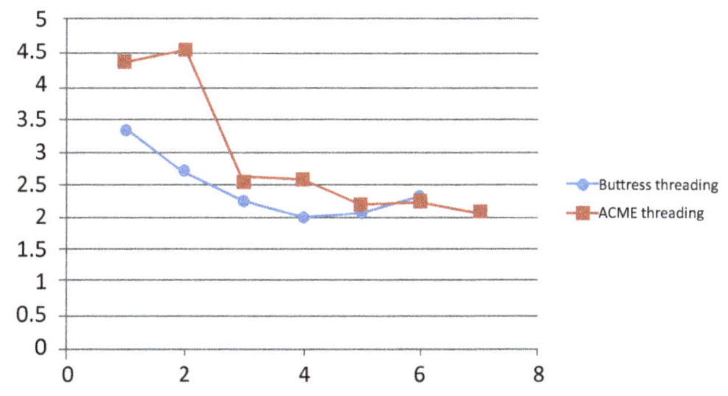

Figure 18: Buttress thread and ACME threads stress concentration factor comparison.

Design Description	Sketch 2D Model	3D Model For ACME Thread	Stress concentration factor	3D Model For Buttress Thread	Stress concentration factor
Model-1 This is the base model which features a 1 – inch diameter bolt with corresponding nut.	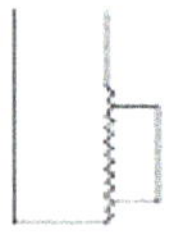	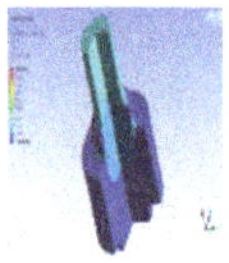	Kt=4.3716	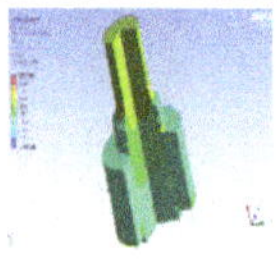	Kt=3.334
Model-2 A groove is added to the face of the nut closer to the head of the bolt.	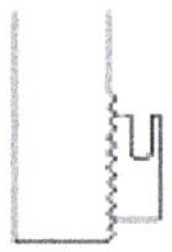		Kt=4.5684	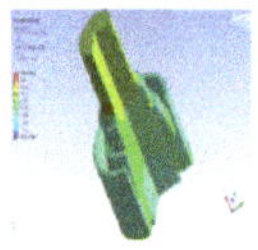	Kt=2.70
Model-3 A groove is added to the lower end of the bolt, in addition to the groove on the nut.	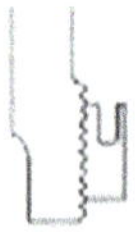		Kt=2.516	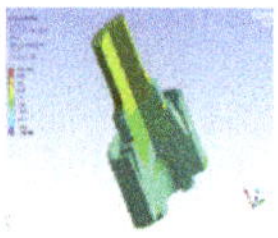	Kt=2.241
Model-4 A step is added to the nut with no grooves on the bolt or the nut.	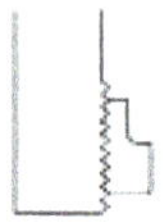	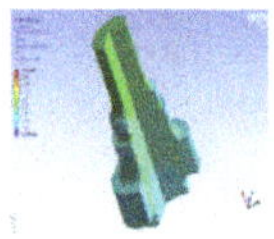	Kt=2.5759	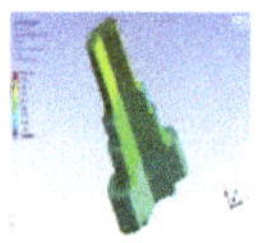	Kt=2.015
Model-5 A step added to the nut and a groove is added to the lower end of the bolt.		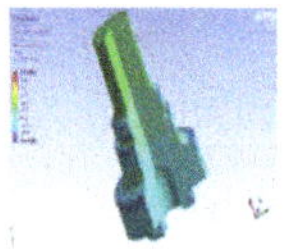	Kt=2.1821	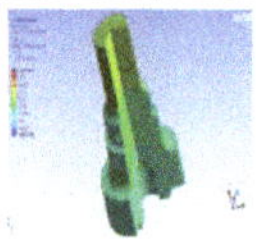	Kt=2.05
Model-6 A step is added to the nut and a groove is added to the lower end of the bolt. the shank diameter of the bolt is reduced.		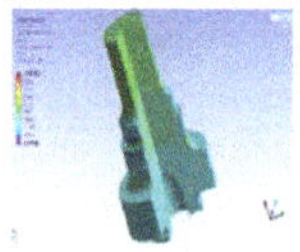	Kt=2.232	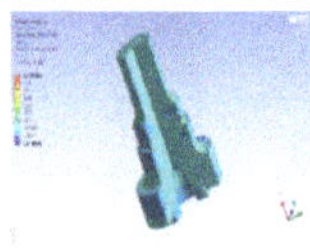	Kt=2.30
Model-7 A taper is added to the nut and a groove is added to the lower end of the bolt.		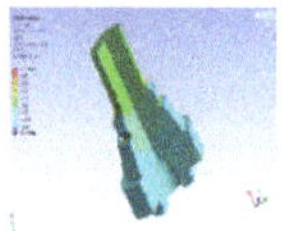	Kt=2.0678	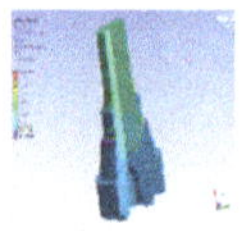	Kt=2.7955

Table 2: 2D and 3D results and comparision; Results from various thread forms compare with validation.

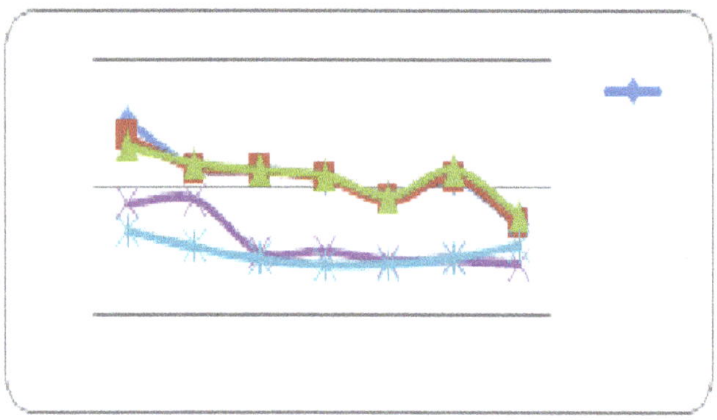

Figure 19: Model Vs Stress concentration factor of various thread forms.

1. Stress concentration factor is maximum at first thread and minimum at fourth thread. It is observed that in an individual thread the highest stress occurred below the deepest point of the thread root.

2. The validated results are found to be reasonably accurate when compared with the published values. The minor error is due change in software used. Here we used Ansys whereas published Journal used Algor for FEM.

3. The stress concentration Factor has been reduced for different models of designed Buttress thread form with respect to base model. So, buttress Thread is effective when compared to 1-8 UNC Thread and it can as well use for structural works.

4. Among the three models done for buttress thread- model in which, a step is added to the nut with no grooves on the bolt and nut has less stress concentration factor. So, Model 4 is best design.

5. Essentially all of the stress reduction methods based on a modification of the threaded section and nut remove material. Therefore, even though the stress concentration changes due to changes in the nut and thread end appear to be less dramatic than those resulting

from reducing the shank diameter, the nut and thread changes may be more efficient from a material utilization and cost standpoint.

Conclusion

3-D analysis gives exact values than 2-D analysis, we can consider the stress concentration factor (**K_t**) values obtained in this study as more exact and standard values for 1-8 UNC, ACME and Buttress threads in nominal dimensions. And these **K_t** values for nominal dimensions can be used in design modifications.

The stress concentration factor is less for Buttress thread form, compared to the other forms of threads. So, Buttress thread form is effective when compared to 1-8 UNC thread form and ACME thread form and it can be used for structural woks. Model 4 is the optimum model for nominal dimensions. This will help in reducing the material required for bolt in the design and cost consideration.

References

1. Muhammad-Tandur KH, Magami AI (2008) Computing the stress concentration factor in bolted joint using FEM. International Journal of Applied Engineering Research 3: 369.
2. Hetenyi M (1943) A photoelastic study of bolt and nut fastenings. Trans ASME 65: 93-100.
3. Seika M, Sasaki S, Hosono K (1974) Measurement of stress concentrations in threaded connections, Bull. JSME 17: 1151-1156.
4. Patterson EA, Kenny B (1987) Stress analysis of some bolt-nut connections with some modifications to the external shape of the nut. J Strain Anal 22: 187-193.
5. Patterson EA, Kenny B (1985) Stress analysis of some bolt-nut connections with modification to the nut thread form. J Strain Anal 20: 35-40.
6. Kenny B, Patterson EA (1985) Load and Stress Distribution in Screw Threads. Exp Mech 25: 208-213.
7. Venkatesan S, Kinzel GL (2006) Reduction of Stress Concentration in Bolt-Nut Connectors. Journal of Mechanical Design 128: 1337-1342.
8. de Koning AU, Henriksen TK (1996) Damage tolerance analyses of bolt/nut assemblies, National Aerospace Laboratory NLR, Amsterdam, The Netherlands.
9. Aryassov G, Strizhak V, Zverkov A, Copeland P, Michael O (2002) Dynamic lightening of bolted joint. Department of Machine Science.
10. Lesniak Mickel JR, Zickel Christopher J, Whelch S, Jhonson DF (2000) An innovative Polaris cope for photo elastic stress analysis.
11. Oberg E (2004) Machinery's Handbook for mechanical engineering (27thedn), Industrial Press Inc, New York.
12. Fukuoka T, Hamada M, Yamasaki N, Kitagawa H (1986) Stresses in Bolt and Nut. Bull JSME 29: 3275-3279.
13. The Finite Element Method And Applications In Engineering Using, Ansys® By Erdogan Madencilbrahim Guventhe University Of Arizona Printed In The United States of America, Springer.Com.
14. Sethuraman R, Sasi Kumar T (2009) Finite Element Based Member Stiffness Evaluation of Axisymmetric Bolted Joints. Journal of Mechanical Design 131: 11-12.
15. NagendraBabu R, Ramana KV, Mallikarjuna Rao K (2008) Determination of Stress Concentration Factors of a Steam Turbine Rotor by FEA.Proceedings of World Academy of Science, Engineering and Technology 1307-6884.
16. Bretl JL, Cook RD (1979) Modeling the Load Transfer in Threaded Connections by Introduction to machine design.
17. Tanaka M, Miyazawa H, Asaba E, Hongo K (1981) Application of the finite element method to bolt-nut joints. Bull JSME 24: 1064-1071.
18. Bickford JH, Nasser S (1998) Handbook of bolts and bolted joints, Marcel Dekker, New York.

XFEM Potential Cracks Investigation using Two Classical Tests

Martínez Concepción ER[1]*, De Farias MM1 and Evangelista F[2]

[1]PPG, Geotechnics, University of Brasilia, DF, Brazil
[2]PECC, Structures, University of Brasilia, DF, Brazil

Abstract

Extended Finite Element Method (XFEM) is used in this work, first to perform the simulation of crack initiation and propagation mechanisms in plane models and then to determine the stress distribution singularities in the closest surroundings of a front fracture inserted in three-dimensional models. The essentials of XFEM is the well-known Finite Element Method (FEM) adding to degrees of freedom and enrichment functions, which serve to describe local discontinuities in the model. In XFEM, the fracture geometry is developed independent of the mesh, allowing it to move freely through the domain, without the need to adapt the mesh to discontinuity. In other words, the XFEM reproduces the discontinuity of the displacement field along the fracture, without discretizing this feature directly in the mesh. XFEM carry out the spatial discretization of two classic models in Fracture Mechanics: the single-edge-notch bending test (SEN (B)); and the disck-shaped compact tension test (CDT). The propagation criterion is based on the proportion of energy released and the stress intensity factors (SIF). The solutions provided by the XFEM numerical model indicated an excellent agreement with the results obtained from the experimental data.

Keywords: Fracture mechanics; Fracture; Extended finite element method; XFEM; Enrichment; Stress intensity factor

Introduction

The demand for a fracture analysis method has conducted important contributions since the 1960s. Research was initially distinguished by empirical, analytical, and semi-analytical fundamentals Gross et al. [1] Rice [2]. These methods can be used in simple geometry problems and under specific boundary and loading conditions. For other more complex problems it is common to appeal to numerical methods due to the need to make several simplifications in the analytical models. Thus, several works were available using the traditional Finite Element Method (FEM) to analyze fracture toughness, but this resulted in a complex mesh which required to be adapted to the fracture surface and also be updated at each time-step to refine the elements size positioned in the surrounding area of the fracture tip. Recently, the Extended Finite Element Method (XFEM) has gained in popularity in its use by the scientific community, it allows developing strategies to analyze fractures without the need for a refinement of the mesh. The XFEM is considered an extension of the conventional FEM and is based on the unit partition concept (i.e., sum of the shape functions must be equal to unity), and were studied by Belytschko and Black [3] and Moës et al. [4] the developers of their initial working algorithm.

Since the publication of Belytschko and Black [3] it is usual to find fracture researches on two-dimensional field using XFEM and highlighted for the explicit description [4]. In addition, the solution of fractures in three-dimensional models is constantly updated with news features for improve the approximations, from the purely explicit, implicit forms or coupling both as studied by Sukumar et al. [5]; Stolarska et al. [6]; Fries and Belytschko [7]; Baydoun and Fries [8].

In this paper, a numerical analysis is developed using the Extended Finite Element Method (XFEM) to study the initiation and propagation process in fractures. The validation of XFEM approximation by numerical simulations in two dimensions of fracture problems with known solutions and the numerically extraction of stress intensity factors (K) in three-dimension solid is performed.

XFEM Approach for Fracture Problems

The XFEM was introduced first by Belytschko and Black [3] and Moës et al. [4] and incorporating enrichment functions and degrees of freedom in addition to the conventional approximation of finite elements in the region where the fracture is placed, to simulate discontinuities and singularities. The type of enrichment functions are described as asymptotic (capture the singularity at the crack tip) and discontinuous (representing the gap between the crack surfaces). The enriched area surrounding the crack tip and over of the fracture are exemplified in Figure 1 (a). The mathematical formulation to approximate the displacement field through an implicit-explicit description was introduce for Baydoun and Fries [3] as follows:

$$u(x)=\sum_{i\in I} N_i(x)u_i + \sum_{i\in I^{cut}} N_i^*(x)\left[H(x)-H(x_i)\right]a_i + \sum_{i\in I^{branch}} N_i^*(x)\left[\sum_{m=1}^{4}\left(B^m(x)-B^m(x_i)b_i^m\right)\right] \quad (1)$$

The first part of equation represents the classical FEM approximation, defined by a continuous shape function $N_i(x)$ and unknowns at nodal points . The enrichments are taken for the shape functions N_i^*. I. is the set of all nodes in the domain. The discontinuity in the field of fracture displacements and the enrichment that captures the special behavior at the crack tip are considered, respectively, by the second and third terms. Two types of enrichment functions are implemented by the XFEM formulation, the Heaviside function, $H(x)$, and the fracture tip asymptotic function $B(x)$, as shown in Figures 1b and 1c.

Through process of investigation in two dimensions fracture problems, Moës et al. [4] explained exactly how discontinuity functions are added in the finite element approximation.

Denoted to Figure 1a the enhanced nodes belonging elements

***Corresponding author:** Martínez Concepción ER, PPG, Geotechnics, University of Brasilia, DF, Brazil
E-mail: edel.rolando@gmail.com

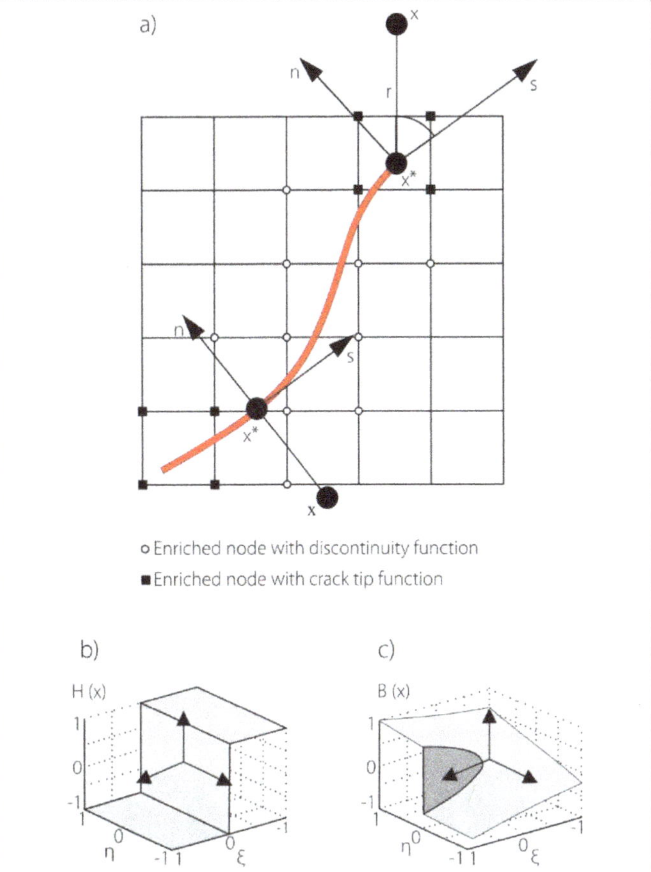

Figure 1: (a) Enriched nodes by XFEM functions (b) Heaviside enrichment function for discontinuity (c) Branch asymptotic function for fracture tip.

crossed by the fracture interface (set of nodes I^{out}), as well the nodes elements located in the crack tip (set of nodes *I branch*). Along the discontinuities, the nodes are enriched function degree entitled Heaviside; in this case the finite element boundaries are cut by the fracture, while the nodes in elements around the crack tip are enriched with the fracture tip functions named Branch. In Figure 1a encircled nodes are enriched with the Heaviside function and nodes with square symbols are enriched with crack tip functions.

The nodes set within a region around the discontinuity tips present a geometric enrichment. Since the origin of XFEM a geometric criterion was defined for the enrichment zone in order to determine the nodes that will be enriched by singularity functions. In particular, the strategy to include enrichment functions is useful for an efficient approximation of crack singularities and discontinuities as well as interface changes. The addition of discontinuous and asymptotic functions to the elements surrounding the crack tip allows to correctly capturing the singularity in this region Moës et al. [4]. In conditions that the crack tip does not end in boundary elements, functions also describe the discontinuity on the fracture surfaces. Adopting a polar coordinates system, with origin on the crack tip and tangential coordinates on the trajectory propagation, as is shown in Figure 1a. Where is the shortest vector length extended from the crack tip and is the angle measured from rectangular to polar coordinates; the fracture tip enrichment functions for an elastic and isotropic material, were presented by Sukumar et al. [5] as follows:

$$\{B(x)\}_{i=1}^{4} = \left(\sqrt{r}\,sen\left(\frac{\theta}{2}\right), \sqrt{r}\cos\left(\frac{\theta}{2}\right), \sqrt{r}sen\left(\frac{\theta}{2}\right)sen(\theta), \sqrt{r}\cos\left(\frac{\theta}{2}\right)sen(\theta)\right) \quad (2)$$

The use of the enrichment function involves the addition of four degrees of freedom to all nodes in the improved region. For each freedom degree, is associated a term of the function. The pair (r, θ) represents local polar coordinates at the crack tip for the interval of $-\pi \leq \theta \leq \pi$. The expression $\sqrt{r}\,sen\left(\frac{\theta}{2}\right)$ is fundamental in the formulation of the function $(x)$for the reason that it describes the discontinuity on the fracture surfaces. Furthermore, this function is responsible for the representation of phenomenon along crack length, lies the fracture approximation occurs. The other functions of $B(x)$ are used to improve the approximation of the solution in the zone proximately to the fracture tip.

For the discontinuity representation, the XFEM can make use a set of levels Sukumar et al. [5], endorsed for describe the crack interfaces geometry and without required to coincide with the element boundary, or that the surface of the interface matches the element faces. Before the element failure, the enrichment function degenerates into the conventional finite element; as soon as the element presents the damage, it activates the Level Set Method (LSM), which is based on enrichment functions that will assume the modeling of the discontinuities. Numerical simulation by XFEM and LSM incorporated allows modeling the movement of curves and surfaces in a fixed mesh. The mathematical difficulty to represents the fracture problem is assumed by the LSM. For the direction of fracture propagation, the XFEM arranged functions of set levels to track the fracture surface at each time step by Stolarska et al. [6].

Among other characteristics, in XFEM is implemented the approach known as phantom nodes by Song et al. [9]; Dassault Systèmes [10]. This mathematical artifice is based on the internal duplication of each enriched element with the addition of phantom nodes.

Benchmark Cases for XFEM Validation

The numerical simulation of two validation cases is target to test the XFEM accuracy through the comparison of the load-opening curve, computing the stress intensity factors and reproducing the direction of the fracture in the propagation development. The fracture cases analyzed are: disk-shape compact tension, CDT, (ASTM D7313, 2013), and the three-point notched bending tests, SEN (B), (ASTM E1820 [11]. The ASTM standards describe the CDT and SEN (B) tests as important ways to measure fracture toughness. The experimental application determine the magnitude and singularity of the stress field originated in the volume surrounding the fracture front through the parameters, identified as stress intensity factors as exposed by Irwin [12]. The crack problems were taken from experiments data on: disk-shape compact tension available by Wagoner et al. and in concrete beams tested in flexion described in the research of Evangelista et al. [13]. This researches describes the geometry and materials of the specimens, the laboratory test configuration and results (load-opening curves) obtained, and the analogous fracture energy values, statistically ensuring reliable results.

Disk-shape compact disk test

The first validation model is represented in Figure 2. The cylindrical sample extracted from a pavement and subjected to uniaxial tensile following the ASTM D7313 (Wagoner et al. in 2005; ASTM D7313, 2013). The specimen has the following dimensions: = 150, = 25, = 25, = 35 and = 110, and the length of the notch is + = 62.5, while the thickness is = 50. In Table 1, are summarized properties of the material used in the disk.

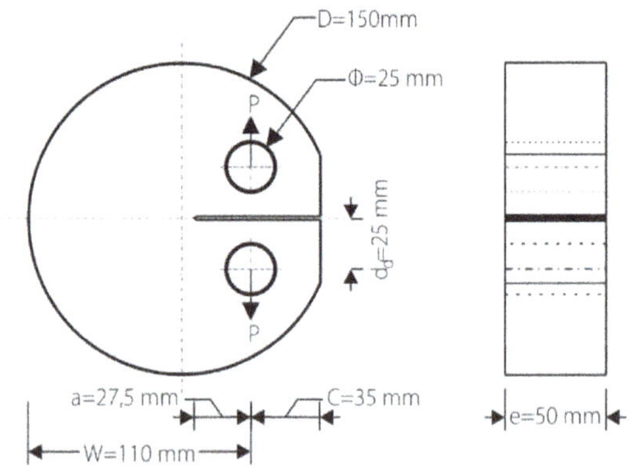

Figure 2: Specimen dimensions for disk-shape compact disk test.

Parameters	Value
Fracture Energy	G_F = 328 *N*/m
Tensile Strength	$f_t^{'}$ = 3.56
Young's Module	*E* = 14.2
Poisson's ratio	v = 0.35

Table 1: Material properties for the compact disk (Wagoner et al., 2005).

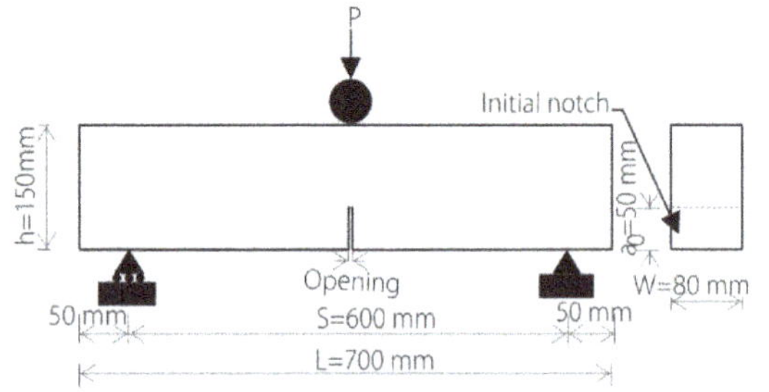

Figure 3: Single-edge notched bending test SEN (B).

Parameters	Value
Fracture Energy	G_F= 99 *N*/ m
Tensile Strength	$f_t^{'}$ = 5.04 MPa
Young's Module	E = 27GPa
Poisson's ratio	v= 0.19
Concrete Density	$\rho conc$ = 2400 kg/m³

Table 2: Material properties for the beam (Evangelista et al., 2013).

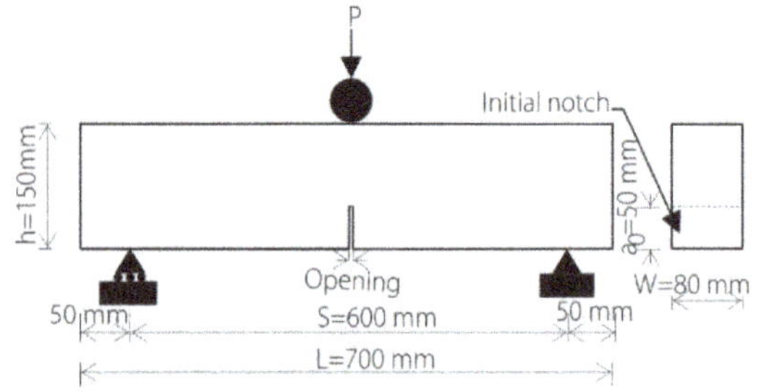

Figure 3: Single-edge notched bending test SEN (B).

Single-edge notched bending test

To obtain a load versus crack opening displacement that is used for determine fracture properties of the material, Evangelista et al. [13] performed this classic test, represented in Figure 3. The material parameters considered in the beams are exposed in Table 2.

Numerical Results and Discussion

Discretization density analysis

The mesh density is analyzed thought in a function of stress intensity factor convergence for a simulation of CDT in the three-dimensional space, while in the SEN(B) model the mesh density convergence is investigated for the two-dimensional propagation of arbitrary crack tests, with base in the approximation of the peak load.

CDT model: For both 2D and 3D models, three mesh configurations were analyzed: coarse, intermediate and fine, as shown in Figure 4 for the plane model. The numerical results for CDT model in 2D and 3D, presented in the research, correspond to the fine discretization.

A convergence study as a function of mesh density, was performed in the 3D fracture models as shown in Figure 5 for the three types of discretization.

Density mesh analysis was started with approximately 5 · 10^5 elements and is reached to 75 · 10^5 elements. Accordingly increasing the mesh density in the simulations, that is, decreased size of the elements, it was observed that the stress intensity factor in the second mode (K_{II}) was converging to a minimum, as shown in Figure 6. In the first simulation a K_{II} of 0.10$MPa\sqrt{m}$ was obtained and in the

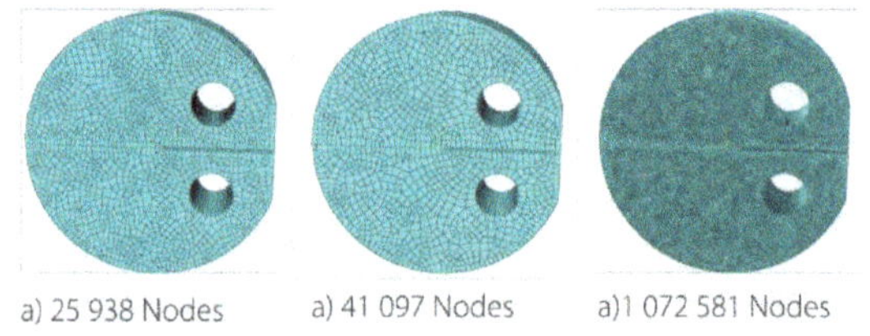

Figure 4: Mesh pattern in bi-dimensional model (a) coarse (b) intermediate (c) fine.

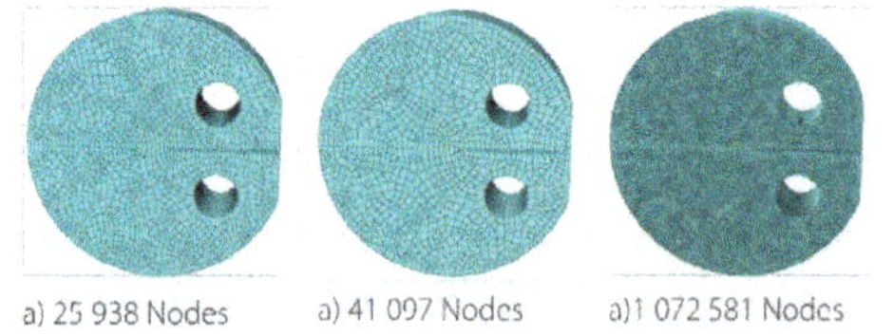

Figure 5: CDT tri-dimensional meshes (a) coarse (b) intermediate (c) fine.

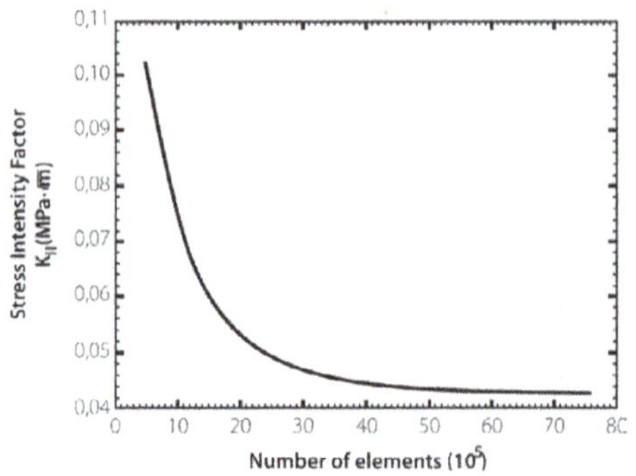

Figure 6: Trend line during the sensitivity analysis for the three-dimensional discretization in the CDT model.

sequential simulations the value decreased to approximately 0.040 $MPa \sqrt{m}$, that is, the increase in the number of elements suggests a better approximation in the magnitudes of the stress intensity factors. This statement is a consequence for how the test is conducted, where dominate a fracture propagation in a pure opening mode (K_I). After approximately $17{\cdot}10^5$ finite elements the value of the stress intensity factor in Mode II of propagation begins to decrease in smaller increments, according to Figure 6, since this amount of elements was considered appropriate to carry out all the following simulations for the three-dimensional CDT fracture problem.

SEN (B) model: In the single-edge notched bending numerical simulation was considered a structured discretization with three different mesh densities, as can be observed in Figure 7. The type element use is the plane strain and the mesh was refined in the fracture region. The maximum loads obtained, for the same applied displacement, is shown in Table 3. In order to evaluate the mesh density performance, the four cases enumerated in Table 3, are offered the discretization configuration, the evolution of the relative error and the processing time for the different simulations.

The error associated to the approximation of the peak load of the test was calculated according to the equation:

$$e = \frac{P_{\max.test} - P_{numerical}}{P_{numerical}} .100 \quad (3)$$

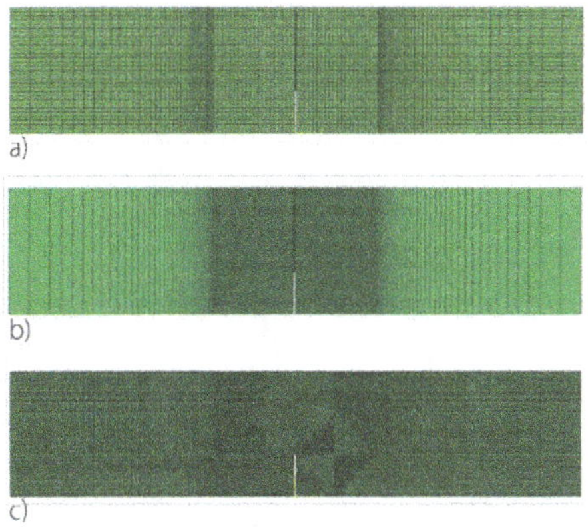

Figure 7: Discretization of the beam in finite elements (a) coarse (12,625 elements) (b) intermediate (22,610 elements) (c) fine (41000 elements).

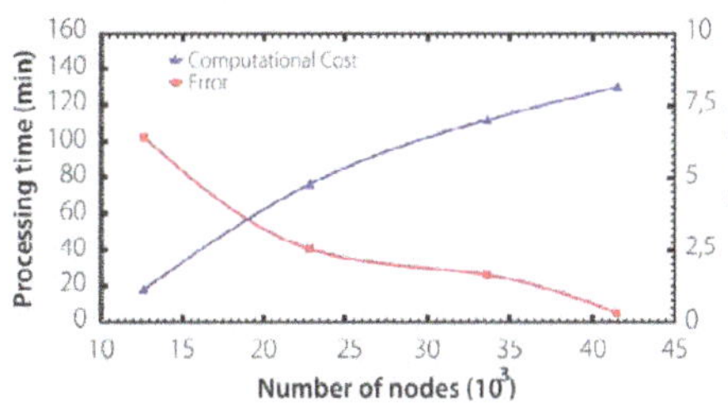

Figure 8: Error representation and the processing time as a function of the SEN B) discretization model.

The Figure 8 shows the magnitude of the XFEM approximation error based on the maximum test load. In the abscissa axis is represented the number of nodes used in the model, variable used to characterize the mesh density, while the axis of ordinates represent the processing time consumed in the simulation, as well as the completed relative error through the XFEM approximation. As can be observed, the increase in the number of nodes implies in extended processing times and a significant reduction of the relative error, which influences are the practically insignificant value of 0.30%. It is reasonable, that XFEM errors in the numerical model are small, since the method implements enrichment functions around the fracture to obtain better approximations in the singularities of reaction forces field to a better-quality mesh model.

Opening displacements and stress intensity factors

The fracture openness results in the bi-dimensional model and the stress intensity factors in the three-dimensional models corresponds to the improved discretization of both tests. Plane models did not have to devote considerable time to shape the geometry and boundary conditions. In the three-dimensional model, the geometry and the variables to assign are in fact more complex. Firstly, contours are established to perform an evaluation of the corresponding contour integral and determine the stress intensity factor, $K_{i,ring}$. In the simulations with inserted fractures it was possible to estimate for each off evaluation point, that is the stress intensity factor in modes I, II and III. Note that only the evaluation points along the fracture front are computed. The approximation is performed from the stress intensity factor value at each contour evaluation point $K_{i,ing}$ by the following expression:

$$K_i = \frac{1}{N-n} \sum_{ring=n+1}^{N} K_{i,ring} \quad (4)$$

where is the number of rings excluded and is the total of contours requested in the contour integral. The variable specifies the number of these elements used in the contour integrals computation. The stress intensity factor is expected as the average value, generally the first contours are diverging values so omitted.

CDT model

The deviation in the fracture path at the CDT test advises the presence of the pure opening mode. The Figure 9 shows the enriched elements that define the fracture faces.

However, Mode I propagation predominate in the bi-dimensional simulations, was found results which deviated fracture trajectory from the horizontal axis of the model. The effect shown in numerical modeling is related to the method capability to detect changes in the fracture energy. It was experimental that the Von Mises equivalent stresses at the Gauss mesh points take up maximum values around the notch vertex. The behavior response represented by the load-opening curve was compared with the experimental results in Figure 10, was observed a good correspondence between XFEM model and the research laboratory results. The curve slope for the loading period in the numerical simulation matches with the experimental results, at the

Model	Number of elements	Number of nodes	CPU time (min)	$P_{máx}$ Nodal (*kN*)	Approximation error to P = 3.56 *kN* (%)
1	12625	12895	18	3.33	6.46
2	22610	22967	76	3.47	2.52
3	33678	33966	112	3.5	1.69
4	41000	41475	130	3.55	0.3

Table 3: Discretization analysis and simulation results for the SEN (B) model.

maximum load obtained the XFEM estimates a slight higher than the results of Wagoner et al. and after the peak, the numerical results are intermediate in relation to the experimental curves.

The 3D model is used to investigate the opening process in a crack inserted to the numerical simulations and performed using the same material parameters of the two-dimensional model. The Figure 11 shows the distribution of stress intensity factors along the fracture length front. For the number of plotted points correspond a number of evaluation points used in the XFEM model, there is a predominance of Mode I fracture propagation.

SEN (B) model

Based on the experiment procedure, it is expected that the fracture simulation will exhibit Mode I dominated propagation, as can be perceived in Figure 12. The total fracture length was verified at the beam mid-span, achieve approximately the half distance between the notch tip and the model top. The symmetrical stresses are due to the geometry and contour conditions in the model, as well as the load position. The stress registered on the projection of the fracture tip at the beam top.

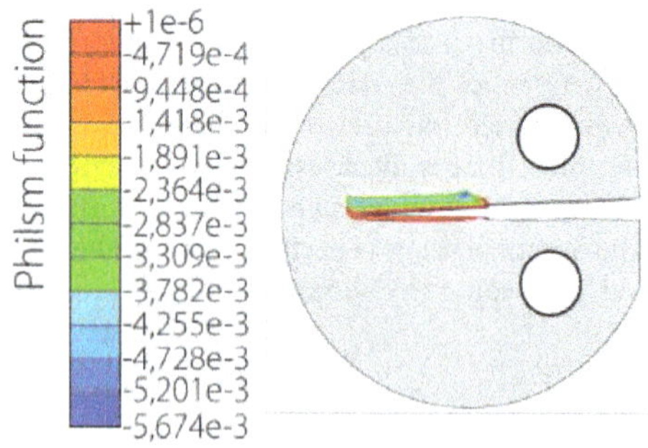

Figure 9: Enriched elements represented by the philsm function along the fracture in the bi-dimensional CDT model.

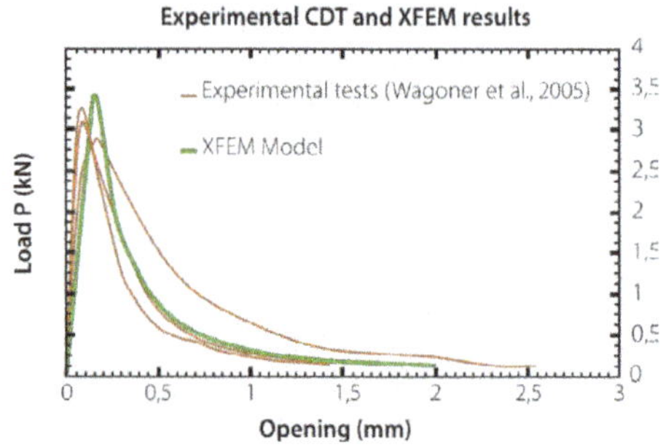

Figure 10: Experimental and numerical results by the use of XFEM in the CDT model.

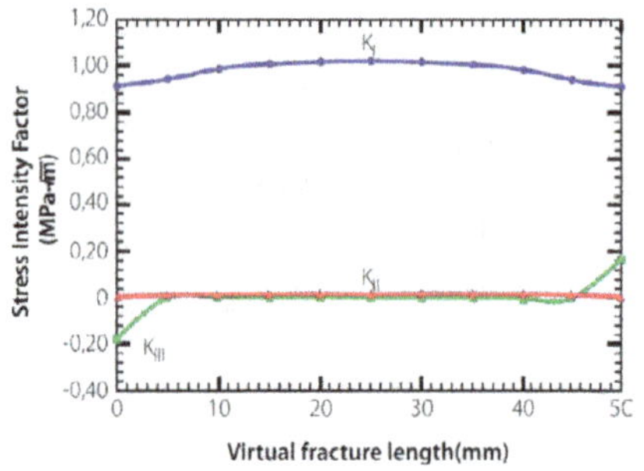

Figure 11: Stress intensity factors at evaluation points distributed along the fracture length.

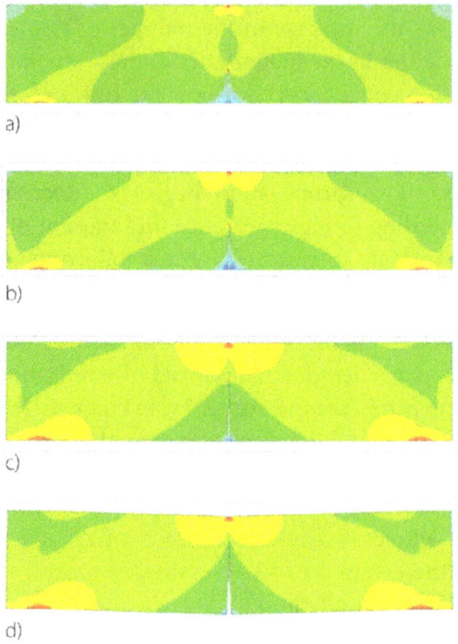

Figure 12: von Mises stress representation for different crack lengths in SEN(B) simulation (a) = 0,1 (b) = 5 (c) = 22 (d) = 54.

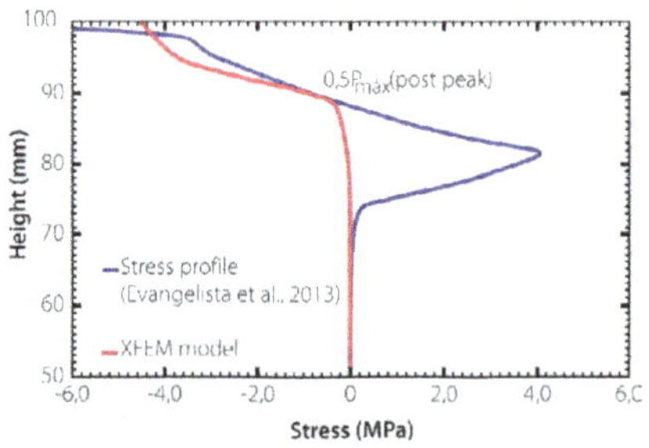

Figure 13: Stress evaluation in fracture contiguous elements.

In a trajectory, close to the fracture line the development of the stress state was investigated, the stresses distribution in the beam mid-span from the notch to the loaded surface in the interval of h (50 ≤ h ≤ 150) was considered. The final stress at the integration points for each crack interface element was plotted in Figure 13 and compared with the experimental results determined by Evangelista et al. [13] for the 0.5 p_{max} pós-peak loading stage. The difference in the 75-85 mm beam height sector is due to cohesive elements considerations in the numerical model used in the reference research. Despite this difference, was perceived a relationship in stress at the load point of the model.

The load-opening curves obtained by the XFEM model are compared with the experimental data in Figure 14. A maximum load achieved by XFEM model represents the reaction force acting at the top center node for a controlled displacement. During the loading path the numerical model present a premature change in the curve slope. This is a consequence numerical computations sensitivity, but in overall, the results are satisfactory. The distribution of the stress intensity factors is determined on a beam 3D model with an inserted vertical fracture, positioned over the notch in the model mid-span and represented by a plane. The 3D model was discretized with an accurate density of elements used in simulation approximately $1 \cdot 10^7$ nodes. The scheming of the stress intensity factor computed by XFEM is shown in Figure 15. The theoretical virtual fracture length is the width of the beam and equal to 80 mm. As expected, was a propagation predominance in the

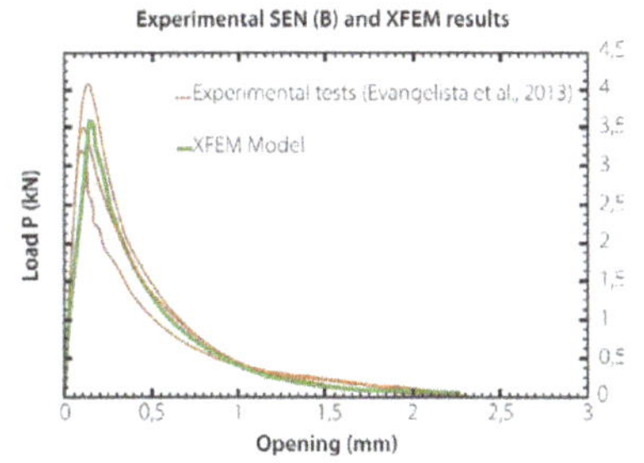

Figure 14: Experimental and numerical simulation results using XFEM in notched beams subjected to bending.

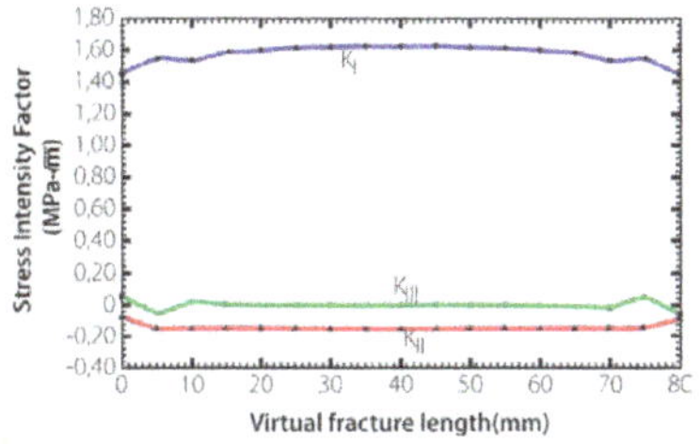

Figure 15: Stress intensity factors for SEN (B) model.

pure opening mode, due to the experimental configuration, which agrees with the results obtained in the three-dimensional simulations [14,15].

Concluding Remarks

Fracture problems were successfully solved using XFEM; eliminating the requirement to re-discretize the fracture surfaces, thus controlling computational costs and mesh projection errors associated with the conventional finite element method. Through the proper use of XFEM and refining the mesh in the fracture region were obtained excellent results in numerical approximated solutions to the experimental data. The two-dimensional model had an important role as an initial step to understand the problem simulation, and was used to learning the crack propagation due to the low processing time. In both cases of validation it was demonstrated that the maximum force reached in the two-dimensional numerical models is close to the experimental values. Convergence analysis over the mesh refinement returned results as accurately as possible. However, the fundamental discretization parameters must be carefully selected, to enhance the calculation speed and reduce the cost of monitoring. In three-dimensional models with inserted fractures the stress intensity factors revealed the predominance of the pure opening Mode I.

References

1. Gross B, Srawley JE, Bronw WF (1964) Stress intensity factors for a singled-edge-notch tension specimen by boundary collocation of a stress function. NASA scientific and technical publications. Washington DC, USA.
2. Rice JR (1968) A path independent integral and the approximate analysis of strain concentration by notches and cracks. J Appl Mech 36: 379-386.
3. Belytschko T, Black T (1999) Elastic crack growth in finite elements with minimal remeshing. Int J Numer Meth Eng 45: 601-620.
4. Moës N, Dolbow J, Belytschko T (1999) A finite element method for crack growth without remeshing. Int J Numer Meth Eng 46: 131-150.
5. Sukumar N, Moës N, Moran B, Belytschko T (2000) Extended finite element method for three-dimensional crack modelling. Int J Numer Meth Eng 48: 1549-1570.
6. Stolarska M, Chopp DL, Moës N, Beltyschko T (2001) Modelling crack growth by level sets. Int J Numer Meth Eng 51: 943-960.
7. Fries TP, Belytschko T (2010) The extended/generalized finite element method: An overview of the method and its applications. Int J Numer Meth Eng 84: 253-304.
8. Baydoun M, Fries TP (2012) Crack propagation criteria in three dimensions using the XFEM and an explicit–implicit crack description. Int J Fracture 178:1-20.
9. Song JH, Areias PMA, Belytschko T (2006) A method for dynamic crack and shear band propagation with phantom nodes. Int J Numer Meth Eng 67: 868-893.
10. Systèmes D (2013) Abaqus 6.13. Example problems, Manual D Systèmes Providence, USA.
11. ASTM (2013) Standard test method for determining fracture energy of asphalt-aggregate mixtures using the disk-shaped compact tension geometry. D7313 West Conshohocken United States ASTM Int 9.
12. Irwin GR (1957) Analysis of stresses and strains near the end of a crack traversing a plate. J Appl Mech 24: 361-364.
13. Evangelista F, Roesler JR, Proença SP (2013) Three-dimensional cohesive zone model for fracture of cementitious materials based on the thermodynamics of irreversible processes. Eng Fracture Mech 97: 261-280.
14. ASTM (2013) Standard test method for measurement of fracture toughness. E1820-13West Conshohocken USA, ASTM Int 9.
15. ASTM (2013) Standard test method for determining fracture energy of asphalt-aggregate mixtures using the disk-shaped compact tension geometry. D7313 West Conshohocken USA, ASTM Int 9.

Comparing the Thermal Stability and Oxidative State of Mineral and Biphenyl Diphenyl Oxide based Heat Transfer Fluids

Wright CI*

Research and Development, Global Group of Companies, Cold Meece, Staffordshire, UK

Abstract

Many industrial processes (i.e., concentrated solar power [CSP] plants) require indirect heating of the product to temperature above ambient temperature. A heat transfer fluid (HTF), such as a Globaltherm™ Omnitech (Global Heat Transfer; Staffordshire, UK), is used in such plants and flows from the heater to source requiring the heat. Not all HTFs are the same, however, and so understanding the difference between fluids is important to ensure that end-users use the correct fluid for the correct operation. This is especially true in CSP plants where HTFs operate at high temperatures for long periods of time and therefore need to be stable under such conditions. Indeed, biphenyl diphenyl oxide (BDO) mixtures are commonly used in CSP plants as they can be heated to 400 degrees Celsius, which is higher than the upper operating temperature for a mineral-based HTF (i.e., ~400 degrees Celsius). It is a fact that all HTFs will thermally degrade over time and so it is important to monitor this to ensure that an early intervention can be taken if a problem starts to appear. The objective of HTF monitoring being to keep the HTF and plant operational for as long as is possible. Routine sampling and chemical analysis is used to assess the physiochemical properties of a HTF. For this to be done effectively, it is important to understand the properties of a virgin HTF and then to assess the rate of thermal degradation over time. Carbon residue, total acid number and closed flash point temperature are routinely measured in the laboratory and the current study proposes their use to assess the extent of thermal cracking and oxidation, two common pathways through which a HTF thermally degrades. The findings of this assessment are presented herein. Future work should consider using a similar approach to assess the condition of other HTFs commonly used in industrial applications.

Keywords: Heat transfer fluid; Thermal heat transfer fluid; Thermal degradation; Oxidation; Sampling and chemical analysis

Introduction

The electricity produced by concentrated solar power (CSP) plants is growing with ~430 MW produced in 2008 [1] and an estimated 20 to 630 GW to be produced by 2020 [2] and 2050 [3,4] respectively. CSP plants use a heat transfer fluid (HTF) to collect heat from the sun. The main types of HTF being: 1) air/other gases; 2) water/steam; 3) mineral based HTFs (e.g., Globaltherm M); 4) synthetic based HTFs; 5) molten salts (e.g., Globaltherm Omnistore); and, 6) liquid metals [3]. Synthetic HTFs that contain a eutectic mixture of biphenyl and diphenyl oxide (BDO), such as Dowtherm A, Therminol VP-1 and Globaltherm Omnitech [4], are commonly used as a heat carrier in CSP plants [5].

The sheer size of CSP plants, for example the first CSP plant built in India used a solar field consisting of three loops with parabolic troughs and measured 1,500 meters in length and covered 8,000 m^2 [6] means that the HTF is an expensive asset [3]. Commercial operations therefore need to work to minimize the cost of the HTF as well as maximize the performance of the CSP plant [3].

It is a fact that HTFs will age with usage and is influenced by a number of factors including elevations in temperature above the HTFs bulk operating temperature, oxidative stress and HTF contamination [7]. Being able to slow the process of degradation is critical to maintaining an effective and efficient operation. Laboratory chemical analysis is routinely used in the condition monitoring of HTFs. Indeed, a HTF that undergoes thermal degradation will form heavy (i.e., carbon formations) and light-chain (flash point components that can be assessed from open and closed flash point temperatures). Organic HTFs operating at high temperatures are susceptible to air oxidation and it is possible to assess the extent of HTF oxidation by looking at its neutralization number (i.e., total acid number [TAN]) (Table 1). Other tests can also be used to assess foreign contaminants and kinematic viscosity [5].

Aluyor and Ori-jesu [8] describes "thermal stability" as the resistance posed by a fluid to either molecular breakdown or the rearrangement of molecules at elevated temperatures in the absence of oxygen (i.e., thermal degradation). Synthetic HTFs, such as eutectic mixtures of BDO, are generally considered to exhibit better thermal stability than mineral-based HTFs [9]. They are also considered to be more resistant to fouling (for example, please see reference [10].

Objectives for This Article

This article aims to compare mineral-based and synthetic (i.e.,

Chemical test	What is assessed?
Carbon residue	Heavy-chain hydrocarbons
Closed flash point temperature	Light-chain hydrocarbons
Open flash point temperature	Light-chain hydrocarbons
TAN	The extent of HTF oxidation

Table 1: What the chemical testing says about the condition of a HTF.

***Corresponding author:** Wright CI, Research and Development, Global Group of Companies, Cold Meece, Staffordshire, UK
E-mail: chrisw@globalgroup.org

BDO-based) HTFs to ascertain the following:

a) Their physicochemical properties.

b) To understand how these HTFs should be maintained and if there are any differences.

c) To compare the thermal stability (from plots of carbon residue versus open and closed flash point temperatures) and anti-oxidative capacity (from plots of carbon residue versus TAN) of these HTFs.

Materials and Experiments

Materials and methods

The sampling of HTFs: A 500 ml of the HTF was sampled from the plant whilst the HTF was in circulation. These samples were then taken to the laboratory for subsequent chemical analysis [7]. This was performed using a closed sampling device to prevent the HTF coming into contact with air and thus ensuring a representative sample of the HTF was collected. This technique has been presented previously [11]. All laboratory analysis was conducted according to ISO14001 [12] and ISO17025 [13].

HTFs sampled and chemically analyzed: Mineral-based (i.e., Globaltherm M) and BDO-based HTFs (i.e., Dowtherm A) were included in this analysis.

Contrast and comparison

Physicochemical properties: The physicochemical properties of virgin mineral and BDO-based HTFs were compared by extracting key data from safety data sheets for Globaltherm M (mineral-based) and Globaltherm Omnitech HTFs (BDO-based) [14].

Comparing the maintenance and management programs required for mineral and BDO-based HTFs: The maintenance guidelines for mineral and BDO-based HTFs were compared based on guidance for usage, routine sampling, chemical analysis and handling. The objective was to determine if these HTFs should be treated any differently in the real-world setting.

Assessments of thermal stability and anti-oxidative capacity: Retrospective test reports were analyzed to assess the thermal stability and anti-oxidative capacity of mineral and BDO-based HTFs. Thermal stability was determined from x-y plots of carbon residue against closed flash point temperature. Likewise, anti-oxidative capacity was determined from x-y plots of carbon residue against TAN.

Comparisons were made by constructing x-y plots and grouped according to carbon residue values: Closed flash point temperatures and TAN were grouped according to carbon residue, which was organized based on 5 groups:

a) Group 1, <0.05% carbon residue and typical of a virgin HTF.

b) Group 2, ≥0.05 and <0.5% and considered a satisfactory rating.

c) Group 3, ≥0.5 and <0.75% and considered a cautionary rating.

d) Group 4, ≥0.75 and <1.0% and action to correct the issue should be considered.

e) Group 5, ≥1.0% and this is considered to be severe and an immediate intervention is required [11].

Results and Discussion

Physicochemical properties

Some typical physicochemical properties for mineral and BDO-based HTFs are compared in Table 2.

It is claimed that eutectic mixtures of BDO have been used in industrial HTF systems for over 60 years [15]. Eutectic mixtures of BDO are commercially available [16,17] and traded under many different brand names. One example is Globaltherm Omnitech [14] which contains a uniform eutectic mixture of biphenyl ($C_{12}H_{10}$) and diphenyl oxide ($C_{12}H_{10}O$) [5,16] with a melting point of roughly 12 degrees Celsius and an upper operating temperature around 400 degrees Celsius [5,18]. This means the operating range is above 12 degrees Celsius as below this temperature the HTF solidifies. This is in contrast to a mineral-based HTF which has a lower operating temperature of -10°C, which means it has the advantage of being able to be used without steam tracing in most cases and less prone to solidifying during storage or cold operating climates. Although, solidification will not present an issue in plants are protected from the weather.

The upper operating temperature of the BDO-based HTF is 400°C and is a reflection of the fluid's superior thermal stability as compared with a mineral-based HTF (i.e., +120°C higher than a mineral-based HTF (Figure 1 and Table 2).

BDO-based HTFs can also be used effectively as a liquid HTF (below boiling point), a vapor phase (above its boiling point) or as a mixture of liquid plus vapor, which is defined by the overall design of the plant [14]. This is compared with a mineral-based HTF which operates in the liquid phased.

BDO-based HTFs also have lower flash point temperatures and higher bulk and auto-ignition temperatures (Figure 1 and Table 2). Indeed, the BDO-based HTF provides a safety margin of 221°C above its upper operating temperature [19] as compared with the 40°C normally provided by a mineral-based HTF [20].

Parameter	Unit	Mineral-based HTF	BDO-based HTF
Other HTF examples	Descriptive	BP Transcal N, Globaltherm M, Shell Thermia B	Dowtherm A, Globaltherm Omnitech, Therminol VP-1
Operating range	°C	-10 to 320	15 to 400
Appearance	Descriptive	Viscous clear-yellow liquid with a mild odor	Clear-to-light yellow liquid with a geranium-like odor
Density at 25°C	kg/m^3	873	1056
Kinematic viscosity (at 40, 100°C)	mm^2/s	29.8, 4.5	2.5, 0.97
Auto-ignition	°C	>320	621
Maximum film	°C	330	425
Boiling point at 1013 mbar	°C	365	257
Open flash point	°C	230	123
Closed flash point	°C	210	113

Table 2: A comparison of the typical physicochemical properties of a mineral and BDO-based HTF.

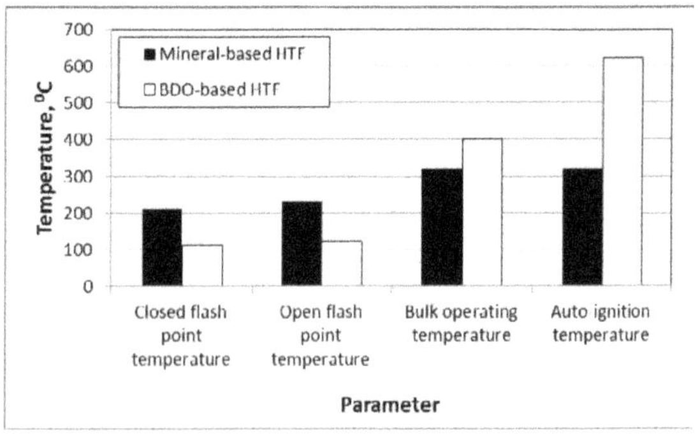

Figure 1: Comparison of flash point, bulk operating and auto-ignition temperatures for mineral and biphenyl and diphenyl oxide (BDO)-based heat transfer fluids (HTFs).

Both mineral and BDO-based HTFs are combustible materials, but the both have relatively high open flash points 230 and 123°C (mineral and BDO-based HTF, respectively) [19]. The flash points for BDO-based HTFs are lower compared with mineral based HTFs, but it is also important to factor the greater thermal stability of BDO-based HTFs.

Table 2 compares the kinematic viscosity for mineral and BDO-based HTFs. For BDO-based HTFs, the kinematic viscosity is lower than mineral-based HTFs and only changes slightly between the lower and upper operating temperatures. For comparative purposes this is seen in Table 2 where the factor of change in kinematic viscosity between 40 and 100°C is 2.6 (0.97 divided by 2.5 mm^2/s) and 6.6 (4.5 divided by 29.8 mm^2/s) for BDO and mineral-based HTFs, respectively. A benefit of the lower viscosity for BDO-based HTFs is that problems on start-up are minimized as the fluid has a lower viscosity than a mineral-based HTF [19].

The carbon residue, water and neutralization number (indicating the TAN for a HTF) are ideally as close to zero as possible. Contamination of a HTF, and potentially subsequent HTF system corrosion and HTF degradation, can be caused by the introduction of foreign particles (e.g., water) or chemicals during system flushing and cleaning [19] as a result of process leaks [19] or during the building of new HTF plants (e.g., welding slag and environmental contaminations) [21,22]. Hence, even at the early stages in the construction of a plant it is important to sample and chemically analyze the HTF to mitigate the exposure to and manage the removal of any identified contaminants [21,22].

Maintenance and management programs for mineral and BDO-based HTFs

The thermal stability of a HTF depends not just on the chemical structure, but also on sound engineering practices [19] and sound maintenance programs to monitor and manage the rate of thermal degradation and degree of oxidation [11].

The advice from manufacturers (from example see [19] when sampling and analyzing HTFs is summarized below:

a) A 500 ml is required to chemically analyze the condition of the HTF.

b) Closed sampling devices should be used to get a representative sample of the HTF.

c) Before sampling, the sampling device should be thoroughly washed.

d) A HTF should be taken from the main circulating line of the HTF system, but samples may be required from other parts of the system if a specific issue is being investigated.

e) Analysis of a HTF provides insights, as discussed above, into the condition of the fluid and is used to identify problems (e.g., contamination) as well as to monitor the state of the HTF and to help define suitable interventions if a correct to the HTF condition is required.

f) Following sampling of a HTF, is should be allowed to cool (to below 40 degrees Celsius) before handling and chemical analysis. This will also work to keep light-ends in solution and stop them escaping from the sample.

g) HTFs should be sampled routinely and research has shown that the condition of mineral-based HTFs improves if sampled every 3 months [7].

h) The chemical analysis conducted on the HTF sample should be defined in consultation with the manufacturer of the fluid.

Assessments of thermal stability and anti-oxidative capacity

Figure 2 plots carbon residue against TAN and closed flash point temperature to provide visual assessments of the thermal stability (bottom panel) and anti-oxidative capacity (top panel) for a mineral and BDO-based HTF. From Figure 2 it is clear that for a mineral-based HTF, carbon residue increases at twice the rate of TAN. Thus demonstrating the need to assess carbon residue closely for mineral-based HTFs as this appears to be an early and sensitive marker for assessing the condition of the HTF. Moreover, once a mineral-based HTF gets to a carbon residue value of ≥1% weight, the HTF will need to be replaced. This can be achieved by monitoring the fluid and this

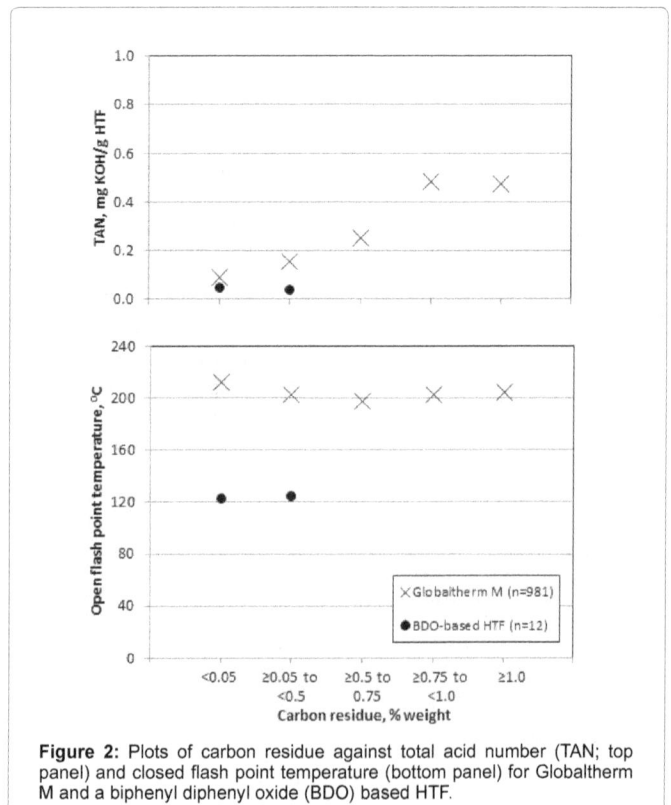

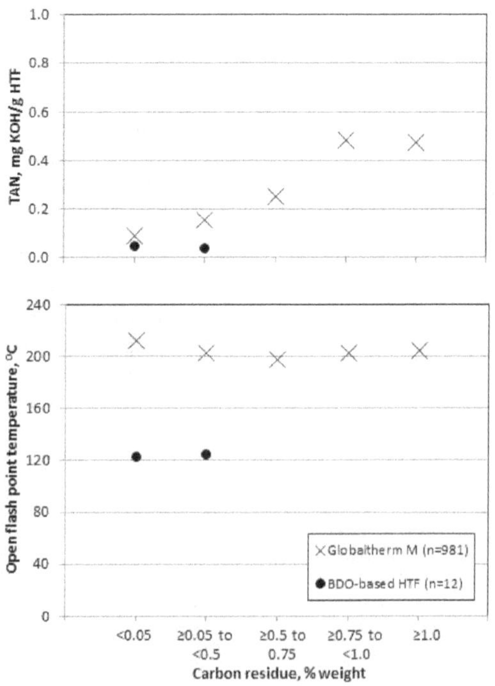

Figure 2: Plots of carbon residue against total acid number (TAN; top panel) and closed flash point temperature (bottom panel) for Globaltherm M and a biphenyl diphenyl oxide (BDO) based HTF.

	Carbon residue, % weight				
Parameter	<0.05	≥0.05 to <0.5	≥0.5 to <0.75	≥0.75 to <1.0	≥1.0
Number of samples chemically analyzed	82	793	46	30	30
Water content, ppm	90.43	84.49	83.60	63.00	182.89
Kinematic viscosity, mm^2/s	30.3	30.2	30.9	33.8	36.2
Elements, e.g., silicon (ppm less than 5 μm)	11.8	14.3	18.3	11.5	16.0
TAN	0.1	0.2	0.3	0.5	0.5
Ferrous debris, insolubles	7.0	15.6	20.9	17.8	29.1
Open flash point temperature, ^0C	212.0	203.1	197.3	202.8	204.2
Closed flash point temperature, ^0C	165.5	155.6	150.4	164.6	160.5
Fire point temperature, ^0C	244.1	243.9	236.6	240.0	240.2

Table 3: The physicochemical properties of the mineral-based HTF for each carbon level.

is an important part of HTF system planned preventative maintenance [23] and offers the opportunity to be able to detect problems and take appropriate action when needed. In the current case this would be maintaining carbon residue as close to the values for a virgin HTF (see typical values presented in Table 1) as possible.

Although, the current data shows that both carbon and TAN were increasing and this raises the question as to whether these changes were primarily driven by oxidation or by thermal cracking. This can be assessed by looking at plots of carbon against closed flash point temperature. Indeed, the bottom panel of Figure 2 shows the thermal stability of the HTF was relatively constant with closed flash point temperature constant at each level of carbon. This data suggests that the degradation of HTF was driven largely by oxidation rather than cracking of the HTF [24]. This is supported by the fact that carbon increases linearly with TAN (i.e., oxidative state) but is stable against closed flash point temperature. This is further supported by Table 3 where kinematic viscosity increases as opposed to decreases. The increase denoting the formation of carbon in solution, as would be occurring through oxidation of the HTF [24].

The comparison of mineral and BDO-based HTFs over the full range of carbon residue values was not possible because of the limited number of samples for the BDO-based HTF. From Figure 2, however, thermal stability and oxidative state are relatively stable over the range of carbon values investigated. The data for BDO-based HTFs highlights the need for further work to allow the comparison of different HTFs Table 3.

Conclusions

The current study highlights three important aspects for every HTF:

i. The importance of referring to the safety datasheet to understand the thermochemical physicochemical properties and performance of a virgin HTF. This is an important reference source to understand a fluid or to compare fluids, such as was done herein to understand the differences between mineral and BDO-based HTFs.

ii. All HTFs need to be maintained as part of the overall maintenance of a plant. All HTFs need to be managed as their condition is directly related to the condition of the overall plant. Indeed, Wagner [5] states that "Relevant regular maintenance services of the system components are necessary and to be carried out by experienced personnel to ensure trouble-free operation of the plant over a longer period." This needs to be done by experienced engineers, such as Global Heat Transfer (www.globalheattransfer.co.uk) using suitable equipment so that a representative sample can be obtained safely whilst the plant is still in operation [22].

iii. Chemical analysis can be used to assess the condition of the HTF and to make early decisions regarding the maintenance of the HTF and the plant. The results of tests for carbon residue, TAN and closed flash point temperature can be used to assess both the extent of thermal cracking (x-y plots of carbon residue versus closed flash temperature) and oxidation (plots of carbon residue versus TAN). The data presented in this paper suggested that the degradation of HTF was driven largely by oxidation as opposed to thermal cracking [24] and these insights were achieved the 11-point test offered by Global Heat Transfer.

Acknowledgement

I would like to acknowledge the engineering support provided by Global Heat Transfer and the writing support provided by Red Pharm communications, which is part of the Red Pharm company (please see @RedPharmCo on Twitter).

References

1. Reports n Reports.com, Global Heat Transfer Fluid Market Applications and Geography-Forecasts To 2017 market research report.
2. Australian Solar Institute (2015) Realising the potential of concentrating solar power in Australia. Report prepared by IT Power (Australia) PTY LTD in 2011.
3. Küser D (2015) Solar Report. Concentrating Solar Power (CSP) Outlook on large potentials and the MENA region.
4. Oetinger J (2011) Troubleshooting heat-transfer fluid systems. Anatomy of a Heat Transfer Fluid Analysis. Chemical Engineering.
5. Wagner WO (1997) Heat transfer technique with organic media. In Heat transfer media 2 edn. Graefelfing Germany Maria-Eich-Straβe 4-58.
6. Biencinto M, González L, Zarza E, Díez LE, Muñoz-Antón J (2014) Performance model and annual yield comparison of parabolic-trough solar thermal power plants with either nitrogen or synthetic oil as heat transfer fluid Energy Conversion and Management 87: 238-249.
7. Wright CI, Picot E, Bembridge T (2015) The relationship between the condition of a mineral-based heat transfer fluid and the frequency that it is sampled and chemically analyzed. Applied Thermal Engineering 75: 918-922.
8. Aluyor EO, Ori-jesu M (2009) Biodegradation of mineral oils-A review. African Journal of Biotechnology 8 : 915-920.
9. Stachowiak GW, Batchelor AW (2005) Engineering Tribology. Butterworth-Heinemann USA.
10. (2015) Therminol SP heat transfer fluid.
11. Wright CI (2014) Thermal heat transfer fluid problems following a system flush with caustic and water. Case Studies in Thermal Engineering 2: 91-94.
12. (2014) Environmental management. ISO 14001:200
13. (2014) General requirements for the competence of testing and calibration laboratories. ISO/IEC 17025: 2005
14. (2015) Globaltherm high temperature thermal fluid range. Global heat transfer

15. (2015) Dowtherm A properties. World of chemicals

16. (2015) Physical properties of DowthermA The Engineering Toolbox.

17. (2015) Diphenyl-heat transfer fluids Lanxess.

18. Free Dictionary (2015) Azeotropic Mixture.

19. (2015) Dowtherm A heat transfer fluid. Product technical data.

20. Dowtherm Q (2015) Heat Transfer Fluid Product Technical Data.

21. Wright CI, Picot E (2015) A case study to demonstrate the value of a system flush and clean prior to filling a plant with virgin heat transfer fluid. Heat Transfer Engineering.

22. Wright CI (2015) Case study in CSP plant performance. Renewable Energy Focus 16: 22-26.

23. Phillips WD (2006) The high-temperature degradation of hydraulic oils and fluids. Journal of synthetic lubrication 23: 39-70.

24. Noria Corporation (2015) The lowdown on oil breakdown. Machinery Lubrication.

Lacking Data Recovery via Partially Overdetermined Boundary Conditions in Linear Elasticity

Abda AB and Khalfallah S*

University of Tunis El Manar ENIT-LAMSIN, Tunisia

Abstract

This work focuses on the sub-Cauchy problem for linear elasticity in two dimentional case. Solving such a problem may be formulated as follows: given the displacement and one component of the traction in a given part of the boundary of the elastic body, reconstruct the displacement field in all the domain. Author propose herein, an iterative method borrowed from the domain decomposition communauty to solve the sub-Cauchy problem. Numerical results highlight the efficiency of the proposed method.

Keywords: Linear elasticity; Shear stress; Cauchy problem; Steklov Poincare operator; Domain decomposition; Inverse problems

Introduction

Many inverse problems in linear elasticity are defined by overdetermined boundary conditions. One can think to the reconstruction of buried flaws such as cracks, voids or inhomogeneities, identification of constitutive law, data completion (that is the recovery of boundary conditions on an inaccessible part of the body boundary) [1].

All the above inverse problems have in common to be defined by overspecified boundary conditions namely the normal stress and the displacement on a part of the boundary which correspond to Cauchy data. Many papers treated this problem, from the numerical view point, this last decade [2-4].

Author would like to mention the work by Bourgeois [5] who applied the Lions-quasi-reversibility method to the data completion. This method leads to a direct inversion process.

Many authors resort to iterative methods based on minimising a least-square type error functional, [6-8]. Marin [9] would like to mention the minimization of an energy-like gap functional in ref. [10] and domain decomposition like method in ref. [11] which are close to what we develop in this work.

Hereafter, Author are concerned by a partially overdetermined boundary conditions. In fact,on a part of the boundary of the domain partially overdetermined boundary data are prescribed, namely one component of the traction and the displacement field. Following ref. [12] author build an energy-gap error functional to recover the lacking boundary data. Author emphasise on the shear stress reconstruction, on the part of boundary where the partial-data is prescribed.

Formulation of Sub-Cauchy Problem as Steklov Poincare Operator

The inverse problem under consideration concerns the recovery of lacking boundary data from the knowledge of partially overdetermined boundary elastic data.

The problem is formulated mathematically as follows : Let Ω be a bounded domain in $\mathbb{R}^2$, the boundary $\Gamma=\partial\Omega$ is split into Γ_c and Γ_i having both non vanishing measure $\Gamma_c \cap \Gamma_i = \varnothing$. Given the displacement U and the normal component of surface traction $\Phi.n$ on Γ_c:

$$\begin{cases} div\sigma(u)=0 & in \quad \Omega, \\ (\sigma(u).n).n=\Phi.n & on \quad \Gamma_c, \\ u=U & on \quad \Gamma_c. \end{cases} \tag{1}$$

where $\sigma = \lambda Tr\varepsilon(u)+2\mu\varepsilon\ (u)$, $\varepsilon = 1/2(\nabla u+\nabla u^T)$ and λ, μ are the Lamé coefficients related to Young's modulus E and the Poisson ratio v via:

$$\mu=\frac{E}{2(1+v)} \qquad \lambda=\frac{Ev}{(1-2v)(1+v)}$$

Our aim is then to reconstruct $(\sigma\ (u).n).\tau$ on Γ_c and both the displacement and traction. To our knowledge, there are no theoretical studies (existence and uniqueness) of this problem despite its great importance in applications. In this paper author treat this problem numerically by solving a data completion problem.

The decomposition of the Cauchy problem (1) is formulated through an unknown function η as follows:

$$(P_D)\begin{cases} div\sigma(u_D)=0 & in \quad \Omega, \\ u_D=U & on \quad \Gamma_c, \\ u_D=\eta & on \quad \Gamma_i. \end{cases} \quad (P_N)\begin{cases} div\sigma(u_N)=0 & in \quad \Omega, \\ (\sigma(u_N).n).n=\Phi.n & on \quad \Gamma_c, \\ u_N.\tau=U.\tau & on \quad \Gamma_c, \\ u_N=\eta & on \quad \Gamma_i. \end{cases} \tag{2}$$

Where η is the virtual control and u is chosen so that u_D and u_N adjust in the best possible on Ω. The solution u_D and u_N are a function of η ($u_D=u_D(\eta)$ and $u_N=u_N(\eta)$).

To express the problem in the framework of virtual control, we introduce the cost functional:

$$J(\eta)=\int_\Omega \sigma(u_D-u_N):\varepsilon(u_D-u_N)\quad d\Omega \tag{3}$$

and consider the minimization problem:

$$\inf_{\eta\in H^{\frac{1}{2}}(\Gamma_i)} J(\eta) \tag{4}$$

***Corresponding author:** Khalfallah S, University of Tunis El Manar ENIT-LAMSIN, Tunisia, E-mail: sinda.khalfallah@yahoo.fr

The solutions u_D and u_N can be written as:

$$u_D = u_D^0 + u_D^* \qquad u_N = u_N^0 + u_N^*$$

Where u_i^0 depends on the data U and $\Phi.n$ where as u_i^* depends on η as follows:

$$(P_D^*)\begin{cases} div\sigma(u_D^*) = 0 & in \quad \Omega, \\ u_D^* = 0 & on \quad \Gamma_c, \\ u_D^* = \eta & on \quad \Gamma_i. \end{cases} \quad (P_D^0)\begin{cases} div\sigma(u_D^0) = 0 & in \quad \Omega, \\ u_D^0 = U & on \quad \Gamma_c, \\ u_D^0 = 0 & on \quad \Gamma_i. \end{cases} \tag{5}$$

Similary, author decompose u_N.

$$(P_N^*)\begin{cases} div\sigma(u_N^*) = 0 & in \quad \Omega, \\ (\sigma(u_N).n).n = 0 & on \quad \Gamma_c, \\ u_N.\tau = 0 & on \quad \Gamma_c, \\ u_N^* = \eta & on \quad \Gamma_i. \end{cases} \quad (P_N^0)\begin{cases} div\sigma(u_N^0) = 0 & in \quad \Omega, \\ (\sigma(u_N).n).n = \Phi.n & on \quad \Gamma_c, \\ u_N.\tau = U.\tau & on \quad \Gamma_c, \\ u_N^0 = 0 & on \quad \Gamma_i. \end{cases} \tag{6}$$

The solution of the problem (4) is recovered if:

$$\sigma(u_D).n = \sigma(u_N).n \quad on\,\Gamma_i \tag{7}$$

With this partition, condition 7 leads to the boundary equation

$$\sigma(u_D^*).n - \sigma(u_N^*).n = \sigma(u_N^0).n - \sigma(u_D^0).n \quad on \quad \Gamma_i \tag{8}$$

Author introduce the Steklov Poincaré operator

$$S\eta = \sigma(u_D^*).n - \sigma(u_N^*).n \quad on \quad \Gamma_i$$

Author define $S_D\eta = \sigma(u_D^*).n$ and $S_N\eta = \sigma(u_N^*).n$.

Author can write the equation (8) according to the Steklov Poincaré operator:

$$S\eta = \xi \quad on \quad \Gamma_i$$

where $\xi = -(\sigma(u_D^0).n - \sigma(u_N^0).n)$.

This operator, borrowed from the domain decomposition community, is widely used in ref. [13].

There are several ways to solve this linear system of equations. Here author use an iterative preconditioned gradient algorithm, which appears to be very efficient. Each iteration of the algorithm is written

$$\eta = \eta + \rho(S_D)^{-1}(S\eta - \xi),$$

where ρ is a relaxation coefficient and S_D is the preconditioning operator.

Thus each iteration requires to compute $S\eta$ by solving the two problems ?? and to solve the system $S_D\chi = S\eta$. This is achieved by solving the following problem:

$$\begin{cases} div\,\sigma(w) = 0 & in \quad \Omega, \\ \sigma(w).n = S\eta & on \quad \Gamma_i, \\ w = 0 & on \quad \Gamma_c. \end{cases} \tag{9}$$

where $\chi = w$ on Γ_i.

Now, author propose an algorithm to approximately solve the sub-Cauchy problem:

Algorithm

1. Choose arbitrary η
2. Solve problems (P_D) and (P_N).
3. solve problem (9).
4. Let $\eta = \eta + \rho\, w$
5. Go back to the first step until the stopping criteria $||u_D - u_N|| \leq \varepsilon$ is reached. (ε is a given tolerance level)

Numerical Results and Discussion

The purpose of this section is to present the numerical implementation of the boundary data recovery process described above.

The numerical implementation is run under FreeFem software [14] based on Finite Element Method. All through this section, author consider an isotropic linear elastic material (Steel $XC10$ to 20° temperature) characterised by the poisson coefficient $\nu = 0.29$ and Young's modulus E = 216 GPa.

Author are concerned by a two dimentional framework corresponding to a square hole domain.

The partially overspecified boundary data is a synthetic one, obtained through the resolution of the following forward problem:

$$\begin{cases} div\sigma(u_0) & = 0 & in \quad \Omega, \\ \sigma(u_0).n & = \sigma(T).n & on \quad \Gamma_c, \\ u_0 & = T & on \quad \Gamma_i. \end{cases}$$

where $T = (Re(\frac{1}{z-a}), Im(\frac{1}{z-a}))$, $z = x + iy$, $a = 1.8$, $\partial\Omega = \Gamma_c \cup \Gamma_i$ and Γ_i being the inner circle.

Notice that we are dealing with a "rough" case, insofar as, the inffered data, are induced by a "near singular" data. The trials used in the litterature come usually from analytical reference solutions.

Preliminary Numerical Test

Our trial concerns the resolution of the sub-Cauchy problem in the following context: We consider a square hole domain: rectangle size: (10 * 20) with inner circle of radius R=2. The internal circle plays the role of the boundary Γ_i and the Cauchy data are donated in the external boundary Γ_c.

Author choose $\varepsilon = 10^{-2}$ in the stopping criteria computation are carried out with "un-noisy" data. Figures 1-3 show the reconstructed displacement and traction on the inner boundary, whereas Figure 4 illustrate the reconstruction of the shear stress in Γ_c. Note that the reconstruction is quite nice in Γ_i and in good agreement with exact for the shear stress.

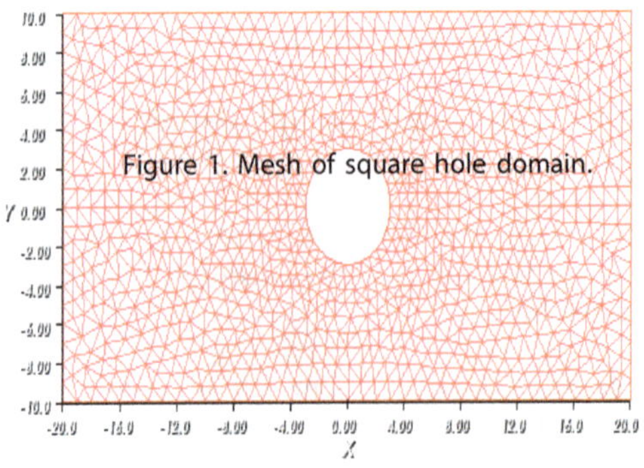

Figure 1: Mesh of square hole domain.

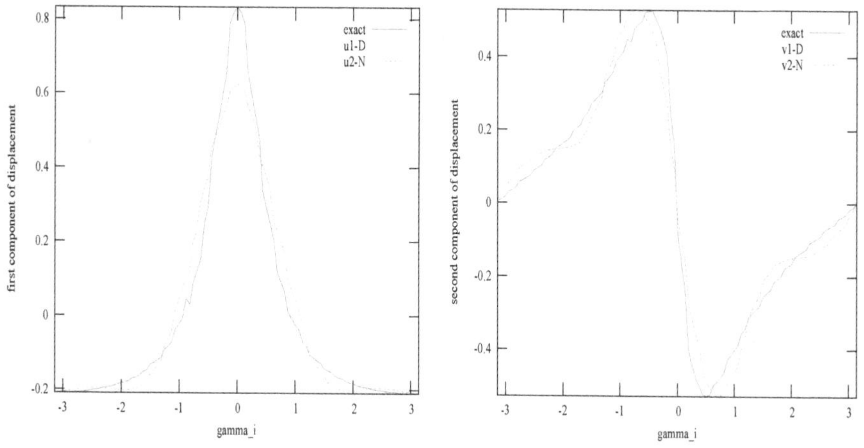

Figure 2: Exact and numerical displacement (left: first component, right: second component) on Γ_i .

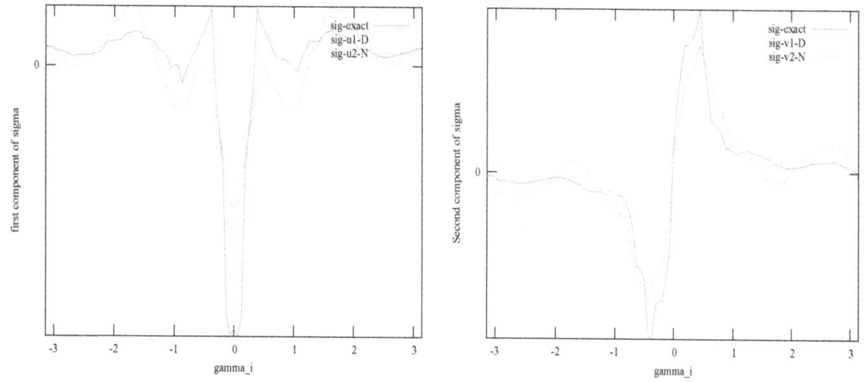

Figure 3: Exact and numerical traction (left: first component, right: second component) on Γ_i .

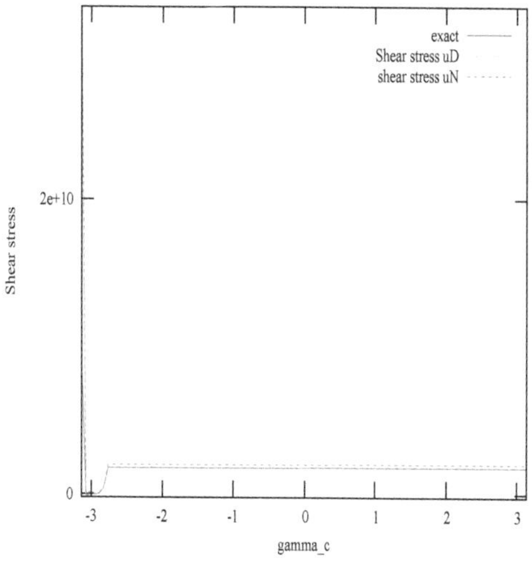

Figure 4: Exact and reconstructed shear stress on Γ_c .

Sensitivity to the Thickness

The following numerical trials are devoted to the influence of the radius of the hole on the reconstructed data.

The results are summerrized in the Table 1. As expected the computed sub-Cauchy problem solution is better when the distance between Γ_c and Γ_i is lower. To confirm the results, Figure 5 where author present the result for the first component of displacement on Γ_i

The same remark is true when we zoom on the shear stress.

Radius	R=2	R=4	R=6	R=8
$\| u_D - u_{ex} \|_{L^2(\Omega)}$	2.029	0.27	0.096	0.055
$\| u_D - u_{ex} \|_{L^2(\Omega)} / \| u_{ex} \|_{L^2(\Omega)}$	0.45	0.098	0.043	0.03
$\| (\sigma(u_D).n).\tau - (\sigma(u_{ex}).n).\tau \|_{L^2(\Gamma_c)}$	$8.4*10-2$	$7,01*10$	$4.41*10^{-2}$	$8.31*10^{-3}$

Table 1: Error between the exact and numerical solution for different radius defined Γ_i .

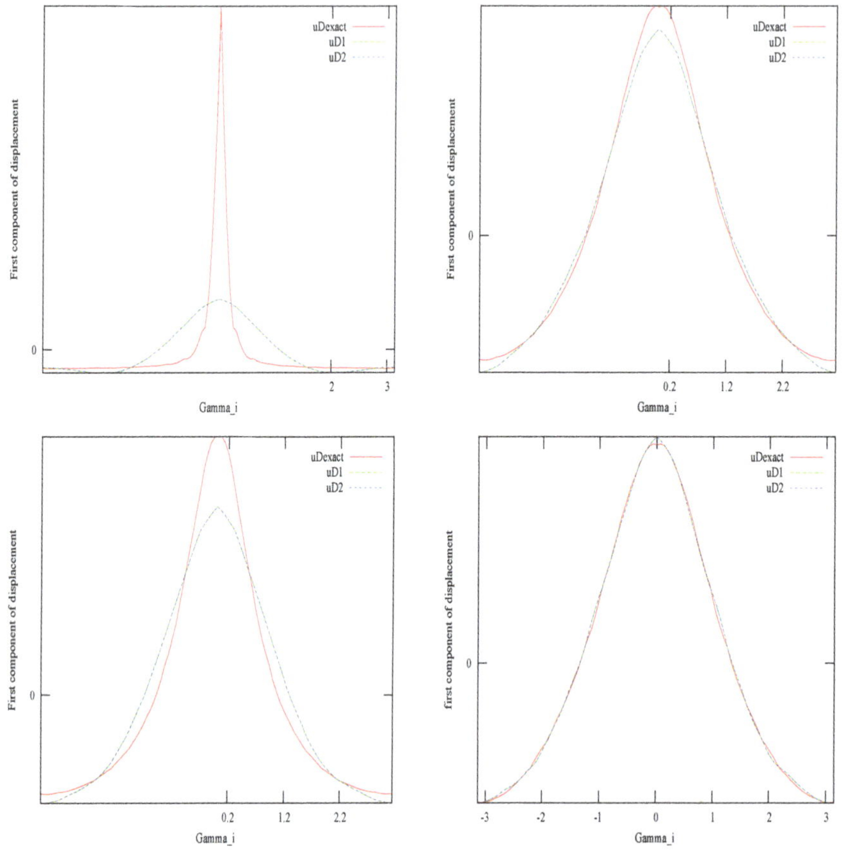

Figure 5: Exact and numerical first component of displacements for different radius: left-top $R=2$, right-top: $R=6$, left-bottom: $R=4$ and right-bottom: $R=8$.

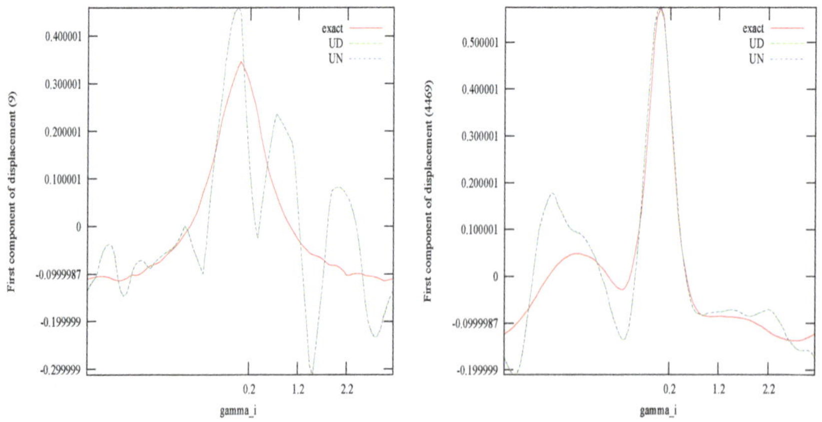

Figure 6: Exact and reconstruted first component of displacement from noisy Cauchy data (4%). Left: The solution in the exact domain. Right: The solution in the extended domain (tolerance $\varepsilon = 2*10^{-2}$).

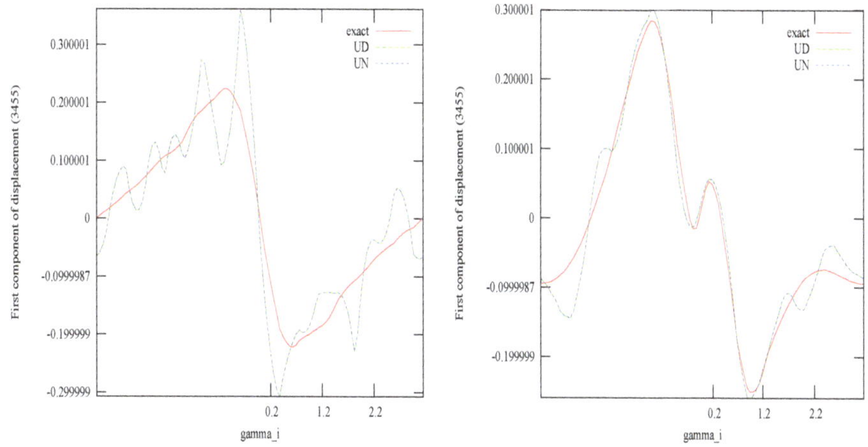

Figure 7: Exact and reconstruted second component of displacement from noisy Cauchy data (4%). Left: The solution in the exact domain. Right: The solution in the extended domain (tolerance $\varepsilon = 2 * 10^{-2}$).

Extended Domain

The following numerical experiments are inspired by Hecht [14]. In ref. [14], the authors resort to un extended domain method to illustrate its regularisation effect on their numerical data completion procedure. Their study was conducted in the framework of Laplace equation.

To our opinion, the proposed method may be used in many practical situations, one can think to the data completion on rough boundary in this situation it is worthfull to extend the domain to a smooth one and to deduce the boundary conditions on the rough boundary.

This trick will avoid the meshing difficulties for instance.

Another possible application may concern a void detection: If one has an apriori knowledge on the void location, the computation may be done on an extended domain, the void being detected by level lines of the displacement field [15] (for the Laplace equation).

Our concern here is to illustrate the deblurring effect of this domain extension procedure.

Of course, it currently happens in practice that the data (on Γ_c) suffer from erroneous measurements, the following numerical experiments illustrates the deblurring effect of the extended domain method.

We consider a random noise of 4% added to the exact data as follows:

$$U = U + \alpha r \qquad \Phi.n = \Phi.n + \beta r$$

where (α , β) denotes the noise level relative to ($\| U \|_{L^2(\Gamma_c)}$, $\| \phi.n \|_{L^2(\Gamma_c)}$), and r is a random function generated by *Freefem*.

The boundary Γ_i is very close to the complete boundary and is then exposed to the noise contamination coming from Γ_c .The possibility of extending domain by a fictitious incomplete bounday can correct this contamination.

The exact domain is defined square by rectangle size (10 * 20) with hole of radius *R=6* while the extended field is defined by the same rectangle, but with a hole of radius *R=4*.

Figures 6 and 7 show the reconstructed displacement for the exact and extended domain. Note that the solution computed in the real domain suffers from hard oscillations. Those obtained in the extended domain seem satisfactory.

Conclusion

In this work the reconstruction of lacking boundary data on a part of the boundary of a body from partially-overspecified boundary conditions on an other part has been investigated numerically.

A domain decomposition like method has been given to describe the reconstruction process. The numerical investigation has been conducted on a "rough" configuration (i.e. the data to be recovered is not extendable on a divergence free stress field outside the domain namely within the hole), it uses FEM.

The numerical section highlights the accuracy of the inverse procedure, as well as the robustness of the inversion process to noisy data as well as its ability to deblur noise.

Acknowledgement

The authors acknowledge the financial support for EPIC (within LIRIMA: http://www.lirima.org). The LAMSIN researchers work is supported on a regular basis by the Tunisian Ministry for Higher Education, Scientific Research and Technology.

References

1. Bonnet M (2005) Inverse problems in elasticity. Inverse Problems 21 R1.
2. Baranger TN, Andrieux S (2008) An optimization approach for the Cauchy problem in linear elasticity. Structural and Multidisciplinary Optimization 35: 141-152.
3. Durand B, Delvare F, Bailly P (2011) Numerical solution of cauchy problems in linear elasticity. Int J Solids and Structures 48: 3041-3053.
4. Bilotta A, Turco E (2009) A numerical study on the solution of the cauchy problem in elasticity. Int J Solids and Structures 46: 4451-4477.
5. Bourgeois L, Dardé J (2010) Quasi-reversibility approach to solve the inverse obstacle problem. Inverse Probl Imaging 4: 351-377.
6. Karageorghis D, Lesnic D, Marin L (2014) The method of fundamental solutions for an inverse boundary value problem in static thermo-elasticity. Computers and Structures 135: 32-39.
7. Belgacem FB, Fekih HE (2005) On cauchy problem: I. A variational Steklov-Poincaré theory. Inverse Problems 21: 1915-1936.
8. Hadamard J (1953) Lectures on cauchy's problem in linear partial differential equations. Dover New York USA.
9. Marin L, Delvare F, Cimetière A (2015) Fading regularization MFS algorithm for inverse boundary value problems in two-dimensional linear elasticity. International Journal of Solids and Structures 78: 9-20.
10. Andrieux S, Baranger TN (2008) An energy error-based method for the

resolution of the Cauchy problem in 3D linear elasticity Comput. Methods Appl Mech Engrg 197: 902-920.

11. Kadri ML (2015) Contact pressures and cracks identification by using the Dirichlet-to-Neumann Solver in Elasticity. J Appl Mech Eng 4: 152.

12. Andrieux S, Baranger T, Ben Abda (2006) A solving cauchy problems by minimizing an energy-like functional. Inverse Problems 22: 115-133.

13. Quarteroni, Valli A (1999) Domain decomposition methods for partial differential equations. Oxford New York Clarendon press.

14. Hecht F, Pironneau O, Le Hyaric A, Ohtsuka K (2011) "FreeFem++" Univ Pierre et Marie Curie Paris.

15. Ben Belgacem F, Du DT, Jlassi F. Extended-domain-Lavrentiev's regularization for the cauchy problem. Inverse prolems journal.

21

Numerical Investigation of Natural Convection Heat Transfer from V-Fin Arrays with Constant Heat Flux

Jebir SK*

Department of Mechanical Engineering, AL-Nahrain University, Iraq

Abstract

Natural convection heat transfer from rectangular V-fins had been investigated numerically with different heat flux values (175, 350, 525, 700 and 875 Watt per square meter). Fin thickness (5) mm, fin high (18) mm, space between fin and other (10) mm, the heat sink base plate was heated by an attached maximum electric heater 2225 W/m2 with an identical size as the base plate. The mathematical model of the base plate and fins are solved numerically using an COMSOL (5.0) after describing the mesh model using the COMSOL (5.0) and assume the properties of air variation with film temperature. After find the numerical results make validation between numerical and the experimental results, where found good agreement between them. Empirical correlations for the overall Nusselt number versus average Rayleigh numbers for these configurations are obtained and compared to other correlations cited in the literature. The range of Rayleigh numbers, Nusselt number and base plate temperature, $1.7*10^7$ to $12*10^7$, (37°C to 83°C) and (25.6°C to 81.34°C).

Keywords: Heat flux; Nusselt number; Rayleigh number

Introduction

Natural convection heat transfer from these heat sinks occur when there is a relative motion between a hot surface and a fluid flowing over the surface and there is a temperature difference between the surface and the air. If the fluid motion is due to density difference caused by temperature variation in the fluid, then it is called natural convection. The convective heat transfer rate depends on the properties of the fluid flow Natural convective heat transfer from a heat sink with rectangular fin has been studied for many years, a comprehensive review of these studies are presented in many heat transfer. By numerical study such as, Senol baskaya and mecil and ozec [1] focused about effect of parameter (length, width, high, spacing and the temperature) on natural convection heat transfer. Fins made from aluminum in rectangular shape. Heat sink in horizontal position solved case by (CFD) this study model.

Abdullatif Ben-Nakhi and Ali J. Chamkha [2] are focused on the analytical study of steady state heat transfer, laminar flow, natural convection in a square base plate enclosure with an inclined thin rectangular fin. Fins material was aluminum the range of Rayliegh (104-108). A numerical solution based on the finite-volume method is obtained. Aularasan R. and veraj R. [3] designed modern heat sink to cooling electronic device in numerical work use (CFD) program to determine the natural convection from rectangular fin.

M. Baris and Mahmetarik [4] studied the effect of much material (copper, aluminum, parotic graphite thermal annealed on fin efficiency where is respect main factor to making electronic devise. Yaclin et al. [5] studied about natural convection heat transfer from a fin array in horizontal position. CFD code used to solve fin model. The range of Rayleigh number is (2×104 - 3.5×107). Ali Al- Qusamy [6] had studied the execution numerical steady of natural convection heat transfer from rectangular fins. Fins made from aluminum. Heat sink in horizontal position the range of Rayleigh number (4×107- 2×108), rang of high (0.1-0.5) m.

Abdullah H and M. AL-Essa [7] focused about natural convection from rectangular fins in horizontal position the fin mad from aluminum material. Ilker Tari and mehdi [8] made comparing between horizontal and incline heat sink with rectangular fins for natural convection heat transfer.

Sam [9] studied natural convection from rectangular interrupted fins in horizontal position. Where the fins made from aluminum, the continuous heat sinks of different designs have been carried through (CFD) simulations. The range of input power (5-25) w and the range of Rayleigh number (104 - 107).

Numerical Analysis

The governing equations, boundary conditions, numerical domain and the corresponding, the assumptions and the mesh independency are discussed. Some of the present numerical results are also presented in this chapter as validation where compared against well-established analytical model available in the literature.

Computational Domain

Different patterns of heat sinks have been modeled. The heat sink geometry can be shown in Figure 1.

Governing Equations

The heat transfer in the heat sink is take place in three ways, conduction, convection, and radiation. The temperature field is obtained by solving the energy equation Maher [10]. The heat conduction in solid is governed by;

$$\rho C_p \frac{\partial T}{\partial t} - \nabla.(k\nabla T) = Q \quad (1)$$

The heat convective from all external surfaces to ambient is governed by;

***Corresponding author:** Jebir SK, Assistant professor, Department of Mechanical Engineering, AL-Nahrain University, Iraq
E-mail: suhakareem@eng.nahrainuniv.edu.iq

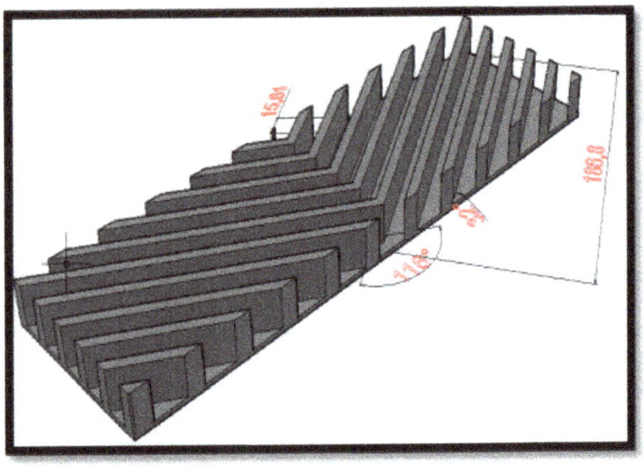

Figure 1: Sketch for the present study.

$$-n.(-k\nabla T) = h(T_{amb} - T) \quad (2)$$

The heat radiation from all external surfaces to ambient is governed by,

$$-n.(-k\nabla T) = h(T_{amb} - T) \quad (3)$$

The following is a summary of the assumptions made to model the fluid flow and heat transfer in a horizontal finned heat sink.

- Steady state, laminar flow, i.e., Rayleigh number Ra <10^9.
- Incompressible flow
- Two-dimensional flow and heat transfer inside the channels.
- Symmetric flow and identical heat transfer in all the channels.
- Iso-heat flux boundary condition for the base plate
- Negligible air velocity entrance in side channels. (The fresh air inflow and outflow from the outmost channels was small compared to the air flow entering from the side of the fin array).

The physical properties of the fluid varied with temperature, density can be shown to follow a simple inverse relationship (ideal gas) with a small correction term:

$$\rho = \frac{351.99}{T} + \frac{344.84}{T^2}\left[\frac{kg}{m^3}\right] \quad (4)$$

$$\mu = \frac{1.4592T^{\frac{3}{2}}}{109.1+T}\left[10^{-6}\frac{(N.s)}{m^2}\right] \quad (5)$$

$$k = \frac{2.334\times10^{-3}T^{\frac{3}{2}}}{164.54+T}\left[\frac{W}{(m.K)}\right] \quad (6)$$

Specific heat follows a quadratic relationship:

$$Cp = 1030.5 - 0.199975T + 3.9743\times10^{-4}T^2\left[\frac{J}{(kg.k)}\right] \quad (7)$$

Boundary Conditions

The boundary conditions employed inside enclosure where the enclosure opens from top. In this case the velocity of air above the heat sink is very low where the velocity of air caused by thermal radiation from fins where the hot air ascent and cold air landing. No slip boundary condition because the air flow over heat sink not have relative velocity to heat sink. For modeling the channel, since the geometry repeats itself, a single channel has been chosen to represent the computational domain According to the flow visualization and velocity measurement of the field flow for a finned plate reported in. Thus, a two-dimensional analysis (instead of three dimensional) is adequate for the purpose of our simulation.

Computational Grid

The governing equations were discretized using a finite-volume method and solved using COMSOL computational fluid dynamic (CFD) package. A computational quadratic meshes ware used for all types of heat sinks. Independent of the grid size has been examined. The coupled set of equations ware solved iteratively, and the solution was considered to be convergent when the relative error was less than 1.0×10^{-6} in each field between two consecutive iterations. We see in group of figures increase of mish even incoming to mesh independence that's give optimum value of measurement factor at 25Watt as a heat flux and (25°C) at an environmental temperature. Complete mesh consists of 36294 domain elements, 21206 boundary elements, and 2622 edge elements Figures 2 and 3.

Validation

The results of the CFD model were verified with experimental results. The results of the average temperature for the heat sinks of CFD model were verified with experimental results in the same condition of the external (ambient) temperature (T_{ext}) for different levels of power (heat flux). The computed average temperature shows in good agreement with the experimental average temperature measured in heat sinks.

Result and Discussion

Figure 4 indicated the variation between temperature different (base plate temperature minus from ambient temperature) in calicoes and heat input in watt for five model of fins with deferent geometry (continues fins,1-inerrupted fins ,4-interrupted fins, inclined fins and V-fins) , as a rule when increase the heat input, increase (ΔT) because of increase the convection and radiation heat transfer, as well as, note (ΔT) of 4-interrupted fins more than rest case becose of small surface area , where the surface area effect on heat transfer, in case of 4-interrupted fins have smallest surface area, lead to weak capacity to carry the heat and Couse high base temperature. They agreed with Senol Baskaya et al. [11] and Salila Ranjan Dixit and Dr. Tarinicharana Panda [12].

Figure 5 shows the variation of heat input and the tip fin temperature for five models of fins. The variation of heat transfer coefficient with length in (mm) shown Figure 6 indicates the variation between heat transfer coefficient and heat input for five configuration of fins. The result show when increase heat input increase heat transfer coefficient.

Figure 7 was prepared for sixth heat input in watts (5, 10, 15, 20, 25, 50 watts). note that the maximum heat transfer coefficient in meddle of heat sink width and decrease whenever approach to end of heat sink width because of the end losses by conduction heat transfer, where is the super heat sink must be insulated from ends and bottom to decrease the different between center temperatures and end temperature as much as possible to decrease heat transfer by conduction. Incropera in 2005 had studied the computational of heat transfer by conduction from bottom of heat sink 3% from heat transfer by conduction and convection so it neglected in heat transfer coefficient compute.

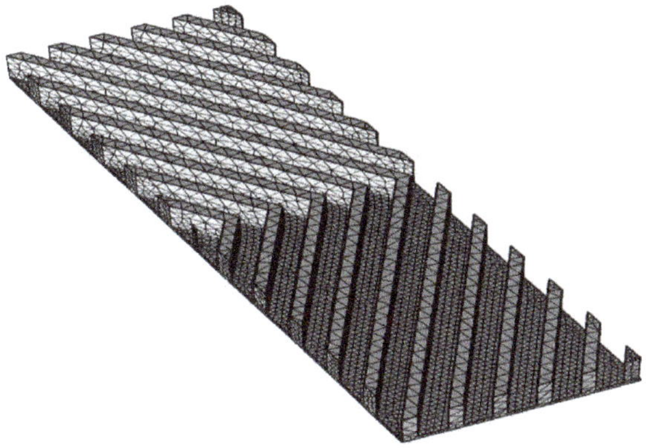

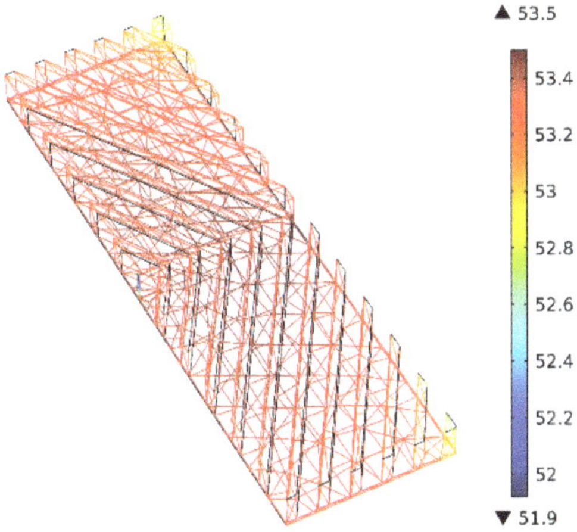

A. Mesh 1592 domain elements, 1038 boundary elements, and 529 edgeelements

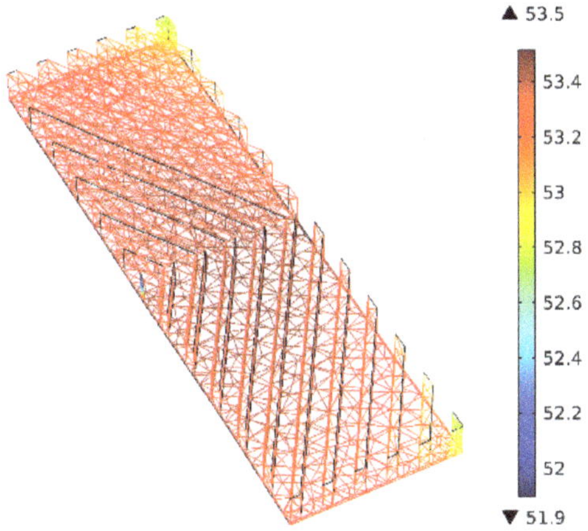

B. Mesh 4636 domain elements, 3228 boundary elements, and 1026 edge elements

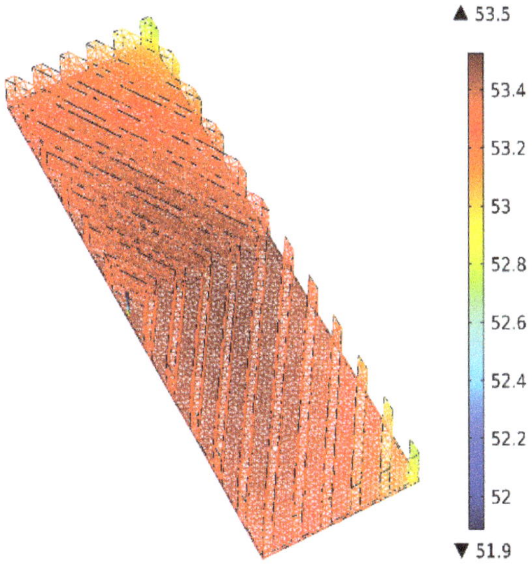

C. Mesh 36294 domain elements, 21206 boundary elements, and 2622 edge elements

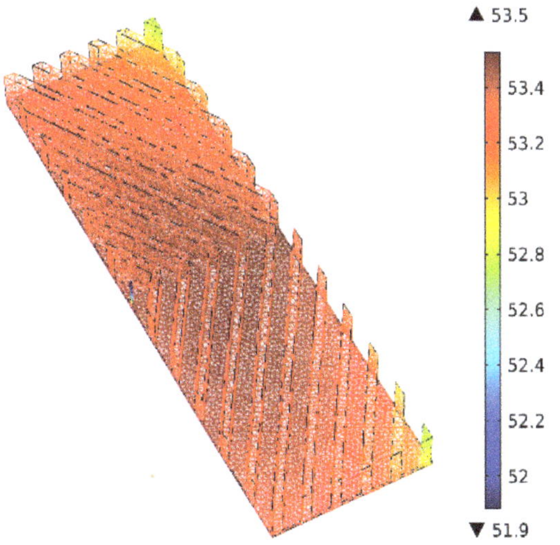

D. Mesh 40069 domain elements, 22718 boundary elements, and 2740 edge elements

Figure 2: Mesh independence of V-fin.

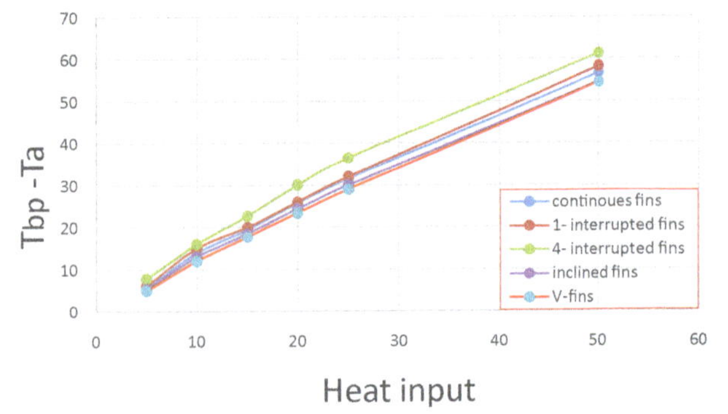

Figure 3: The variation between temperature different in calicoes and heat input.

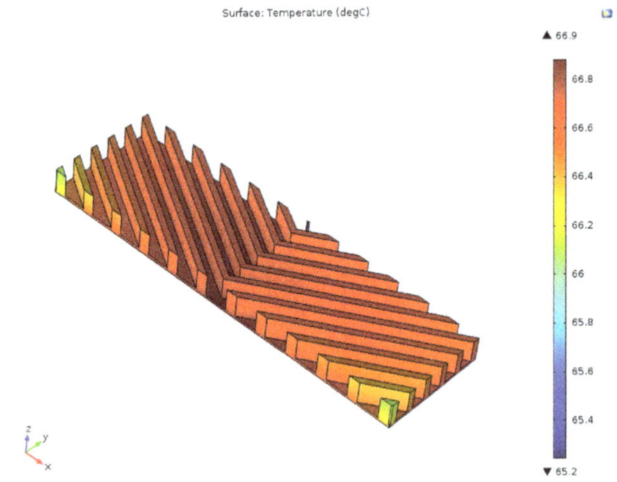

Figure 4: Base temperature distribution of V- fin.

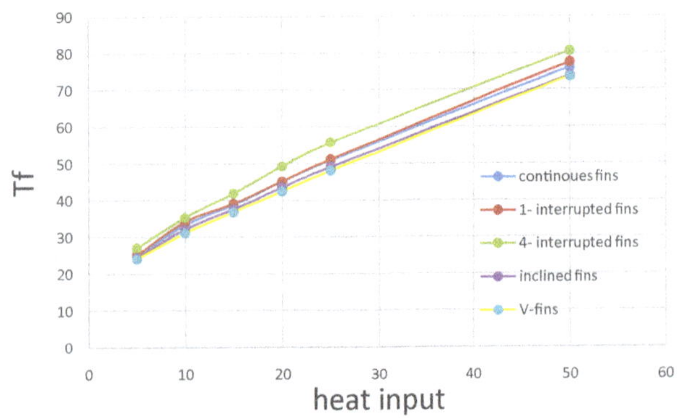

Figure 5: The variation of heat input and the tip fin temperature.

Conclusions

1. For the present research, for the V-fin arrays, it can be concluded that Nu as a function for Ra is: $Nu= 1.1303\ Ra^{0.2142.}$

2. The less average difference between base plate temperature and tip fin temperature in V-fin heat sink had higher efficiently from other cases and equal (78%).

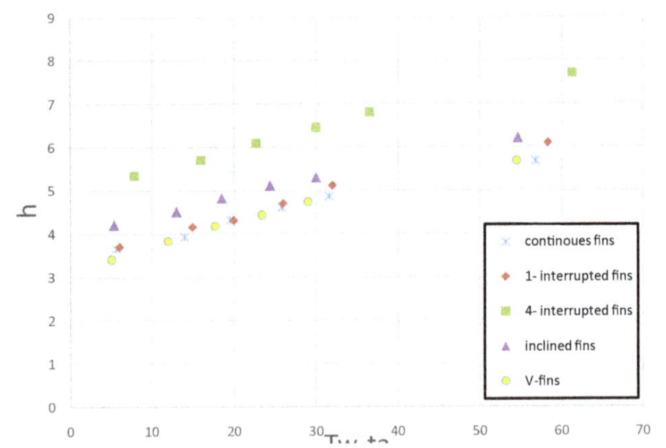

Figure 6: The variation between heat transfer coefficient and heat input.

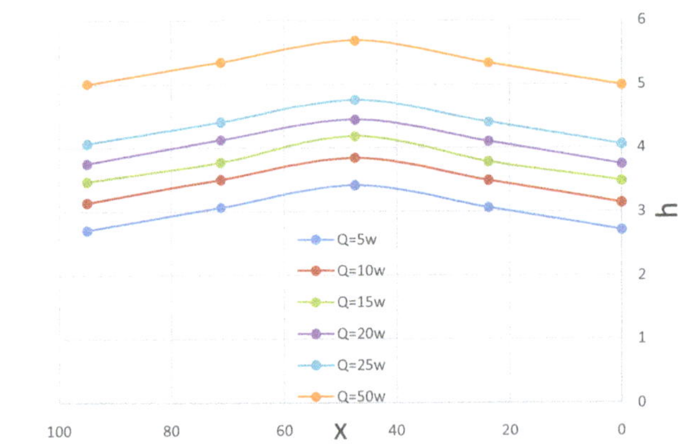

Figure 7: The variation of heat transfer coefficient with length in (mm) for V-fins.

Average Nusselt numbers decrease with increasing the finning factor for same heat input, in this case must V-fin the lowest but heat transfer coefficient depended also on configuration of air flow.

References

1. Jaluria Y (1980) Natural convection: Heat and Mass Transfer. (1stedn), Pergamon Press United States.
2. Bejan A (1995) Convection Heat Transfer (2ndedn), John Wiley & Sons.
3. Kumar S (2007) CFD analysis of electronic chip cooling. Department of Mechanical Engineering National Institute of Technology, Rourkela.
4. Naik S, Probert SD, Wood CI (1987) Natural convection characteristics of a horizontally based vertical rectangular fin-array in the presence of a shroud. Applied Energy 28: 295-319.
5. Sobhan CB, Venkateshan SP, Seetharamu KN (1989) Experimental analysis of unsteady free convection heat transfer from horizontal fin arrays. Wfirmeund Stoffiibertragung 24: 155-160.
6. Sobhan CB, Venkateshan SP, Seetharamu KN (1990) Experimental studies on steady free convection heat transfer from fins and fin arrays. Wfirmeund Stofffibertragung 25: 345-352.
7. Sable MJ, Jagtap SJ, Patil PS, Baviskar PR, Barve SB (2010) Enhancement of natural convection heat transfer on vertical heated plate by multiple v- fin. IJRRAS 5: 255-262.
8. Kharche SS, Farkade HS (2012) Heat transfer analysis through fin array by using natural convection. Int J Emerging Technology and Advanced Engineering 2: 314-322.
9. Wankar VM, Taji SG (2012) Experimental investigation of flow pattern on

rectangular fin arrays under natural convection. Int J Modern Engineering Research 2: 4572-4576.

10. Taji SG, Parishwad GV, Sane NK (2014) Enhanced performance of horizontal rectangular fin array heat sink using assisting mode of mixed convection. Int J Heat and Mass Transfer 72: 250-259.

11. Baskaya S, Sivrioglu M, Ozek M (2000) Parametric study of natural convection heat transfer from horizontal rectangular fin arrays. Gazi University Ankara, Turkey.

12. Dharma Rao V, Naidu SV, Govinda Rao B, Sharma KV (2006) Heat transfer from a horizontal fin array by natural convection and radiation a conjugate analysis. Int J Heat and Mass Transfer 49: 3379-3391.

The Potential Impact of Biofield Treatment on Physical, Structural and Mechanical Properties of Stainless Steel Powder

Mahendra Kumar Trivedi[1], Gopal Nayak[1], Shrikant Patil[1], Rama Mohan Tallapragada[1], Omprakash Latiyal[2] and Snehasis Jana[2*]

[1]*Trivedi Global Inc., 10624 S Eastern Avenue Suite A-969, Henderson, NV 89052, USA*
[2]*Trivedi Science Research Laboratory Pvt. Ltd., Hall-A, Chinar Mega Mall, Chinar Fortune City, Hoshangabad Rd., Bhopal-462026, Madhya Pradesh, India*

Abstract

Stainless steel (SS) has gained extensive attention due to its high corrosion resistance, low maintenance, familiar lustre, and superior mechanical properties. In SS, the mechanical properties are closely related with crystal structure, crystallite size, and lattice strain. The aim of present study was to evaluate the effect of biofield treatment on structural, physical and mechanical properties of SS powder. SS (Grade-SUS316L) powder was divided into two parts denoted as control and treatment. The treatment part was received Mr. Trivedi's biofield treatment. Control and treated SS samples were characterized using particle size analyzer, X-ray diffraction (XRD), and Fourier transform infrared (FT-IR) spectroscopy. Result showed that biofield treatment has significantly reduced the particle size d_{10}, d_{50}, d_{90}, and d_{99} (size, below which 10, 50, 90, and 99% particles were present, respectively) of SS powder up to 7.42, 12.93, 30.23, and 41.38% respectively, as compared to control. XRD result showed that the unit cell volume of SS was altered after biofield treatment. Moreover, crystallite size was significantly reduced upto 70% in treated SS as compared to control. The yield strength calculated using Hall-Petch equation, was significantly increased upto 216.5% in treated SS, as compared to control. This could be due to significant reduction of crystallite size in treated SS after biofield treatment. In FT-IR spectra, intensity of the absorption peak at wavenumber 1107 cm^{-1} (control) attributing to Fe-O-H bond was diminished in case of treated SS. These findings suggest that biofield treatment has substantially altered the structural, physical and mechanical properties of treated SS powder.

Keywords: Biofield treatment; Austenitic stainless steel; X-ray diffraction; FT-IR; SUS316L

Introduction

Stainless steel (SS), invented in the beginning of the 20th century, is known for high resistance to corrosion and staining. It primarily consists of iron (Fe), nickel (Ni), chromium (Cr) and molybdenum (Mo). Based on microstructure, SS is classified into three categories: austenitic, ferritic, and martensitic. The austenitic SS is mainly responsible for corrosion resistance properties and nonmagnetic behaviour. It exist in the form of face centred cubic (FCC) crystal structure with nickel (12-15 wt.%), chromium (16-18 wt.%). Due to high content of Cr, it is suitable for high corrosion resistance applications [1]. Beside this, the superior mechanical properties of austenitic SS is very useful for nuclear fuel clad tubes and fuel assembly [2]. Nano crystalline austenitic SS is mainly consist of large volume fraction of crystallite and crystallite boundaries, which significantly alters their physical and mechanical properties [3]. Further, it is well known fact that the crystallite size of metals are inversely proportional to its yield strength and hardness [4]. Additionally, the mechanical properties of austenitic SS strongly depends on the chemical composition and lattice strain i.e. higher the lattice strain, higher is yield strength. Thus, it is possible to change the mechanical properties of metals by modulating the crystallite size and lattice strain. Currently, in steel industries, mechanical properties of austenitic SS are mainly controlled through various heat treatment process such as annealing, normalizing and quenching etc [5-7]. In heat treatment process, crystallite refinement is strongly required by steel industries in order to increase the strength of material [8]. Furthermore, the heat treatment processes require costly equipment set up and high power supply, to modulate the mechanical properties. Due to this, it becomes important to study an alternative and economically safe approach that could be utilized to modify the physical and structural properties of SS powder.

Recently, several researchers have reported that human body functions as macroscopic quantum system [9-13]. The famous Physicist Feyman had explained the scientific aspects behind quantum biology using quantum-electrodynamics and quantum-chromo dynamics [14]. In other words, each quantum system consists of quantum-domains that have some oscillators within, which generate the potential field. Due to this, a human has ability to harness the energy from environment/universe and can transmit into any object (living or non-living) around the Globe. The object(s) always receive the energy and responded into useful way that is called biofield energy. This process is known as biofield treatment.

Mr. Trivedi's biofield treatment has known to alter the characteristics in various things at atomic, molecular and physical level in many fields such as material science [15-22], microbiology [23-25], biotechnology [26,27] and agriculture [28-30]. The biofield treatment has also shown significant results in graphite carbon, for instance, the unit cell volume was decrease by 1% and crystallite size was increased by 100% after treatment [16]. In the present study, we evaluated for the first time, an impact of biofield treatment on physical, structural and mechanical properties SS powder.

***Corresponding author:** Snehasis Jana, Trivedi Science Research Laboratory Pvt. Ltd., Hall-A, Chinar Mega Mall, Chinar Fortune City, Hoshangabad Rd., Bhopal-462026, Madhya Pradesh, India
E-mail: publication@trivedisrl.com

Experimental

The SS powder (Grade-SUS316L) was purchased from Alfa Aesar, USA. The sample was equally divided into two parts, considered as control and treated. Treated group was in sealed pack and handed over to Mr. Trivedi for biofield treatment under laboratory condition. Mr. Trivedi provided the treatment through his energy transmission process to the treated group without touching the sample. The control and treated samples were characterized using X-ray Diffraction (XRD), surface area analyzer, and Fourier transform infrared (FT-IR) spectroscopy.

Particle size analysis

For particle size analysis, laser particle size analyzer SYMPATEC HELOS-BF was used, which had a detection range of 0·1-875 μm. The particle size data was collected in the form of a chart of particle size *vs.* cumulative percentage. Four parameters of particle sizes *viz.* d_{10}, d_{50}, d_{90}, and d_{99} (size below which 10%, 50%, 90%, and 99% particles are present, respectively) were calculated from the particle size distribution curve. The percent change in particle size were calculated using following equation:

$$\% \text{ change in particle size, } d_{10} = \frac{[(d_{10})_{\text{Treated}} - (d_{10})_{\text{Control}}]}{(d_{10})_{\text{Control}}} \times 100$$

Where, $(d_{10})_{\text{Control}}$ and $(d_{10})_{\text{Treated}}$ are the particle size, d_{10} of control and treated samples respectively. Similarly, the percent change in particle size d_{50}, d_{90} and d_{99} were calculated. For particle size analysis treated part was divided into four parts, referred as T1, T2, T3, and T4.

X-ray diffraction study

XRD analysis was carried out on Phillips, Holland PW 1710 X-ray diffractometer system, which had a copper anode with nickel filter. The radiation of wavelength used by the XRD system was 1.54056 Å. The data obtained from this XRD were in the form of a chart of 2θ vs. intensity and a detailed table containing peak intensity counts, d value (Å), peak width (θ⁰), relative intensity (%) etc. Additionally, PowderX software was used to calculate lattice parameter and unit cell volume.

The crystallite size (G) was calculated by using formula:

$$G=k\lambda/(b\cos\theta),$$

Here, λ is the wavelength of radiation used and k is the equipment constant (=0.94). However, the percentage change in all parameters such as lattice parameter, unit cell volume and crystallite size was calculated using the following equation:

Percent change in lattice parameter $=[(a_t-a_c)/a_c]\times100$

Where, a_c and a_t are lattice parameter value of control and treated powder samples respectively

Percent change in unit cell volume $=[(V_t-V_c)/V_c]\times100$

Where, V_c and V_t are the unit cell volume of control and treated powder samples respectively

Percent change in crystallite size $=[(G_t-G_c)/G_c]\times100$

Where, G_c and G_t are crystallite size of control and treated powder samples respectively. XRD analysis was carried out for control, T1, T3, and T4.

FT-IR Spectroscopy

To study the impact of biofield treatment at atomic bonding level in SS the FT-IR analysis was carried out using Shimadzu, Fourier transform infrared (FT-IR) spectrometer with frequency range of 300-4000 cm^{-1}. FT-IR analysis was carried out for control and T1.

Results and Discussion

Particle size analysis

Particle size analysis result of SS powder are presented in Table 1, Figures 1 and 2. In order to study the effect of biofield treatment on various sizes of particles, four kind of particle size (d_{10}, d_{50}, d_{90}, and d_{99}) were analyzed. Data result showed that smaller particle size d_{10}, was reduced from 18.58 μm (control) to 17.74, 17.36, 17.20, and 17.55 μm in T1, T2, T3, and T4 respectively (Table 1). It indicates that d_{10} was reduced up to 7.42% (T3) as compared to control (Figure 1). Medium particle size, d_{50} was reduced from 44.15 μm (control) to 39.64, 38.53, 38.44, and 38.65 μm in treated T1, T2, T3 and T4 respectively. It suggests that average particle size, d_{50} was reduced up to 12.93% (T3) as compared to control. Further, large particle size, d_{90} was decreased from 93.61 μm (control) to 68.80, 65.99, 65.95, and 65.31 μm in T1, T2, T3, and T4 respectively. Data showed that d_{90} was significantly reduced upto 30.23% (T4), as compared to control. In addition, larger particle size d_{99} was reduced from 152 μm (control) to 99.15, 95.47, 94.93, and 89.11 μm in T1, T2, T3, and T4 respectively. It suggests that d_{99} was substantially reduced upto 41.38% (T4), as compared to control (Figure 2). Overall, the particle size result indicates that particles of each size i.e. d_{10}, d_{50}, d_{90}, and d_{99} were reduced in all treated samples T1, T2, T3, and T4. It could be due to breaking down of all kind of powder particles into smaller particles. It is assumed that an energy might be transferred to SS powder through biofield treatment. This energy might induce milling in SS powder and that resulted into breaking down of large particles to smaller [16-18]. Furthermore, average percent changes in particle size are illustrated in Figure 3. It was found that average of percent change in particle sizes d_{10}, d_{50}, d_{90}, and d_{99} were reduced by 6.01, 12.08, 28.9, 37.7%, respectively, as compared to control in treated

Group	d_{10} (μm)	d_{50} (μm)	d_{90} (μm)	d_{99} (μm)
Control	18.58	44.15	93.61	152
T1	17.74	39.64	68.8	99.15
T2	17.36	38.53	65.99	95.47
T3	17.20	38.44	65.95	94.93
T4	17.55	38.65	65.31	89.11

T1, T2, T3, and T4 are biofield treated stainless steel samples.

d_{10}, d_{50}, d_{90}, and d99 are the sizes below which 10%, 50%, 90%, and 99% particles are present, respectively.

Table 1: Particle size of stainless steel (SS) powder.

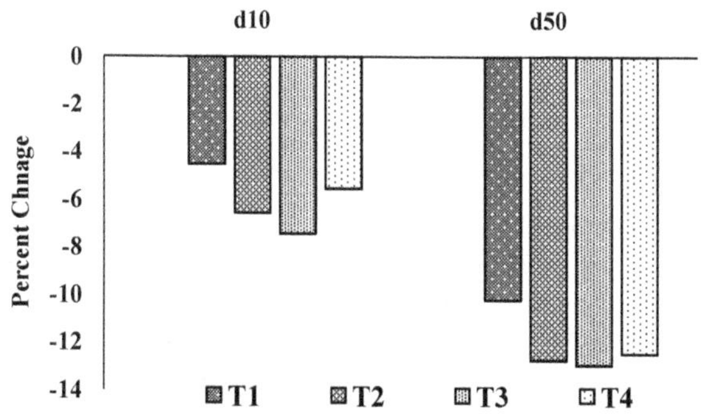

Figure 1: Effect of biofield treatment on percent change in particle size d_{10} and d_{50} of stainless steel powder.

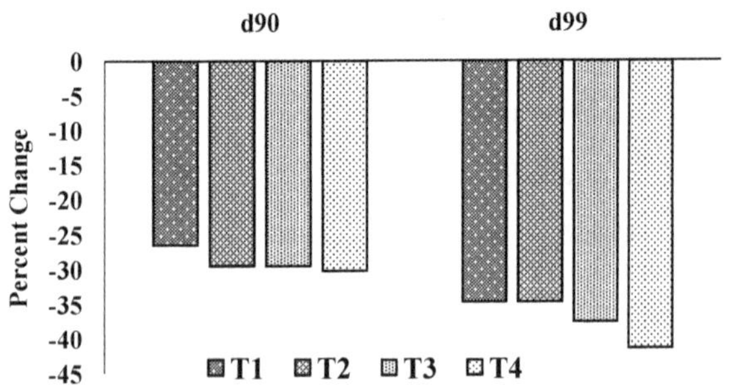

Figure 2: Effect of biofield treatment on percent change in particle size d_{90} and d_{99} of stainless steel powder.

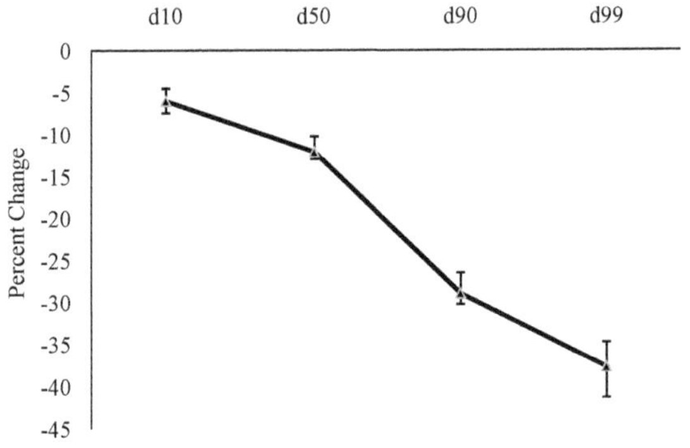

Figure 3: Effect of biofield treatment on percent change of particle sizes of stainless steel powder with respect of control.

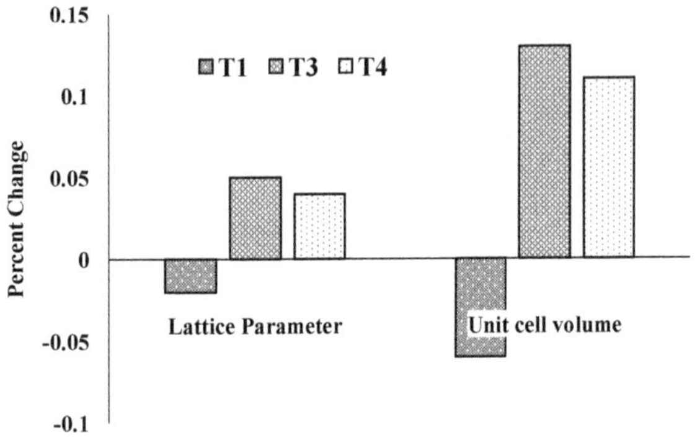

Figure 4: Effect of biofield treatment on lattice parameter and unit cell volume of stainless steel powder.

SS powder. Similar results of particle size reduction in titanium and antimony had been reported by our group in previous studies [15,17].

X-ray diffraction (XRD)

XRD results of control and treated SS samples are depicted in Figures 4-6. It was found that the lattice parameter of unit cell slightly altered in biofield treated samples (T1:-0.02%, T2: 0.05%, T3: 0.04%) as compared to control. This change in lattice parameter led to alter the unit cell volume slightly by -0.06%, 0.13%, and 0.12% in treated T1, T3, and T4 respectively as compared to control (Figure 4). It indicates that both kind of stress (compressive and tensile) might present in treated SS powder, after biofield treatment [15,16]. Thus, it is hypothesised that the high-energy milling induced through biofield treatment may lead to generate tensile and compressive stress in SS powder that resulted into alteration of lattice parameter and unit cell volume. Besides this, the crystallite size was computed using Scherrer formula is presented in Figure 5. It was found that crystallite size was 148.44 nm in control, whereas crystallite size of treated samples was 74.2, 44.53, and 63.61 nm in T1, T3 and T4, respectively. It indicates that crystallite size was significantly reduced by 50, 70, and 57.15% in treated T1, T3 and T4 respectively, as compared to control (Figure 6). The existence of severe lattice strains are evidenced by the change in lattice parameters (Figure 4). Thus, it is assumed that presence of these internal strain may leads to fracture the grains into sub grains and decrease the crystallite size [21]. On the other hand, the relation between strength of material and crystallite size is given by Hall-Patch equation as given below:

$$\sigma = \sigma_o + k / \sqrt{G} \qquad (1)$$

Where, σ is strength of the material, σ_o is a material constant for the starting stress for dislocation movement, k is the strengthening coefficient, G is crystallite size.

Singh et al. reported the k=575 MPa $\mu m^{1/2}$, σ_o=150 MPa for true strain less than 0.02 [31]. Yield strength was computed using these constants and results are shown in Figures 7 and 8. It was found that yield strength of 2086.8 Mpa in control, which increased to 4024.6, 6606.3, and 4669.7 MPa in treated SS samples T1, T3 and T4, respectively. This indicates that the yield strength was significantly enhanced by 92.86, 216.5, and 123.7% in treated SS samples T1, T3, and T4 respectively as compared to control. It is already reported that the strength of materials can be modulated by changing the crystallite size. The decrease in crystallite size in treated SS powder results into increase

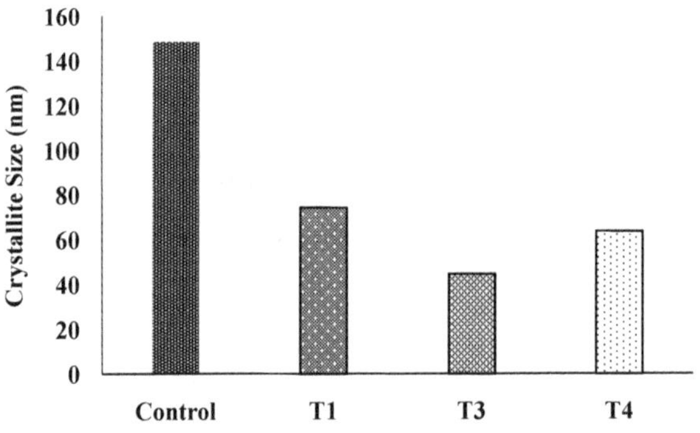

Figure 5: Effect of biofield treatment on crystallite size of stainless steel.

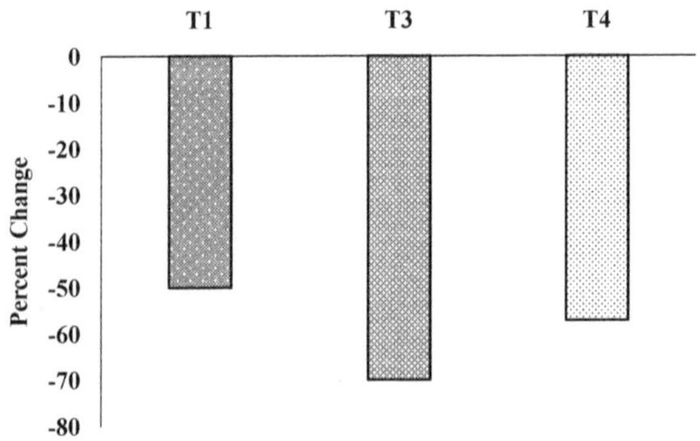

Figure 6: Effect of biofield treatment on percent change in crystallite size of stainless steel.

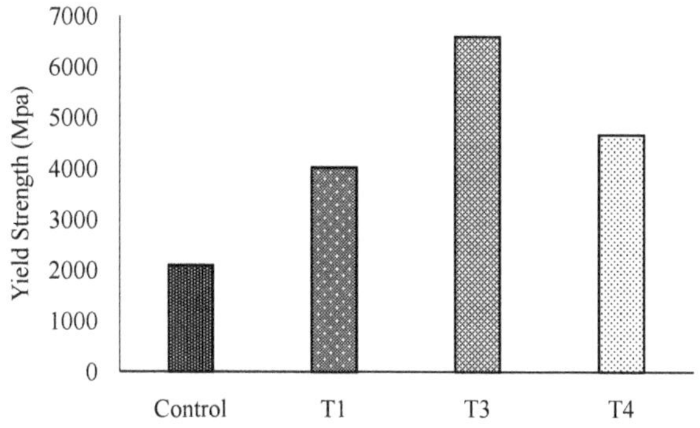

Figure 7: Effect of biofield treatment on yield strength of stainless steel.

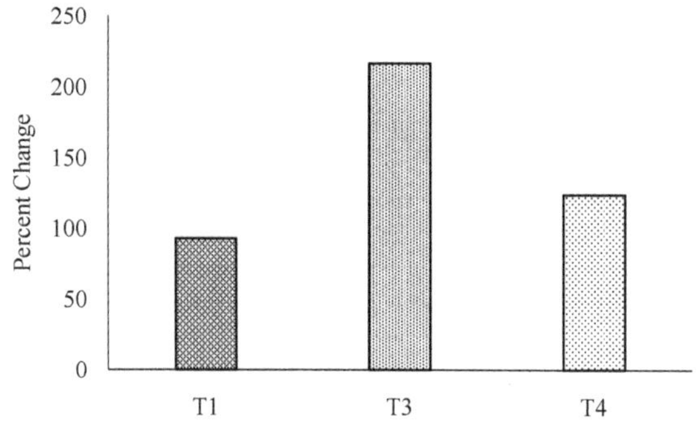

Figure 8: Effect of biofield treatment on percent change in yield strength of stainless steel as compared to control.

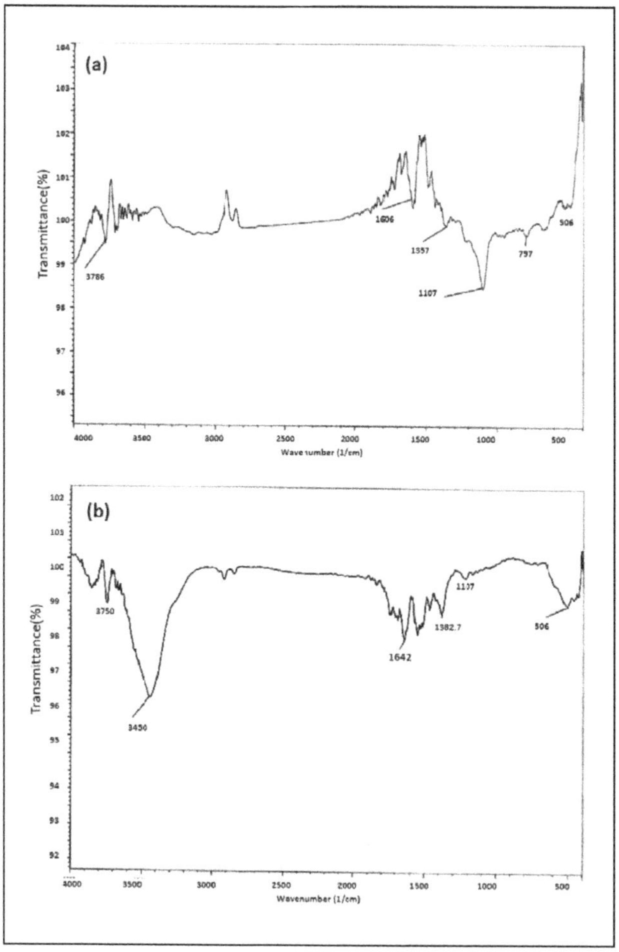

Figure 9: FT-IR spectrum of stainless steel powder (a) control and (b) treated (T1).

the crystallite boundaries. Thus, higher crystallite boundaries in treated SS powder hindered the dislocation movement gliding along the slip planes and thereby increased yield strength [4]. Thus, it is postulated that biofield treated SS powder could be more useful in automobile parts and nuclear reactor applications.

FT-IR spectroscopy

FT-IR spectrum of control and treated SS powder are shown in Figure 9. The absorption peaks observed at wavenumber 3786 and 1606 cm^{-1} (control) and 3759, 3450, and 1542 cm^{-1} (treated) were assigned to bonding vibration of water molecules due to moisture absorption by sample. Another peak observed at wavenumber 506 cm^{-1} in control and treated SS sample were due to Cr-O bond vibrations [32]. Peak found at wavenumber 1107 cm^{-1} (control) attributed to Fe-O-H bond vibrations, was completely diminished in treated SS. It may be due to alteration in F-O-H bond at atomic level through biofield treatment [32]. Thus, it is hypothesized that biofield treatment may be acting at atomic level to cause these alteration.

Conclusion

In summary, the biofield treatment has significantly reduced the particle size and crystallite size in SS powder. Average particle size was reduced upto 12.93% in treated SS powder as compared to control. In addition, the reduction in crystallite size upto 70% after biofield treatment led to increase in yield strength by 216.57% as compared to control (Hall-Petch effect). This could be due to increase in crystallite boundaries after biofield treatment, which hindered the dislocation movement and thereby increased yield strength. FT-IR spectra showed peak at wavenumber 1107 cm^{-1} in control, which assigned to Fe-O-H was significantly reduced in treated SS. It might be due to alteration of bond properties in treated SS after biofield treatment. Based on these promising results, it is expected that biofield treatment could be applied to improve the mechanical properties of SS powder for nuclear reactor, appliances, and automobile.

Acknowledgement

Authors gratefully acknowledged to Dr. Cheng Dong of NLSC, Institute of Physics, and Chinese academy of Sciences for providing the facilities to use PowderX software for analyzing XRD data.

References

1. Silva G, Baldissera MR, Triches EDS, Cardoso KR (2013) Preparation and characterization of stainless steel 316L/HA biocomposite. Mater Res 16: 304-309.
2. Desu RK, Krishnamurthy HN, Balu A, Gupta AK, Singh SK (2015) Mechanical properties of austenitic stainless steel 304L and 316L at elevated temperatures. J Mater Res Technol (In press).
3. Meyers MA, Mishra A, Benson DJ (2006) Mechanical properties of nano crystalline materials. Prog Mater Sci 51: 427-556.
4. Callister WD (2008) Fundamentals of materials science and engineering: An integrated approach. (3rdedn), John Wiley and Sons.
5. Tanwar AK (2015) Effect of various heat treatment processes on mechanical

properties of mild steel and stainless steel. AIJRSTEM 57-61.

6. Tukur SA, Dambatta MS, Ahmed A, Muaz NM (2014) Effect of Heat treatment temperature on mechanical properties of the AISI 304 stainless steel. Intl J Innov Res Sci, Eng Technol 3: 9516-9520.

7. Jithin M, Hameed AA, Jose B, Jacob A (2015) Influence of heat treatment on duplex stainless steel to study the material properties. Intl J Sci Technol Res 4: 291-293.

8. Nakashima H (2008) Trends in materials and heat treatments for rolling bearings. Technical Review- 76. NTN corporation Japan.

9. Del Giudice E, Doglia S, Milani M (1989) Magnetic flux quantization and josephson behavior in living systems. Phys Scrip 40: 786-791.

10. Nobili R (1985) Schrodinger wave holography in brain cortex. Phys Rev A 32: 3618-3626.

11. Popp FA (1989) Electromagnetic Bio-Information. Munchen, Baltimore: Urban & Schwarzenberg.

12. Smith CW (1998) Is a living system a macroscopic quantum system?. Frontier Perspect 7: 9-15.

13. Bohm DA (1952) Suggested interpretation of quantum theory. Phys Rev 85: 166-178.

14. Feynman RP (1949) Space-time approaches to quantum electrodynamics. Phys Rev 76: 769-782.

15. Trivedi MK, Tallapragada RM (2008) A transcendental to changing metal powder characteristics. Met Powder Rep 63: 22-28, 31.

16. Trivedi MK, Tallapragada RM (2009) Effect of super consciousness external energy on atomic, crystalline and powder characteristics of carbon allotrope powders. Mater Res Innov 13: 473-480.

17. Dhabade VV, Tallapragada RM, Trivedi MK (2009) Effect of external energy on atomic, crystalline and powder characteristics of antimony and bismuth powders. Bull Mater Sci 32: 471-479.

18. Trivedi MK, Patil S, Tallapragada RM (2012) Thought intervention through bio field changing metal powder characteristics experiments on powder characteristics at a PM plant. Proceeding of the 2nd International conference on future control and automation 173: 247-252.

19. Trivedi MK, Patil S, Tallapragada RM (2013) Effect of biofield treatment on the physical and thermal characteristics of silicon, tin and lead powders. J Material Sci Eng 2: 125.

20. Trivedi MK, Patil S, Tallapragada RM (2013) Effect of biofield treatment on the physical and thermal characteristics of vanadium pentoxide powder. J Material Sci Eng S11: 001.

21. Trivedi MK, Patil S, Tallapragada RM (2014) Atomic, crystalline and powder characteristics of treated zirconia and silica powders. J Material Sci Eng 3: 144.

22. Trivedi MK, Patil S, Tallapragada RM (2015) Effect of biofield treatment on the physical and thermal characteristics of aluminium powders. Ind Eng Manage 4: 151.

23. Trivedi MK, Patil S, Bhardwaj Y (2008) Impact of an external energy on Staphylococcus epidermis [ATCC –13518] in relation to antibiotic susceptibility and biochemical reactions – An experimental study. J Accord Integr Med 4: 230-235.

24. Trivedi MK, Patil S (2008) Impact of an external energy on Yersinia enterocolitica [ATCC –23715] in relation to antibiotic susceptibility and biochemical reactions: An experimental study. Internet J Alternat Med 6.

25. Trivedi MK, Patil S, Bhardwaj Y (2009) Impact of an external energy on Enterococcus faecalis [ATCC – 51299] in relation to antibiotic susceptibility and biochemical reactions – An experimental study. J Accord Integr Med 5: 119-130.

26. Patil S, Nayak GB, Barve SS, Tembe RP, Khan RR (2012) Impact of biofield treatment on growth and anatomical characteristics of Pogostemon cablin (Benth.). Biotechnology 11: 154-162.

27. Altekar N, Nayak G (2015) Effect of biofield treatment on plant growth and adaptation. J Environ Health Sci 1: 1-9.

28. Shinde V, Sances F, Patil S, Spence A (2012) Impact of biofield treatment on growth and yield of lettuce and tomato. Aust J Basic Appl Sci 6: 100-105.

29. Lenssen AW (2013) Biofield and fungicide seed treatment influences on soybean productivity, seed quality and weed community. Agricultural Journal 8: 138-143.

30. Sances F, Flora E, Patil S, Spence A, Shinde V (2013) Impact of biofield treatment on ginseng and organic blueberry yield. Agrivita J Agric Sci 35.

31. Singh KK, Sangal S, Murty GS (2002) Hall–Petch behaviour of 316L austenitic stainless steel at room temperature. Mater Sci Technol 18: 165-172.

32. Junqueira RMR, Loureiro CRDO, Andrade MS, Buono VTL (2008) Characterization of interference thin films grown on stainless steel surface by alternate pulse current in a sulphochromic solution. Mat Res 114: 421-426.

23

Comparative Study of Different with Al-Sic for Engine Valve Guide by using FEM

Srivastva HK[1*], Chauhan AS[2], Raza A[3], Kushwaha M[3] and Bhardwaj PK[3]

[1]*Department of Mechanical Engineering, Naraina Group of Institutions, Kanpur 208020, India*
[2]*Department of Mechanical Engineering, Maharana Pratap Engineering College, Kanpur 209217, India*
[3]*Department of Mechanical Engineering, Naraina Group of Institutions, Kanpur 208020, India*

Abstract

In this Work an effort has been designed to raise the reliability of engine using Al-Sic composites with other alternatively materials for the engine valve guides. Aluminum matrix composites have found the most suitable inside automotive, aerospace and aircraft industries and contain the greatest promise for future year's growth. The finite element analysis of the Al-Sic composite with Titanium alloy (Ti-834), Copper Nickel Silicon alloys (CuNi3Si), and aluminum bronze alloy as an alternative material for engine valve guide was done using Ansys 13.0 software. The stress analysis of engine valve guide under the different pressure and temperature is considered, the pressure is taken as from 10 MPa to 100 MPa with Different temperatures varying from 600 °C to 650 °C. The temperature, principal stress and principal strain distribution on the entire surface area of the engine valve guide were obtained. The stresses were observed to be well below the permitted stress for all the materials but the Al-Sic composites found the most suitable one. Valve guide is modeled in pro-engineer software and analysis is carried out in Ansys 13.0. The deformations and stresses induced due to structural and thermal loading is illustrated and discussed.

Keywords: Engine valve; Materials

Introduction

In reaction to increasing worldwide competition and developing concern for environment the auto manufacturers have already been encouraged to meet up the conflicting demands of increased power and performance, lower fuel consumption, lower pollution emission and reduced vibration and noise. To be able to fulfill these newer and emerging needs the Automobile industry has acknowledged the necessity for materials substitution. Metal matrix composites are providing outstanding properties in a true number of automotive components such as for example piston, cylinder liner, engine valves, brake discs, brake drums, clutch discs, linking rods etc. Several function has been documented on the substitution of presently utilized materials by the aluminum matrix composites for various automotive parts viz. piston, cylinder liner, engine valves, valve seat inserts etc. Al-Sic composite engine valve guides have already been fabricated through powder metallurgy and casting processes. The radial crushing strength, hardness, and wear resistance of the Al-SiC cast and composite iron engine valve guides were measured and compared. Al-SiC composites with 5 to 30 wt.% of SiC were discovered to have increased Rockwell hardness and radial crushing strength compared to the cast iron engine valve guides. Al-SiC composite engine valve guides with 20 and 30 wt. % of SiC were discovered to possess higher wear level of resistance compared to the cast iron engine valve guides. Existing work incorporates the finite element strategy to include the prospects of Al-SiC composites just as one alternative materials for the engine valve guides. The finite element Analysis of the engine valve guides was done making use of Ansys 13.0 software program. Temperature, displacement and pressure boundary conditions were used and the temperature, stress and stress distribution on the entire surface area of the engine valve guides was obtained. Aluminum alloys have discovered greater adoption while potential matrix materials in comparison to some other alloys. In 2000, Katsunao Chikahata and Koichiro Hayashi (Publication numberUS6012703 A) [1] presented valve guide and process for manufacturing, they say that valve guide which is manufactured by a sintered alloy and used for an engine, and more particularly, to a valve guide which has high wear resistance and can prevent appearance of scuffing on a surface of a valve stem associated with the valve guide. In 2009, Suhael Ahmed [2] presented the Development and Characterization ofAl7075 Based Hybrid Composites. Aluminum based Metal matrix composites have been emerged as an important class of materials for structural, wear, thermal, transportation and electrical applications. Srinivasa R. Bakshi, Di Wang, Timothy Price, ArvindAgarwal in 2009,[3] presented Microstructure and Wear properties of Aluminum/Aluminum–Silicon Composite coatings prepared by cold spraying. Manoj Singla, D. Deepak Dwivedi, Lakhvir Singh, Vikas Chawla in 2009, [4] developed a conventional low cost method of producing MMCs and to obtain homogenous dispersion of ceramic material. In 2011, Dunia Abdul Saheb [5] presented Aluminum silicon carbide and aluminum graphite Particulate composites. Ramanpreet Singh, Er. Rahul Singla in 2012 [6] presented Tribological Characterization of Aluminium-Silicon Carbide Composite Prepared by mechanical alloying aluminum metal composites is combinations of materials in such a way that the resulting materials have certain design properties on improved quality.

Material and Properties

Al SIC composites

Shown in Tables 1-4

Problem Specification

Issue of sustainability, stability, durability and reliability of present

***Corresponding author:** Srivastva HK, Associate Professor, Department of Mechanical Engineering, Naraina Group of Institutions, Kanpur 208020, India
E-mail: hemendra28@rediffmail.com

Material	Elastic Modulus (GPa)	Poisson ratio	Rockwell Hardness(HRC)
Al 10% wt. of SiC	77.4	0.33	46.5
Al 20% wt. of SiC	86	0.32	51.5
Al 30% wt. of SiC	92	0.31	53

Table 1: Mechanical properties.

Material	Coefficient of thermal expansion (x10^{-7} /^{0}k)	Thermal Conductivity W/m.K
Al 10% wt. of SiC	2.0	173
Al 20% wt. of SiC	1.75	168
Al 30% wt. of SiC	1.55	164

Table 2: Thermal properties.

Modulus of elasticity X-Direction (Pa)	1.2e+011
Thermal expansion coefficient X-Direction	1.2e-006
Major Poisson's ratio Z-Plane	0.31
Minor Poisson's ratio Z-Plane	0.31
Density (kg/m^3)	4550
Thermal conductivity X-Direction (w/m.k)	25

Table 3: Titanium alloy (Ti-834) properties.

Modulus of elasticity X-Direction (Pa)	1.3e+011
Thermal expansion coefficient X-Direction	1.6e-006
Major Poisson's ratio Z-Plane	0.32
Minor Poisson's ratio Z-Plane	0.32
Density (kg/m^3)	8870
Thermal conductivity X-Direction (w/m.k)	230

Table 4: Copper Nickel Silicon alloy (CuNi3Si) properties.

material for engine guide valve at high temperature about 600°C to 1000°C under very high pressure about 10 mpa-100 mpa back ground of the problem.

At the moment the engine valve guides are made of iron-based materials, which result in a true amount of problems within an automotive engine-

• During cold start condition the viscosity of oil is high and also sufficient lubricant is not available therefore high wear of the valve stem / valve guide takes place. In the adverse conditions the valve may jam in the guide.

• During running condition of the engine the temperature of the valve stem and valve guide increases to about 500°C. Therefore at high temperature the clearance between valves stem and valve guide decreases due to thermal expansion, which results in high wear of the valve stem and the guide.

• The superimposed rocking motion in addition to the sliding of the engine valve causes high wear at the ends of the valve guide called "bell-mouthing", generally more pronounced in the rocker arm actuation mechanism.

These problems call for a high wear resistant material with low coefficient of thermal expansion and engine valve guides based on the Al-SiC composites are expected to provide a better solution.

Temperature Distribution

To understand the problem of wear and decreasing the clearance the clearance due to temperature change at various load condition, firstly have to analyze the temperature distribution in the valve and valve guide.

Typically we take example of exhaust valve temperature distribution in spark ignition engine of V-8 car with full throttle (3500 r/min). During expansion stoke the temperature of the burnt gases is about 600-800°C. These hot burnt gases get impact with head of poppet valve and head of poppet valve conduct the heat through various portion of poppet valve.

- Small part about 30% of the heat is conducted through valve seat and remaining 70% heat conduct through valve stem.
- 25% heat conducted through valve stem is further conduct through valve guide.

Due to this impact of hot gases, the temperature at the center of head is maximum about 700 °C and further distributed at various element, nodes and section as shown in figure below.

Temperature at the starting section of valve stem is 613 °C and at the open end of valve guide is about 580 °C and further temperature distribute in different nodes and elements as shown in the figure below and at other end of valve guide the temperature is about 390 C.

Problem Analysis

At this temperature range, materials which are currently used for guide valve expands due to which the clearance between valve stem and valve guide get reduced which creates a lot of problems in the valve operating system such as:

- Valve stems breakage
- Burnt Valve seat
- Valve Stem to Valve Guide Seizure
- Mechanical Damages
- Excessive Valve Stem & Valve Guide Wear
- Valve Tip Breakage etc.

Analysis of Engine Guide Valve

Finite element method has become one of the most widely used techniques, for analyzing mechanical loading characteristics in modern engineering components. Traditional analysis techniques can only be satisfactorily applied to a range of conventional component shapes and specific loading conditions. Unfortunately, the majority of engineering loading situations are not simple and straight forward therefore the traditional techniques often need to be modified and compromised to suit situations for which they were not intend. The uncertainty thus created, commonly leads to the designer applying excessively high factor of safety to the mechanical loads and so to over design components by specifying either unnecessarily bulky cross section or high quality materials, inevitably the cost of the product is adversely affected.

Before using Valve guide in the I.C. Engine first of all we have to check its deformation, Vonmises stresses-Strain value, Maximum & Minimum principle stress-strain and its failure point at every load conditions. FEA is the best method of determining the deformation, Vonmises stresses-Strain value, Maximum & Minimum principle stress-strain and its failure point at each and every load conditions. Main advantage of FEA is that it converts the problem into a number of elements and nodes and then solve problem and give the result at every element and node. It also State that at which node the material is going to fail at which load condition. Therefore this can make our material safe in all load condition. In this way, FEM is applied (using Ansys software) on Valve guide for making a comparative study of Aluminium silicon carbide of different compositions with other materials (which are currently in use):

- Titanium alloy (Ti-834) (used in racing Cars)
- Copper Nickel Silicon Alloy (Cu Ni3Si) (used in locomotive)

At different load conditions such as:

Different temperatures varying from 550°C to 1000°C.

Pressure varying from 10 MPa to 100 MPa.

Modeling has been carried in pro-engineering software. The Engine guide valve is drawn in Pro-E (Figure 1). Sketch of the guide valve is done in the sketcher part of the Pro-Engineering. The section of the guide valve is revolved to 360^0 about the central axis to obtain 3D model of guide valve. The inner wall and outer wall is inclined angle 92°.

Finite Element Model of Guide valve

Analysis of guide valve is carried over in following steps Guide valve model has been modeled in the pro-engineer as shown in Figure 2. For performing the analysis over the guide valve a finite element model is necessary. In order to get the good quality mesh and to maintain the tetrahedron elements, the cap cavity plate is meshed in hyper mesh 12.0 Hyper mesh 12.0 is the product from Altair Hyper Works is a commercially available software package. It mainly used for the finite element modeling of the components. Guide valve as shown in Figure 2 is retrieved in hyper mesh using solid 10 node 92 element type(From Ansys library) is used to mesh the Guide valve and a converged mesh is shown in Figure 3. A Solid 10 node 87 element type is used for thermal analysis.

Result and Discussion

From the resultant variation (Figures 4 and 5) the maximum deformation is 0.0169mm and 0.01436mm is at inside of guide valve when the material is Al 10% Sic and Al 20% Sic and subjected pressure is 10 MPa at top and 1MPa at bottom.

From the resultant variation (Figures 6 and 7) the maximum deformation is 0.0778 mm and 0.124 mm is at inside of guide valve when the material is titanium alloy and CuNi3si alloy and subjected pressure is 10 MPa at top and 1MPa at bottom. The variation of deformation as material is changed is shown in Table 5.

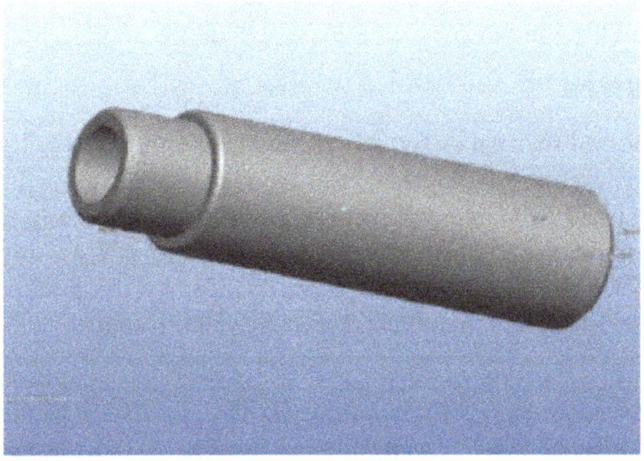

Figure 1: Geometric modeling.

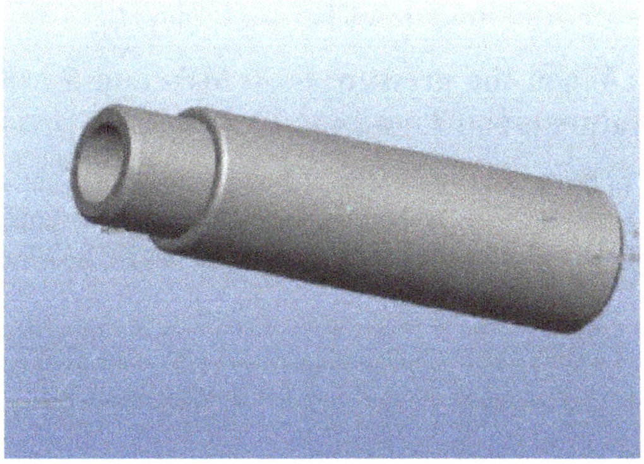

Figure 2: Guide valve.

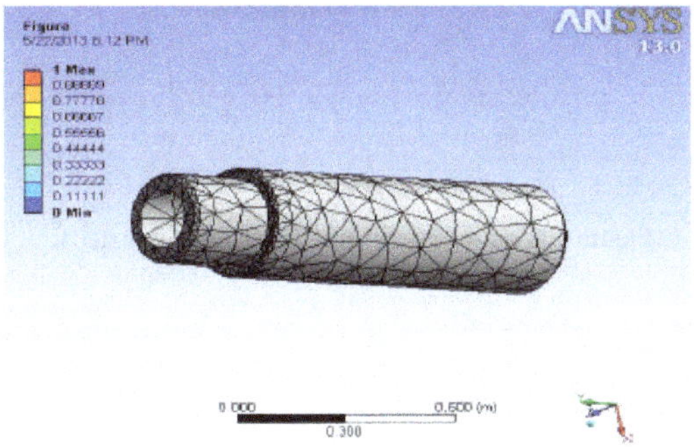

Figure 3: Meshed model of guide valve.

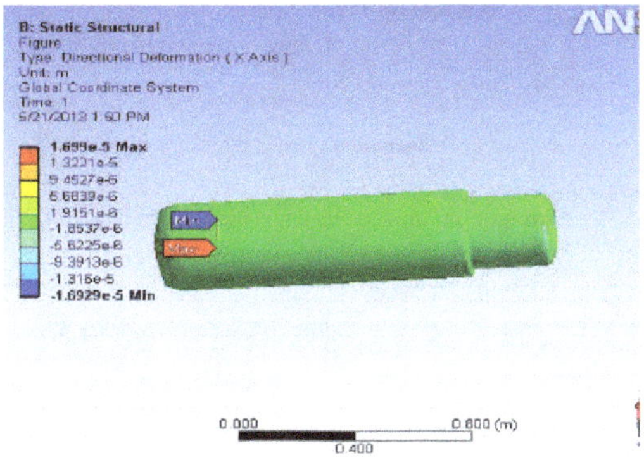

Figure 4: Deformation variation for Al 10% Sic material.

From the table it is clear that the deflection is varying with change of material it is minimum with Al 30% composites and maximum with CuNi3si Alloy that is 0.124mm. However the guide valve is safe due to very less deformation (Figures 8 and 9).

Table 6 shows the resultant displacement and resultant stress of the guide valve under the different material conditions. under the constant load condition of 10 MPa, and 1 Mpa the safe stress is 23.06 MPa, 21.79MPa, 20.09MPa, 165.39MPa, 286.31MPa, when the guide valve is made up of Al 10% composites, Al 20% composites, Al 30% composites, Titanium Alloy (Ti 834), CuNi3si Alloy, respectively . Which is not exceeded the material yield stress respectively. It is concluded that when the von mises stress are less than the yield stress there the guide valve will be safe. And when the von mises stress are more than the yield stress than the structure will be failed (Figures 10-15).

Case I: When the pressure is 50 MPa and 5 MPa and temperature is 600 °C and 400 °C

The pressure load of 50Mpa at bottom and 5MPa, at top and temperature of 600⁰ c, at bottom and 400⁰c at top is applied and deformations and stresses are computed on guide valve (Table 7).

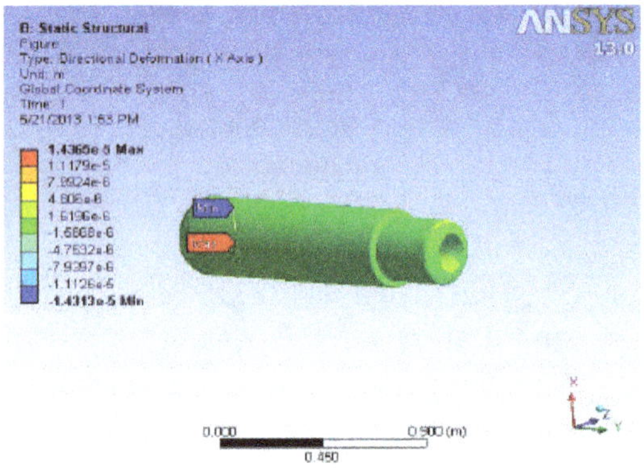

Figure 5: Deformation variation for Al 20% Sic material.

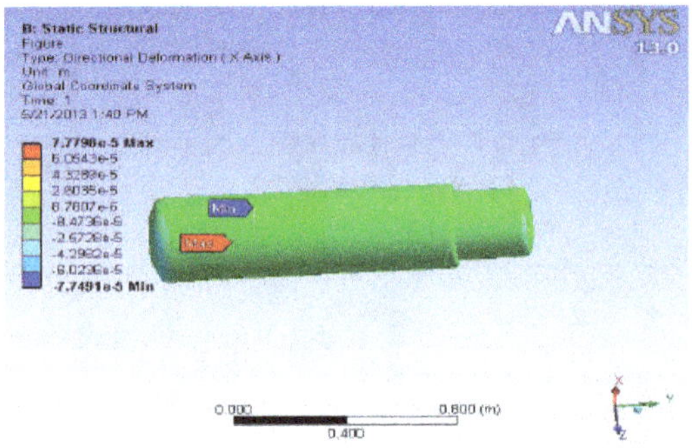

Figure 6: Deformation variation for titanium alloy.

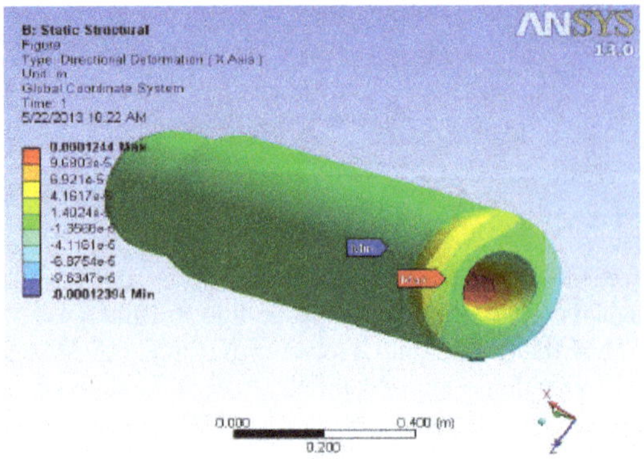

Figure 7: Deformation variation for CuNi3si Alloy.

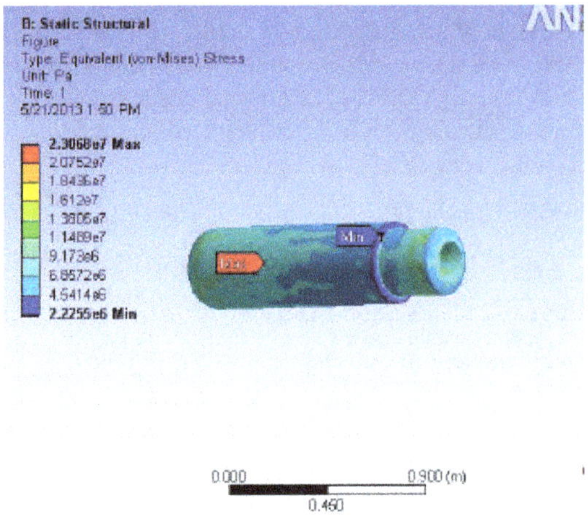

Figure 8: Von mises stress for Al 10% Sic material.

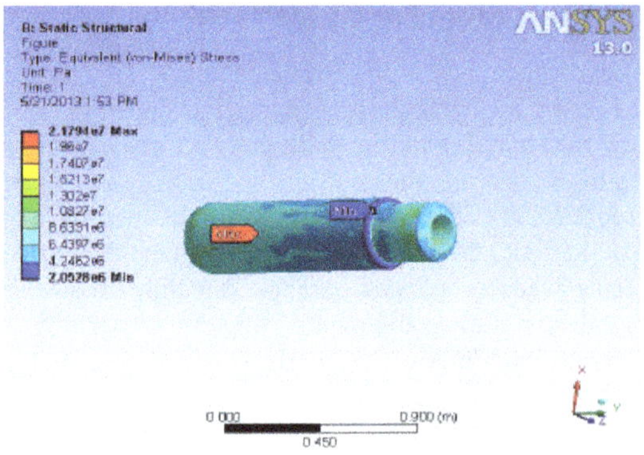

Figure 9: Von mises stress for Al 20% Sic material.

Case II: When the pressure is 100 MPa and 10 MPa and temperature is 600 °C and 400 °C

The pressure load of 100Mpa at bottom and 10MPa, at top and temperature of 600 °C, at bottom and 400°c at top is applied and deformations and stresses are computed on guide valve (Table 8).

Case III: When the pressure is 10 MPa and 1 MPa and temperature is 650°C and 450°C

The pressure load of 10Mpa at bottom and 1MPa, at top and temperature of 650°c, at bottom and 450°c at top is applied and deformations and stresses are computed on guide valve (Table 9).

Conclusion

Static and thermal analysis is carried over Guide valve. The Guide valve is analyzed for different pressure load cases (Tables 1-5). At a pressure of 10 MPa and 1 MPa and temperature 600° c and 400°c the maximum deformation is 0.124mm and maximum stress is found to be 286.31MPa (CuNi3si Alloy) . Stresses developed for this pressure for all the materials are less than their respective yield strength.

Result/ Material	Al 10% composites	Al 20% composites	Al 30% composites	Titanium Alloy (Ti 834)	CuNi3si Alloy
Deformation	0.0169mm	0.01436mm	0.0123mm	0.0778mm	0.124mm

Table 5: Deformation with material.

Load (MPa)		Temp °c						
Max (bottom)	Min (Top)	Max (bottom)	Min (Top)	Material	Max. deformation, mm	Von-mises, MPa	Safety/Failure	Material yield stress MPa
10	01	600	400	Al 10% composites	0.0169mm	23.06 MPa	Safe	257
10	01	600	400	Al 20% composites	0.01436mm	21.79MPa	Safe	263
10	01	600	400	Al 30% composites	0.0123mm	20.09MPa	Safe	269
10	01	600	400	Titanium Alloy (Ti 834)	0.0778mm	165.39MPa	Safe	910
10	01	600	400	CuNi3si Alloy	0.124mm	286.31MPa	Safe	550

Table 6: Resultant displacement and resultant stress of the guide valve under the different material conditions.

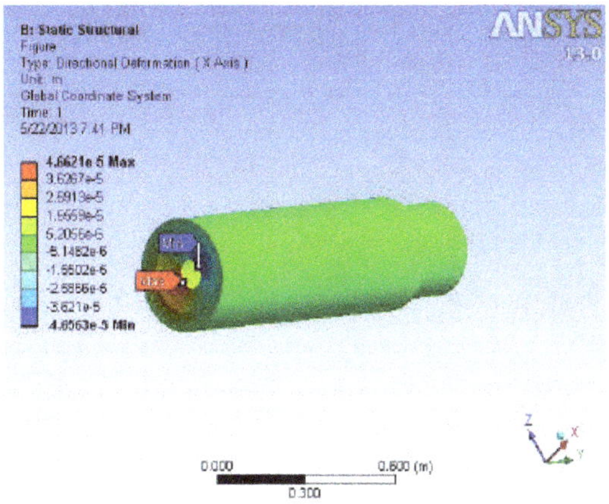

Figure 10: Deformation variation for Al 30% Sic material.

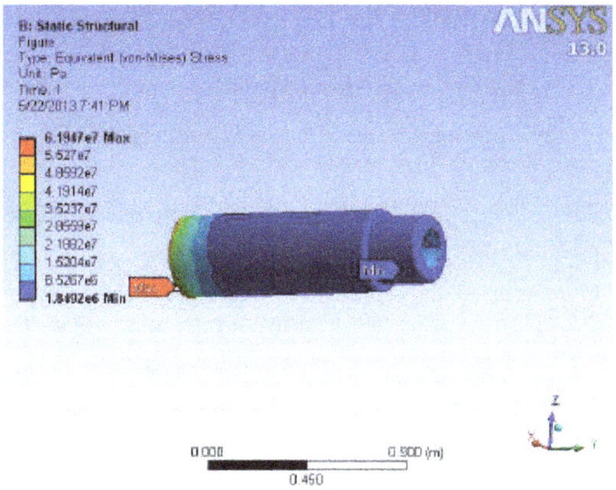

Figure 11: Vonmises stress for Al 30% Sic material.

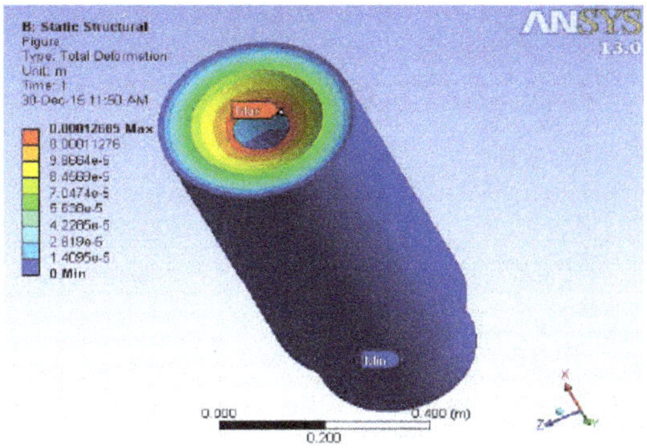

Figure 12: Deformation variation for Al 30% Sic material.

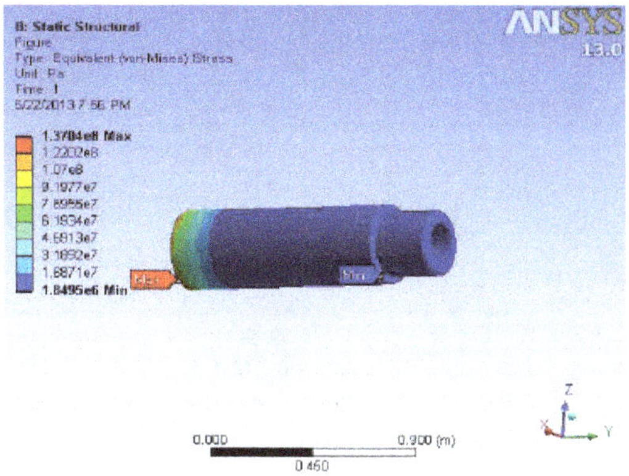

Figure 13: Von mises stress for Al 30% Sic material.

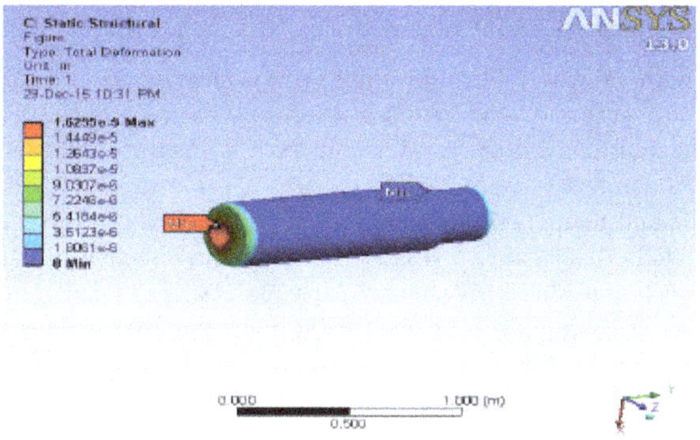

Figure 14: Deformation variation for Al 30% Sic material.

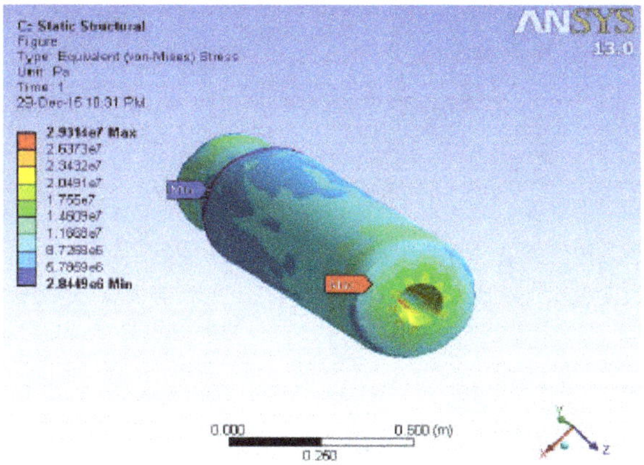

Figure 15: Von mises stress for Al 30% Sic material.

Load (MPa)		Temp °c						
Max (bottom)	Min (Top)	Max (bottom)	Min (Top)	Material	Max. deformation, mm	Von-mises, MPa	Safety/Failure	Material yield stress MPa
50	5	600	400	Al 10% composites	0.0564 mm	60MPa	Safe	257
50	5	600	400	Al 20% composites	0.05034 mm	59.79MPa	Safe	263
50	5	600	400	Al 30% composites	0.04662 mm	61.94MPa	Safe	269
50	5	600	400	Titanium Alloy (Ti 834)	0.12907 mm	200.33 MPa	Safe	910
50	5	600	400	CuNi3si Alloy	0.2268 mm	338.30 MPa	Safe	550

Table 7: Deformation and Von mises stress according to respective materials.

Load (MPa)		Temp °c						
Max (bottom)	Min (Top)	Max (bottom)	Min (Top)	Material	Max. deformation, mm	Von-mises, MPa	Safety/Failure	Material yield stress MPa
100	10	600	400	Al 10% composites	0.192 mm	129.65	Safe	257
100	10	600	400	Al 20% composites	0.16189 mm	131.94	Safe	263
100	10	600	400	Al 30% composites	0.1268 mm	132.93	Safe	269
100	10	600	400	Titanium Alloy (Ti 834)	0.00986 mm	202.87	Safe	910
100	10	600	400	CuNi3si Alloy	0.1816 mm	340.94	Safe	550

Table 8: When the pressure is 10 MPa and 1 MPa, and temperature is 6500c and 4500c.

Load (MPa)		Temp °c						
Max (bottom)	Min (Top)	Max (bottom)	Min (Top)	Material	Max. deformation, mm	Von-mises, MPa	Safety/Failure	Material yield stress MPa
100	10	600	400	Al 10% composites	0.020315 mm	25.968	Safe	257
100	10	600	400	Al 20% composites	0.01734 mm	26.183	Safe	263
100	10	600	400	Al 30% composites	0.01625 mm	29.314	Safe	269
100	10	600	400	Titanium Alloy (Ti 834)	0.0778 mm	216.61	Safe	910
100	10	600	400	CuNi3si Alloy	0.1244 mm	367.18	Safe	550

Table 9: Deformation and Von mises stress according to respective materials.

In the same way when the guide valve is analyzed for the different set of pressure and temperature parameters (i.e., pressure 50 MPa and 5 MPa and temperature is 600°c and 400°c),(pressure 100 MPa and 10MPa and temperature is 600°c and 400°c), (pressure is 10 MPa and 1 MPa and temperature is 650°c and 450°c) the max deformation is found to be 0.2268 mm , 0.1816 mm , 0.1244 mm respectively and max stress is found 338.30 MPa, 340.94 Mpa , 367.18Mpa in CuNi3si Alloy . However it has been observed that at high temperature the stresses produced in Al-SiC composites are very less as compared to other materials at different parameters (pressure & temperature both) From the results obtained finally, it is figured that Al/sic_p is suitable materials for high temperature applications i.e., Turbo charged engines. Racing Cars, Diesel Loco Engines, Air craft engines where cost is not a major factor.

References

1. Chikahata K, Hayashi K (2000) US patent Publication number US6012703 A.
2. Ahmed S (2013) Development and characterization of Al7075 based hybrid composites. Paripex Indian j research.
3. Bakshi SR, Di Wang, Price T, Agarwal A (2009) Composite Material-Aluminium Silicon Alloy: A Review. Paripex Indian j research.
4. Singla M, Dwivedi DD, Singh L, Chawla V (2009) Development of Aluminium based Silicon carbide particulate metal matrix composits. 8: 455-467.
5. Saheb DA (2011) Aluminum Silicon Carbide and Aluminum Graphite Particulate Composites ARPN J Engineering and Applied Sciences 6: 41-46.
6. Singh R, Singla ER (2012) Tribological characterization of Aluminium-Silicon carbide composite prepared by mechanical alloying. Int J Applied Engineering Research.

24

Analysis of Changing Wear Assembly Screw-Engine with the use of Automatic Systems

Epifantsev K* and Nikulin A

Saint Petersburg Mining University, 199106, Saint Petersburg, 21 Line of Vasilievskiy Island, Russia

Abstract

Waste recycling in Russia merely reaches 5% to 7% versus up to 60% of MSW in the EU, and over 90% of waste in Russia is delivered to waste landfills and unauthorised dumps so that waste accumulation is increasing. This environmental situation is a national priority. The Decree of the President of the Russian Federation dated 5 January 2017 announces 2017 the Year of Ecology in Russia. Most environmental reforms stipulated by amendments to laws are enacted from 1 January 2017. These are primarily measures aimed at emissions and discharges control using best available technologies and breakthrough provisions of the Industrial and Consumer Waste law. The Clean Country, priority project of the Russian Government, will be implemented from 2017 to 2025 with the key aim of reducing the environmental footprint from municipal solid waste disposal and mitigating environmental risks of an accumulated environmental damage. The priority project involves the construction of five environment-friendly facilities for the thermal processing of municipal solid waste (waste incineration plants), four of them to be built in the Moscow Region and one facility to be built in the Republic of Tatarstan. An alternative to waste incineration is municipal waste recycling by moulding in extrusion machines to make pellets to be further used in the fuel or construction industries. The profitability of a waste recycling facility is dependent on a sound choice of extrusion equipment with the best value for money.

Keywords: Automatic systems; Screw-engine

Introduction

An increase in the reliability of extrusion machines used to shred and extrude refractory materials (plastics, hard food waste, and couched paper) is subject to rational use of the extrusion force inside the machine frame. It requires automatic gauges and mechanisms capable of seamlessly increasing or reducing the effort and rotation speed at the right time to preserve the engine life and prevent failures [1-5].

When producing fuel pellets or construction materials from waste, an extrusion machine is exposed to significant loads on its main operating elements such as: inside the frame, on the auger shaft, coupling, connecting auger to the engine shaft, and inside the matrix.

The most common extruder failure in the extrusion process is matrix clogging with a moulded stock (Figure 1), formation of a tight clog impeding auger rotation, loss of efficiency, and engine emergency shutdown. A failure response takes one to two hours to dismantle and clean the auger.

The above loads need to be reduced to not only minimise machine assembly wear but to reduce power consumption, and enhance the reliability of the engine which accounts for 30% of the machine value. Currently, the set-up of a waste recycling facility requires a thorough review of power consumption per 1kg of product. Energy saving on the machine makes extrusion equipment more affordable in regions with high energy cost.

Figure 1: Clogged matrices reduce palletising to 50% matrix holes.

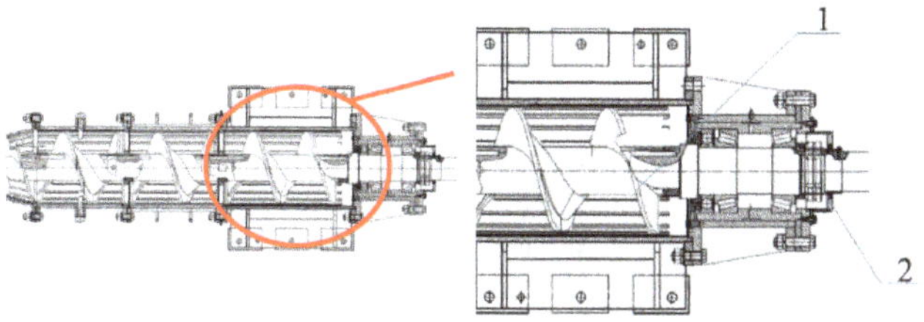

Figure 2: The most common deformed elements of the auger machine when clogged: 1 – weld location of an auger head and shaft, 2 – box coupling on the rotation shaft.

An operating body of the auger machine is an auger rotated by the engine through a box coupling. Clogging increases pressure from a moulded feed both on a welded auger head (Figure 2) and on the box coupling. These are main machine operating elements, and a reduction in their strength will result in the production of defective pellets (less tight and more friable).

***Corresponding author:** Epifantsev K, Saint Petersburg Mining University, 199106, Saint Petersburg, 21 Line of Vasilievskiy island, Russia
E-mail: epifancew@gmail.com

Statistically, without engine emergency shutdown, these elements develop microcracks and are deformed. Twisting deformation can be assessed as twisting of a beam with a non-round cross-section. If we consider an auger to be a circle with a flattened surface, the maximum effort is applied at the middle of the flat cut: τ_{max} = h/2 [6,7], i.e., the maximum effort is in the middle point at the location where the auger diameter changes and the auger has a cone shape (Figure 3).

We will calculate the moment of resistance for the maximum overload force at the shaft when it is clogged and a contingency situation develops [8].

$$W_k = \frac{b^3(2,6\frac{h}{b}-1)}{8(0,3\frac{h}{b}+0,7)} = \frac{113^3(2,6\frac{115}{113}-1)}{8(0,3\frac{115}{113}+0,7)} = 297\,кН \cdot м \quad (1)$$

Therefore, estimates were sufficient to determine the moment of resistance as 238.04 kN/m. When exceeded, the auger would be exposed to microcracks and deformation.

Materials and Methods

The extrusion moulding process was studied using waste from dumps and wastewater treatment facilities, non-marketable TPP products and wood processing waste [9]. Clogging processes were researched using a newly built auger-type machine (MN-3) with heating elements (Figure 4).

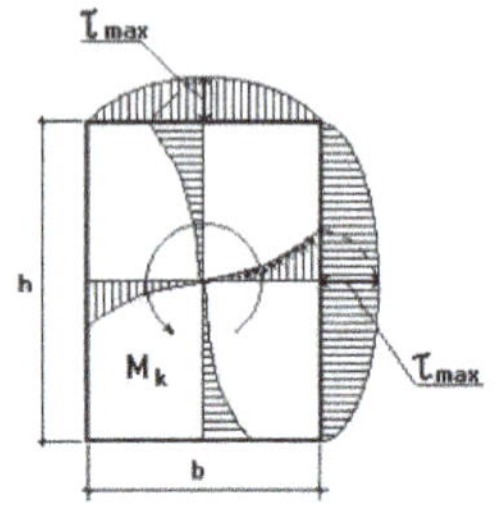

Figure 3: Curve showing shaft moment of resistance with auger narrowing, h=115 mm, b=113.

Figure 4: MN-3 extruder.

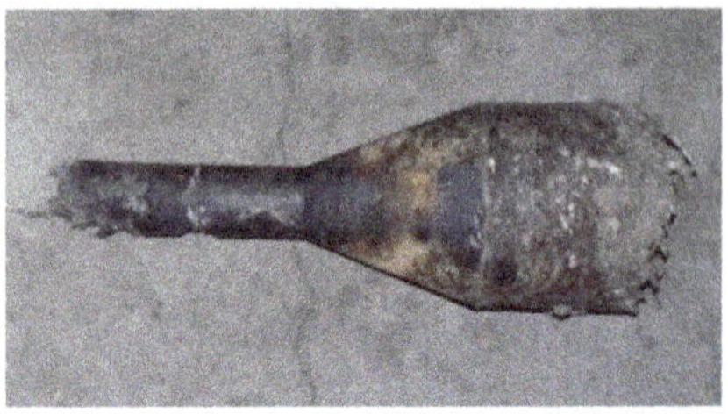

Figure 5: Stopper clog removed from the machine when the frame was cleaned.

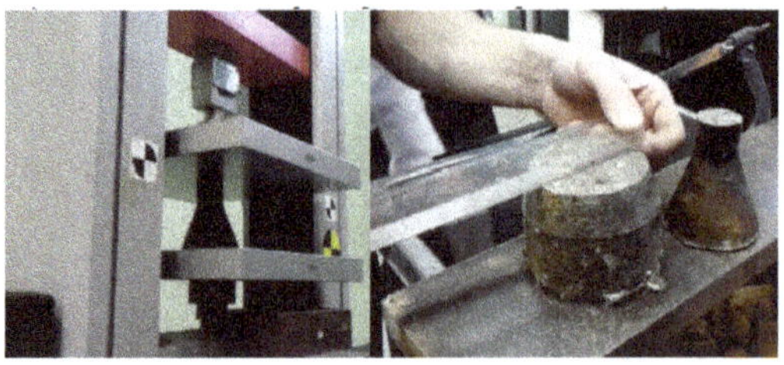

Figure 6: Uniaxial compressive strength test of the clog.

Test No.	Max. load (N)	Destruction shift (mm)	L, mm	P, kg/m²
Sample 1	3,907.600	2.902	115	56.38050057
Sample 2	4788.900	14.090	125	69.09626859

Table 1: Results of uniaxial compressive strength test of the clog.

The ingredients of the mixture to be palletised with a moisture content of 45% were as follows:

1. 20% of cardboard.
2. 15% of plastic.
3. 40% of coal charge.
4. 10% of sawdust.
5. 15% of waste from the wastewater treatment facilities (sludge).

The feed was loaded into the hopper of MN-3 auger-type machine which started to produce pellets 20 minutes after pre-heating of its heating elements and ramp-up. The process was going for two hours when machine performance began to decrease. The auger rotation speed was increased using a frequency converter, but that raised power consumption by 20%. After three hours of operation, a decision was made to shut the machine down since its performance was minimum with the maximum rotations per minute of the engine.

During the experiment, the frequency converter helped to control engine and, therefore, auger rotation speed. However, speed was adjusted manually by an operator depending on pellet output. Clogging of the auger was not controlled since it was mounted in a metal frame and the process could not be seen. Such actions were random and chaotic. Temperature was not switched off, either, heating bundles were operated in the normal mode so that the edges of the clog in the auger melted and became harder. When the frame was dismantled, the cause of the loss of efficiency could be identified. It was a clog consisting of a compacted waste feed which blocked the auger and resulted in the emergency shutdown (Figure 5).

A decision was made to study the clog at the destructive press:

1) For uniaxial compressive strength test;

2) For split test; and

3) To determine a normal effort at the auger shaft in ParaView software by comparing it to the estimated effort of 238.04 kN and to visually monitor the clogging process.

This data was needed to determine the range of action of a pressure gauge to be later mounted in the extruder frame to alarm of the start of the clogging process. By interacting with the frequency converter, the gauge will prevent emergency shutdown which reverses the engine through a change in poles. A shaker may also be used inside the auger to reduce stock adhesion to the auger and prevent clogging.

Results

The uniaxial compressive strength test required to split the clog mechanically into two parts to prepare for testing at Testometric 500 press in the Geomechanics and Mining Issues Centre of the Mining University. The diameter and height of produced samples were duly measured (Figure 6).

A cone sample (Sample 1) was compressed at the press first followed by a cylinder sample (Sample 2) (Table 1).

Figure 7 shows graphically how changes in deformation (mm) depend on the load (N). Sample 2 apparently has better performance (by 881.3 N) as compared to Sample 1.

These were followed by the uniaxial compressive strength tests of produced pellets (Figure 8 and Table 2).

Clogging was analysed then in a 3D simulation model of the extrusion machine in ParaView software. The simulation included similar machine parameters with its matrix split using the finite elements method. A type of an approximation function is randomised per element. The simplest case is linear polynomial (Figure 9).

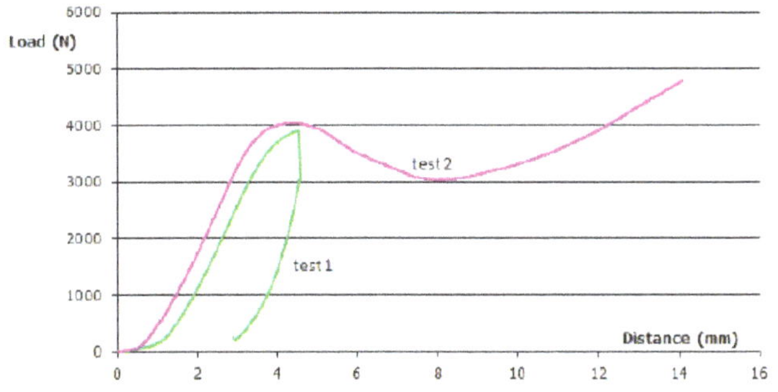

Figure 7: Deformation of Samples 1 and 2 during uniaxial compressive strength test.

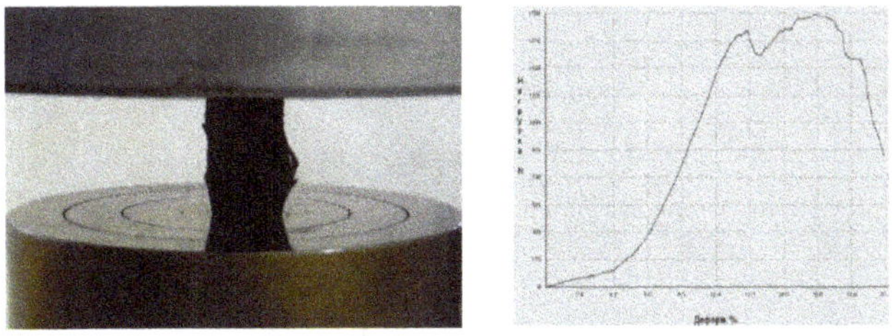

Figure 8: Uniaxial compressive strength test of the pellets.

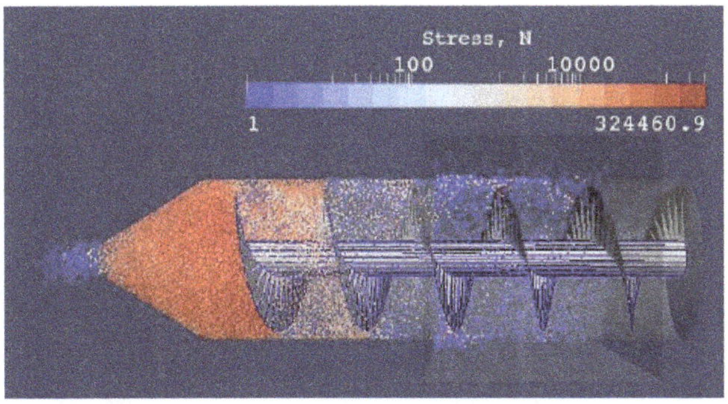

Figure 9: Simulation of the clogging process.

Item No.	D, mm	N, mm	m, g	Max. load (N)
1	33.5	53.5	40.65	1720
2	35.5	51.2	36	1300
3	31.5	38	28	1400

Table 2: Results of uniaxial compressive strength test of the pellets.

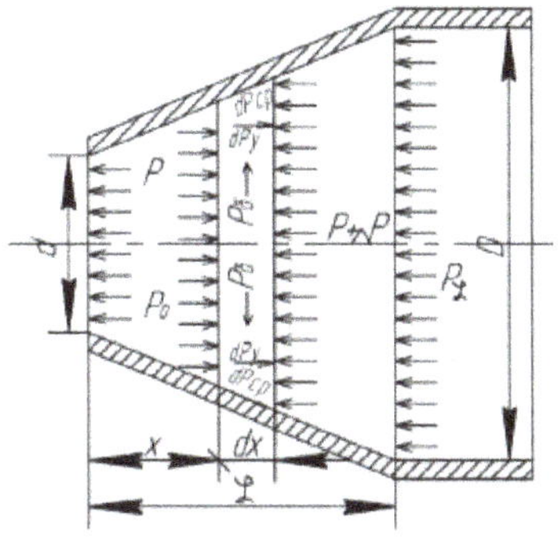

Figure 10: Visual representation of pressure summation inside the matrix.

Discussion

The studies completed revealed several important parameters required to set automatics cut-off values:

1) Moment of resistance of the auger according to Formula 1 is 297 kN/m. When it is exceeded, the auger is subject to microcracks and deformation. When simulated in ParaView software, the moment at the auger shaft during clogging reaches 324.46 kN. In this case we will take the effort range between 297 and 324.460 kN;

2) Uniaxial compressive strength test shows the difference between the cone and cylinder parts of the auger as 3907.6 N and 4788.9 N which demonstrates the difference of 800N and supports I.N. Chisty's theory [10,11] of pressure summation in the middle point of the cone output (P+▲P) after which the clog is hard to destroy (Figure 10).

This experiment enabled to determine the pressure summation point as 115 cm – the length of the cone part of the clog (Table 1). It is the point where a leap certifying to clogging takes place.

Conclusion

The experiments facilitated the collection of information on clog location and hardness in different parts. The studies formed the basis for the upgrade of the existing MN-3 extruder which was re-equipped with pressure gauges 115 cm off the matrix edge and a strain gauge capable of recognising an effort on the auger between 297 kN and 324.5 kN. The main function of the two gauges is to start auger in the reverse for the clog to "roll away' and destroy. The auger was also upgraded with a vibrating element reducing feed stuck to the coils and minimising clog development and matrix clogging.

This area of research will be effective means of creating "transparent" extruder control systems for automatic monitoring of the clogging process and protection against machine breakdown and loss of reliability.

Acknowledgement

The results of this work were marked on the Competition for the best scientific work about Earth (Tomsk Polytechnic University, 2012); Grant Competition of scientific works of graduate students (Government of St. Petersburg, 2012); Grant of the German Academic Exchange Service (DAAD 2012); The grant for young scientists (Government of St. Petersburg, 2016)

References

1. Environmental Protection (2012). National Programme of the Russian Federation / Ministry of Natural Resources and Ecology of the Russian Federation (2012-2020).
2. IFC (2013) Final report on the waste in Russia: Garbage or a valuable resource? Scenarios of development of the municipal solid waste handling sector. International Financial Corporation (IFC, World Bank Group).
3. Federal Law (1998) 24 June 1998, No. 89-FZ. Industrial and consumer waste, Consultant Plus.
4. Clean Country (2016) Priority project profile. Government of the Russian Federation.
5. Nikulin AN (2014) The research of possibility to use the machine for biofuel production as a mobile device for poultry farm waste recycling. Life Sci J 11: 464-467.
6. Kocserha I, Kristály F (2010) Effects of extruder head's geometry on the properties of extruded ceramic products. Material Sci Forum 659: 499-504.
7. Ottino JM (1995) Paste flow and extrusion: Oxford series on advanced manufacturing. AIChE J 41: 741.
8. Bogatov BA (1985) Managing peat bog development. Mn Higher School.
9. Epifancev K, Nikulin A, Kovshov S, Mozer S, Brigadnov I (2013) Modeling of peat mass process formation based on 3D analysis of the screw machine by the code YADE. American J Mech Eng 1: 73-75.
10. Chisty IN (1980) Granulated peat production. Minsk pp: 420.
11. Kosov VI (2005) Peat and organic slime as a powerful energy and geo-ecological potential of Russia: Exploration, production, and refining of minerals. Interment Eng.

The Influence of Corrosion Damage on Low Cycle Fatigue Life of Reinforcing Steel Bars S400

Koulouris K[1], Konstantopoulos G[1], Apostolopoulos Alk[2], Matikas T[2] and Apostolopoulos Ch[1]*

[1]Department of Mechanical and Aeronautical Engineering, University of Patras, Greece
[2]Department of Material Science, University of Ioannina, Greece

Abstract

In this study the influence of corrosion and seismic load (low-cycle fatigue LCF) in the mechanical performance of reinforcing steel bars of S400 grade with 10 mm nominal diameter was investigated. There took place 140 tensile, LCF and salt spray tests which performed on reinforcing bars in different conditions. The results show that the corrosion level and surface conditions are the main parameters which affect to the low-cycle fatigue life of reinforcing bars. Moreover, through a non-linear regression analysis of the experimental data, a model of predicting the expectancy life of the corroded rebars was conducted.

This prediction was based on two models: the first model was about an imposed of (total) strain amplitudes (ε_α) and the second model on predicting the strength degradation per cycle of fatigue in correlation with the plastic strain amplitudes (ε_p). Both the experimental study and the prediction modeling conducted for the same steel grade S400 with and without ribs. The model prediction of non-linear regression analysis, show a good agreement with the observed experimental results and adequately confirmed the experimental results showing that from the first levels of corrosion, the degradation of their life expectancy was obvious as well the rebars without ribs (smoothed) which present more advanced mechanical behavior and life expectancy against to the respective ribbed rebars.

Keywords: Steel Reinforcement bar; Fatigue Behavior; Corrosion; Nonlinear Analysis; Low-Cycle Fatigue model

Introduction

It is well known that corrosion effect is an electrochemical nature phenomenon which constitutes one of the basic factors of degradation of reinforcing concrete structures. In past, lots of studies [1-5] have presented the negative circumstances of corrosion effect, such as the local decrease of cross section and the respective mass loss. Meanwhile, corrosion effect has an impact on the mechanical behavior of steel bar due to the reduction of strength properties, the ductility and the bonding between the concrete and the steel bar. The corrosive factor in correlation with the effect of seismic loads plays an important role in the mechanic performance of structures. Sheng and Gong [6] studied and showed that the effect of seismic loads can be simulated, in a laboratory, in low cycle fatigue conditions. This effect can induce a reduction of steel bar's loading ability as well as their failure.

Corrosion effect appears to begin with chlorides through the pores of concrete, through the action of capillary voids of water or a combination of them. An important percentage of chloride concentration, on corrosion effect, is about 0.4% of concrete's weight [7]. In case of corrosion effect, occurring through pits (chlorides), the tension rate of stress and also the concentration rate of stress increase and as a result the formulation and the development of micro - cracking which, in combination with the fatigue, cause the material's failure. Although a significant number of researchers [8-10] presented the consequences of mechanical degradation of steel bar due to seismic loads and corrosion effect, the international design regulations of structures, except for the Portuguese and Spanish regulations [11,12], did not include similar technical requirements for the reinforcing steel bars. Moreover, special well known life expectancy predicting models of metal materials belong to Coffin-Manson [13,14] and Koh–Stephens [15]. Based on the above models, in this study there is an effort of predicting the life expectancy of steel bar S400. In more detail, based on the results of an extensive experimental study, in which steel bars in various seismic loads (Low Cycle Fatigue) were examined, before and after several periods of time exposed to an artificial imposed of accelerated corrosion effect in salt spray chamber. Steel bars S400 with and without ribs have undergone some fatigue tests in monotonic sinusoidal loading of 0.5 Hz frequency in various deformation range values such as ± 1%, ± 2.5%, and ± 4% [16]. S400 steel bar category, even though today has a limited usage, the last decades constituted the main material of many structures in Mediterranean countries (Greece - Italy - Turkey). Therefore, the potential for predicting the life expectancy of steel bars in already existing structures (of various level of corrosion) is really interesting fact for the engineer researchers because it contains useful information about the level of steel bars' mechanical performance and for the level of reliability of crucial structural elements of constructions (such as the columns).

Experimental Procedure

The experiments were conducted on S400 steel grade reinforcing steel, specially produced for the needs of the current investigation by a Greek steel mill. Chemical composition of steel S400, is shown on Table 1. S400 steel (widely known as StIII or BSt 420) has officially been withdrawn since the late 1990's from production, it still holds as the backbone of reinforced structures aging from 20 to 50 years. The material was delivered in the form of 10 mm nominal diameter ribbed bars according to postolopoulos and Pasialis [16] study. Specimens

***Corresponding author:** Apostolopoulos CH, Professor, Department of Mechanical and Aeronautical Engineering, University of Patras, Greece
E-mail: charrisa@mech.upatras.gr

with 170 mm total length and 60 mm in gauge length were cut for the LCF tests. The gauge length was equal to six times the nominal diameter of steel specimens. Prior to the tests, the specimens were pre-corroded using accelerated laboratory corrosion tests in salt spray environment. Salt spray tests were conducted according to the ASTM B117-94 specification. For the tests, a special apparatus, model SF 450 specially designed by C and W. Specialist Equipment Ltd. was used. The salt solution was prepared by dissolving 5 parts by mass of sodium chloride (NaCl) into 95 parts of distilled water. The duration times of exposure were 10, 20, 30, 45, 60 and 90 days. Upon completion, the specimens were washed with clean running water to remove the remaining salt deposits from their surfaces and then were dried. The oxide layer was removed using a bristle brush, according to the ASTM G1-90 specification. In order to make a more comprehensive study of the mechanical behavior of the steel, except of LCF tests, additionally tensile tests on ribbed bars were performed, before and after, corrosion. The mean value tensile test results (corroded and non-corroded) are shown in Figure 1 and Table 2. Table 3 presents the low cycle fatigue test results (in different amplitudes of deformation ± 1, ± 2.5, and ± 4%).

C %	Mn %	S %	P %	Si %	Ni %	Cr %	Cu %	V %	Mo %	N %
0.35	0.94	0.026	0.013	0.26	0.26 0.10	0.16	0.42	0.002	0.023	0.01

Table 1: Chemical composition of S400.

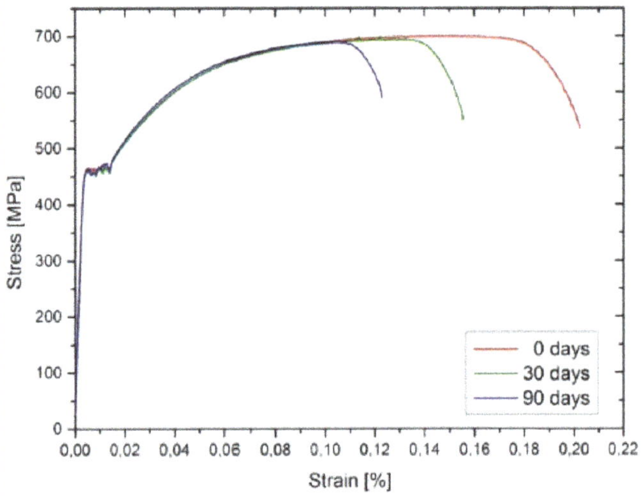

Figure 1: Stress-strain curves of ribbed bars.

	0 days	30 days	90 days
Yield Stress [MPa]	454,86	452,53	437,61
Tensile Strength [MPa]	695,12	695,29	674,93
Plastic Strain Ag [%]	15,53	12,88	9,00
Total Strain Agt [%]	19,73	15,33	10,53
Energy Density [MPa]	126,56	97,98	63,98

Table 2: Tensile test results.

Modeling low-cycle fatigue life of steel bar

It is well known that the low-cycle fatigue life of reinforcing bars without the effect of corrosion has been studied by several researchers. The current study examined the life prediction, based on Coffin-Manson's and Koh-Stephen's models, which are more popular among researchers. Coffin-Manson model relates the plastic strain amplitude (εp) to fatigue life.

$\varepsilon_p = \varepsilon'_f\,(2N_f)^c$, where ε'f is the ductility coefficient i.e., the plastic fracture strain for a single load reversal, c is the ductility exponent and $2N_f$ is the number of half-cycles (load reversals) to failure. Koh-Stephen extended the Coffin-Manson's model for modeling the low-cycle fatigue life of materials based on the total strain amplitude (elastic strain + plastic strain) as described in the following equation.

$\varepsilon_a = \varepsilon_f\,(2N_f)^{\alpha}$, where ε_f is the ductility coefficient i.e., the total fracture strain for a single load reversal, α is the ductility exponent and $2N_f$ is the number of half-cycles (load reversals) to failure.

Between these two models, Koh-Stephen's model is used for the analysis and the prediction of low-cycle fatigue life of reinforcing bars. Furthermore, the influence of corrosion on fatigue material constants εf and α is also explored. The Koh-Stephen equation is fitted with the observed experimental data of each exposure time to calibrate the fatigue material constants (ε_f and α). In a similar way, using the Coffin-Manson model, the prediction of strength loss of hysteresis loops was conducted. The prediction of strength degradation of reinforcing bars is made by using a type expression, $\varepsilon_{pl} = \varepsilon_d\,(f_{SR})^{\alpha}$, where, εd and α are material constants and f_{SR} is the strength loss factor per cycle as measured in a fatigue test at a constant plastic strain amplitude of ε_{pl}. The results of Fatigue Life material and Strength loss material coefficients are shown in Tables 4 and 5.

Days of corrosion	Strain	Ribbed bars		Smoothed bars	
		Cycles to Failure	Dissipated Energy [MPa]	Cycles to Failure	Dissipated Energy [MPa]
0	± 1.0%	1280	7103	1435	7420
	± 2.5%	40	1059	51	1334
	± 4.0%	11	537	12	579
30	± 1.0%	509	2902	750	3905
	± 2.5%	26	694	27	705
	± 4.0%	9	423	9	423
90	± 1.0%	349	1862	365	2040
	± 2.5%	24	587	24	626
	± 2.5%	7	272	7	344

Table 3: Low cycle fatigue test results.

Results and Discussion

Mass loss measurements for several periods of time exposed to salt spray chamber 10, 20, 30, 45, 60 and 90 days led to 1.58%, 2.50%, 3.77%, 5.18%, 7.23% and 8.48% percentage mass loss respectively. The results of mechanical tensile tests of ribbed bars are presented in Table 2. They show that the decrease of strength properties is (about) equivalent to mass loss decrease, opposite to the ductility properties where a dramatic decrease is presented in. It is known that corrosion of embedded steel

Days of corrosion	Mass loss	Smoothed Bars			Ribbed Bars		
		ε_f	α	R^2	ε_f	α	R^2
0	0	0,10363	-0,296	0,994	0,10405	-0,314	0,986
10	1,58	0,11075	-0,33	0,989	0,11103	-0,351	0,952
20	2,5	0,10602	-0,337	0,973	0,11388	-0,367	0,985
30	3,77	0,10635	-0,339	0,985	0,11531	-0,372	0,981
45	5,18	0,09286	-0,331	0,987	0,12121	-0,391	0,975
60	7,23	0,11727	-0,388	0,971	0,10321	-0,361	0,998
90	8,48	0,11511	-0,393	0,999	0,10389	-0,363	0,998
Mean*		0,10806	-0,353		0,11142	-0,368	

*(not including 0 days of corrosion)

Table 4: Fatigue life material coefficients.

bar initially (for mass loss rates 1,5% to 2%) has a positive impact on bonding between concrete and steel bar. As a consequence, the prediction of seismic loads behavior (low cycle fatigue) will have as a reference some experimental results with higher level percentage of mass loss. However, the analysis of experimental tests of low cycle fatigue results, through the statistical regression analyses showed that there are fatigue life prediction models for 0-10% mass loss.

Figures 2 and 3 present the curves of prediction model with dashed line. Table 3 represents the low cycle fatigue test results. In Table 4, the calibrated constants of fatigue material ε_p, a in consequence of regression analysis are presented in. The results of modeling, show high convergence reliability (values of R2). The analysis of empirical constants (mean) εf and a of life prediction models reflects the influence of the corrosive factor in corrosion levels of concrete. For both types of steel (ribbed and smoothed), based on these mean values the prediction model was resulted from. The curves of the two prediction models are in a good agreement with the experiment results as they take into account the fatigue phenomena and corrosion damage. As it was expected, the corrosion affected negatively the life expectancy of steel specimens.

It is obvious, from Figures 2 and 3, that increasing exposure time, the life expectancy of specimen material is steadily decreased. From the first exposure times of specimens, in smaller strain amplitude (mainly in ± 1%, ± 2.5%), shorter life expectancy is recorded. On the contrary, in larger strain ranges (± 4%) Kashani's study results are confirmed [17], in which additional negative phenomena highlighted due to the effect of inelastic buckling.

Days of corrosion	Smoothed bars			Ribbed bars		
	ε_d	α	R^2	ε_d	α	R^2
0	0,01762	0,369	0,934	0,01693	0,411	0,979
10	0,01696	0,421	0,958	0,01617	0,444	0,981
20	0,01465	0,490	0,992	0,01455	0,507	0,978
30	0,01479	0,466	0,974	0,01383	0,516	0,981
45	0,01291	0,531	0,994	0,01353	0,477	0,972
60	0,01235	0,593	0,991	0,01314	0,554	0,973
90	0,01189	0,535	0,986	0,01241	0,521	0,972
Mean*	0,01392	0,506		0,01394	0,503	

Table 5: Strength loss material coefficients.

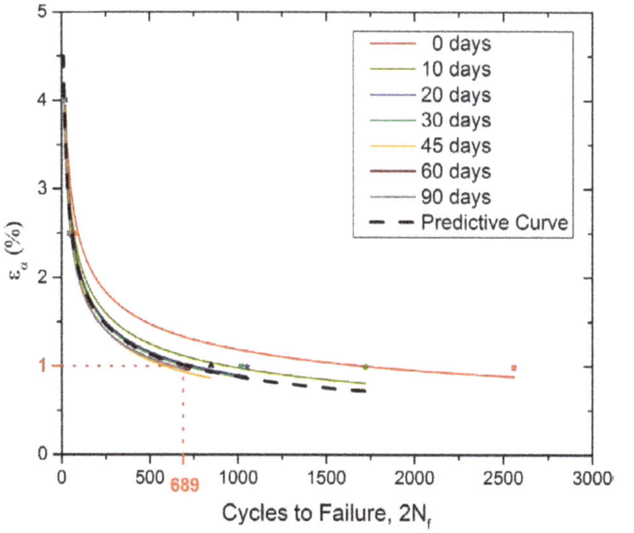

Figure 2: Fatigue life of ribbed bars.

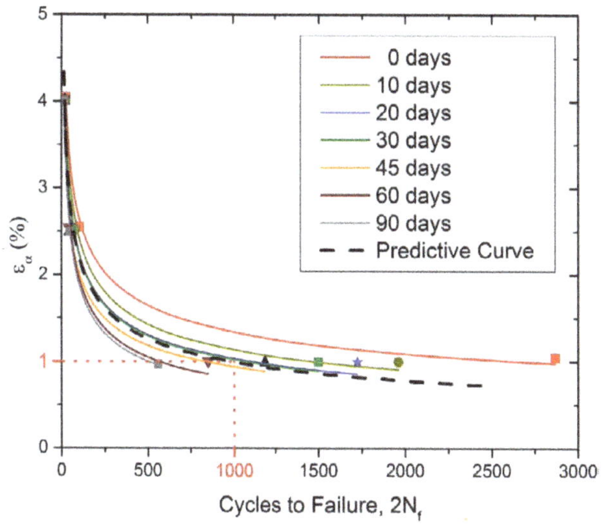

Figure 3: Fatigue life of smoothed bars.

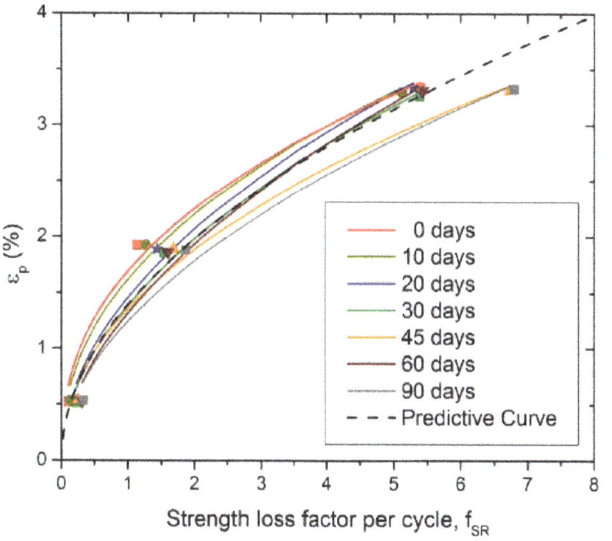

Figure 4: Strength loss of ribbed bars.

Very interesting results came from exploring the roles of steel ribs. Comparing Figures 2 and 3, it is observed that for the same number of reversals ($2N_f$), the maximum deformation of specimens without ribs are higher than the ribbed specimens. On the other hand, for a given range deformation (e.g. ± 1.0%) the number of reversals of the ribbed bars is presented lower than the smoothed bars.

According to the modified Coffin-Manson's equation that is referred to terms of plastic deformation and the coefficient strength loss per cycle of fatigue [18], comes up a prediction model of strength loss which is related to loading cycles. In equation, $\varepsilon_{pl}=e_d(f_{SR})^{\alpha}$, e_{pl} is the plastic strain amplitude, f_{SR} is strength loss coefficient per cycle and εd and α, are the empirical coefficients which are based on the material. In this analysis, the f_{SR} measurement was calculated by deriving the total strength loss from the total number of failure cycles. The experimental procedure led to prediction modeling of strength loss per cycle fatigue at S400 steel with and without ribs. The diagrams of Figures 4 and 5 are related to this, showing the prediction curves (dashed lines). In the

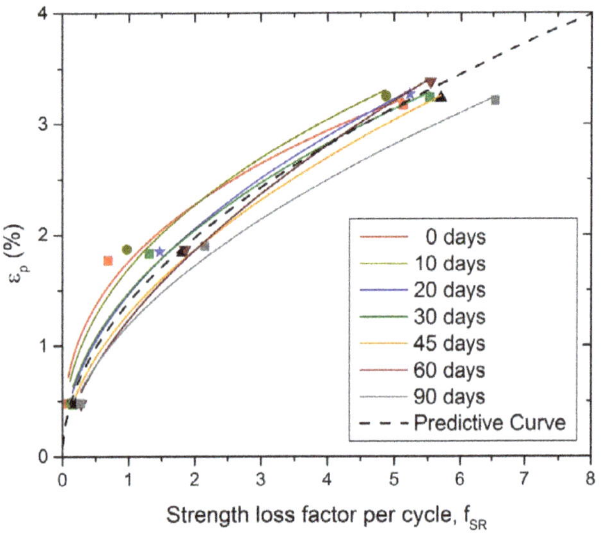

Figure 5: Strength loss of smoothed bars.

following diagrams, (Figures 4 and 5) the modified Coffin-Manson equation is displayed fitted to the experimental data using non-linear regression analysis. The results of regression analyses are summarized in Table 5. Herein, it observed that the impact of corrosion causes a shift in curves, in ribbed and also in smoothed steel bars, as the value of strength loss factor increases. The increase of strength loss factor rate is combined with lower percentage increase of ε_{pl} (plastic strain). The dashed line represents a mean curve condition of corroded bars.

Conclusions

1. The effect of corrosion has significant impact on low-cycle fatigue behavior of S400 reinforcing bar. As the duration of exposure increased the LCF life decreased and therefore the energy dissipation capacity of the bar under cyclic loading reduced.
2. The non-corroded bars show a ductile failure mechanism compare to corroded bars. This is also observed in case of smoothed compared to ribbed bars. However, as the strain amplitude increases the influence of ribs are reduced and the fracture of bars is mainly governed by the stress concentration of buckling phenomena.
3. The predictive models combine the effect of corrosion (concerning mass loss), the morphology of outer surface of rebars (ribbed-smoothed) and low-cycle fatigue degradation of S400 steel rebar. These results of prediction refer to mass loss rate less than 10% because after this rate, the strength bonding loss is too high. At these circumstances (mass loss > 10%) the study of mechanical performance and durability of RC structures serves no purpose.

References

1. Ma Y, Wang L, Zhang J, Xiang Y, Liu Y (2014) Bridge remaining strength prediction integrated with Bayesian network and in situ load testing J Bridge Eng 10: 1061.
2. Apostolopoulos C.A, Papadakis VG (2008) Consequences of steel corrosion on the ductility properties of reinforcement bar. Construction and Building Materials 22: 2316-2324.
3. Apostolopoulos C.A (2008) The influence of corrosion and cross section diameter on the mechanical properties of B500c steel. Journal of Materials Engineering and Performance 18: 190-195.
4. Apostolopoulos C.A, Demis S, Papadakis VG (2013) Chloride-induced corrosion of steel reinforcement -mechanical performance and pit depth analysis. Constr Build Mater 38: 139-46.
5. Zhang J, Gardoni P, Rosowsky D (2009) Stiffness degradation and time to cracking of cover concrete in reinforced concrete structures subject to corrosion. J Eng Mech 136: 209-19.
6. Sheng GM, Gong SH (1997) Investigation of low cycle fatigue behavior of building structural steel under earthquake loading. Acta Metallur Sin (Engl Lett) 10: 51-55.
7. Shi X, Xie N, Fortune K, Gong J (2012) Durability of steel reinforced concrete in chloride environments: An overview. J Constr Build Mater 30: 125-38.
8. Apostolopoulos C.A (2007) Mechanical behavior of corroded reinforcing steel bars S500s tempcore under low cycle fatigue. Constr Build Mater 21: 1447-56.
9. Apostolopoulos C.A, Papadopoulos M (2007) Tensile and low cycle fatigue behavior of corroded reinforcing steel bars S400. J Constr. Build Mater 21: 855-64.
10. Zhanga W, Songa X, GuaX, Li S (2012) Tensile and fatigue behavior of corroded rebars. J Constr Build Mat. 34 409-17.
11. (2000) UNE Norma Espanola experimental barras corrugadas de acero soldable con caracteristıcas especiales de ductilidad para armaduras de horigon armado UNE 36065 EX 2000.
12. (2008) LNEC Varoes de ac, A400 NR de ductilidade especial para armaduras de betao armado caracteristıcas, ensaios e marcac, AO, LNEC E455-2008.
13. Manson SS (1953) Behavior of materials under conditions of thermal stress. Heat Transfer Symp University of Michigan Engineering Research Institute Ann Arbor MI: 9-75.
14. Coffin LF Jr (1954) A study of the effects of cyclic thermal stresses on a ductile metal. Trans Am Soc Mech. Eng New York NY 76: 931-950.
15. Koh SK, Stephens RI (1991) Mean stress effects on low cycle fatigue for a high strength steel. Fatigue Fract Eng Mater Struct 14: 413-428.
16. Apostolopoulos C.A, Pasialis VP (2010) Effects of corrosion and ribs on low cycle fatigue behavior of reinforcing steel bars S400. Journal of Materials Engineering and Performance 19: 385-394.
17. Kashani MM (2015) Influence of inelastic buckling on low-cycle fatigue degradation of reinforcing bars. Construction and Building Materials 94: 644-655.
18. Kunnath SK (2009) Nonlinear uniaxial material model for reinforcing steel bars. Journal of Structural Engineering 135: 335-343.

26

Radiative Heat Transfer on Nanofluids Flow over a Porous Convective Surface in the Presence of Magnetic Field

Kandasamy R* and Mohamad R

Research Centre for Computational Mathematics, Faculty of Science,Technology and Human Development, UniversitiTun Hussein Onn Malaysia, Johor, Malaysia

Abstract

Nanofluids have been considered for applications as advanced heat transfer fluids for almost two decades. Nanofluid is a new kind of heat transfer medium, containing nanoparticles which are uniformly and stably distributed in a base fluid.Convection in nanofluids plays a key role in enhancing the rate of heat transfer either for heating or cooling nanodevices.In this paper, we investigate theoretically the impact of a convective surface on the heat transfer characteristics of water-based nanofluids over a static or moving wedge in the presence of magnetic field with variable stream condition. The governing nonlinear partial differential equations are made dimensionless with the similarity transformations. Numerical simulations are carried out through the very robust computer algebra software MAPLE 18 to investigate the effects of various pertinent parameters on the flow field. The results show that the temperature distribution in a nanofluid in the presence of thermal radiation with magnetic field significantly depends on the surface convection parameter. This finding is new and has not been reported in any open literature.

Keywords: Radiative heat transfer; Convective surface; Magnetic field; Nanofluids; Porous static or moving wedge

Introduction

The novel and advanced concepts of nanofluids offer fascinating heat transfer characteristics compared to conventional heat transfer fluids. There are considerable researches on the superior heat transfer properties of nanofluids especially on thermal conductivity and convective heat transfer. Applications of nanofluids in industries such as heat exchanging devices appear promising with these characteristics. The increases in effective thermal conductivity are important in improving the heat transfer behavior of fluids. Nanofluids, defined as fluids in which nanosized particles are suspended in a base liquid, are a novel types of working fluids used for heat transfer and cooling. Many of the publications on nanofluids about understand their behavior so that they can be utilized where straight heat transfer enhancement is paramount as in many industrial applications, nuclear reactors, transportation, electronics as well as biomedicine and food. Choi [1] first proposed the concept of using nanofluids to enhance thermal conductivity, these working fluids have shown great promise in heat transfer applications. Metal, metal oxides, ceramics and nonmetals such as carbon nanotubes and graphene can be used as nanoparticles in nanofluids, whereas water, ethylene glycol, oil and polymer solutions are generally used as base fluids. Several studies on nanofluids have been conducted by a number of researchers with respect to thermal conductivity enhancement and viscosity [2-5], phase change heat transfer [6-9] and convective heat transfer [10-13].Ghasemi and Aminossadati [14], Congedo et al. [15] and Ho et al. [16] and [17] presented some numerical studies on the modeling of natural convection heat transfer in nanofluids using finite volume techniques or traditional finite difference methods.Magnetic nanofluid is a unique material that has both the liquid and magnetic properties. Many of the physical properties of these fluids can be tuned by varying magnetic field. In addition, they have been wonderful model system for fundamental studies.

Researchers have shown that nanofluids have not only better heat conductivity but also greater convective heat transfer capability than that of base fluids. Based on the radiative motion properties of nanoparticle, the utilization of nanofluids in thermal system becomes the new study hotspot. Radiative transport in porous media has important engineering applications in thermal energy collectors and the porous medium acts as a means to absorb or emit radiant energy that is transferred to or from a fluid. Generally, the fluid itself can be assumed to be transparent to radiation, because the dimensions for radiative heat transfer among the solid structure elements of the porous medium are usually much less than the radiative mean free path for scattering or absorption in the fluid. In this new age of energy awareness, our lack of abundant sources of clean energy and the widespread dissemination of battery operated devices, such as cell-phones and laptops, have accented the necessity for a smart technological handling of energetic resources.

Due to many engineering applications in aerodynamics, geothermal systems, crude oil extractions, ground water pollution, thermal insulation, heat exchanger, storage of nuclear waste, etc., convective flows over wedge shaped bodies have been extensively studied since the early formulation of the problem in 1931 by Falkner and Skan. They first studied two-dimensional flow of viscous incompressible fluid over a wedge. Since then many investigators have studied and reported results on wedge flow considering various flow conditions. Lin and Lin [18], Yih [19], Pantokratoras [20], Martin and Boyd [21], Rahman and Eltayeb [22], Rahman and Al-Hadhrami [23] and Muhaimin et al. [24]. Three properties that make nanofluids promising coolants are: (i) increased thermal conductivity, (ii) increased single-phase heat transfer, and (iii) increased critical heat flux. Research has shown that relatively small amounts of nanoparticles (5% or less volume fraction), can enhance thermal conductivity of the base fluid to a large extent. Therefore, exploiting the unique characteristics of nanoparticles, nanofluids are created with two features very important for heat

***Corresponding author:** Kandasamy R, Research Centre for Computational Mathematics, Faculty of Science, Technology and Human Development, Universiti Tun Hussein Onn Malaysia, Johor, Malaysia, E-mail: ramasamy@uthm.edu.my

transfer systems: (i) extreme stability, and (ii) ultra-high thermal conductivity. The effect of thermal radiation becomes important when a device operates in a high-temperature field. Radiative flows have important industrial, technological and geothermal applications such as high-temperature plasmas, cooling of nuclear reactors, liquid metal fluids, MHD accelerators, and power generation systems. A wide variety of industrial processes involve the transfer of heat energy. Throughout any industrial facility, heat must be added, removed, or moved from one process stream to another and it has become a major task for industrial necessity. These processes provide a source for energy recovery and process fluid heating/cooling.

In view of the above discussion the objective of the present investigation is to analyze the impact of a convective surface on the flow dynamics and radiative heat transfer characteristics of nanofluids over a static or moving wedge in the presence magnetic field. For a moving wedge its surface may be stretched or may be contracted. Thermal radiation, permeability of the porous medium, convective surface parameter, solid volume fraction of the nanofluid, magnetic parameter and suction/injection parameter on velocity and temperature fields for nanofluids using the thermophysical properties nanoparticles of the base fluid (water).

Mathematical Analysis

Consider the steady laminar two- dimensional flow of an incompressible viscous nanofluid of density ρ_{f_n} past a moving or static porous wedge sheet with uniform velocity $U_0(x)=ax^m$ driven by the pressure gradient of the corresponding inviscid flow solution (Figure 1). We also consider influence of a constant magnetic field of strength B_0 which is applied normally to the sheet. The temperature at the lower surface of the wedge is heated by convection from a hot fluid (different from nanofluid)at a temperature T_0 which provides the heat transfer coefficient $h$while the temperature of the moving or static porous wedge sheet T_∞. It is further assumed that the induced magnetic field is negligible in comparison to the applied magnetic field (as the magnetic Reynolds number is small). The porous medium is assumed to be transparent and in thermal equilibrium with the fluid. Both the fluid and the porous medium are opaque for self-emitted thermal radiation. Also, the solar radiation is a collimated beam that is normal to the plate. The fluid is a water based nanofluid containing different types of nanoparticles: Cu, Al_2O_3, Ag and TiO_2. As mentioned before, the working fluid is assumed to have heat absorption properties. It is assumed that the base fluid and the nanoparticles are in thermal equilibrium and no slip occurs between them. $\frac{dp}{dx}=-a^2m\rho_{nf}x^{2m-1}$. The coefficient a is a function of the flow geometry. The exponent m is a function of the wedge angle parameter $\beta=\frac{\Omega}{\pi}$ such as $m=\frac{\beta}{2-\beta}$. It is worth mentioning that the caseβ = 0corresponds to a horizontal plate, while β=1 corresponds to a vertical plate. In the present work, we consider only $0 \le \beta \le 1$. Under the above assumptions, the boundary layer equations governing the flow and thermal field can be written in dimensional form as

$$\frac{\partial u}{\partial x}+\frac{\partial v}{\partial y}=0 \tag{1}$$

$$\bar{u}\frac{\partial \bar{u}}{\partial \bar{x}}+\bar{v}\frac{\partial \bar{u}}{\partial \bar{y}}=\frac{1}{\rho_{f_n}}\left[U\frac{dU}{dx}\rho_{f_n}+\mu_{f_n}\frac{\partial^2\bar{u}}{\partial \bar{y}^2}+(\rho\beta)_{f_n}\bar{g}(T-T_\infty)\cos\Omega-(\sigma B_0^2+\frac{\nu_f}{K}\rho_{f_n})(\bar{u}-U)\right] \tag{2}$$

$$\bar{u}\frac{\partial T}{\partial \bar{x}}+\bar{v}\frac{\partial T}{\partial \bar{y}}=\alpha_{f_n}\frac{\partial^2 T}{\partial \bar{y}^2}-\frac{1}{(\rho c_p)_{f_n}}\frac{\partial q''_{rad}}{\partial y} \tag{3}$$

Using Rosseland approximation for radiation (Sparrow and Cess [25], Rapits [26] and Brewster [27]) we can write $q''_{rad}=-\frac{4\sigma_1}{3k^*}\frac{\partial T^4}{\partial y}$ where σ_1 is the Stefan–Boltzman constant is, k^* is the mean absorption coefficient.

The boundary conditions of these equations are

$$u=U_w(x)=bx^m, v=V_0, -k_{f_n}\frac{\partial T}{\partial y}=h(T_0-T)\ at\ y=0;$$

$$\bar{u}\to U_0(x)=ax^m, T\to T_\infty\ as\ \bar{y}\to\infty \tag{4}$$

Here a and b are constant, u and v are the velocity components in the x and y directions, $B_0$is uniform magnetic field strength,$\sigma$is the electric conductivity,T is the local temperature of the nanofluid, $\bar{g}$ is the acceleration due to gravity, V_0 is the velocity of suction/injection, K is the permeability of the porous medium, ρ_{f_n} is the effective density of the nanofluid, q''_{rad} is the applied absorption radiation heat transfer, μ_{f_n} is the effective dynamic viscosity of the nanofluid, α_{f_n} is the thermal diffusivity of the nanofluid which are defined as (see Aminossadati and Ghasemi [28]),

$$\rho_{f_n}=(1-\zeta)\rho_f+\zeta\rho_s,\ \mu_{f_n}=\frac{\mu_f}{(1-\zeta)^{2.5}},\ (\rho\beta)_{f_n}=(1-\zeta)(\rho\beta)_f+\zeta(\rho\beta)_s, \alpha_{f_n}=\frac{k_{f_n}}{(\rho c_p)_{f_n}},$$

$$(\rho c_p)_{f_n}=(1-\zeta)(\rho c_p)_f+\zeta(\rho c_p)_s; \frac{k_{f_n}}{k_f}=\frac{k_s+2k_f-2\zeta(k_f-k_s)}{k_s+2k_f+2\zeta(k_f-k_s)}; \tag{5}$$

where ζ is the nanoparticle volume fraction, μ_f is the dynamic viscosity of the base fluid, ρ_f and $\rho_s$are the densities of the base fluid and nanoparticle, k_{f_n} is the thermal conductivity of the nanofluid, $(\rho c_p)_{f_n}$ is the heat capacitance of the nanofluid and k_f, k_s are the thermal conductivities of the base fluid and nanoparticle.

Following the lines of Kafoussias and Nanousis [29], the changes of variables are

$$\eta=\sqrt{\frac{(m+1)U_e}{2\nu_f x}}\,y, \psi=\sqrt{\frac{2\nu_f x U_e}{m+1}}\,f\ and\ \theta=\frac{T-T_\infty}{T_0-T_\infty} \tag{6}$$

Where ν_f is the kinematic viscosity of the base fluid.By introducing the stream function ψ, which defined as $u=\frac{\partial\psi}{\partial y}$ and $v=-\frac{\partial\psi}{\partial x}$

Equations (1)-(3) take the non-dimensional form

$$f'''-\frac{2}{m+1}(1-\zeta)^{2.5}\left[(M+\frac{\delta}{(1-\zeta)^{2.5}})(f'-1)-\left\{(1-\zeta+\zeta\frac{(\rho c_p)_s}{(\rho c_p)_f})\right\}\gamma\cos\frac{\Omega}{2}\theta\right]$$
$$-(1-\zeta+\zeta\frac{\rho_s}{\rho_f})(1-\zeta)^{2.5}\left[\frac{2m}{m+1}(f'^2-1)-f\,f''\right]=0 \tag{7}$$

$$\theta''+\frac{4}{3}\frac{k_f}{k_{f_n}}N\left\{(C_T+\theta)^3\theta'\right\}'-\Pr\left\{1-\zeta+\zeta\frac{(\rho c_p)_s}{(\rho c_p)_f}\right\}\frac{k_f}{k_{f_n}}\left[\frac{2}{m+1}f'\theta-f\theta'\right]=0 \tag{8}$$

where $\Pr=\frac{\nu_f}{\alpha_{f_n}}$ is the Prandtl number, $\delta=\frac{\nu_f}{aK}$ is the porous media parameter, $\gamma=\frac{g(\rho\beta)_f\Delta T}{\rho_f a^2b^4L}$ is the buoyancy or natural convection parameter, $M=\frac{\sigma B_0{}^2}{\rho_f a k^2}$ is the magnetic parameter, $N=\frac{4\sigma_1\theta_w^3}{k_f k^*}$ is the conductive radiation parameter and $C_T=\frac{T_\infty}{T_0-T_\infty}$ is the temperature ratio where $C_T$assumes very small values by its definition as T_0-T_∞ is very large compared to T_∞. In the present study, it is assigned the value

0.1. It is worth mentioning that γ>0 aids the flow and γ<0 opposes the flow, whileγ=0 i.e., $(T_0 - T_\infty)$ represents the case of forced convection flow. On the other hand, if γ is of a significantly greater order of magnitude than one, then the buoyancy forces will be predominant. Hence, combined convective flow exists when $\gamma = O(1)$

The boundary conditions take the following form

$$f'(0)=\lambda, f(0)=S, \theta(0)=1+\frac{1}{\sqrt{2-\frac{2m}{m+1}B_s\left\{1-\zeta+\zeta\frac{(\rho c_p)_s}{(\rho c_p)_f}\right\}}}\theta'(0) \; and \; f'=1, \theta \to 0 \; as \; \eta \to \infty \tag{9}$$

where suction (or injection) parameter $S=\frac{-V_0}{\sqrt{\frac{m+1}{2}\left(\frac{U_0 v_f}{x}\right)}}$ and relative velocity parameter $\lambda = \frac{U_w}{U_0} = \frac{b}{a}$ and the surface convection parameter $B_S = \frac{hx}{k_f}\text{Re}_x^{-\frac{1}{2}}$.

For practical purposes, the functions $f(\eta)$ and $\theta(\eta)$allow us to determine the skin friction coefficient and the Nusselt number as follows

$$C_f = \frac{\mu_{fn}}{\rho_f U^2}\left(\frac{\partial u}{\partial y}\right)_{at\ y=0} = -\frac{1}{(1-\zeta)^{2.5}}(\text{Re}x)^{-\frac{1}{2}} f''(0) \tag{10}$$

$$Nu_x = \frac{x k_{f_n}}{k_f (T_w - T_\infty)}\left(\frac{\partial T}{\partial y}\right)_{at\ y=0} = -(\text{Re}x)^{\frac{1}{2}}\frac{k_{f_n}}{k_f}\theta'(0)\left[1+\frac{4}{3}N(C_T+\theta(0))^3\right] \tag{11}$$

respectively. Here, $\text{Re}_x = \frac{U_w x}{v_f}$ is the local Reynolds number.

The relative velocity parameter signifies the ratio of the wedge velocity to the velocity of the external free stream. The value λ≠0 corresponds to a moving wedge, whereas λ=0 corresponds to a static wedge. It can further be mentioned that λ>0 implies a stretching wedge surface in which both the surface and the free stream are moving along the same direction, whereas λ<0 implies the opposite case. The surface convection parameter or Biot number B_S is a ratio of the hot fluid (the fluid supplying heat to the wedge surface) side convection resistance to the cold fluid (nanofluid) side convection resistance on a surface. For fixed cold fluid properties and fixed free stream velocity, the surface convection parameter at $B_S$any x station is directly proportional to the heat transfer coefficient h associated with the hot fluid. The thermal resistance on the hot fluid side is inversely proportional to h. Hartee pressure gradient parameter $(m=\frac{\beta}{2-\beta})$ is a function of the wedge angle parameter β where the total angle of the wedge is $\beta\pi$. The value m>0 corresponds to accelerating flows whereas m<0 corresponds to decelerating flows. In the present investigation, we considered only accelerated flows.

Results and Discussions

The set of Equations (7) and (8) is highly nonlinear and it cannot be solved analytically and numerical solutions subject to the boundary conditions (9) are obtained using the very robust computer algebra software Maple 18. This software uses a fourth-fifth order Runge-Kutta-Fehlberg method as default to solve boundary value problems numerically using the dsolve command. For the benefit of the readers the Maple worksheet is listed in Appendix A. The transformed system of coupled nonlinear ordinary differential equations (7) and (8) including boundary conditions (9) depend on the various parameters. The numerical results are represented in the form of the dimensionless velocity and temperature. Numerical computations are carried out for several sets of values of the governing parameters, namely, magnetic parameter (M), thermal radiation parameter (N), convective surface parameter (B_S), porosity parameter (δ), relative velocity parameter (λ), nanoparticle volume fraction parameter (ζ) and suction if S>0 and injection if S<0. In order to illustrate the salient features of the model, the numerical results are presented in the following figures with fixed parameters $N=0.5, M=0.5, N=1.0, \lambda=1.0, \delta=0.5, Bi=1.0, \Pr=0.6.2, S=1.0$ .The thermophysical properties of the base fluid and solid nanoparticles are given in Table 1, Oztop and Abu-Nada [30].

In Table 2, the present results of $f''(0)$ is compared with those of Yacob et al. [31] and Rahman et al. [32]. The results show a very good agreement with their results since the errors are found to be very less. This may be due to the fact that we have used Runge-Kutta-Fehlberg method (Maple 18 software) which has fifth-order accuracy. Thus the present results are more accurate than their results.

It is also observed from the Figure 2 that the agreement with the theoretical solution of velocity and temperature profiles for different values of convective surface parameter B_S is excellent compared with Figure 3b of Rahman et al. [32]. Figures 3a and b depict the influence of the suction S on velocity and temperature profiles with uniform magnetic field in the presence of nanofluid flow over a convective surface. It is shown that the velocity of the nanofluid increases and the temperature of the nanofluid decreases with increase of suction strength$(S = 0.0, 0.5, 1.0, 2.0, 3.0)$. The physical explanation for such a behavior is interesting to note that the heated fluid is pushed towards the convective surface where the buoyancy forces can act to retard the fluid due to high influence of the viscosity of the nanofluid. In the case of suction, the convective surface (B_S=0.5) plays a physically significant role on the nanofluid flow and also the nanofluid is brought closer to the surface and enhances the hydrodynamic and reduces the thermal boundary layer thickness. This is due to the fact that the removal of the decelerated nanoparticles through the porous convective surface reduces for the growth of the thermal boundary layer. Decrease in thickness of the thermal boundary layer was caused in two ways: (i) the direct action of suction and (ii) the indirect action of suction of the nanofluid causing a thinner thermal boundary layer which corresponded to a higher temperature gradient, a consequent increase in the buoyancy force and a higher temperature gradient whereas the exact opposite behavior is observed by imposition of wall fluid injection.

Physical properties	Fluid phase (water)	Cu	Al_2O_3	TiO_2
C_p (J/kg K	4179	385	765	686.2
ρ (kg/m³)	997.1	8933	3970	4250
κ (W/m K)	0.613	400	40	8.9538

Table 1: Thermophysical properties of fluid and nanoparticles (Oztop and Abu-nada [30]).

m	ξ	Yacob et al. [31]	Rahman [32]	Present results
0.0	0.1	0.7179	0.7181	0.718094351
	0.2	0.9992	0.9993	0.999265398
0.5	0.1	1.5881	1.5882	1.588163987
	0.2	2.2105	2.2106	2.105852436
1.0	0.1	1.8843	1.8843	1.884257298
	0.2	2.6226	2.6227	2.622647295

Table 2: Comparison of the present results for $f''(0)$ with published works.

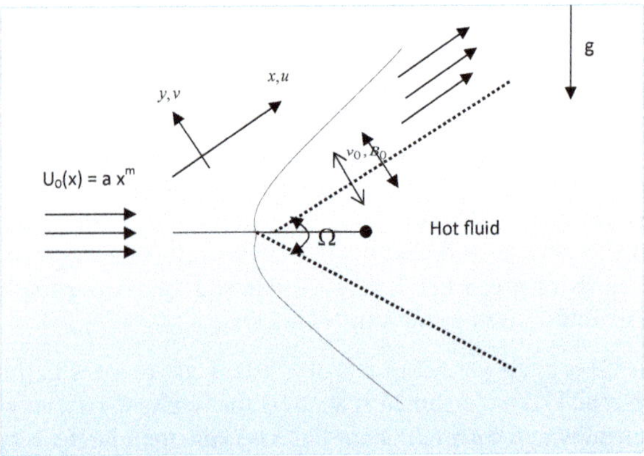

Figure 1: Physical flow model and Coordinate system over a porous wedge sheet.

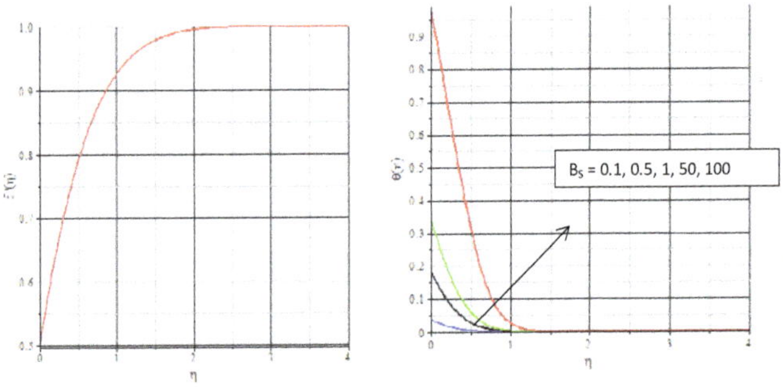

Figure 2: Coparison of convective surface effects on velocity and temperature profiles with Rahman et al. [32].

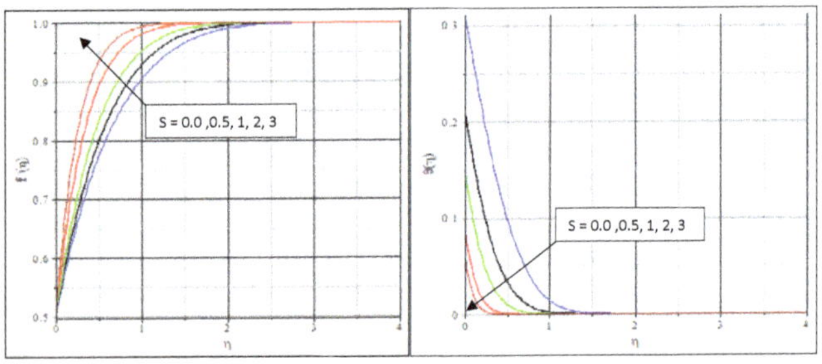

Figure 3: Suction effects on velocity and temperature profiles.

Illustrative velocity and temperature profiles of the nanofluid for five typical angles of inclination ($\Omega = 0.0^0, 10.5^0, 23.5^0, 120^0$ and 180^0) are presented in Figures 4a and 4b. It is revealed that the temperature of the nanofluid is reduced by increasing the angle of inclination of convective wedge surface approximately corresponds to Hartee pressure gradient parameter ($S = 0.0, 0.5, 1.0, 2.0, 3.0$) because the angle of inclination increases the effect of the buoyancy force due to thermal diffusion decreases by a factor of $\cos\Omega$. Consequently the driving force to the fluid increases as a result temperature profiles decrease. From the Figures 4a and 4b, it is observed that the hydrodynamic boundary layer thickness of the nanofluid increases whereas thermal boundary layer thickness decreases as the angle of inclination increases. Similar to the effects of magnetic field, increasing the inclination angle makes it harder for the fluid to flow along the wedge plate and causes it to become warmer. This is due to the reduction in the thermal buoyancy effect caused by decrease in temperature and increase in velocity of the nanofluid. Due to the wedge plate is inclined from horizontal to vertical, the buoyancy effect on the velocity increasing and temperature of the nanofluid profiles decreasing. This is because the incident thermal radiation is initially absorbed by the absorbing fluid-matrix system which, in turn, heats up the ideally transparent plate. This operation of passing the absorbing fluid through an absorbing porous medium is believed to enhance thermal radiation collection by direct absorption in which heat losses are reduced as a result of plate temperature of the nanofluid. All these physical behavior are due to the combined effects of the strength of inclination of the wedge plate in the presence nanofluid with magnetic field.

In the presence of nanofluid flow along the convective wedge surface, the velocity profiles for different values of magnetic and porosity strength are shown in Figures 5 and 6. It is observed from the Figure 5 that the velocity of the nanofluid increases with increase of the magnetic strength because of Lorentz force comes into action. This lorentz force acts in the opposite direction to the flow of nanofluid, opposing the motion of the nanofluid and tends to accelerates with increase of M. This is due to the combined effect of the increase of strength of the electrical conductivity and decrease of the density of the nanofluid in the presence of convective wedge surface. Due to the combined effect of decreasing of the permeability of the porous medium and increasing of the kinematic viscosity of the nanofluid, it is predicted from the Figure 6 that the velocity of the nanofluid increases with increase of porosity parameter.

Figure 7 presents the characteristic velocity and temperature

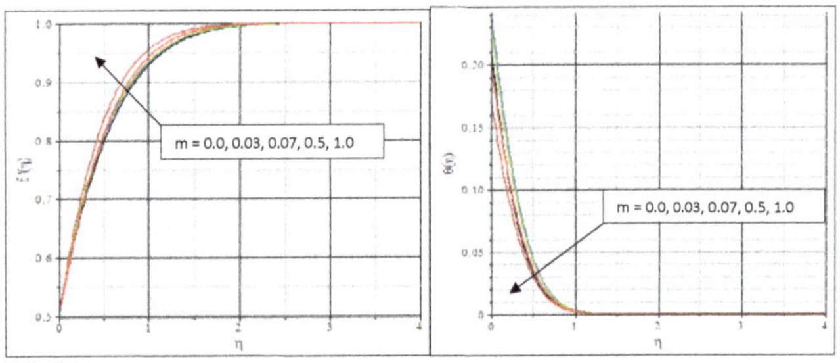

Figure 4: Hartee pressure gradient effects on velocity and temperature profiles.

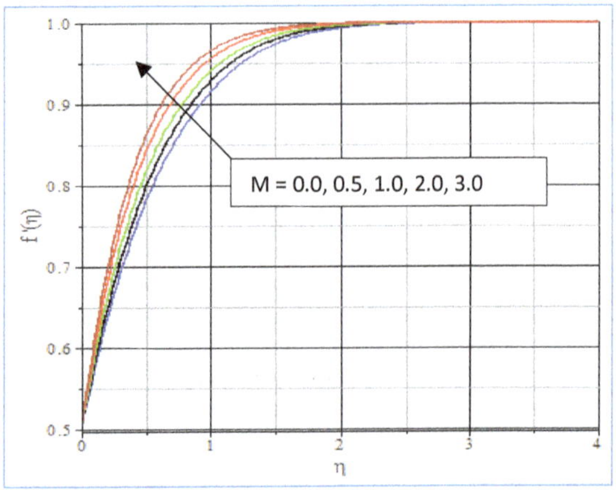

Figure 5: Magnetic effects on velocity profiles.

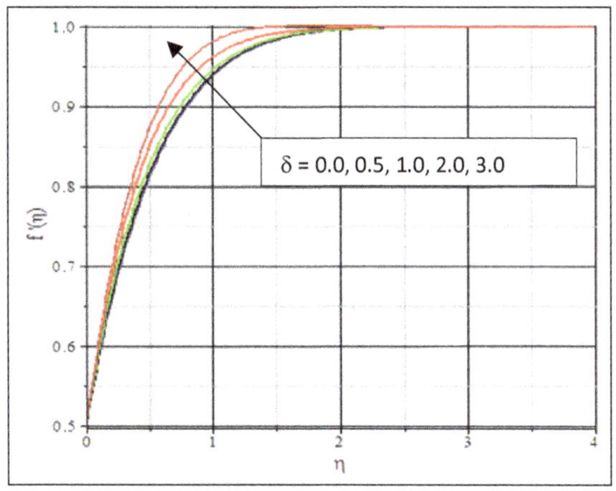

Figure 6: Porosity effects on velocity profiles.

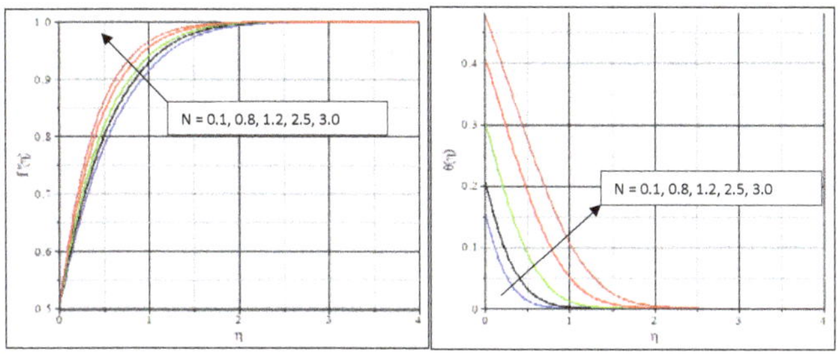

Figure 7: Thermal radiation effects on velocity and temperature profiles.

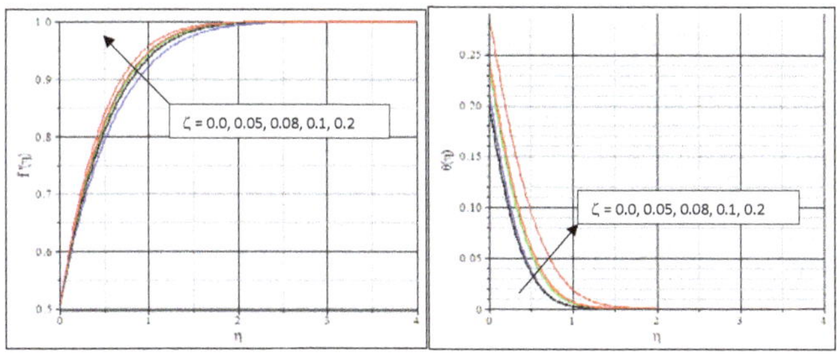

Figure 8: Nanoparticle volume fraction on velocity and temperature profiles.

profiles in the presence of nanofluid flow along the surface for different values of the thermal radiation parameter. According to Equations (2) and (3), the divergence of the radiative heat flux increases as thermal conductivity of the nanofluid (k_f) raises which in turn increases the rate of radiative heat transfer to the nanofluid and hence the nanofluid velocity and temperature of the nanofluid accelerate.N is to enhance the velocity and temperature of the nanofluid significantly in the flow region because of the combined effect of absorption and porosity of the convective surface in the presence magnetic field. This result can be explained by the fact that an increase in the values of $N = \dfrac{4\sigma_1\theta_w^3}{k_f\, k^*}$ forgiven $k^*\ and\ T_\infty$ means a decrease in the Rosseland radiation absorptive k^*. In view of this explanation, the effect of convective thermal radiation becomes more significant as $N \to 0$ ($N \neq 0$) and can be neglected when $N \to \infty$. It is noticed that the velocity and temperature of the nanofluid increases with increase of radiation parameter N. The increase in radiation parameter means the release of heat energy from the flow region and so the fluid temperature increases as the thermal boundary layer thickness becomes thicker. All these physical behavior are due to the combined effect of porous strength of the convective wedge surface and size of the nanoparticles.

Figure 8 displays the effects of volume fraction of nanoparticles on velocity and temperature distribution. In the presence of uniform thermal conductivity of the nanofluid along the convective porous wedge surface, it is note that the velocity and the temperature of the nanofluid increases with increase of the nanoparticles volume fraction parameter ζ. It is also observed that the hydrodynamic and thermal boundary layer increase with increase of nanoparticle volume fraction parameter. This agrees with the physical behavior that when the volume fraction of nanoparticles increases the thermal conductivity and then the thermal and hydrodynamic boundary layer thickness increases. Variability of the size, shape, material, and volume fraction ofthe nanoparticles allows for tuning to maximize spectral absorption of thermal radiation throughout the fluid volume. Enhancement in the thermal conductivity can lead to efficiency improvements, although small, via more effective fluid heat transfer. Vast enhancements in convective surface area due to the extremely small particle size, which makes nano fluid-based thermal radiation systems attractive for thermochemical and photocatalytic processes.

Figure 9 illustrates typical profiles for velocity and temperature for different strength of convective surface in the presence of uniform magnetic field. In the presence of uniform porosity of the convective surface of the wedge, it is clearly shown that the velocity and temperature of the nanofluid increase with increase of the strength of convective surface parameter because of the combined effects of increase of heat transfer coefficient associated with the hot nanofluid and decrease of thermal conductivity of the nanofluid. In particular, the velocity and the temperature of the nanofluid gradually changes from lower value to the higher value only when the strength of convective surface parameter increases in the presence of nanofluid flow along the porosity of the wedge convective surface. For heat transfer characteristics mechanism about the connective surface, important result is the large distortion of the temperature field caused for $1 < B_S \leq 50$. It is interesting to note that the characteristic of strength of convective surface plays a dominant role on heat transfer of the nanofluid. All these physical behaviors are due to the combined effects of the ratio of the hot fluid side convection resistance to the cold fluid (nanofluid) side convection resistance on a surface.

Conclusions

In the present paper, we have examined the performance of radiative heat transfer on water based nanofluid flow over a porous

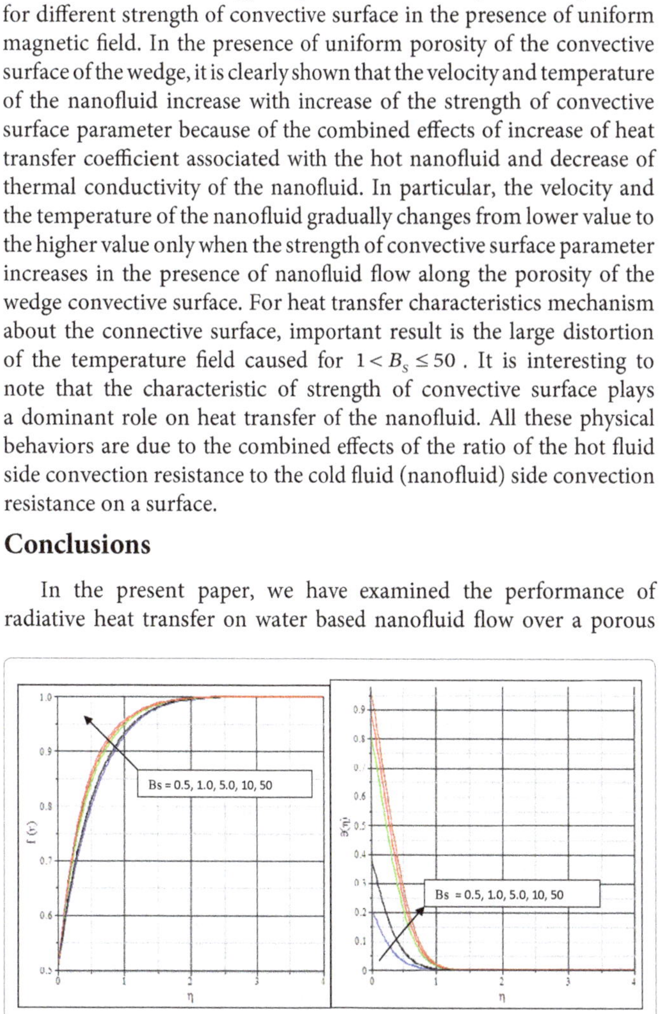

Figure 9: Thermal radiation effects on velocity and temperature profiles.

static or moving convective wedge surface in the presence magnetic field with variable stream condition. The results of this study lead to the following conclusions:

- In the presence of uniform porosity of the convective surface of the wedge, it is clearly shown that the velocity and temperature of the nanofluid increase with increase of the strength of convective surface parameter because of the combined effects of increase of heat transfer coefficient associated with the hot fluid and decrease of thermal conductivity of the nanofluid. It is interesting to note that the characteristic of strength of convective surface plays a dominant role on heat transfer of the nanofluid.

- In the presence of uniform magnetic field, the velocity enhances and the temperature of the nanofluid reduces with increase of angle of inclination of the convective wedge surface approximately corresponds to Hartee pressure gradient. The fact is that as the angle of inclination increases the effect of the buoyancy force due to thermal diffusion increases by a factor of $\cos\Omega$ and consequently the driving force to the fluid increases as a result temperature decreases in the presence of water based nanofluid.

- Due to convective wedge surface, it is important to note that the velocity of the nanofluid increases whereas the temperature of the nanofluid decreases with increase of suction strength because the heated fluid is pushed towards the convective surface where the buoyancy forces can act to retard the fluid due to high influence of the viscosity of the nanofluid.

- Increase of thermal field due to increase in nanoparticle volume fraction parameter shows that the temperature increases gradually as we replace Copper by Alumina, and Titanium in the said sequence.

- It has been shown that mixing nanoparticles in a liquid (water based nanofluid) has a dramatic effect on the liquid thermophysical properties such as thermal conductivity. The porous convective wedge surface plays a very significant role on thermal boundary layer regime.

- It is interesting to predict that the velocity of the nanofluid increases with increase of the magnetic and porosity strength of the convective wedge surface because of the combined effect of increase of the strength of electrical conductivity and decrease of permeability of the porous medium and density of the nanofluid.

- It is noticed that the velocity and temperature of the nanofluid increase with increase of thermal radiation parameter. The increase in radiation parameter means the release of heat energy from the flow region and so the fluid motion and temperature of the nanofluid increases as the momentum and thermal boundary layer thickness enhance in the presence convective wedge surface.

- Nanofluids due to thermal radiation energy in the presence of magnetic field along the convective surface are important because they can be used in numerous applications involving heat transfer and other applications such as in detergency, solar collectors, drying processes, heat exchangers, geothermal and oil recovery, building construction, etc. The interdisciplinary nature of nanofluid research presents a great opportunity for exploration and discovery at the frontiers of nanotechnology. It is hoped that the present work will serve as a stimulus for needed experimental work on this problem.

References

1. Choi SUS (1995) Enhancing thermal conductivity of fluids with nanoparticles. Developments and applications of non-newtonian flows, FED-231 66: pp. 99-105.
2. Kole M, Dey TK (2010) Viscosity of alumina nanoparticles dispersed in car engine coolant. Experimental Thermal and Fluid Science 34: 677-683.
3. Lee S, Choi SUS, Eastman JA (1999) Measuring thermal conductivity of fluids containing oxide nanoparticles, ASME Journal of Heat Transfer 121: 280-289.
4. Xuan Y, Li Q (2000) Heat transfer enhancement of nanofluids.International Journal of heat and fluid flow 21: 58-64.
5. Xie H, Wang J, Xi T, Liu Y, Ai F, et al. (2002)Thermal conductivity enhancement of suspensions containing nanosized alumina particles.Journal of Applied Physics 91: 4568-4572.
6. Das SK, Putra N, Roetze W (2003) Pool boiling characteristics of nano-fluids. International Journal of Heat and Mass Transfer 46: 851-862.
7. Das SK, Putra N, Roetzel W (2003) Pool boiling of nano-fluids on horizontal narrow tubes. International Journal of Multiphase Flow 29: 37-47.
8. You SM, Kim JH, Kim KH (2003) Effect of nanoparticles on critical heat flux of water in pool boiling heat transfer.Applied Physics Letters 83: 3374-3376.
9. Vassallo P, Kumar R, Amico SD (2004) Pool boiling heat transfer experiments in silica–water nano-fluids.International Journal of Heat and Mass Transfer 47: 407-411.
10. Kang C, Okada M, Hattori A, Oyama K (2001) Natural convection of water-fine particle suspension in a rectangular vessel heated and cooled from opposing vertical walls (classification of natural convection in the case of suspension with a narrow-size distribution). International Journal of Heat and Mass Transfer 44: 73-82.
11. Putra N, Roetzel W, Das SK (2003) Natural convection of nano-fluids. International Journal of Heat and Mass Transfer 39: 775-784.
12. Khanafer K, Vafai K, Lightstone M (2003) Buoyancy-driven heat transfer enhancement in a two-dimensional enclosure utilizing nanofluids. International Journal of Heat and Mass Transfer 46: 39-53.
13. Kim N, Kang YT, Choi CK (2004) Analysis of convective instability and heat transfer characteristics of nanofluids. Physics Fluids 16: 395-401.
14. Ghasemi B, Aminossadati SM (2009) Natural convection heat transfer in an inclined enclosure filled with a water-Cu nanofluid, Numer. Heat Transfer Part A Appl 55: 807-823.
15. Congedo PM, Collura S, Congedo PM (2009) Modeling and analysis of natural convection heat transfer in nanofluids.In: Proc. ASME Summer Heat Transfer Conf 3: 569-579.
16. Ho CJ, Chen MW, Li ZW (2008) Numerical simulation of natural convection of nanofluid in a square enclosure: Effects due to uncertainties of viscosity and thermal conductivity. Int J Heat Mass Transfer 51: 4506-4516.
17. Ho CJ, Chen MW, Li ZW (2007) Effect of natural convection heat transfer of nanofluid in an enclosure due to uncertainties of viscosity and thermal conductivity.In: Proc. ASME/JSME Thermal Eng. Summer Heat Transfer Conf. HT 1: 833-841.
18. Lin HT, Lin LK (1987)Similarity solutions for laminar forced convection heat transfer from wedges to fluids of any Prandtl number. Int J Heat Mass Transfer 30: 1111-1118.
19. Yih KA (1998) Uniform suction/blowing effect on the forced convection about a wedge: Uniform heat flux. Acta Mech 128: 173-181.
20. Pantokratoras A (2006) The Falkner-Skan flow with constant wall temperature and variable viscosity. Int J Thermal Sci 45: 378-389.
21. Martin MJ, Boyd ID (2010) Falkner-Skan flow over a wedge with slip boundary conditions. J Thermophys Heat Transf 24: 263-270.
22. Rahman MM, Eltayeb IA (2011) Convective slip flow of rarefied fluid over a wedge with thermal jump and variable transport properties. Int J Thermal Sci 50: 468-479.
23. Rahman MM, Al-Hadhrami AMK (2012) Nonlinear slip flow with variable transport properties over a wedge with convective surface.Stavrinides SG (Ed.), Chaos and complex systems, 21 Springer, Berlin.
24. Muhaimin I, Kandasamy R, Khamis AB, Roslan R (2013) Effect of thermophoresis particle deposition and chemical reaction on unsteady MHD mixed convective flow over a porous wedge in the presence of temperature-dependent viscosity.Nucl Eng Des 261: 95-106.

25. Sparrow EM, Cess RD (1978) Radiation heat transfer, Washington: Hemisphere.

26. Raptis A (1998) Radiation and free convection flow through a porous medium. Int Comm Heat Mass Transfer 25: 289-295.

27. Brewster MQ (1972) Thermal Radiative Transfer Properties, John Wiley and Sons.

28. Aminossadati SM, Ghasemi B (2009) Natural convection cooling of a localized heat source at the bottom of a nanofluid-filled enclosure. Eur J Mech B/Fluids 28: 630-640.

29. Kafoussias NG, Nanousis ND (1997) Magnetohydrodynamic laminar boundary layer flow over a wedge with suction or injection. Canadian Journal of Physics 75:733.

30. Oztrop, Abu-nada (2008) Numerical study of natural convection in partially heated rectangular enclosures filled with nanofluids. Int J Heat Fluid Flow 29: 1326-1336.

31. Yacob NA, Ishak V, Pop I (2011) Falkner-Skan problem for a static or moving wedge in nanofluids. Int J Thermal Sci 50: 133-139.

32. Rahman MM, Al-Mazroui WA, Al-Hatmi FS, Al-Lawtia MA, Eltayeb IA (2014) The role of a convective surface in models of the radiative heat transfer in nanofluids. Nuclear Engineering and Design 275: 382-392.

Ultrasound Tomography for Spatially Resolved Melt Temperature Measurements in Injection Moulding Processes

Hopmann C and Wipperfürth J*

Institute of Plastics Processing (IKV), RWTH Aachen University, Seffenter Weg 201, 52074 Aachen, NRW, Germany

Abstract

In injection moulding processes, the measurement of the temperature distribution is very important for the validation of models used for simulative part design due to the high influence on shrinkage and warpage of the moulded part, but is also very challenging to measure. During the injection moulding process high mould pressures occur and the cavity is not easily accessible. Therefore, contact sensors cannot be used since they induce shear stress into the melt, which changes the flow behaviour of the melt and thus the temperature field. In this work, we present a method for the contactless determination of the temperature distribution of a moulded part during injection moulding using ultrasound tomography. With time-of-flight ultrasound measurements from different directions it is possible to reconstruct the distribution of ultrasound velocity in the cross-section of a moulded part. With this distribution, the temperature field can be calculated using additional material characteristic properties. Based on this concept, an injection mould was designed, that allows performing ultrasound tomography with 20 ultrasound transducers radially arranged around a cylindrical shaped cavity. This allows the temperature determination under real process conditions with a spatial resolution of 3.5 mm^2. A highly parallelised measurement device allows recording of several complete datasets before no more signals can be detected due to shrinkage of the moulded part. During several injections moulding-cycles all sensor positions were able to detect noticeable signals. Due to internal signal processing of the measurement device, it is not yet possible to calculate arrival times of the ultrasound signal but amplitude-scans show the general feasibility of ultrasound tomography during injection moulding..

Keywords: Temperature; Injection moulding; Tomography; Ultrasound

Introduction

The quality of moulded plastics parts, such as shrinkage, warpage; mechanical and thermal properties are highly influenced by the injection moulding process [1]. In the last years, the increasing requirements for quality and accuracy of moulded parts lead to a significant need of simulation and are routine tools for the engineer in the meanwhile to predict accurately the properties of the final parts [2,3]. A lot of commercial software is available on the market, such as Moldflow, Sigmasoft, Moldex3D and Cadmould. Each of the software uses different models and approaches to describe parts of the injection moulding process and their effect on the moulded part. In the injection moulding process the main parameters are injection time, pressure, mould temperature and melt temperature [2]. While injection time, pressure and mould temperature can be described accurately via theoretical models and validated under injection moulding conditions, the melt temperature is still a challenging parameter. The temperature history of a moulded part is complex to describe theoretically due to high cooling rates, complex heat transfer mechanisms to the mould and the forming microstructure especially with respect to semi-crystalline thermoplastics. However, the knowledge of the temperature field of a polymer in the cavity during the injection moulding process is required to perform precise simulations for the prediction of shrinkage and warpage of the moulded part. Today's measurement techniques, such as thermocouples, are still based on the direct contact between sensor and melt, which influences the melt flow and induces shear stresses. Furthermore, only punctual information can be obtained with high response times, which prevent precise temperature measurements and thus a detailed process understanding. Therefore, contact measurement techniques are not suitable for the determination of the real temperature. Other sensor techniques, such as IR-Sensors, do not require direct contact but special sapphire windows to measure the punctual temperature at or near the surface [4]. This is adverse, e.g. for the validation of newly developed models used in part design, since the temperature is measured under a different contact behaviour at the interface of the melt and the cavity wall. In this context, an increasing importance for non-invasive measurement techniques is caused by these disadvantages. Several different research approaches were carried out in the last years, such as pyrometry [5], fluorescence spectroscopy [6] or ultrasound [7-9]. Recent approaches base on the use of tomographic methods to measure the local temperature distribution. There are two current approaches: the ultrasound tomography [8,9] and the electrical capacitance tomography (ECT) [10]. Both methods have in common that signals can be introduced into the melt from different directions. Due to interactions of the radiation with the melt, various sensors, around the measurement region, can measure a response. Using this, distribution ultrasound velocities can be calculated and correlated with the local temperature. In the case of ultrasound tomography first the distribution of ultrasound velocities is calculated based on the time of flight measurements and then the temperature field is calculated with the help of further material properties such as the bulk modulus and the shear modulus. The ultrasound tomography used in this work is based on the procedure of Praher et al. [8] who suggests using ultrasound signals to measure the temperature field in the screw antechamber of an injection-moulding machine and assumed a radial symmetric temperature distribution to realise a fast temperature

***Corresponding author:** Jens Wipperfürth, Institute of Plastics Processing (IKV), RWTH Aachen University, Seffenter Weg 201, 52074 Aachen, NRW, Germany
E-mail: jens.wipperfuerth@ikv.rwth-aachen.de

measurement with only five ultrasound transducers. In this work, we present an extended approach to measure the temperature distribution of a moulded part during injection moulding and solidification that will focus on the spatial resolution. The advantage of this measurement technique is the stability against the high cooling rates and complex heat transfer mechanisms prevailing during the injection moulding process, since the acquired data include this information. In future, the knowledge of the temperature distribution and its evolution during the process will considerably improve the simulative part design.

Principle of Ultrasound Tomography

In ultrasound-time-of-flight (TOF) tomography, an ultrasound transducer emits an ultrasound wave that propagates through the melt. A set of receiving transducers radially arranged around the region of interest measures the arrival time of the emitted wave that is used for the reconstruction of the distribution of ultrasound velocity. Typically, tomographic methods require consecutive measurements from different directions. This requires a rotation of either the sample or the sensor-detector-arrangement. In our approach of ultrasound tomography, a rotation is not required since the transducers are able to both emit and detect ultrasound signals.

Between an ultrasound emitter and an ultrasound receiver one can assume that an emitted ultrasound wave propagates along the path g between the emitter and the receiver. The ultrasound wave can be detected at the opposite receiver after an arrival time $t_{tot} = t_E - t_0$, which is the difference of the start time of the propagation t_0 and the detection time t_E. The time required for the ultrasound wave to traverse the melt depends on the ultrasound velocity c that is a temperature and pressure dependent material property. However, along the path g many different transit times t_{ik} that arise from different ultrasound velocities due to temperature inhomogeneity, determine the arrival time t_{tot}. That means by discretising the measurement region into j× l arbitrary subareas s the arrival time can be expressed as the amount of the propagation times t_{ik} in an individual subarea s_{ik} (equation (1)). Further, each transit time t_{ik} can be expressed in terms of the ultrasound velocity c_{ik}. For this purpose,the length of each path $\vec{g}$ in a subarea s_{ik} has to be known. This can be done by introducing a factor b_{ik} that scales the length of $\vec{g}$ in S_{ik}.

$$t_{tot} = \sum_{s_{i=l,k=1}}^{s_{jl}} t_{ik} = \sum_{k=1}^{l}\sum_{i=1}^{j} \frac{b_{ik}}{c_{ik}(T)} \qquad (1)$$

From equation (1), it is clear that the transit times in every subarea s_{ik} are needed to solve the equation for c_{ik}. In practice, this is realised by measuring the transit times from different directions.

Furthermore, equation (1) is valid for n emitter/receiver-paths $\vec{g}_n$. Considering this, a system of linear equations can be set up as shown in equation (2) where in,

$$an, ik = b_{ik}(\vec{g}_n)$$

$$\begin{pmatrix} a_{11} & \cdots & a_{1,jl} \\ \vdots & \ddots & \vdots \\ a_{n1} & \cdots & a_{n,jl} \end{pmatrix} \begin{pmatrix} 1/c_1 \\ \vdots \\ 1/cjl \end{pmatrix} \qquad (2)$$

The Matrix in equation (2) includes the information of the individual path lengths $b_{ik}(\vec{g}_n)$. This matrix is multiplied with the corresponding reciprocal ultrasound velocity c_{ik} in the element s_{ik} and the result is a vector, containing the measured transit times for every path $\vec{g}$. Equation (2) can be expressed in a general form:

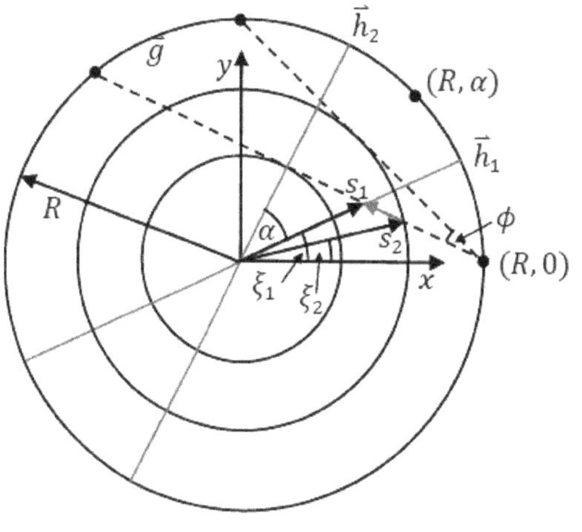

Figure 1: Division of the measurement area (inner circles and grey lines) for the determination of the distance (grey arrow) of an individual path (dashed line) in an area element.

$$A\vec{c} = \vec{t}_{tot} \qquad (3)$$

The matrix A is often called design matrix or system matrix in the literature [11]. This should clarify that A is constructed using information about the structure of the observed measurement region. However, to solve equation (3) for c_{ik} first the system matrix A has to be determined and subsequently an inversion of equation (3) is required.

To construct A, the measurement (Figure 1) region has to be divided into subareas, so that more than one individual path crosses each subarea. This ensures that for each area, sufficient information is available to reconstruct the ultrasound velocity in this element. For this purpose, a Cartesian coordinate system is defined at first, which origin coincides with the centre of the measurement region (Figure 1). In a first step, the measurement region is divided in different circular subareas. The circuits are designed so that each circuit is spanned tangentially to a path g_n. The radii r_n of each circular subarea can be described by equation (4):

$$r_i = R\cos\left(\frac{\alpha_n}{2}\right) \qquad (4)$$

In equation (4) R is the radius of the measuring region and α_n is the angle between two sensors and the origin of the coordinate system and results for a given sensor n from the total number of sensors

$$\mathrm{N}: \alpha_n = \frac{2\pi}{N}(n-1), n = \{1, ..., N\} \qquad (5)$$

The division of the measuring region however, can be further divided (Figure 1). To this end, additional lines h are defined (Figure 1, grey lines) which are passing through the origin of the coordinate system and through the centre of the distance between two sensors. This can be expressed as a function of α, where v is the scaling factor for the direction vector:

$$\vec{h}: v\begin{pmatrix} \cos\left(\alpha + \frac{\Delta\alpha}{2}\right) \\ \sin\left(\alpha + \frac{\Delta\alpha}{2}\right) \end{pmatrix} \qquad (6)$$

To set up the system matrix A successfully, in a next step, the path length of the ultrasound wave has to be calculated in an area segment. A line g that represents the path of the ultrasound wave from an emitter to any receiver, can be described by equation (7):

$$\vec{g}: R\begin{pmatrix}\cos(\varphi)\\ \sin(\varphi)\end{pmatrix} + \mu\begin{pmatrix}\cos(\pi-\alpha)\\ \sin(\pi-\alpha)\end{pmatrix} \quad (7)$$

In equation (7) μ is a scaling factor and ϕ is the angle between two paths and can be calculated using equation (8):

$$\varphi = \frac{\pi}{2}\left(1-\frac{2n}{N}\right) \quad (8)$$

From the intersection of $\vec{g}$ and $\vec{h}$ and from the intersections of $\vec{g}$ and the circuits, which are described by r_i, the length of the path can be calculated in a subarea by (Figure 1, grey arrow):

$$d = \left|\vec{g}(s_1) - \vec{g}(s_2)\right| = r_i^2 + r_{i+1}^2 - 2\cos(\xi_1 - \xi_2) \quad (9)$$

To calculate the matrix A, the relations described above have been implemented in a routine in the software Matlab 2016a (The MathWorks Inc., MA; USA). For an efficient calculation, it is sufficient to describe the radiation extinguished from a specific sensor. Furthermore, the routine has to calculate the paths only for one semi-cycle, the other semi-cycle can be achieved by mirroring (Figure 2).

In the routine, the resulting distances for one ray path are stored in a table of the form shown in Figure 2.

For each ray path at one sensor position such a table is calculated. By moving each row to the row beneath, the next sensor position can be fully described. From this, the distances of all ray paths for each area element can be determined for a specific measurement region (Figure 3, left) that has a defined radius and a specified number of ultrasound transducers (Figure 3, white circles). To create the system matrix the rows of one table are concatenated into one row. Subsequently, the rows are strung together and form the system matrix (Figure 3, right). The system matrix thus has N(N-1) rows and the number of columns describes the number of area elements 1 - 32. It can be followed that the system matrix includes the information of the distance for each ray in every area element. The entries represented in black in Figure 3 are zero-entries, which mean that the specific ray does not pass that area element. However, values greater than zero represent a distance of a beam through that area element and the higher the grey value the longer the distance. Since A has many zero entries, it is suitable to store the matrix as a sparse matrix, which is only for the efficiency of the routine.

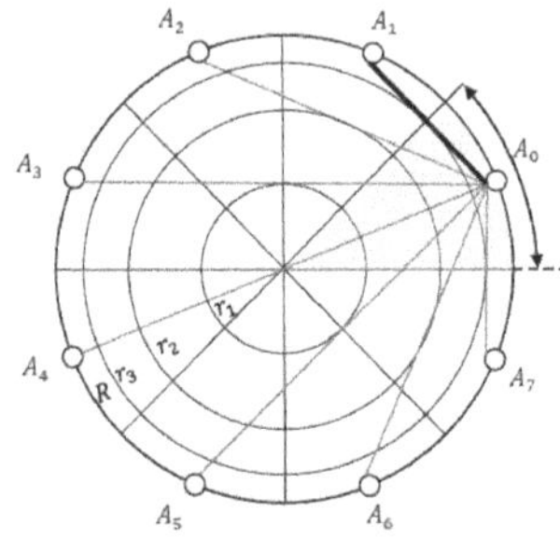

	r_1	$\Delta_{r_{2,1}}$	$\Delta_{r_{3,2}}$	$\Delta_{r_{R,3}}$
A_0	0	0	0	d
A_1	0	0	0	d
A_2	0	0	0	0
A_3	0	0	0	0
A_4	0	0	0	0
A_5	0	0	0	0
A_6	0	0	0	0
A_7	0	0	0	0

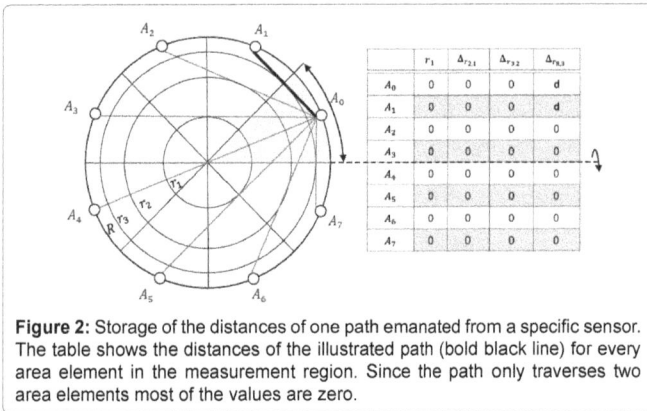

Figure 2: Storage of the distances of one path emanated from a specific sensor. The table shows the distances of the illustrated path (bold black line) for every area element in the measurement region. Since the path only traverses two area elements most of the values are zero.

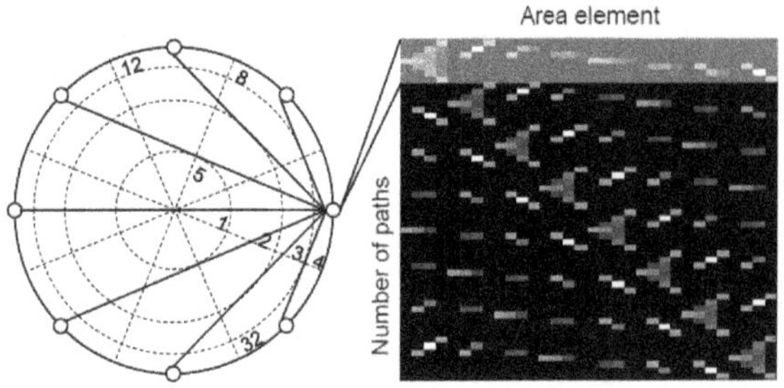

Figure 3: The system matrix (right) that describes the length of a specific path in each area element of the measurement region (left). The grey box in the system matrix marks the paths of the ultrasound beams emitted by one emitter. A Radius R = 0.015 m was assumed.

To calculate the individual ultrasound velocities from the measured transit times, the system of equations in equation (3) has to be inverted. Due to A, this system of equations is overdetermined so that no direct solution can be obtained. Therefore, iterative reconstruction techniques like ART (Algebraic Reconstruction Technique), SIRT (Simultaneous Iterative Reconstruction Technique) or SART (Simultaneous Algebraic Reconstruction Technique) have to be applied. The procedure of each iterative reconstruction technique is explained and discussed in the literature in detail [12]. In this work, the SART algorithm is used [13] that provide high accuracies [12]. The algorithm is part of the Matlab open source package AIRtools of the Technical University of Denmark [13]. The input parameters for the algorithm are the system matrix *A*, a vector t that contains the measured arrival times and the number of iterations *k* that should be performed.

Spatial resolution

The dimension of an area element defines the spatial resolution. This means, that each segment of a cycle has another spatial resolution. Mathematically the dimension of an area element is defined by equation (10).

$$A(r,\alpha) = \int_{r_{n-1}}^{r_n \le R} \int_0^{2\pi/N} r\,d\alpha\,dr \quad (10)$$

Since the ultrasound transducers are equidistantly arranged around the radial measurement region it follows $d\alpha = \Delta\alpha = const$ and equation (10) becomes:

$$A_\alpha(r) = \Delta\alpha \int_{r_{n-1}}^{r_n \le R} r\,dr \quad (11)$$

From equation (10) and (11) it is clear, that only the total amount of ultrasound transducers has an influence on the spatial resolution. This causes a huge difference to conventional tomography, like x-ray tomography, where the spatial resolution is a function of both the rotation step and the total amount of detector elements. In our approach of ultrasound tomography, the rotation step is only a function of the number of detector elements. Therefore, with a static sensor arrangement one resolution-influencing parameter is lost. However, this is only a minor restriction, since rotations would be hardly realisable under injection moulding conditions and would cause high measurement times. The spatial resolution of an area element in an assumed measurement region with a radius of 0.015 m and eight sensors is about 20 mm². Assuming 20 transducers in a measurement region with the same radius, the spatial resolution increases to 3.5 mm².

Dependence of temperature and ultrasound velocity

The inversion of equation (3) is directly related to the desired goal of a non-invasive determination of the temperature distribution in a polymer melt channel. However, up to this point the temperature is still an unknown quantity. To determine the temperature distribution, it is necessary to convert each reconstructed ultrasound velocity at a specific location into a local temperature. The coherence shown in equation (12) of the ultrasound velocity c and the temperature T can be attached on the density ρ of the polymer [14].

$$c(x,y)=\left[\frac{1}{\rho(x,y)}\left(K+\frac{4G}{3}\right)\right]^{1/2} \tag{12}$$

In equation (12), K is the adiabatic bulk modulus and G is the shear modulus of the polymer melt. From equation (12), it is clear that for the determination of the temperature distribution, additional material-specific characteristic data has to be known. The density and the compression modulus can be described via Tait approach [5]. In this work, the Tait approach for a semi-crystalline polypropylene (PP505P, Saudi Basic Industries Corporation, Riyadh, Kingdom of Saudi Arabia) is used (Figure 4).

From the pvT-relationship the bulk modulus can be calculated using the adiabatic compressibility κ (equation (13)) [8]:

$$K^{-1}=\kappa=-\frac{1}{v}\left[\left(\frac{\partial v}{\partial p}\right)_T+\frac{T}{c_p}\left(\frac{\partial v}{\partial T}\right)_p^2\right] \tag{13}$$

The shear modulus relates to the bulk modulus and Young's modulus *E* via the fundamental equation (14).

$$G=\frac{3KE}{9K-E} \tag{14}$$

In the melt state the shear modulus can typically be neglected [8], but for the solid-state Young's modulus has to be described in dependence of the temperature. For this, tensile tests (Z100, Zwick Roell AG, Ulm, Germany) in a climatic chamber from 30°C to 80°C with 10°C per step were performed with polypropylene (PP). To provide a sample with homogenous temperature, the tensile bars were stored for 16 h in the climatic chamber for each temperature. We use the optical system LIMESS Messtechnik und Software GmbH, Krefeld/Germany, to measure the strain of five tensile bars at each temperature (Figure 5). In order to compare the experimentally determined data, Young's modulus of Sabic PP505P is presented at room temperature from literature [15].

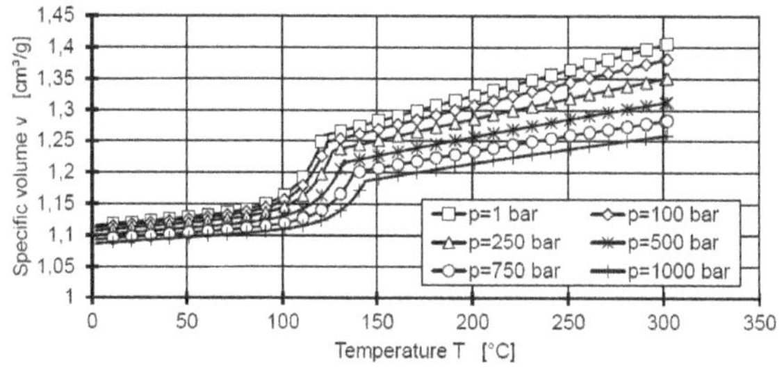

Figure 4: Calculation of the pvT-relation of a polypropylene via Tait-model.

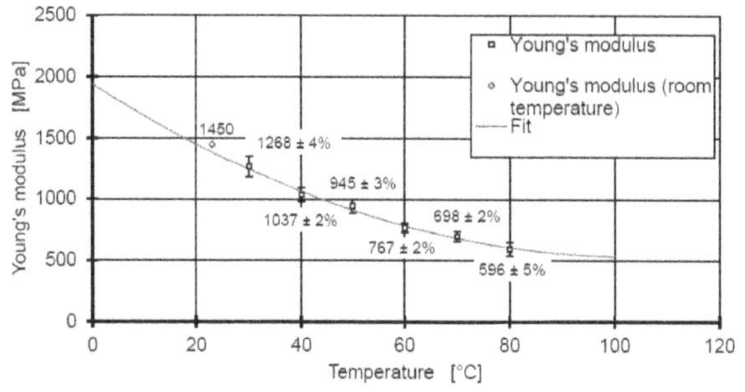

Figure 5: Experimental determination of Young's modulus in dependence of the temperature. The data points represent the mean values of five tensile tests at each temperature. The error bars represent the doubled standard deviation for a better illustration. For this reason, the single relative standard deviation for each data point is specified numerically.

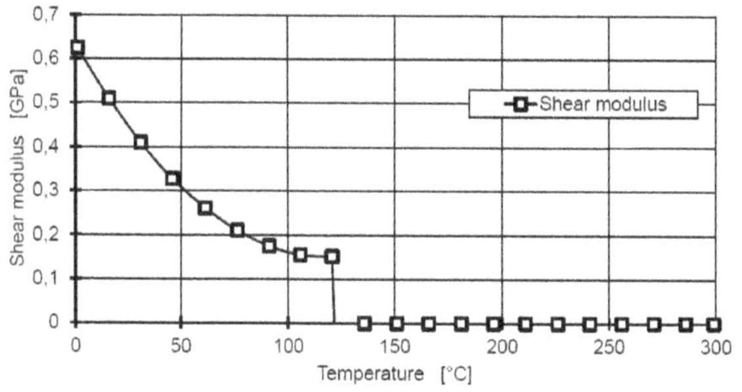

Figure 6: Determination of the temperature dependent shear modulus.

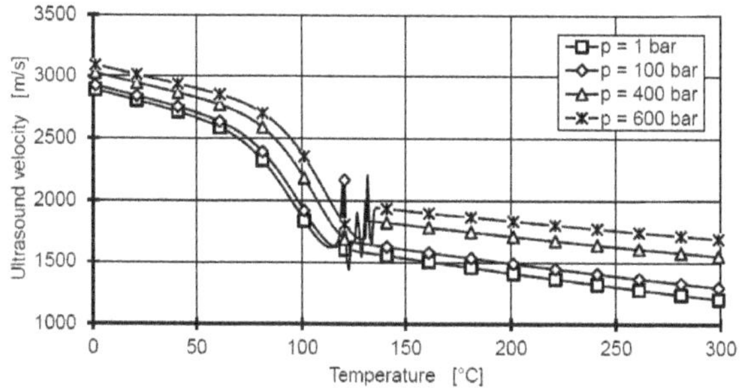

Figure 7: Calculated ultrasound velocity in dependence of the temperature.

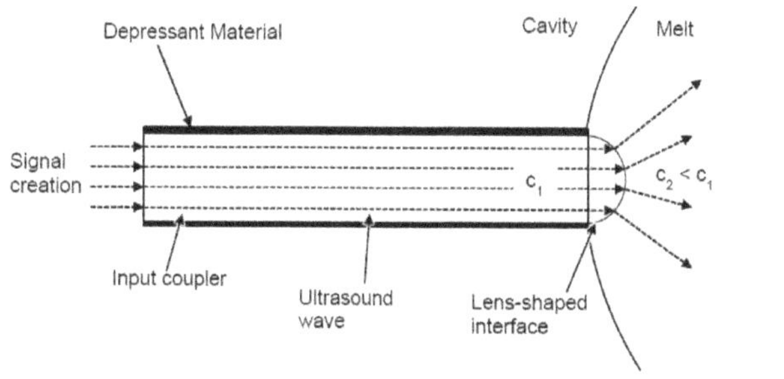

Figure 8: Concept of the buffer rods and the behaviour of ultrasound at curved boundary surfaces.

With experimentally determined Young's modulus and the assumption that the shear modulus is neglectable in the melt the shear modulus can be determined (Figure 6).

From the equations (12) - (14) the ultrasound velocity in dependence of the temperature can be calculated (Figure 7). The discontinuity near the melt temperature arises from the pvT behavior of semi crystalline thermoplastics, since the transition from solid to melt state cannot be described by linear relations.

The ultrasound velocity depends on the pressure and temperature conditions prevailing in the melt. In this context, the pressure is always assumed constant in the measurement region so that only local variations in temperature result in local changes in the ultrasound velocity. This results from the intended application to integrate the ultrasound-tomographic measurement concept into an injection mould. Here, no pressure gradient within a cross section of the flow channel is assumed, which prevails perpendicular to the flow direction [16]. Of course, the pressure in the measurement area has to be known and thus has to be monitored during the injection moulding process.

Properties of ultrasound transducers and constructional challenges

At this point, the theory of ultrasound tomography is fully described. However, for practical realisation analytics on the real behaviour of ultrasound transducers has to be done. In the assumption made above, a fan shaped propagation of the ultrasound wave was assumed. Depending on the sensor geometry the real propagation profile of an emitted ultrasound wave can be divided into to two parts: The near field or Fresnel-zone and the far-field or Fraunhofer-zone [17]. The propagation behaviour in the Fresnel-zone is simple straight, while in the Fraunhofer-zone the ultrasound wave begins slightly to diverge. This behaviour is adverse for the usage in tomography. However, this problem can be solved, using lens-shaped interfaces. If an ultrasonic beam applies to an interface of two media, the beam is reflected, scattered and refracted, but if the surface is curved, the interface behaves similar to a lens in geometrical optics.

Mathematically, a modified lens equation describes the behaviour of ultrasound waves at a curved interface [18]:

$$\frac{1}{b}+\frac{c_1/c_1}{g}=\frac{1}{f} \tag{15}$$

In equation (15), b is the image distance, c_1 and c_2 are the ultrasound velocity of two media, the distance of the transmitter to the apex of the curvature and f describes the focal point.

The ultrasound velocities of the two media and the direction of the curvature play a determining role. Depending on the ratio of the two ultrasound velocities the ultrasound wave either is focused or diverges in terms of a particular direction of the curved interface. This behaviour changes if the direction of the curvature is changed. For the use of ultrasound tomography in the injection moulding process, one can expect that the ultrasound velocity of the melt is much smaller than the ultrasound velocity of the mould, which is made of steel. From this follows, that the curvature must have a convex form (Figure 8). For this reason, additional buffer rods have to be used that direct the ultrasound wave through the mould in the melt (Figure 8). To avoid any circulations of the ultrasound wave within the mould these input couplers must be coated with a depressant material. One end of the buffer rod, which directs to the melt, has a convex form; the other end is a planar area, which is used to mount the ultrasound transducer.

The theory of scattering the ultrasound wave at curved interfaces confirms the results of Praher et al. who tried different geometries to scatter an ultrasound wave and identified the lens-shaped geometry as the best working one [8]. The usage of input couplers is beneficial since they shield the temperature sensitive ultrasound transducers.

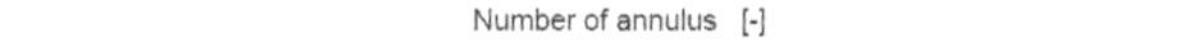

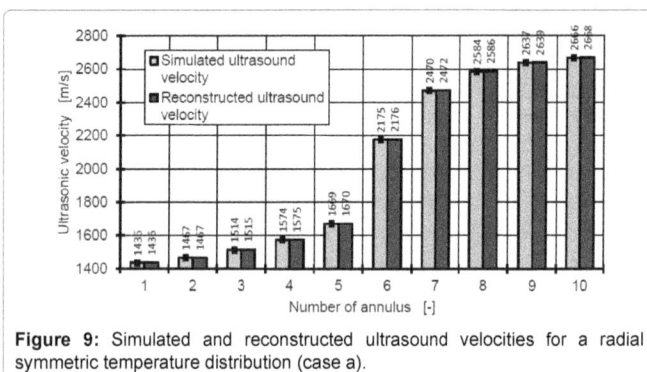

Figure 9: Simulated and reconstructed ultrasound velocities for a radial symmetric temperature distribution (case a).

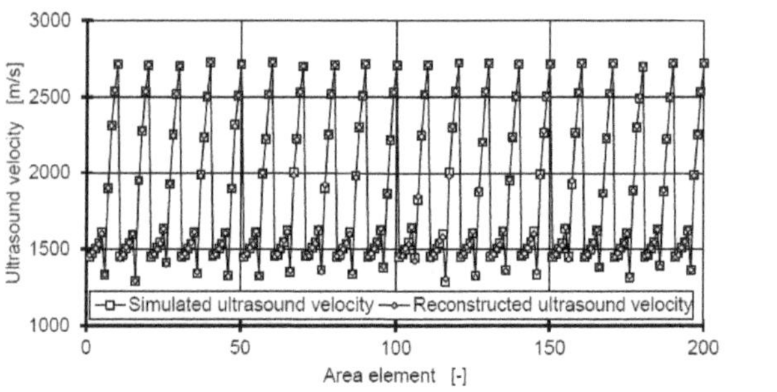

Figure 10: Simulated and reconstructed ultrasound velocities for a random radial symmetric temperature distribution (case b).

Results and Discussion

Virtual test problems

For later reconstructions of the temperature distribution, it is important to evaluate the accuracy of the algorithm first. Less importance is attributed to the calculation speed of the algorithm, since the measurement data need to be available completely.

To carry out the system, first the matrix A is calculated. For this purpose, a radius R = 0,015 m and a number of sensors N = 8 is assumed. Then equation (3) is used to calculate the time-of-flight data that represent the measured values. The number of iterations is set to k = 100. From the resulting set of solutions the one having the least squares with respect to the predetermined values is selected.

Two virtual test problems are generated: (a) radial symmetric ultrasound velocity profile (Figure 9), a random radial symmetric ultrasound velocity profile, with a temperature gradient of 5°C towards the centre of the measurement region (Figure 10). In addition, the temperature randomly varies between the upper and the lower limit of each circle. The numeration of the area elements is illustrated in Figure 3.

In both cases, the ultrasound velocities can be reconstructed very accurately. The standard deviation is 1 m/s (case a) and 8 m/s (case b). From this follows that the used models for ultrasound tomography depict a reliable method to accurately determine the temperature distribution during injection moulding. However, the used model underlay restriction in the near of the melt temperature due to non-linear behaviour of the plastics. For a clear correlation of ultrasound velocity with the temperature, time-of-flight measurements should be performed under defined laboratory conditions. For this, a measurement device is required that allows measurements without any cooling effects. Furthermore, for calculating the temperature additional information of the material, e.g. pvT-relations, must be known in detail, since process parameters like cooling rate influence this behaviour [19]. In addition, the models, especially the designed system matrix and thus the propagation behaviour of the ultrasound wave, underlay no disturbing effects in the virtual test problems. In view of the real process conditions, one must expect that such effects will appear in time-of-flight measurements, such as refractions, scattering and reflection. In this case the system matrix can be extended by beam propagation methods (BPM) or ray-tracing that can be handled with the SART-Algorithm [20].

Figure 11: Image of mould shows the measurement region inside the cavity (left) and the radially arranged ultrasound transducer (right).

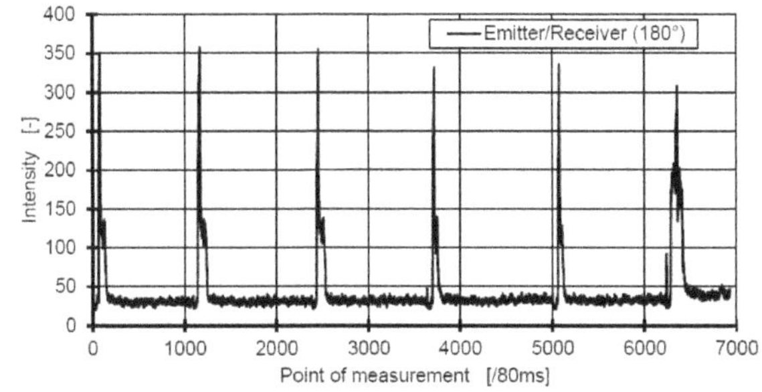

Figure 12: Amplitude-scan during six injection moulding cycles of a receiver opposite to an emitter.

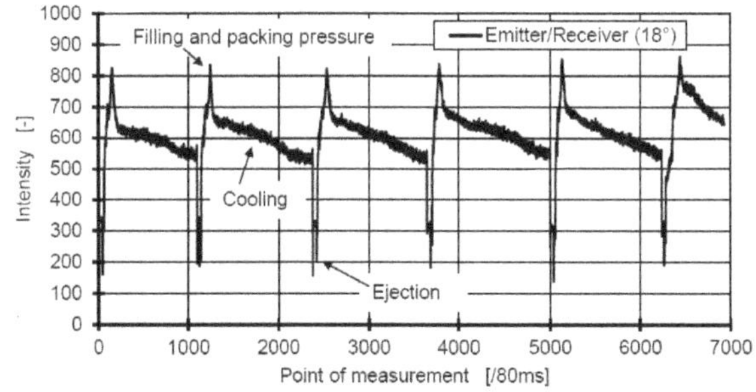

Figure 13: Amplitude-scan during six injection moulding cycles of a receiver next to a receiver.

Pilot experiments

For the non-invasive measurement of the temperature distribution a mould was built containing with 20 ultrasound transducers radially arranged around a cylindrical shaped cavity (r = 15 mm, h = 25 mm). Buffer rods shield the ultrasound transducers from the high temperature and their lens-shaped tip scatter the signal within the melt. To provide a tomographic measurement the buffer rods are pushed minimally into the melt by a mechanical mechanism during the measurement and are pulled back before the moulded part is ejected. To avoid circulations of the ultrasound wave within the mould the buffer rods are coated with PEEK. To provide a homogenous pressure within the cross section the mould was designed as compression mould so that a plane pressure can be performed on the moulded part (Figure 11).

To measure the temperature distribution during the injection moulding process a fast detection of the signals is required, since high cooling rates lead to a fast solidification of the moulded part. Furthermore, shrinkage during the solidification leads to separating of the moulded part from the buffer rods and prevent injection of the ultrasound signal into the melt. For this reason, a customised measurement device was designed in cooperation with MACEAS GmbH, Bexbach/Germany that provides a parallelised control per three transducers. Therewith, a complete measuring cycle (20 × 19 measurements) can be realised within 80 ms with an accuracy of 10 ns and a standard deviation of 3 ns. The deployed ultrasound transducers are FABP-MSWQC-ALP 2.25 × 25 (GE Sensing and Inspecting Technologies GmbH, Hürth/Germany) that are excited with a frequency of 2 MHz. For a suitable transmission of the ultrasound signal from the transducer into the buffer rod, petrolatum is used as coupling medium.

First experiments are carried out with the injection moulding machine KraussMaffei 160-1000 CX (KraussMaffei Technologies GmbH, Munich/Germany). Because of the temperature sensitivity of the ultrasound transducer of around 60°C the mould temperature was set up to 50°C, which is adverse for the measurement. The high difference between melt temperature (220°C) and mould temperature leads to a fast solidification and thus shrinkage and warpage. During the process the amplitudes at each receiver were detected. Figures 12 and 13 show the resulting amplitudes of six injection-moulding cycles for an opposite receiver and a next-neighbour receiver, respectively. The signal heavily increases after the filling process and during the packing pressure phase. Especially in Figure 13, the decrease of the signal intensity during the cooling phase can be seen. Since the shrinkage of the moulded part between two neighbouring transducers is affected by the sticking out buffer rods, the connectivity between the moulded part and the transducers is guaranteed. This information is lost in in the measurement shown in Figure 12 due to shrinkage of the moulded part, so that no signal can be detected. In Figure 13 the ejection of the moulded part can be seen, as the amplitude drops to around 200. The relative high noise from neighbouring receiver can be explained by ultrasound waves that traverse through the mould. The measurements are highly reproducible with every transducer combination over different injection moulding cycles. In general, this shows the feasibility of ultrasound measurements during the injection moulding process. Even next-neighbour transducer are able to detect noticeable signals, which is beneficial with respect to the spatial resolution, since the number of signals that are detectable produce the maximum spatial resolution of around 3.5 mm^2 with respect to a diameter of the cavity of 30 mm. Due to internal signal processing of the measurement device, it is not yet possible to calculate arrival times of the ultrasound signal. An explanation for this is the complex path of buffer rod, solidifying melt and again buffer rod. This causes a broadening of the ultrasound pulse that is not detectable by the internal software, since calibration of the software was not performed on solidified and molten PP and a much higher signal to noise ratio was expected. New calibration is on-going with the data from the pilot experiment. In future, also signal averaging will be possible to reduce noise.

Conclusion and Outlook

In this work, we present the principle and feasibility of the ultrasound tomography for the spatially resolved temperature measurements of a moulded part during the injection moulding process. Measuring the arrival times of an ultrasound wave emitted by a transducer from different direction allow to reconstruct the distribution of ultrasound velocity in a cross section of a cavity. From this distribution, the temperature field can be calculated using additional material characteristic properties, which can be either determined through simulation or through experiments. Based on the concept of ultrasound tomography, an injection mould was designed, that allows time-of-flight measurements with 20 ultrasound transducers radially arranged around a cylindrical shaped cavity. Lens-shaped buffer rods enable the possibility the scatter the ultrasound wave within the melt, to provide the necessary information density that is required for the temperature calculation. Furthermore, the buffer rods shield the temperature-sensitive ultrasound transducer from the high temperatures of the melt. In this work, we show that the maximum possible resolution of 3.5 mm^2 with respect to the described measurement area could be reached, since it was possible to detect noticeable signals at each sensor position. A highly parallelised measurement device allows the acquisition of 380 arrival times within 80 ms. For this reason, several complete datasets can be recorded before, due to shrinkage of the moulded part, no more signals can be detected. Internal signal processing of the measurement device actually prevents to calculate arrival times of the ultrasound signal but amplitude-scans show the general feasibility of ultrasound tomography during injection moulding.

The used model to calculate the temperature from the ultrasound velocity requires several additional material specific data, which should be determined over a wide temperature and pressure range. It could be shown that, using standard tests and models, the used model leads to high inaccuracies, especially in the area of the melt temperature, because of the non-linear behaviour of the material. Due to the high accuracy of the ultrasound measurement device the highest error for the determination of the temperature is thus produced by the used model. Therefore, it is suitable to correlate the ultrasound velocity with the temperature directly under defined temperature and pressure states. This requires the construction of a special measurement device, which prevents cooling of the material. These data would also be less sensitive to the high cooling rates that occur during injection moulding.

Acknowledgements

All presented investigations were conducted in the context of the Collaborative Research Centre SFB1120 “Precision Melt Engineering” at RWTH Aachen University and funded by the German Research Foundation (DFG). For the sponsorship and the support, we wish to express our sincere gratitude.

References

1. Ospald F (2014) Numerical simulation of injection molding using Open FOAM. Proceedings in applied Mathematics and Mechanics 14: 673-674.
2. Shelesh-Nezhad K, Siores E (1997) An intelligent system for plastic injection molding process design. J of Materials Processing Technology 63: 458-462
3. Zhou H, Li D (2005) Integrated simulation of the injection molding process with stereolithography molds. Int. J Adv Manuf Technol 28: 53-60.
4. Bendada K, Cole K, Lamontagne M, Simard Y (2003) A hollow waveguide infrared thermometer for polymer temperature measurement during injection moulding. J Opt A Pure Apllied Opt 5: 464-470.
5. Bur AJ, Roth SC, Spalding MA, Baugh DW, Koppi KA, et al. (2004) Temperature gradients in the channels of a single-screw extruder. Polym Eng Sci 44: 2148-2157.
6. Migler KB, Bur AJ (1998) Fluorescence based measurement of temperture profiles during polymer Processing. Polym. Eng Sci 38: 213-221.
7. Tzu-Fang C, Ky TN, Szu-Sheng LW, Cheng-Kuei J (1999) Temperature measurement of polymer extrusion by ultrasonic techniques. Meas Sci Technol 10: 139.
8. Praher B, Straka K, Steinbichler G (2013) An ultrasound-based system for temperature distribution measurements in injection moulding: system design, simulations and off-line test measurements in water. Meas Sci Technol 24: 84004.
9. Halmen N, Kugler C, Hochrein T, Heidemeyer P, Bastian M (2017) Ultrasound tomography for inline monitoring of plastic melts. J Sensors Sens Syst 6: 9-18.
10. Yang Y, Yang W, Zhong H (2008) Temperature distribution measurement and control of extrusion process by tomography. IEEE International Workshop on Imaging Systems and Techniques.
11. Buzug T (2008) Computed tomography. Springer-Verlag, Heidelberg, Berlin.
12. Kak AC, Slaney M (2001) Principles of computerized tomographic imaging. Society for Industrial and Applied Mathematics, Philadelphia.
13. Hansen PC, Saxild-Hansen M (2012) AIR tools - A MATLAB package of algebraic iterative reconstruction methods. J Comput Appl Math 236: 2167-2178.
14. Oakley BA, Barber G, Worden T, Hanna D (2003) Ultrasonic parameters as a function of absolute hydro-static pressure. I. A Review of the Data for Organic Liquids. J. Phys. Chem. Ref. Data, 32: 1501-1533.
15. Drummer D, Ehrenstein GW, Hopmann C, Vetter K, Meister S, et al. (2012) Analysis and comparative assessment of different process technologies for manufacturing polymer micro-elements. Mater Sci Eng B 2: 347-362.
16. Hopmann C, Poppe E, Spekowius M, Wipperfürth J, Spina R, et al. (2016) Integrative Kunststofftechnik 2016. Shaker Verlag, Aachen.
17. Hendee WR, Ritenour ER (2002) Medical imaging physics. (4th edn), Wiley-Liss Inc., New York.
18. Krautkrämer J, Krautkrämer H (1961) Werkstoffprüfung mit Ultraschall. Springer-Verlag OHG, Berlin, Göttingen, Heidelberg.
19. Zuidema H, Peters GWM, Meijer HEH (2001) Influence of cooling rate on pVT-date of semi-crystalline polymers. J Appl Polym Sci 82: 1170-1186.
20. Andersen AH, Kak AC (1984) Simultaneous algebraic reconstruction technique (SART): A superior implementation of the art algorithm. Ultrason. Imaging 6: 81-94.

An Approach to Grid-based Fire Frequency Analysis for Design Accidental Loads in Offshore Installations

Seo JK[1]* and Bae SY[2]

[1]*The Ship and Offshore Research Institute, Pusan National University, Busan, Korea*
[2]*Electric and Electronic Research Division, Korea Marine Equipment Research Institute, Busan, Korea*

Abstract

This paper describes the approach for establishing the Design Accidental Load (DAL) fire based on a grid-based fire risk analysis. Representative cases are screened via an initial fire risk analysis where the leak frequencies, ignition probabilities and inventories are combined to determine the cases with the highest risk. The fire risk analysis is subsequently performed based on the consequence results and the fire frequencies. Although many initiates for risk assessment were taken, there are many limitation and uncertainties on frequency analysis. Especially, calculation of ignition probability for an accidental hydrocarbon release on an offshore platform is a complex issue. To overcome these limitations of historical accident data, time dependent ignition model is developed a model for probability of ignition of hydrocarbon gas leakages on offshore platforms on based of ignition model presented by some JIPs for offshore risk assessment and improved understanding of ignition mechanisms. In this paper, we reviewed the existing probabilistic risk assessment method, such as ignition models, fire and explosion models, and selected the ones most suitable for offshore conditions. Then applied grid-based fire frequency analysis in the risk assessment. Two main revisions were incorporated: a grid-based approach was adopted to enable better consequence/impact modelling and analysis of radiation, and an enhanced onsite ignition model was integrated in the consequence assessment process to obtain better results. This study will be useful for the fire frequency analysis on offshore platform topsides as one of procedures of quantitative risk assessment.

Keywords: Risk assessment; Fire frequency; Design accidental loads; Offshore installation; Grid-based method

Introduction

The past few decades have seen a wide range of major accidents with a number of fatalities, economic losses and damage to the environment. Examples of accidents in the offshore oil and gas industry [1] include the structural failure and loss of platform, the flooding, a blowout, the process leak leading to fires and explosions on Piper Alpha in the UK, the explosion and sinking of production semisubmersible, and the recent helicopter accident on offshore platform. Experience shows that operational failures and human errors are the most common initiating events for accidents offshore. While operational failures could be arrested by safety-instrumented systems (through monitoring and restriction to the desirable limits of safety integrity level. Recently DiMattia et al. [2] have developed a unique human error probability calculation index for offshore mustering. The operational failures can be mainly attributed due to design faults or improper inspection and maintenance. An offshore development can never be completely safe, but the degree of inherent safety [3] can be increased by selecting the optimum design in terms of the installation configuration, layout and operation. This is done in an attempt to reduce the risk to a level that is as low as reasonably practicable (ALARP) without resorting to costly protective systems. This requires the identification and assessment of major risk contributors, which can be accomplished using quantitative risk assessment (QRA) techniques early in the project's life. If a structured approach of identification and assessment is not carried out early in the project, it is possible that the engineering judgement approach will fail to identify all of the major risks, and that loss prevention expenditures will be targeted in areas where there is little benefit. This may result in expensive remedial actions later in the life of the project [4].

Leakage or spillage of flammable material can lead to a fire that is triggered by any number of potential ignition sources (sparks, open flames, and so on). Depending on the types of release scenarios in the offshore environment, fires are mainly classified into four types: pool fires, jet fires, fireballs and flash fires. Although there are additional ones such as flares, fires on the sea surface and running liquid fires, they are in one way or the other modelled as one of the four defined types For example, flares can be treated as jet fires, and fires on the water surface and running liquid fires can be treated as pool fires. A review of the fire models and analysis of their characteristics has been carried out in our earlier work [5,6]. Jet fires usually occur due to immediate ignition of continuous high pressure releases. They represent a significant element of risk associated with major incidents on offshore installations, with the fuels ranging from light flammable gases to two-phase crude oil releases. Between horizontal and vertical jet fires, the former is more dangerous because of the high probability of impingement on objects downwind. This can lead to structural, storage vessel, and pipe-work failures, and can cause further escalation of the event. These are considered to be the most dangerous among all the fires, and hence need considerable attention.

The main aim of frequency and consequence analysis is to identify the personnel, equipment, plant and structure exposed to the initial and escalating events, and to assess the likely effects and failures. Frequency of fires is expressed in terms of leak probability and ignition probability. Also, the consequences of fires are usually expressed in terms of thermal radiation intensity, smoke concentration. The

***Corresponding author:** Seo JK, The Ship and Offshore Research Institute, Pusan National University, Busan, Korea, E-mail: seojk@pusan.ac.kr

analysis of consequences resulting from a small process leak leading to major fires and explosions is shown in Figure 1. For an unwanted release event, the first step involved in analysing the consequences is to select an appropriate ignition, leak sources model based on the type and phase of release. The second step is to select a dispersion model to estimate the dimensions and concentration of the gas cloud. The third step is to select an ignition model to estimate the probability of ignition and the frequency of fire. The final step is to estimate the heat dose, and flame length from fires, and to evaluate their impacts (Figure 1).

In this paper, we reviewed the existing consequence models, such as source models, fire frequency models, ignition models for fire risk assessment, and selected the ones most suitable for offshore conditions. These models were then used to perform a consequence assessment for an offshore platform by simulating four different scenarios. Two main revisions were incorporated: a grid-based approach was adopted to enable better consequence/impact modelling and analysis of radiation, and an enhanced onsite ignition model was integrated in the consequence assessment process to obtain better results compare with current industrial practices.

The Proposed Approach

The present method is a combination of the conventional method and heuristic approach. The method consists of the following steps.

The probabilistic grid based assessment procedure for estimating jet fire loads is divided into two main sections as illustrated in Figure 2. First step describes the grid based approach for establishing the Design accidental Load (DAL) jet fire. Second step is determination of jet fire load. This step describes the approach for establishing the Design Accidental Load (DAL) jet fire based on a fire risk analysis. Representative cases are screened via an initial fire risk analysis where the leak frequencies, probability of immediate ignition and grids information are combined to determine the cases with the highest risk. These representative cases are further considered in the consequence assessment involving detailed fire load modelling with CFD and/or analytical solutions. The jet fire risk analysis is subsequently performed based on the consequence results and the fire frequencies. The output from the jet fire risk analysis is the jet fire exceedance plot which is used to assess the DAL jet fire scenario based on modelling of girds (Figure 2).

Grid-based modelling

A grid-based approach is one of the most straightforward ways to numerically compute a discrete probability distribution that approximate the continuous probability density function. This approach has found increasing popularity in mathematical, numerical and practical methodologies for complex systems due to their simplification of gird or segments divided from complex system. For example, the FPSO topside, an offshore installation is very compact with high degree of congestion and confinement due to space limitations and environmental conditions. In order to get the exceedance curve and the DAL jet fire of offshore installation, it should be determined by analysing all probable fires in the module, and obtaining the degree of consequence and the frequency for each jet fire. Generally, the dynamic fire loads can be calculated using CFD modelling and include all probable fire scenarios for target module or system [7-10]. However, the number of analyses required to do so will become too large. In practice, a selection of critical fires from the initial fire risk assessment can be applied.

In this study, the grid-based approach can be employed to enable

Figure 1: Fire and explosion accidents in offshore installations.

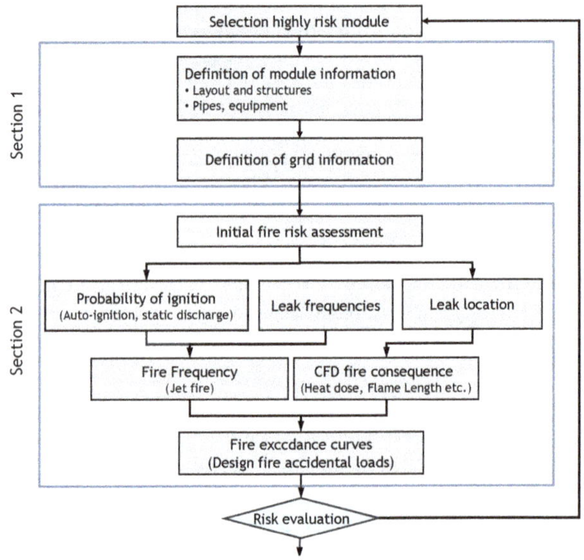

Figure 2: Proposed flow diagram of probabilistic fire accidental load procedure based on grid information.

better modelling of probability of ignition, leak frequencies, and analysis of fire consequence (flame length, heat dose etc.) at grids in the FPSO topside module. The grid modelling is need to details of number of pipes, equipments and layout of module information etc. of described the targe module in order to definition of grid information.

This approach is a grid of charge arithmetical values defined on an orthogonal lattice for calculation of fire frequency and fire consequence in Figure 3. The charge arithmetical values of grid can be obtained from numerical simulation, accidental data and probabilistic analysis based on grid modelling. The shape of grid in grid modelling appropriates orthogonal lattice then non-orthogonal lattices in order to easier modelling applied and combined to whole module information. If it is difficult to make an orthogonal lattice, the generalization to non-orthogonal lattices should be implemented considered with specific boundary conditions, significant complication, and detail of geometric properties (Figure 3).

Additionally, the analysis of radiation heat and the blast waves is uncertain if the presence of obstacles (partial barriers, e.g. process equipment and solid barriers, e.g. passive fire protection walls, explosion proof walls) is not taken into account. This issue does not arise when using CFD models, as application of appropriate boundary conditions will solve the problem. However, it appears to be difficult to resolve this issue while using semi-empirical models unless a grid-based approach is used [6]. The capability of this approach is apparent from Figure 4, which show the effect of partial barriers and solid barriers on the heat radiation from a jet fire. It is clear that the solid barrier totally blocks the radiation while the partial barrier reduces the effect to some extent. Also, the damage contours can be obtained permit the development of a clear picture of potential impact zones. This can facilitate proper selection and specification of safe separation distances to prevent injury to people and damage to nearby equipment. The grid heat load and overpressure load obtained from the analysis can also facilitate the design of protective layers. In addition, a grid-based approach seems to be most valuable for modelling of dispersing or expanding clouds (i.e., in dynamic simulation), and in defining risk (Figure 4).

A case is defined as a leak in a sub-segment where the sub-segment is the part of a segment which is located in one area. As a segment can be located in more than one area, it is often necessary to divide different segments into sub-segments. In this particular case, an area is defined as the area where a leak can have a direct influence. The area is limited by solid walls and decks and borders to free air. For larger areas e.g. a large area with smaller modules within the area, it is necessary to divide the area into sub areas. It is a good practice to make a schematic diagram showing the segments and the areas of the platform. This diagram can be made by combining the General Arrangement drawings and the Process and Instrument/Process Flow drawings of the main hydrocarbon carrying process equipment. When all segments and sub-segments are identified and located in the right area, a case list can be made. Often, a large segment has a small sub-segment in one area e.g. a riser segment which ends with an ESD valve in the import/export area [10]. This small sub-segment will have a large inventory and a small leak frequency, hence the risk can be significant and it is important to include it in the analysis.

Intial fire risk assessment

Fire frequencies

Leak frequencies: Leak frequencies and ignition probabilities are normally calculated as a part of a QRA and details are described in guidelines and standards such as references. It should be noted that leak frequencies need to be detailed for each sub-segment.

The inventory is defined as the mass of hydrocarbon (HC) in the segment (kg) and needs to be calculated for each segment. The volume applied is the volume inside the process equipment, bounded by the closed ESD valves. The density is calculated at the operating pressure. In a segment with different pressures, the settle-out pressure should be applied. Risk ranking of cases, when all cases are defined, the cases with the highest combination of fire frequency and inventory have the highest potential risk. These are the cases which are applied as representative for the detailed consequence analysis. The parts of the fire case regarding location and leak rates and wind conditions are considered in the next section. It is noted that the simplifications arising from modelling only a few cases with simulation tools and using those cases as a basis for representing all fires is justified partly because the analysis method applied leak rates and not whole sizes to represent the cases. Using this approach, a case with a low pressure and a large hole will be defined in a similar way to a case with high pressure and smaller leak hole.

Probabilities of ignition: Source models or release models [11-13] are used to estimate the amount of fuel released, or the rate of release of fuel. These models play a crucial role in the risk assessment process as the release rate and quantity of fuel released determine the size of the resulting cloud and hence the probability of ignition. Furthermore, these models are also used to find the initial sizes of fires and explosions. The initial release rate through a leak depends mainly on the pressure inside the equipment, the size of the hole and the phase of release. Offshore hydrocarbon releases are usually gaseous, liquid and two-phase. Among these, the gases can be hydrocarbons ranging from C1 to C4, while liquids can be crude oil, diesel oil, aviation fuel, and others. Condensate is considered to be two-phase as it is a mixture of hydrocarbons (mainly C4 to C6) that condenses from the gas during compression. This material is liquid while it is held under pressure but becomes gas if the pressure is released. Identification of the appropriate phase and its corresponding model is essential, as, this being the initial step for risk assessment; it may prove to be highly sensitive to the risk estimated.

The probability of immediate ignition depends upon both the potential for auto-ignition (Pai) and the potential for static discharge (Psd). In this model, the former is related to the release temperature (T) relative to the auto-ignition temperature (AIT), and the latter to the minimum ignition energy (MIE) and "release energy" for the material being released. The "release energy" may be considered a function of the process pressure, or release rate, or yet some other surrogate parameter that expresses the often observed result that the more "energy" behind a release, the more likely it is to ignite. Expert Opinion e Subject matter experts were solicited to predict the expected probability of immediate ignition for a variety of hypothetical release scenarios, based on their experience with similar events in process plants. This resulted in an algorithm which was based on MIE, T and AIT. The available static "energy" was the fourth parameter, and was independently developed as a function of the square root of the process pressure. Another work in this area [14] by three recognized experts proposes that the immediate ignition probability is roughly

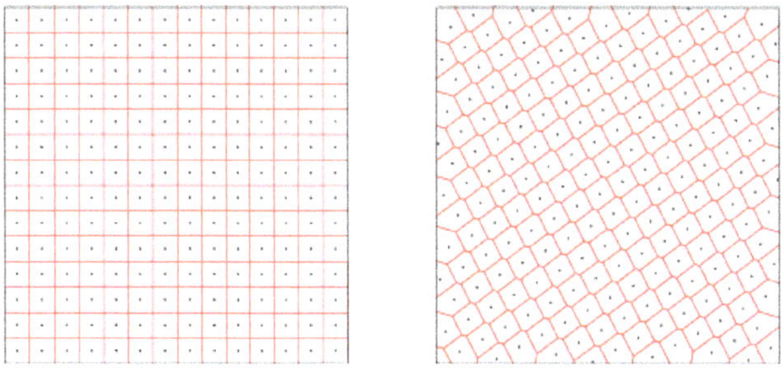

(a) Orthogonal Lattices (b) Non-orthogonal lattices

Figure 3: The typical gird modelling as a two-dimensional matrix form.

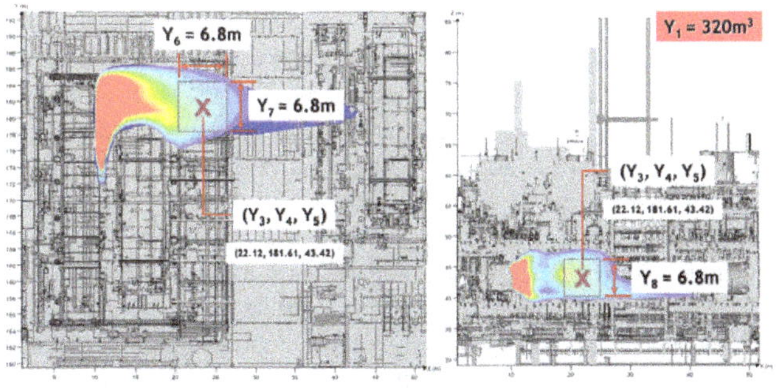

Plan view Elevation view

Figure 4: Jet fire heat radiation in the presence of solid and partial barriers.

proportional to the cube root of the flow rate of the release. In contrast, the square-root-of-pressure relationship proposed independently is equivalent to indicating that this probability is linearly proportional to the release rate. Given this conflict, a compromise is proposed in which the probability of immediate ignition is proportional to the cube root of the source pressure.

The approaches above were combined, resulting in the following relationship [13]:

$$P_{imm.ign} = Pai + Psd \tag{1}$$

$$P_{imm.ign} = (1-5000e^{-9.5(T/AIT)}) + (0.0024 \times \frac{P^{1/3}}{MIE^{2/3}}) \tag{2}$$

Where AIT and T are in degrees AIT and T are in degrees Fahrenheit, P is in psig, and MIE is in mJ. The following constraints are place on this equation: a minimum value of 0 is allowed for T, For T/AIT < 0.9, Pai = 0, For T/AIT > 1.2, Pai = 1, and $P_{imm.ign}$ cannot be greater than 1.

In this method, it is assumed that there is always some probability of non-immediate ignition if T is no more than 200° higher than the AIT (this 200° factor being necessary because AIT values in the literature vary widely for many chemicals, depending on the investigator/apparatus used to measure it). If this condition is met, then $P_{imm.ign}$ is prevented from having a value greater than 0.98. Table 1 lists representative MIE and AIT values for four common chemicals. Values for many other chemicals are provided in Bond [15-17].

Figure 5 shows ignition source parameters for determining the probability of ignition at each grids by Equations 1 and 2. The ignition probability will be calculated using potential probability of the counted equipment (Figure 5).

Leak rate: Leak frequency per offshore platform can be identified on the basis of historical data and presented as a function of the leak amount, where this amount is the product of leak rate and leak duration. Currently, actual data are available for a ten-year period, 1996–2005, for all installations on the Norwegian Continental Shelf. The quality of the data is good for the 2001–2005 periods, and this should be the main data source. Figure 2 shows the trend of hydrocarbon leaks with a flow rate above 0.1 kg/s, normalized against installation years, for fixed production installations. These data are taken from the PSA annual report for 2008 (Figure 2) [18].

In addition, the Health and Safety Executive (HSE) has compiled a database of floating production, storage and offloading unit (FPSO) leak frequencies in association with leak amounts (amounts of fuel released) and the number of FPSOs. This report briefly examines release trends over the data range of interest and provides the total number of releases that occurred each year between 1980 and 2003 [19].

The definitions of the three groups are categorised based on the amount of fuel released are as follows into major, significant, and minor categories. Major releases have a rapid and significant impact on the local area, and typically affect temporary refuges, block escape routes, escalate to other areas of the installation, and cause serious injuries or fatalities. Significant releases cause serious injuries or fatalities among personnel within the local area and escalate within that local area, such as by causing structural damage, secondary leaks or damage to safety systems. Minor releases cause serious injury to personnel in the immediate vicinity, but are very unlikely to escalate or cause multiple fatalities.

Fire location: The number of fire locations depends on the size of the process area. For an average sized process area, two or three locations should be appropriate. Typically, one location near the centre, and others near the edges should be selected. In addition, the leak location needs to be placed where there are pipes and equipment belonging to the representative cases.

Leak positions comprise one of the most important parameters of gas detector placement. On production platforms with complex process trains, leaks from process equipment are inevitably relatively frequent events. In most cases, they are small and readily controlled. Occasionally, they develop into explosions and fires. Very infrequently, they may escalate into major accidents such as the one that occurred on the Piper Alpha oil platform. Leak positions are difficult to consider [20].

For convenience, it can be assumed that the range of leak positions in x-, y and z-directions is based on the confinement of the piping and vessels causing a leak inside the production platform. Also, leak directions can be considered with regard to the leak points.

Jet direction: The jet direction should be selected so that both long jets and impinging jets are simulated. The long jet needs to be directed in the horizontal direction where it is not hitting large equipment and where it is still within the process area. Impinging jet directions are down/up into solid decks, or horizontally hitting a firewall or a large piece of equipment. The impinging, diffusive gas fire could be represented by a low velocity release, although the most realistic diffusive fire is obtained with a high speed release impinging on a solid structure. The number of jet directions can be reduced by not simulating jets where the jet fire will be directed out of the area. Typically 2 or 3 jet directions need to be simulated, one diffusive impinging jet, and two non-impinging jets which cause the majority of the flames to be located inside the area.

Wind conditions: Wind speed and wind direction have less influence on the flame location and the extreme heat loads than the previously mentioned parameters. The strong momentum and buoyancy in the

Material	MIE (mJ)	AIT (F)
Hydrogen	0.011 [15]	752-1085 [15] 752 [16]
Methane	0.28/0.3[15]	999-1103 [15] 999 [17] 1004 [16]
Propane	0.12/0.48[15]	842-912 [15] 842 [16, 17]

Table 1: Minimum ignition energies and auto-ignition temperatures for some common chemicals.

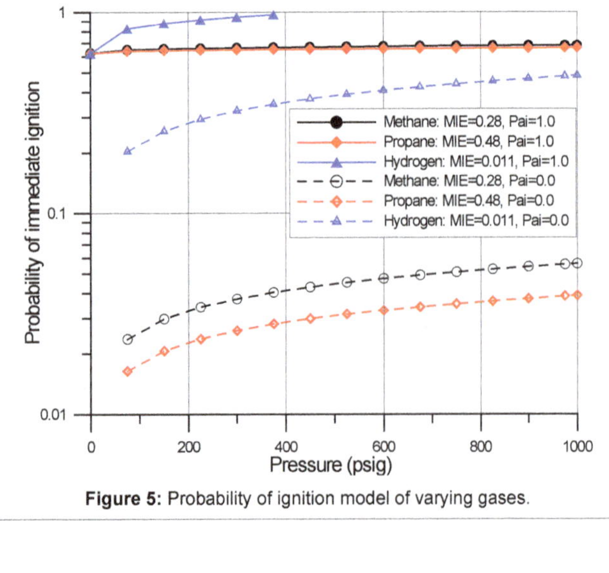

Figure 5: Probability of ignition model of varying gases.

upstream parts of a fire are dominating over the wind momentum; hence wind conditions have a lesser influence. For diffusive, large fires the flames will generally follow the wind direction through the process area; hence, the wind direction should be selected so that it blows the flames mostly into the process area. The wind direction should be selected together with the fire location and jet direction so that most of the fire burns inside the process area. Symmetry conditions can be utilised where appropriate and simulations can be performed on one side, applying the same results on the symmetric opposite side without simulating it. In cases where the wind speed and direction have a significant influence on the fire loads, the wind rose can be applied to assess the impact probability. Note that smoke movement is mostly driven by the wind and in cases where the fire simulations are used for smoke dispersion patterns, the wind conditions need also be considered.

Fire consequence modelling: Fire load modelling using a simulation tools can be performed after the initial fire risk analysis described above and is an input to the probabilistic analysis. Heat loads to be used in the analysis are obtained based on CFD and/or analytical models that are validated against research and experiments. The probabilistic procedure involves a finer distribution of heat flux as a function of leak rate and time compared to traditional methods. The procedure hence accounts for higher heat loads for larger fires as well as dynamic effects. The upper cut-off should reflect the maximum credible initial leak rates. Normally, this would reflect a pipe rupture scenario. Note that the full bore scenarios are in most cases of a very transient nature, and the initial leak rate should represent a typical, average leak rate which occurs in the period before ESD valves are closed. When the QRA gives leak frequencies in only 3 categories, the sub-division of the leak frequency in the smaller categories should be documented [7-9].

Fire exceedance curves: Establishing the fire exceedance plot involves the combination of consequences, probabilities and frequencies to determine the fire risk. The fire exceedance plot is the main result from the probabilistic fire analysis. The generation of the plot for a complex process area and dynamic fires is described. Results from the above sections are used. The fire load to be used on the x-axis of the plot is defined as the heat dose, Temperature received at a spot value. The fire exceedance plot is also used to pick there preventative dynamic fire scenario which is used in the structural response analysis. The requirement for the x-axis is that all fires with a lower x-value than the DAL fire value must be considered in the design. Fires with a higher x-value than the DAL fire need not be designed against because these have too low frequency. When the x-axis is defined correctly, it will be enough to design the mitigating measures using the DAL fire only, and it is not necessary to design against any fires with smaller severity.

Determination of design of fire load: This dynamic behaviour of the fire load is more important for pipes and equipment than for structures. This is because the pipe stress is strongly dependent on the pressure in the pipes and hence pipes are more vulnerable to high heat loads at the start of the fires before the pressure is relieved by blow-down. For pipes and equipment, one can use a set of DAL fires which has the same heat dose when performing the heat-up and response calculations. The set can be selected from the CFD simulations as realistic heat flux vs. time curves. Separate exceedance curves for different durations of the heat flux can also be made so that the separate frequency for the different durations can be found. The extent of the fire and the average heat flux in the area is also important for pipes and equipment as this determines the general heating the piping and equipment receives.

The main output form the first part of this probabilistic procedure is the exceedance curve and the DAL fire. This is determined by analysing all probable fires in the module, and obtaining the degree of severity (consequence) and the frequency for each fire. Ideally, the dynamic fire loads are calculated using CFD modelling and/or analytical solutions and include all probable fire scenarios. However, the number of analyses required to do so will become too large. In practice, a selection of critical fires from the initial fire risk assessment can be applied. Such critical fires are called cases in this procedure.

Application Examples

A simple generic case study is provided to illustrate the benefits of the developed approach. The topside of a hypothetical FPSO is considered as an applied example, and a three-dimensional view of the topside is depicted in Figure 6 [8]. The topside module is divided into a process area and a utility area. The former includes space for hydrocarbon-containing equipment, the flare tower, and compression and separation equipment, whereas the latter includes space for utility and power-generation equipment. In this application example, the one of process module on FPSO topside are selected for calculation of fire frequency analysis and consequence analysis. The process decks on FPSO topside are usually either grated decks or plated decks. The former are more convenient for draining rainwater, and they ventilate gas clouds in fires by minimizing the fire-induced temperature, whereas the latter are more useful in preventing the escalation of fire from the lower to the upper deck (Figure 6). In separation module under consideration here, the lowest process deck in plated and the middle and top decks are grated (Figure 7).

For grid-based computation, the process area under study is divided into a specific number of computational or/and manual grids, and the hazard potentials and consequences are then evaluated at each

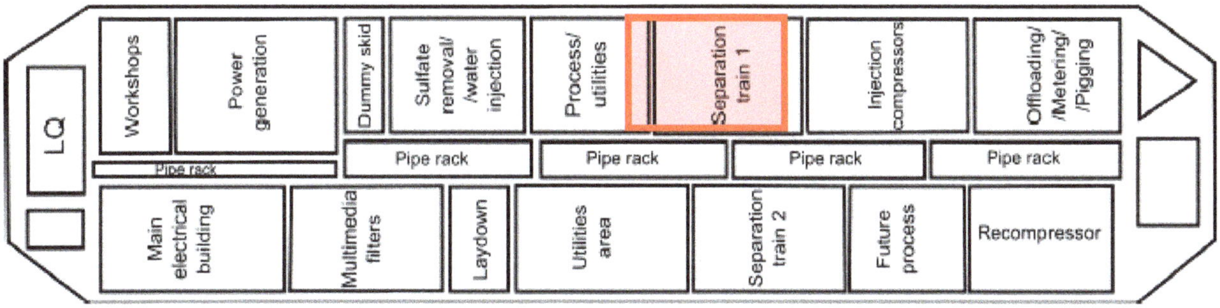

Figure 6: Layout of the object FPSO topsides.

end of the grid. This leads to a two-dimensional or three-dimensional matrix of hazard and consequence results which can be finally plotted as contours. Contour plotting is a more user-friendly representation than the ordinary line plots obtained by other software packages. It can be performed extensive trial simulation runs on process area to study the effect of number of grids.

To obtain the leak frequency should be identified and determined by a two-dimensional or three-dimensional space. In this study, length, breadth, and height are divided by uniformly spaces (Figure 8). Sensitivity of course grids space should be bigger affected then fine grids. Therefore, the computational time and the precision of results for plotting were found to be the most important as well as highly sensitive parameters on the application of a grid system.

Table 2 shows the definition of leak source parameters for determining the frequency of leak at each grid. In the case study, leak frequency for selected equipment is obtained by HSE database [19]. Then it can be counted number of equipment in each grid. Table 3 shows the definition of the location grid matrix used to identify the location of each grid for leak frequency analysis. The results of leak frequency show that Area C is the highest value due to numbers of pipes and values equipment located in Area C. The initial potential gird areas can be revised by increasing the grid numbers, it will be more accurate results for leak frequency.

Table 4 shows Ignition source parameters for determining the probability of ignition at each grids by equations 1 and 2. Probability

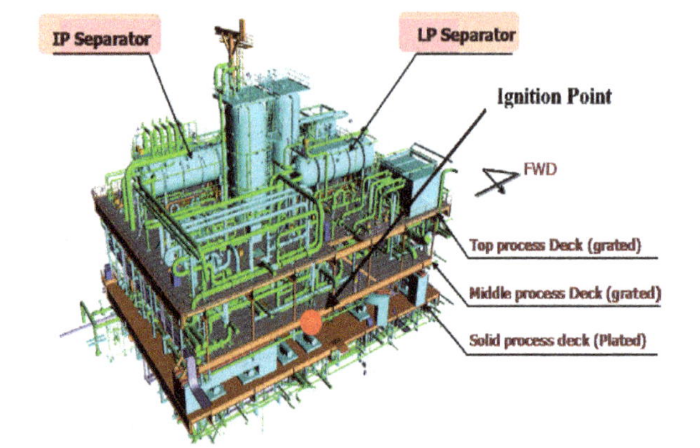

Figure 7: Isometric view of a typical separation module.

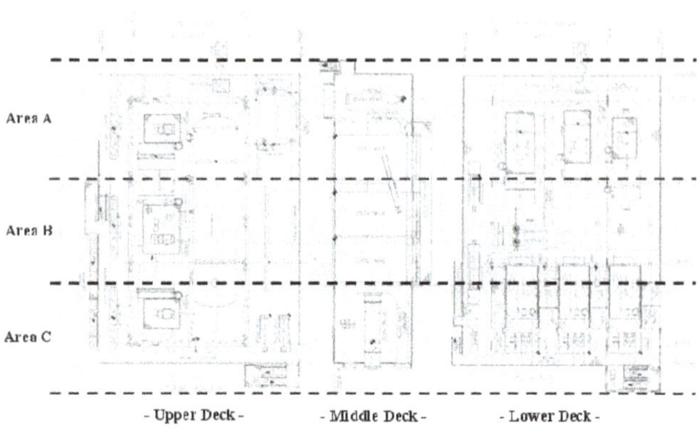

Figure 8: Plan view of separation module.

Equipment type	Index Name	Location	Frequency	Numbers
Separator (HP, LP, IP, Desalter)	SEP.	A,B,C	0.0026	4
Pump (Produced water booster, Desalter water, Crude oil transfer)	PUM	A,B,C	0.0073	9
Crude oil transfer cooler	COL	B,C	0.0053	1
Crude oil heater and exchanger	EXC.	C	0.0060	3
Hydro-cyclone package and Produced water gas flotation unit	UNI	A,B,C	0.0007	4
Piping	PIP	A,B,C	0.000068	5500
Flange	FLA	A,B,C	0.000055	5500
Valves	VAL	A,B	0.0001	5000

Table 2: Leak frequencies per equipment item and numbers of equipment.

Grid	Leak parameters and number								Leak Frequency	
	SEP	PUM	COL	EXC	UNI	PIP	FLA	VAL		
A × U	1	0	0	0	1	400	400	200	0.0718	0.1915
A × M	0	0	0	0	0	300	300	150	0.0519	
A × L	0.5	2	0	0	0	300	300	150	0.0678	
B × U	0	0	0	0	2	800	800	400	0.1398	0.2437
B × M	0	0	0	0	0	600	600	300	0.1038	
B × L	0.95	4	1	0	0	600	600	300	0.000065	
C × U	1	0	0	3	1	1,100	1,100	500	0.2066	0.4721
C × M	0	0	0	0	0	700	700	350	0.1211	
C × L	0.55	3	0	0	0	700	700	350	0.1444	

*notes: H.E. is Heavy equipment, M.E. is Medium equipment and L.E. is Light equipment.

Table 3: The calculation of leak frequencies for Area A, B, and C.

Grid number	Potential probability					Probability of ignition	Rank
	Auto-ignition			Static discharge			
	T(F)	AIT(F)	T/AIT	P(psig)	MIE(mJ)		
A × U	150.98	919	0.164	50.70	0.25	0.02238	9
A × M	150.98	919	0.164	50.70	0.25	0.02238	8
A × L	123.30	999	0.123	66.25	0.28	0.02269	7
B × U	123.30	999	0.123	50.70	0.28	0.02075	6
B × M	123.30	999	0.123	50.70	0.28	0.02075	5
B × L	123.30	999	0.123	127.9	0.28	0.02825	4
C × U	150.98	1,085	0.139	50.70	0.011	0.17959	3
C × M	150.98	1,085	0.139	50.70	0.011	0.17959	2
C × L	150.98	1,085	0.139	118.48	0.011	0.23832	1

*notes: H.E. is Heavy equipment, M.E. is Medium equipment and L.E. is Light equipment.

Table 4: Ignition source parameters for determining the probability of ignition at each grids.

of ignition was estimated using the developed ignition model when the cloud concentration is within the flammability limits. The ignition probability can be calculated using potential probability of the counted equipment. As would expected, Area C is also highly ranked compared with others.

To obtain the characteristics of the gas clouds in each fire frequency and consequence of jet fire, in this study uses and verifies the DNV PHAST [20] results when simplified release modelling is sufficient. DNV PHAST was used to provide time-dependent descriptions of the gas clouds for the gas leak scenarios considered in the study. The study considered only cloud dispersion in two dimensions. The data provided by DNV PHAST included the time-based history of the gas cloud length, width and size at 20%, 50% and 100% LFL.

Scenario 1: jet fire due to ignition of a gas cloud formed by an instantaneous release (IR) of 100 kg of methane gas over the process area; a continuous release (CR) of gas at 1.0, 1.5, and 3 kgs due to leak in a storage tank. The gas that is released (instantaneously or continuously) forms a cloud and disperses with initial momentum in low wind conditions. The dispersion of these clouds and jet fire were simulated using a DNV PHAST dispersion model and jet fire model to estimate the dimension of the cloud and jet fire. These results are presented in Figure 9 and Table 5 for all scenarios (Figure 9).

Flame length contours obtained as a result from the scenario show that most of the Area C is affected, whereas the contours from the Area A and B scenario show that relatively a very small part of the plant is affected. With this we can conclude that jet fire from current simple scenarios is the more dangerous event.

The example of the initial fire risk analysis is to get an overview of the fire cases and the potential risks involved. This is used to obtain a preliminary risk ranking of the cases which are used to select representative cases for further detailed consequence analysis.

Conclusion and Remarks

This paper describes the approach for establishing the Design Accidental Load (DAL) fire based on a grid-based fire risk analysis. Representative cases are screened via an initial fire risk analysis where the leak frequencies, ignition probabilities and inventories are combined to determine the cases with the highest risk. In this paper, we reviewed the existing probabilistic risk assessment method, such as ignition models, fire and explosion models, and selected the ones most suitable for offshore conditions. Then applied grid-based fire frequency analysis in the risk assessment. Two main revisions were incorporated: a grid-based approach was adopted to enable better consequence/impact modelling and analysis of radiation, and an enhanced onsite ignition model was integrated in the consequence assessment process to obtain better results. This study will be useful for the fire frequency analysis on offshore platform topsides as one of procedures of quantitative risk assessment.

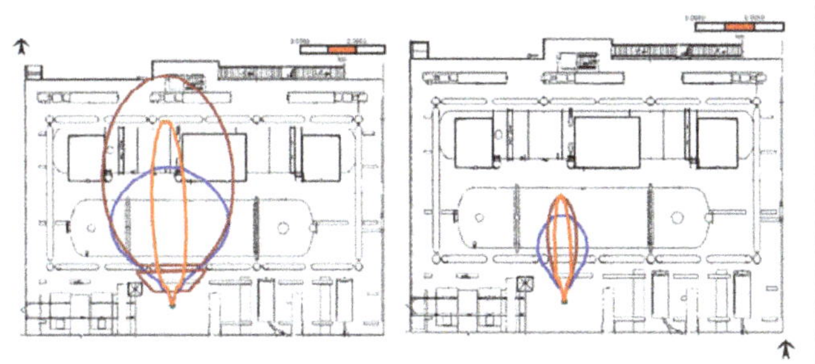

Figure 9: Plan view of gas clouds (left: Area C, right: Area A).

Grid number	Probability of ignition (/year)	Leak Frequency (/year)	Fire Frequency (/year), F	Flame Length (m), C	Risk =F × C	Rank
A × U	0.02238	0.0718	0.001607	7.2	0.0115704	6
A × M	0.02238	0.0519	0.001162	7.2	0.0083664	8
A × L	0.02269	0.0678	0.001538	7.2	0.0110736	7
B × U	0.02075	0.1398	0.002901	13.5	0.0391635	4
B × M	0.02075	0.1038	0.002154	13.5	0.029079	5
B × L	0.02825	0.0000646	1.82E-06	13.5	0.00002457	9
C × U	0.17959	0.2066	0.037103	19.8	0.7346394	1
C × M	0.17959	0.1211	0.021748	19.8	0.4306104	3
C × L	0.23832	0.14443	0.034421	19.8	0.6815358	2

*notes: H.E. is Heavy equipment, M.E. Medium equipment and L.E. is Light equipment.

Table 5: Fire frequency and fire consequence results.

Acknowledgment

This research was financially supported by the Ministry of Trade, Industry and Energy (MOTIE) and Korea Institutes for Advancement of Technology (KIAT) through the Promoting Regional Specialized (Grant no.: A010400243).

References

1. Spouge J (1999) A Guide to quantitative risk assessment for offshore installations The Center for Marine and Petroleum Technology (CMPT) UK.
2. DiMattia DG, Amyotte PR, Khan FI (2005) Determination of human error probabilities for offshore platform musters. Int conference on Bhopal and its effects on process safety IIT-Kanpur India 18: 488-501.
3. Khan FI and Abassi SA (1999) Assessment of risks posed by chemical industries - Application of a new computer automated tool MAXCRED-III. J of Loss Prevention in the Process Industries 12: 455-469.
4. Vinnem JE (1998) Evaluation of methodologies for QRA in offshore operations. Reliability engineering and system safety 61: 39.
5. Pula R, Khan FI, Veitch B, Amyotte PR (2004) Revised fire consequence models for offshore quantitative risk assessment. Int Conference on Bhopal and its effects on process safety IIT Kanpur India 18: 443-454.
6. Pula R, Khan FI, Veitch B, Amyotte PR (2006) A Grid based approach for fire and explosion consequence analysis. Safety and Environmental Protection 84: 79-91.
7. Paik JK, Czujko J (2009) Explosion and fire engineering and gas explosion of FPSOs (phase I) hydrocarbon releases on FPSOs review of HSE's accident database.
8. Paik JK, Czujko J (2010) Explosion and fire engineering of FPSOs (phase II) definition of fire and gas explosion design loads.
9. Paik JK, Czujko J (2011) Explosion and fire engineering of FPSOs (EFEF JIP), definition of design fire loads. Berkshire, UK: Newsletter of Fire and Blast Information Group (FABIG).
10. Paik JK, Czujko J, Kim BJ, Seo JK, Ryu HS et al. (2011) Quantitative assessment of hydrocarbon explosion and fire risks in offshore installations. Marine Structures 24:73-96.
11. CCPS (2000) Guidelines for chemical process quantitative risk analysis (American Institute of Chemical Engineers New York USA).
12. Crowl DA, Louvar JF (2002) Chemical process safety: fundamentals with applications. Prentice Hall Int series in the physical and chemical engineering sciences.
13. Mike M (2011) Development of algorithms for predicting ignition probabilities and explosion frequencies. Journal of Loss Prevention in the Process Industries 24: 259-265.
14. Cox AW, Lees FP, Ang M L (1990) Classification of hazardous locations. IChemE.
15. Bond J (1991) Sources of ignition Flammability characteristics of chemicals and products. Oxford Butterworth-Heinemann.
16. Zebetakis MG (1965) Flammability characteristics of combustible gases and vapors. US Bureau of Mines Bulletin.
17. (2002) CRC handbook of chemistry and physics (83rdedn) CRC Press.
18. PSA (2008) Risk level on the Norwegian continental shelf main report 2008. PSA Report.
19. HSE (2005) Accident statistics for floating offshore units on the UK continental shelf (1980-2003) HMSO RR 353 London Health and Safety Executive.
20. DNV PHAST (2015) User's manual. Norway DNV GL Software.

Mechanical Technique for Raising an Obelisk

Shiells JE*

US Government, 1401 Clearview rd, Edgewood MD, USA

Abstract

A simple technique is outlined which may answer the mystery of how the Egyptian obelisks were raised to their upright position. This technique relies on a mechanical ram which is characterized by equations that are derived in this article. The mechanical ram is basically a leverage device which would enable the ancients to apply a very large force with only a minimal input force. A paper analysis is developed to show how this technique may have raised the heavy 130,000 kg Thutmose obelisk upright, although the procedure would have been painstakingly slow. Note however, that no case is made here that the Egyptians actually used this technique.

Keywords: Mechanics; Obelisks; Archaeology

Introduction

A paper analysis is presented of a proposal to erect the moderate size Thutmose obelisk. This paper compares the turning moment to set Thutmose upright with the theoretical available force that could be obtained from the force multiplier method described. This technique depends on an unusual lever arrangement which will hereafter be referred to as a mechanical ram. The hydraulic ram could be considered as the modern functional equivalent of this mechanical ram. However, an important disadvantage of the mechanical ram is the minute working stroke as compared to a hydraulic ram. Therefore such a mechanical ram would have to be repositioned (reset) many times in order to ratchet the obelisk centimetre by centimetre to the final upright position.

First, some background material is presented which motivated this study including a brief description of the current thinking about this problem. Then the mechanical ram is described in detail, followed by an estimate of force required to raise Thutmose. Finally, some practical considerations are explored including the application of a finite element analysis Bathe [1] to establish the practically of the theoretical proposal.

Background

The PBS television program NOVA in "Secrets of Lost Empires: Obelisks" Barnes [2] made the public aware of the mystery of how the Egyptians raised the massive obelisks to the upright position. Levers could raise the obelisks to a certain height but eventually the raising height limits the availability of pivot points as well as the amount of force that can be applied to the end of the lever. Archeologists Englebach [3] were well aware of this problem many years ago. Later Chevrier [4] a method was proposed by which an earthen (sand) ramp was used. The idea was to shove the obelisk (backwards) up the gradual sloping end of the ramp and then have the base of the obelisk go over the steep end of the ramp under control of ropes until it settled on the pedestal rock. This method is sometimes referred to as the "sand funnel" method which is said to be unproven Arnold [5].

A very interesting discovery was made during the production of the PBS video Barnes [2]. When the crew went to the "obelisk graveyard" in Tanis, Egypt, they found pedestal stones with a turning groove along with associated abandoned obelisks. Figure 8 shows a drawing of such a stone supporting the edge of the obelisk base. This kind of stone would be absolutely necessary to prevent slipping or twisting if a method were known for pushing the obelisk upright from the horizontal position like the technique described here. The "sand funnel" technique envisions that the obelisk is set down at a high angle in which slipping would be unlikely, hence no need for a turning groove. In chapter 3 of his book, Wirsching [6] present a number of mechanical proposals for solving this erection problem. None of these methods are remotely similar to the technique outlined in this paper. Most of these methods depend on brute force with little attention to mechanical advantage except for one. This one is similar to the "sand pit" method which represents the latest thinking. This idea Cort [7] is that the obelisk is slowly lowered while turning into the upright position by allowing the supporting sand under the obelisk base to escape in a controlled manner. This technique however requires the obelisk to be contained in a giant sand box and is not universally accepted.

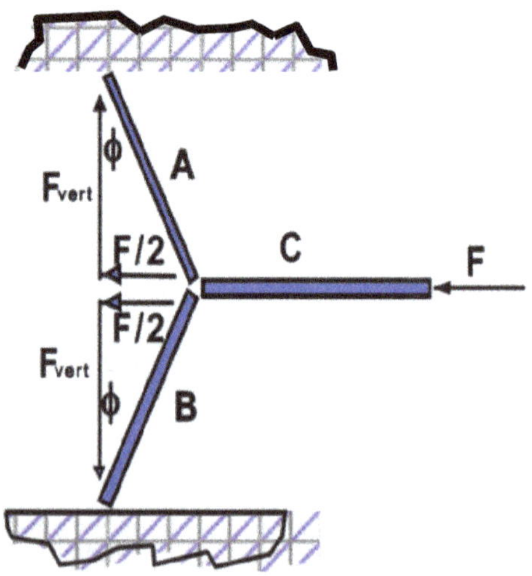

Figure 1: Force vector diagram for mechanical ram

***Corresponding author:** James E Shiells, US Government, 1401 Clearview rd, Edgewood MD, USA, E-mail: jespd109@gmail.com

Mechanical ram description

Figure 1 show the essential features of the mechanical ram along with the associated force vectors. Two rigid beams (A and B) are situated between two stationary surfaces about which they are free to rotate but not slide (pivot only). The beams are oriented at the angle (phi) ϕ off the vertical as shown. Where these beams meet is the point where beam C is used to apply the force F. This force F compresses both beams A and B equally. From an examination of the force diagram it follows that each of these beams must resist with a horizontal force component of magnitude F/2. With further analysis of the force diagram it also follows that the vertical force component is given by equation.

$$\mathbf{Fvert=(\ F/2\ )\ cot}\ (\phi) \quad (1)$$

The mechanical advantage (MA) of the system is defined as the ratio of the vertical force component to the applied force F as shown by equation 2.

$$\mathbf{MA=Fvert/F} \quad (2)$$

Combining equation 1 with equation 2 yields equation 3 which defines the mechanical advantage in terms of the angle ϕ.

$$\mathbf{MA=.5}\ \cot(\phi) \quad (3)$$

From equation 3 it becomes apparent that very high values of mechanical advantage (MA) may be obtained with this system if small values of angle ϕ are used. However, the lower value of the starting angle ϕ implies that the useful working stroke (WS) becomes smaller. Figure 2 illustrates how the working stroke for a unity length beam is calculated.

Since two beams are involved in the mechanical ram, the maximum working stroke is given by equation 4.

$$WS=2\ \{1.0-\cos(\phi)\} \quad (4)$$

Table 1 list several values of mechanical advantage (MA) and working stroke (WS) as a function of the angle phi.

For example, if two 6.1 m beams are set at the initial angle of ϕ=4 degrees it follows from Table 1, that a mechanical advantage of 7.1 with a working stroke of (6.1×100× .00488) or 3 cm is available. Referring to Figure 1, this means that a force F of 10000 N will result in a vertical force Fvert of 71000 N with a maximum lift of 3 cm. When the ram advances the value of phi decreases which means the MA becomes higher and the required input force for a constant load drops accordingly. While moderately impressive for a force multiplier it clearly falls short of the task of raising Thutmose which is approximately 130,000 kg. However, the situation changes dramatically when consideration is directed toward a serial mechanical ram configuration. Figure 3 show two mechanical rams placed in series.

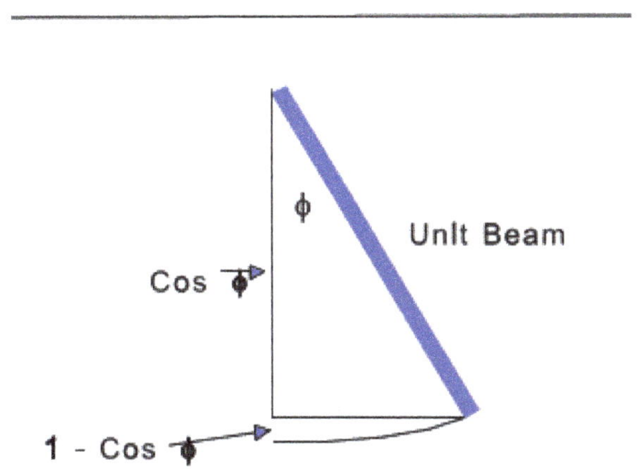

Figure 2: Unit beam working stroke (WS).

In this arrangement a force F on the first ram (beams C and D) is magnified by MA before being applied to the second ram (beams A and B) where it is magnified again by MA. In order for the second ram to complete its full working stroke it is necessary for the first ram to be used several times. The advance of the second ram must be locked temporarily else it would snap back when resetting the first ram. In raising an obelisk, this would mean locking each incremental increase of elevation with a temporary support or "follower" beam.

Table 2 is a repeat of Table 1 except that the values of MA are squared to describe the features of the serial mechanical ram.

As an example of the capability of the serial mechanical ram, consider the case of four 5 m beams (A, B, C, D) in Figure 3 with a starting angle ϕ equal to 5 degrees. From Table 2 the mechanical advantage (MA) is now 32.5 and the working stroke (WS) is (5× 100× .00762) or 3.8 cm. Referring to Figure 3, this means that if a force (F) of 5000 N is applied, then a lift Fvert of 162,500 N can be expected. The first ram (C/D) would have to be reset several times so that the second ram (A/B) could complete the 3.8 cm working stroke. The order of magnitude of force available from such a device suggests it may be up to the challenge of raising Thutmose.

Obelisk force estimations

The analysis of the force required to turn the obelisk to the upright position follows: The Thutmose I obelisk Wirsching [6] is approximately 20 m long and a mass estimated at 130,000 kg. Clearly, the force required to upright the obelisk depends on where the force is applied as well as the angle of inclination theta (θ) of the obelisk. The angle θ being 0 degrees when the obelisk is on its side and θ is 90 degrees when it is upright. In order to make these estimates it is first necessary to calculate the turning moment (Mo) of the obelisk as a function of the inclination angle θ. Figure 4, represents the obelisk at some intermediate angle θ. For modelling purposes, a simplifying assumption is made that the 130,000 kg mass is distributed uniformly along the 20 m length. In reality, the actual obelisk tapers slightly toward

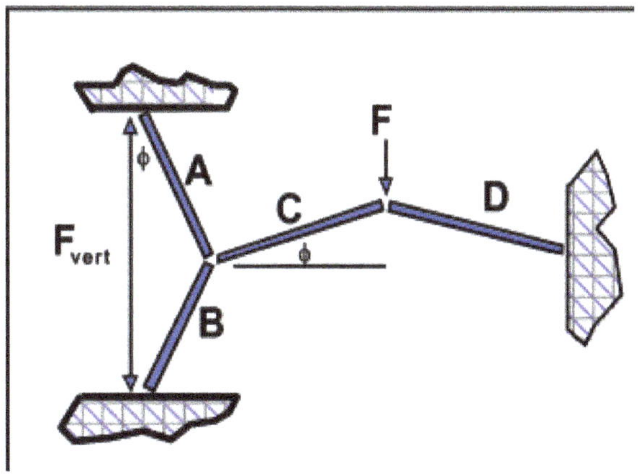

Figure 3: Serial mechanical ram configuration

phi (degrees)	mechanical advantage (MA)	(WS) working stroke (m)
2	14.3	0.00122
3	9.5	0.00274
4	7.1	0.00488
6	4.7	0.01096
8	3.55	0.01946

Table 1: Mechanical ram parameters.

phi (degrees)	mechanical advantage (MA)	(WS) working stroke (m)
2	204.5	0.00122
3	90.2	0.00274
4	50.4	0.00488
6	22.1	0.01096
8	12.6	0.01946

Table 2: Serial mechanical ram parameters.

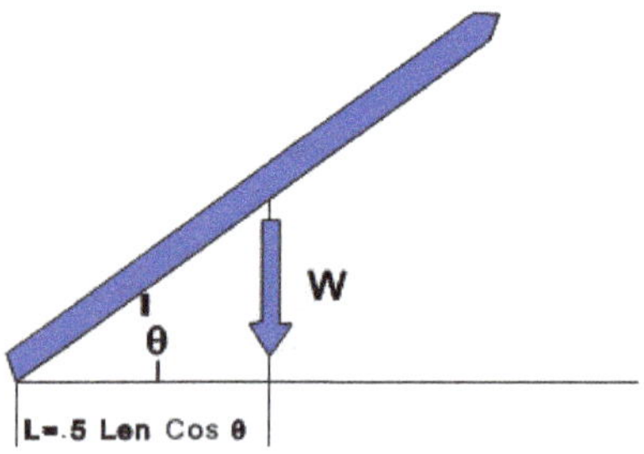

Figure 4: Obelisk turning moment arm.

its top and its centre of gravity would therefore be closer to its base. This assumption of uniform density would make the modelled obelisk more difficult to turn upright than the actual Thutmose I. Figure 4 illustrates the moment arm length L about which the weight W acts hence the turning moment Mo is given by equation 5 where Len is the length and W is the weight of the obelisk.

$$\mathbf{Mo} = \mathbf{.5}\ \text{Len W Cos}\ \theta \quad (5)$$

Table 3 list the turning moments as calculated by equation 5 for various values of the inclination angle However, or very high angles, (i.e. 75 degrees or greater) the turning moment is overstated by equation 5. This is because an increasing fraction of the obelisk mass acts to counter-balance the mass on the other side of the pivot point.

Once the turning moment (Mo) is established, it becomes convenient to calculate the single force (F) required to balance the obelisk weight depending on the location where the force is applied. This force (F) is calculated by dividing the turning moment (Mo) by the turning arm (L) length. Figure 5 show the relationship between the arm length (L), the point of force application (x) and the angle of inclination θ.

Table 4 list the results obtained for the case of single force acting at a point 19-1/6 m from the base of the obelisk. Ideally, the force should be applied, as nearly as possible, perpendicular to the side of the obelisk. In practice, the point of force application would likely have to be lowered as the obelisk rises. This would allow for shorter and more reasonable length beams.

Single serial ram application

The above tables show the order of magnitude of force (600,000 N) required by a mechanical device to upright Thutmose. The serial mechanical ram example Figure 3 described earlier would be capable of about 487,500 N lift with only 15,000 N of input force. Figure 8 illustrates how the serial mechanical ram might be employed to upright the obelisk. The obelisk is shown pivoting on its turning block at an angle of 20 degrees. The rigid beams (A,B,C,D) which make up this serial mechanical ram are shown at an exaggerated starting angle (ϕ=10) for clarity. Likely some initial obelisk angle of perhaps 10 degrees or so could have been achieved through the use of regular lever beams. This is because a number of levers could be operated at the same time given the obelisks near horizontal position. For the postulated obelisk it is seen from Table 4 that a force of 625,033 N would be required to lift the obelisk when it is at 20 degrees. From Table 2 if the mechanical ram were set up with a starting angle (ϕ) of 4 degrees, then a mechanical advantage (MA) of 50.4 could be expected. This means that an input force of (625,033/50.4) or 12,400 N need be applied manually by a group of men pulling on a rope. In this example the beams shown to scale are about 4 m long. The working stroke (WS) may be calculated by using Table 2 and is found to be (4 × 100 × 0.00488) or only about 2 cm.

Multiple serial ram application

When multiple mechanical rams are used in parallel then a substantially longer stroke is possible. For example, if four mechanical rams replaced the single ram then each ram would require a mechanical advantage (MA) of one fourth of that of the single ram. In this case that would be a mechanical advantage of 50.4/4 or 12.6. From Table 2 or thru the use of equation 3 the starting angle would now be 8 degrees rather than the original 4 degree starting angle for the single ram. From Table 2 the new working stroke (WS) for the unit beam is found to be 0.01946. Therefore the working stroke for the quad system is (4 × 100 × 0.01946) or about 8 cm. This quad system would require four separate

Theta(degrees)	Turning Moment(Mo)N-m
0	12,748,640
20	11,979,803
40	9,766,025
60	6,374,320
80	2,213,778

Table 3: Obelisk turning moments for various inclination angles.

Theta(degrees)	Force (F) N at x = 19-1/6 m
0	665,146
20	625,033
40	509,532
60	332,573
80	115,501

Table 4: Force to lift obelisk at 19-1/6 meters from base.

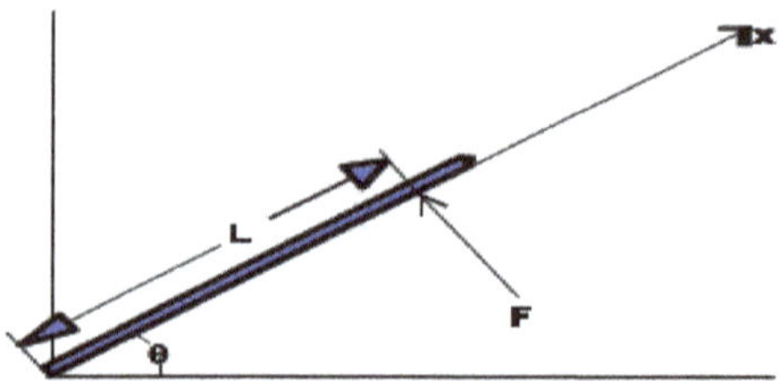

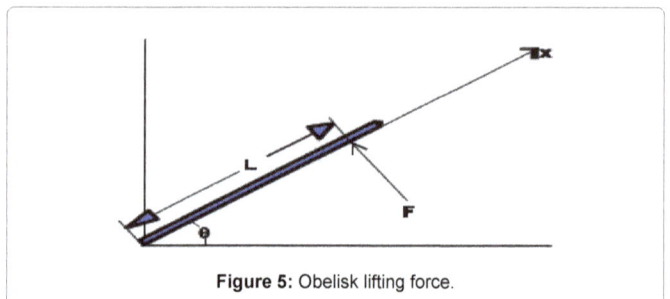

Figure 5: Obelisk lifting force.

applications of the 12,400 N force. The width of Thutmose would easily accommodate such a quad system of parallel serial rams.

Finite Element (FE) Simulation

One of the rams in the quad system is examined with the finite element program ADINA in order to check the above theory. The purpose of the FE analysis is not only to verify the predicted force against the obelisk but also to assess the stress on the beam. In the FE simulation it was assumed that cedar beams make up the mechanical ram. This is because in Arnold [5] it was stated that cedar beams were used to prop up a wall in a pyramid during ancient construction. The mechanical properties of yellow cedar (12% moisture) were taken from ref. 8. In Figure 8 beam A (like beam B) is overwhelmingly subjected to the most stress when lifting the obelisk. The FE analysis was therefore directed to beam A. Figure 6, shows (not scaled for clarity) the physical configuration for the analysis of beam A. The load is one half of the input force (12400 N) times the mechanical advantage of the first mechanical ram (C/D). This works out to be ($0.5 \times 12400 \times \sqrt{12.6}$) or 22010 N. The cedar beam has a square cross-section of 30×30 cm and is 4 m long. The boundary conditions are the same that were assumed for Figure 1.

In the earlier development of the ram equations, it was assumed that rigid beams connected the contact points. The red line in Figure 6 represents such a one dimensional beam between the peak pressure points. The configuration was designed so that the angle of this equivalent beam was 8 degrees above the horizon, although the cedar beam is only 4 degrees.

FE Simulation Results

The color coded stress results of the FE simulation are shown in Figure 7a. This confirms that the maximum contact force (color light green) occurs as indicated by the red arrow heads in Figure 6. It also shows that the cedar beam is not at all over stressed. The maximum stress is listed as 8,375 kPa while the maximum compression limit for cedar is more than 5 times greater at 43,500 kPa.

The FE program also provided the nodal contact forces between the end of the cedar beam and the obelisk surface i.e.CS1. Figure 7b is a listing of these nodal forces which resulted from the 22010N load. The non-zero nodal forces were, as expected, on the lower face of the beam end as depicted by the red arrow head in Figure 6. The sum of these nodal forces added up to 15.4 E04 N. The mechanical advantage (MA) readily follows by dividing this force by twice the input load of 22010 N. This value of 3.49 compares favorably with Table 1 (phi=8), which predicted 3.55 based on the elementary vector analysis (Figure 7c).

Conclusion

In conclusion, it has been demonstrated through the use of vector

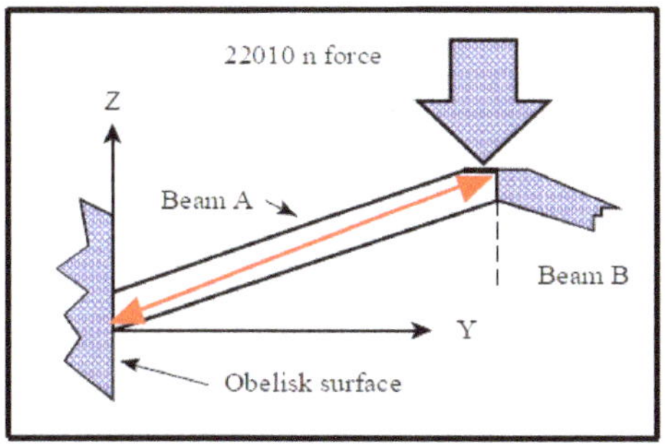

Figure 6: FE problem set-up.

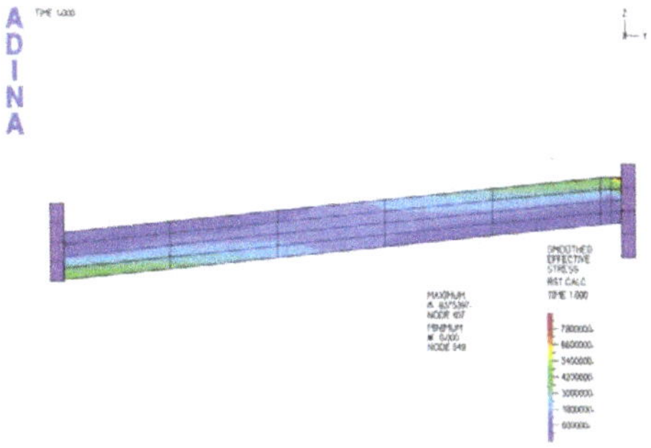

Figure 7a: AFE results for beam A.

ADINA: AUI version 8.8.1, 1 April 2013: SOLID-RAM-101 yellow cedar
Licensed from ADINA R&D, Inc.
Finite element program ADINA, response range type load-step:
Listing for zone CS1:

POINT	CONTACT_FORCE-Y
Node 6	0.00000E+00
Node 12	0.00000E+00
Node 18	0.00000E+00
Node 24	0.00000E+00
Node 30	0.00000E+00
Node 30	0.00000E+00
Node 42	0.00000E+00
Node 48	0.00000E+00
Node 54	0.00000E+00
Node 60	0.00000E+00
Node 66	0.00000E+00
Node 72	0.00000E+00
Node 78	1.49403E+03
Node 84	2.98805E+03
Node 90	2.98805E+03
Node 96	1.49403E+03
Node 102	1.85176E+03
Node 100	3.70351E+03
Node 114	3.70351E+03
Node 120	1.85176E+03
Node 236	0.00000E+00
Node 237	0.00000E+00
Node 238	0.00000E+00
Node 254	0.00000E+00
Node 255	0.00000E+00
Node 256	0.00000E+00
Node 272	0.00000E+00
Node 273	0.00000E+00
Node 274	0.00000E+00
Node 290	5.97611E+03
Node 291	5.97611E+03
Node 292	5.97611E+03
Node 308	7.40703E+03
Node 309	7.40703E+03
Node 310	7.40703E+03
Node 331	0.00000E+00
Node 332	0.00000E+00
Node 333	1.52244E+03
Node 334	3.66903E+03
Node 355	0.00000E+00
Node 356	0.00000E+00
Node 357	3.04489E+03
Node 358	7.33806E+03
Node 379	0.00000E+00
Node 380	0.00000E+00
Node 381	3.04489E+03
Node 302	7.33006E+03
Node 403	0.00000E+00
Node 404	0.00000E+00
Node 405	1.52244E+03
Node 406	3.66903E+03
Node 622	0.00000E+00
Node 623	0.00000E+00
Node 624	0.00000E+00
Node 625	0.00000E+00
Node 626	0.00000E+00
Node 627	0.00000E+00
Node 628	6.08977E+03
Node 629	6.08077E+03
Node 630	6.08977E+03
Node 631	1.46761E+04
Node 632	1.46761E+04
Node 633	1.46761E+04

Figure 7b: ADINA (R) FE output; **R:** ADINA is a registered trademark of K.J. Bathe/Adina R & D Inc.

analysis and computer simulation how the ancients may have brought to bear the tremendous force necessary to upright the 130,000 kg obelisk. The technique outlined in this paper required only a direct force of approximately 12,400 N. Figure 8 shows the basic serial ram set-up although the required "follower" beam which locks each elemental increase in elevation of the obelisk is not shown. An advantage of this

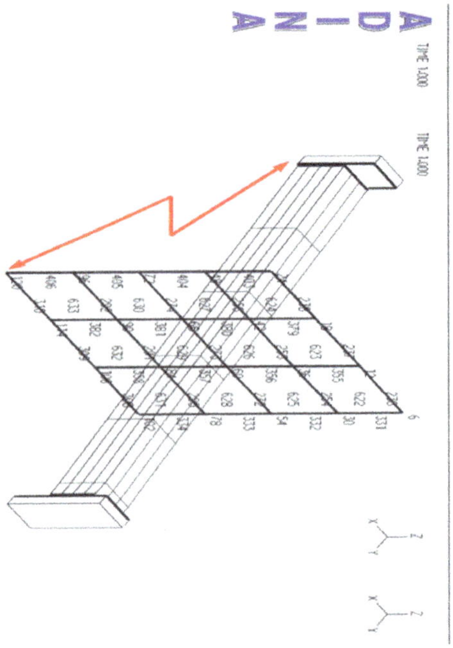
Figure 7c: FE Contact Surface (CS1) Node Identification.

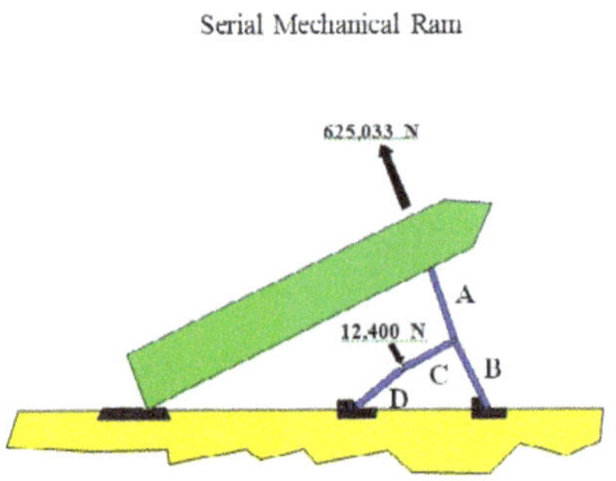

Figure 8: Serial ram set-up for raising the obelisk.

slow gradual procedure is that it allows time for care that the obelisk does not flip to either side during erection. When the obelisk is nearly upright, a team controlling ropes Cort [7] attached to the top has proven effective in the final stage of the ascent. The technique outlined in this paper may also have been used by the ancients to move massive building blocks. While this technique required incredible patience, this task would pale by comparison to the job of shaping the obelisk from solid granite with only primitive tools.

References

1. Bathe K (2006) Finite Element Procedures, Prentice Hall, New Jersy.
2. Barnes M (1997) Secrets of Lost Empires: Obelisks VGBH Video, Boston.
3. Englebach R, George HD (1923) Problem of the Obelisks. NewYork.
4. Chevrier H (1952) Note on Lerection of Obelisks. Annals of the Department of Antiquities, Kairo.
5. Arnold D (2003) Encyclopaedia of Ancient Egyptian Architecture, Tauris, London.
6. Wirsching (2007) Obelisks transporting and erecting in Egypt and in Rome. Production and Publishing, Norderstedt.
7. Cort J. Secrets of Lost Empires II: Pharaoh's Obelisk, VGBH Video, Boston.
8. Green D,Winandy J, Kretschmann D (2007) The Encyclopedia of Wood Skyhorse Publishing Inc New York NY: 4-7.

30

Delivering Medical Products to Quarantined Regions using Unmanned Aerial Vehicles

Otiede ED[1], Odeyemi MM[1], Nwaguru I[1], Onuoha C[2], Awal SI[2], Ajibola O[2], Adebayo VK[3], Itoro E[3], Ejejigbe TR[3], Ejofodomi OA[4*] and Ofualagba G[1]

[1]*Department of Electrical and Electronics Engineering, Federal University of Petroleum Resources (FUPRE)*

[2]*Department of Mechanical Engineering, Federal University of Petroleum Resources (FUPRE)*

[3]*Department of Marine Engineering, Federal University of Petroleum Resources (FUPRE)*

[4]*RACETT Nigeria Ltd., 32 Agric Road, GRA, Effurun, Nigeria*

Abstract

Unmanned aerial Vehicles (UAVs) possess the capacity to transport goods quickly, safely, and inexpensively across both accessible and inaccessible terrains such as to stranded mountain climbers or boats adrift. Medical supplies are typically delivered by ground transport as well as aircraft, both fixed and rotor wing. During emergencies, the availability of blood products and pharmaceuticals is often limited at critical access hospitals, and conventional channels of supply may become disrupted. This article aims to demonstrate the feasibility of using small UAVs to transport pharmaceutical products to epidemic-stricken regions.

A Syma X5-c quadcopter was loaded with a sample drug. Using its accompanying transmitter, the drone was flown from a base location to a pre-determined delivery location. Upon successful delivery of the drug, the drone was navigated back to base by the operator. Future improvements to the drone model include facial identification system to verify recipients before delivery, auto-pilot control of drone with the aid of an on-board computer program and a GPS unit, real-time picture, and video relay/transmission back to control station, and autonomous delivery of drug without requiring physical contact from the recipient.

Keywords: Unmanned Aerial Vehicles; Automated pharmaceutical delivery; Quadcopters

Introduction

Unmanned Aerial Vehicles (UAVs) are aerial vehicles or drones that operate without a human operator on-board. They are either controlled through radio frequency from a remote control or through a computer program installed on a computer on-board the drone. Initially used primarily by the military sector (e.g. for missile deployment during war), applications of UAV have extended to the civilian sector. Some of the current applications for UAV include: agriculture [1] surveying [2], wildlife monitoring and conservation [3], real estate assessments [4], surveillance, meteorological studies, such as taking geographical samples at extreme heights, and courier services, such as Amazon CEO Jeff Bezo's announcement in 2013 that the company would use drones for package delivery [5]. Advances in technology and decreasing costs have led to an increased use of unmanned aerial vehicles (UAVs) by the military and civilian sectors. The use of UAVs in commerce is restricted by US Federal Aviation Administration (FAA) regulations. The FAA continues to demand that drones be controlled by a human pilot and stay within that pilot's sight line. These regulations forestall implementation of a large-scale drone delivery fleet. The FAA also has established a no-fly zone, which lies between 400 and 500 vertical feet, to ensure that drones remain distinct on radar from human-piloted aircraft. However, such restrictions do not necessarily apply to other countries [5]. In fact, 57 countries and 270 companies were producing UAVs in 2013 [6].

The use of UAVs in the medical industry is relatively new. UAVs have been used for medical product transport [7]. Drones delivered care packages in the aftermath of the 2010 Haitian earthquake and have been deployed in various disasters, such as the category 5 cyclone that struck the islands of Vanuatu, the Nepal earthquake, and Super-storm Sandy. In 2014, a medical drone delivery of medication was undertaken by Deutsche Post DHL AG from Norddeich, Germany to Juist, a remote, car-free island in the North Sea. The drone flew for approximately 12 km off the German coast and landed on the island. Doctors without Borders also used drones to transport dummy tuberculosis (TB) test samples from a remote village to the city of Kerema, Papua New Guinea in 2014. The country has a significant TB burden and an increasing burden of multidrug-resistant TB, as well as inaccessible roads and generally poor weather conditions that moor transport boats 5 months of the year [6,7].

Singer et al. [8] have proposed the use of small UAVs for transporting laboratory samples for early infant diagnosis of HIV in Malawi. Transportation of laboratory samples for HIV (EID and Viral Load), tuberculosis and other diseases is a tremendous barrier in Malawi due to limited transportation infrastructure, poor roads (some of which are impossible during the rainy season), and the high cost of fuel (approximately USD 1.40 per liter). There are several companies specializing in UAVs for the health sector and a consortium of non-profit, for-profit, academic, and international organizations working together on developing UAVs for health [9] UAVs have been considered for delivering health commodities, and have been tested for transporting laboratory samples in Haiti, Lesotho, Papua New Guinea, and South Africa [10].

***Corresponding author:** Ejofodomi OA, RACETT Nigeria Ltd., 32 Agric Road, GRA, Effurun, Nigeria, E-mail: tegae@yahoo.com

The goal of this research project is to test and explore the possibility of using UAVs to provide medical assistance to remote, inaccessible regions undergoing epidemic outbreaks, such as the Ebola virus that affected over four countries in the West African sub region. Under such circumstances, the virus responsible for the disease can be transmitted by physical contact either in person or body fluids of an infected person, with death almost certain to occur within days. This makes it very hard for medical attention to be given to people in the epidemic stricken region. In the case of the Ebola epidemic in West Africa, several health-care providers were infected while trying to offer medical assistance, while some lost their lives. The use of UAVs for drug delivery to quarantined regions experiencing this epidemic outbreak could potentially reduce the cases of infection and mortality among health care workers wishing to provide medical assistance to these regions. This project explores and demonstrates the practical feasibility of utilizing UAVs in the transportation of medical drugs to quarantined regions experiencing cases of epidemic outbreaks.

The drugs required by the quarantined region are loaded on the drone at the base station (a region free from the epidemic outbreak). The drone is then navigated to the quarantined region by an operator until it gets to the quarantined region. After successfully arriving at the quarantined region, the drone lands at a specific location where the intended recipient for the drug is waiting. The recipient retrieves the drug from the drone. After successful delivery of the drug to the intended recipient in the quarantined region, the drone is then navigated back to the base station by the operator. Details about the drone utilized in this study, and the methodology that would be employed in aerial drug delivery for quarantined regions are presented in the Materials and Methods Section. The results section show the outcome obtained after implementing the previously described methodology. The challenges encountered and possible improvements to the system are explained in the results and discussion section.

Materials and Methods

Materials

Syma X-5c drone: The Syma X-5c is a remote-controlled quadcopter with an on-board camera (Figure 1). A quadcopter, also called a quadrotor helicopter or quadrotor, is a multi-rotor helicopter that is lifted and propelled by four rotors. Control of vehicle motion is achieved by altering the rotation rate of one or more rotor discs, thereby changing its torque load and thrust/lift characteristics. There are several advantages to quadcopters over comparably-scaled helicopters. Quadcopters do not require mechanical linkages to vary the rotor blade pitch angle as they spin. This simplifies the design and maintenance of the vehicle. Second, the use of four rotors allows each individual rotor to have a smaller diameter than the equivalent helicopter rotor, allowing them to possess less kinetic energy during flight. This reduces the damage caused should the rotors hit anything. For small-scale UAVs, this makes the vehicles safer for close interaction. Some small-scale quadcopters have frames that enclose the rotors, permitting flights through more challenging environments, with lower risk of damaging the vehicle or its surroundings. Due to their ease of construction and control, quadcopter aircraft are frequently used as amateur model aircraft projects. Moreover, due to the fact that they are unmanned, there is a reduced risk of loss of lives in the event of a crash. These characteristics make drones very useful and would therefore be harnessed by the project to fulfil its aim. Some key features of the Syma X-5c quadcopter include:

1. 360° roll.

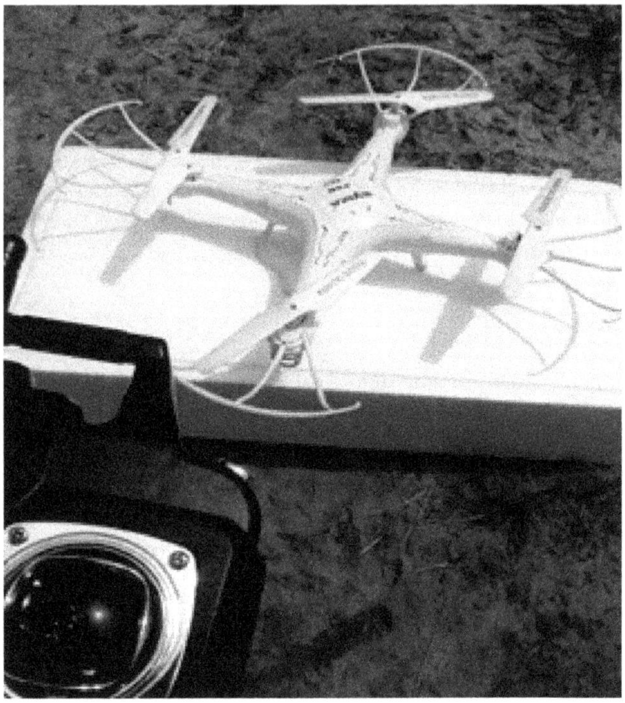

Figure 1: Syma X-5c and transmitter.

2. 6-axis gyroscope.
3. Hover capability.
4. HD camera for video/image capture.
5. 100x faster radio control.
6. Further remote distance with spectrum technology.
7. Modular design.
8. Indoor /outdoor flight.
9. 2 pounds in weight.
10. 50 meter remote distance.

Transmitter/remote controller: The drone is accompanied with a remote controller (Figure 2). The remote controller interfaces with an on-board receiver that in turn controls the motion of the rotors and the speed of rotation as the case may be. It works with a radio frequency with a range of 2.4 GHz.

High/low speed switch: This button regulates the speed of directional transit of the drone. Basically it throttles the speed of sideways transition of the drone.

Left control accelerator lever: The accelerator lever is used to the control the speed of rotation of the fan. The speed of rotation is directly proportion to the height of flight of the drone; therefore the accelerator is used to control the height of flight of the drone. The acceleration is controlled by moving the joystick upwards and downwards. The accelerator is also used to synchronize the signal of the transmitter with that of the drone.

Hover up and down trimmer: This trimmer is used to fine-tune the default height of flight of the drone. In some cases when the drone is switched on, it moves higher without controlling it to do so. In cases

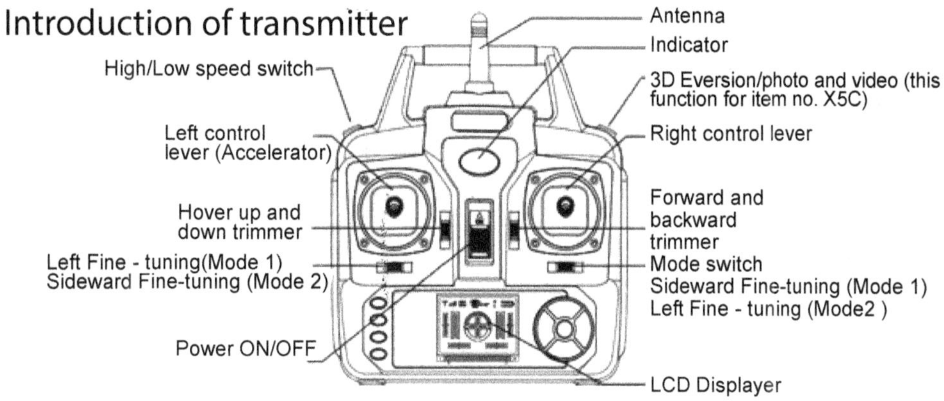

Figure 2: Transmitter for Syma X5-c Quadcopter.

like that the trimmer can be used to ensure that the drone is stabilized and restore it to a default height of hover as specified by the position of the accelerator.

Side-ways fine-tuning button: This applies in cases where the drone flies sideways without controlling it to do so. The side-ways fine-tuning button is used to adjust the default sideways movement of the drone.

Antenna: This is used to transmit radio frequency that controls the drone and also receives feedback signal from the drone as the need be.

Indicator: This indicates when the transmitter is powered on and also blinks when the transmitter operates.

Power button: The power button is used to switch the transmitter On/Off.

3d eversion/photo and video button: This button controls the drone to display some stunt in air. The photo/video button is used to actuate the on-board camera for photo and video capture.

Forward and backward trimmer: Like the other trimmers it is also used to adjust the default position of the drone.

Direction control lever: This is used to direct the drone forward, backward, leftward and rightward.

LCD display: This gives a visual display of the controls and operational state of the drone, showing the acceleration and direction.

Methodology

Figure 3 shows the flow chart operation of the drone for drug delivery in an epidemic-stricken area. An order is made for medical supplies that are needed urgently in the region. The GPS coordinates of the delivery point is made available. The drone is loaded with the required drug. For demonstration purposes, a single tablet of a malaria drug is secured to the drone. The navigation of the drone is a basic responsibility of the operator, i.e. it is done with no programming code involved. The drone was controlled in an open field with sufficient space to prevent collision with humans or structures. Once the drone reached the delivery point, it was lowered to ground level. A picture of the recipient was snapped using the on-board camera. The recipient retrieved the drug. The drone was then flown back to the base. It should be noted that FAA regulations demand that drones be controlled by a human pilot and stay within that pilot's sight line. However, these regulations do not exist in the African countries where the Ebola epidemic outbreak occurred, and so the navigation of the drone could

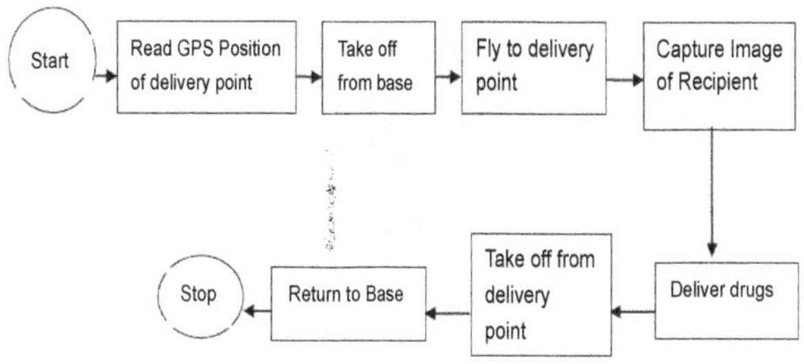

Figure 3: Flow chart for drug delivery using UAV.

eventually be done autonomously, enabling the drone to travel further distances and away from the operator's line of sight. It would also require less input from the operator.

Discussion and Conclusion

The drone was successfully controlled to deliver the specified goods to the delivery point and to fly back to its base. Figure 4a shows the drone prior to take off. The malaria drug is being loaded on the drone. Figure 4b shows the operator manning the drone during flight. Figure 4c shows the drone navigating to the delivery point. Figure 4d shows the recipients retrieving the drug from the drone at the delivery location. Figure 4e shows the drone navigating back to base. Figure 4f shows the drone back at the base.

Several challenges were encountered during this project. The mechanical force imposed by direction of breeze distorted to an extent the positioning of the drone while hovering in air. The effect of the wind was taken into consideration during the control of the drone. The battery life of the drone has a rating of 3.5v 500mAh, which is quite low. The drone takes about 3 hours to get fully charged but only has an operational time of about 10-15 minutes. However, this drone was sufficient to demonstrate the feasibility of using UAVs for drug delivery.

There are still many things that can be added to this drone make it more efficient. Future improvement to the drug delivery drone includes inclusion of a facial identification system to verify recipients before delivery. Control of the drone will be upgraded from the transmitter to auto-pilot mode with the aid of an on-board. computer program and a GPS unit. Real-time picture relay/transmission back to control

Figure 4: Drug delivery using UAV (a) Loading drug on UAV at base.(b) Controller manning UAV after take-off from base. (c) UAV en route to delivery point. (d) Recipients retrieving drug from UAV at delivery point. (e) UAV departing after delivering drug (f) UAV arriving at home base.

station would also be included in the upgraded model. The battery life of the drone will be upgraded to increase the flight time of the drone. Additionally, an automated drug release mechanism (such as a small robotic gripper) could be integrated into the drone so that the recipient has no cause to physically touch the drone. With this feature, once the drone arrives at the destination point, it would automatically deposit the drugs on the ground before taking off and flying back to the base station. This would ensure that the epidemic does not spread beyond the quarantined region.

This research project has successfully demonstrated that UAVs can be used for drug delivery in quarantined regions undergoing epidemic outbreaks. Future work will include autonomous navigation of the UAV, developing an efficient power supply for the UAV, photo identification of recipient prior to drug delivery, and automatic delivery of drug at destination point without requiring physical contact from the recipient.

References

1. The Drone News (2013) Northrop Grumman wants to sell unmanned drones to farmers?
2. Handwerk B (2013) 5 Surprising drone uses. A National Geographic update.
3. WWF Global (2012) Unmanned aerial vehicle to aid Nepal's conservation efforts.
4. Australian UAV (2016) Australian UAV provides professional aerial mapping, survey and inspection services throughout the country using our cutting-edge drone technology.
5. Mullins N (2016) The drones of medicine. SMA, Southern Medical Blog.
6. Office for the Coordination of Humanitarian Affairs (OCHA) (2014) Policy and studies series, Unmanned aerial vehicles in humanitarian response. United Nations.
7. Thiels CA, Aho JM, Zietlow SP, Jenkins DH (2015) Use of unmanned aerial vehicles for medical product transport. Air Med J 34: 104e108.
8. Singer D, Sherman J, Saka E, Bancroft E (2015) Cost analysis and feasibility of using unmanned aerial vehicles to transport laboratory samples for early infant diagnosis of HIV in Malawi.
9. School of Public Health (2016) How can drones improve global health?
10. Mendelow B, Muir P, Boshielo BT, Robertson J (2007) Development of e-Juba, a preliminary proof of concept unmanned aerial vehicle designed to facilitate the transportation of microbiological test samples from remote rural clinics to National Health Laboratory Service Laboratories. SAMJ 97: 1215-1218.

Mechanical Behavior of Al7025-B_4C Particulate Reinforced Composites

Nagaral M[1]*, Attar S[2], Reddappa HN[2], Auradi V[3], Suresh Kumar S[4] and Raghu S[5]

[1] *Aircraft Research and Design centre, HAL, Bangalore-560037, Karnataka, India*

[2] *R&D Centre, Department of Mechanical Engineering, BIT, Bangalore-560004, Karnataka, India*

[3] *R&D Centre, Department of Mechanical Engineering, SIT, Tumkur-572103, Karnataka, India*

[4] *Cornelius Technology Centre India Private Limited, Bangalore, Karnataka, India*

[5] *PES College of Engineering, Mandya, 571401, Karnataka, India*

Abstract

In comparison to unreinforced alloy, Al7025 reinforced with hard ceramic particles possesses higher strength, hardness,wear resistance and low coefficient of thermal expansion. They can be used for automotive components and aircraft structures. The aim of the present work is to investigate the effects of adding micro size-B4C particales to Al7025 alloy on the mechanical properties of the composites.The Al7025 alloy reinforced with 6 wt. % of B_4C particulate composites were fabricated by stir casting method.Microstructure and mechanical properties such as hardness, ultimate tensile strength, yield strength, percentage elongation and density of the composites were examined. Microstructure of the samples has been investigated by using optical microscopy to know the uniform distribution of reinforcement particulates in the matrix. It was observed that the hardness, ultimate tensile strength and yield strength of Al7025 alloy increased with the addition of 6 wt. % B_4C particulates.From the study, it was revealed that the elongation and density of Al7025-6wt.% B_4C composite decreased in comparison to that of the base Al7025 alloy.

Keywords: Al7025 Alloy; B_4C Particulates; Hardness; Ultimate tensile strength; Yield strength; Stir casting

Introduction

Metal matrix composites (MMCs) offer designers requirements, they are particularly suited for applications requiring high strength to weight ratio at high temperature, good structural rigidity, dimensional stability, and light weight. The inadequacy of metals and alloys in providing both strength and stiffness to a structure has led to the development of various composites particularly metal matrix composites (MMCs) [1-3]. Composite materials are used extensively as their higher specific properties (properties per unit weight) of strength and stiffness, when compared to metals, offer interesting opportunities for new product design. MMCs are metals reinforced with other metal, ceramic or organic compounds. They are made by dispersing the reinforcements in the metal matrix [4]. Reinforcements are usually done to improve the properties of the base metal like strength, stiffness, conductivity, etc. The particle distribution plays a very vital role in the properties of the Al MMC and is improved by intensive shearing [5]. Addition of hard ceramic particles like SiO_3, SiC, Al_2O_3, TiB_2, B_4C etc. to Al matrix lead to strengthening of the matrix with improved properties. Ceramic particles such as Al_2O_3 and SiC are the most widely used materials for reinforcement with aluminium. Boron carbide (B_4C) could be an alternative to SiC and Al_2O_3 due to its high hardness (the third hardest material after diamond and boron nitride). Boron carbide has an attractive properties like high strength, low density (2.52 g/cm^3), extremely high hardness, good wear resistance and good chemical stability. Hence reinforcing the aluminium with boron carbide particles confers high specific strength, elastic modulus, good wear resistance and thermal stability [6]. From the Literaure survey it can be concluded that, most of the studies on aluminium based MMCs are devoted to SiC and Al_2O_3 particulate reinforcements; however, use of B_4C particulates as reinforcements in aluminium matrix is relatively limited. B_4C is considered to be the third hardest material and is an extremely promising material for a variety of applications like bullet proof vests, armor tanks and as neutron absorber material.

Cun-Zhu Nie et al., [7] studied the Boron carbide particulates reinforced 2024 Aluminum matrix composites were fabricated by mechanical alloying–hot extrusion technology successfully. A clean interface of B_4C between aluminum was obtained in their experiment, the yield strength and Young's modulus values were improved significantly over the monolithic 2024 alloy. Vijaya et al., [8] investigated the mechanical properties of aluminium alloy alumina boron carbide metal matrix composites

The present investigation focuses on fabrication and evaluation of mechanical behavior of Al7025 alloy matrix reinforced with B_4C particles.

Materials and Experimental Details

Matrix material

The matrix material used in the experimental investigation is aluminium 7025 alloy whose chemical composition is listed in Table 1. Al7025 alloy is one type wrought aluminium alloy, containing zinc as a major alloying element. The theoretical density of Al7025 is taken as 2.80 g/cm^3.

Reinforcement material

The main advantage of introducing reinforcement material to

***Corresponding author:** Madeva Nagaral, Design Engineer, Aircraft Research and Design centre, HAL, Bangalore-560037, Karnataka, India
E-mail: madev.nagaral@gmail.com

Element	Symbol	Wt. Percentage
Zinc	Zn	5
Copper	Cu	0.1
Manganese	Mn	0.6
Magnesium	Mg	1.5
Ferrite	Fe	0.4
Chromium	Cr	0.35
Silicon	Si	0.3
Titanium	Ti	0.1
Aluminium	Al	Bal

Table 1: Chemical composition of Al7025 alloy.

base metal or alloy is to increase the properties there by enhancing the mechanical and tribological properties of composites. In the current research Boron Carbide particulates of size 70-80 microns (μm) were used as a reinforcement material, which was procured from Speedfam (India) Pvt. Ltd., Chennai. The density of B_4C is 2.52 g/cm^3 which is lower than the base Al matrix, contributes in weight saving.

Preparation of Al7025-B_4C composites

In the engineering materials, the MMC's can be manufactured by a unique technique such as casting, as it is inexpensive and suitable for mass production of components. The synthesis of metal matrix composite used in the study was carried out by liquid metallurgy route in particular stir casting technique. Initially B_4C particulates were preheated for 300-400°C. In the present work, an attempt has been made to study the mechanical properties of as cast Al7025 alloy and Al7025- B_4C particulate composites. The composites containing 6 wt.% of B_4C particulates were prepared. Initially required amount of charge or matrix material was placed in a graphite or silicon carbide crucible, which was placed in electric resistance furnace at a temperature of around 730 degree Celsius. After complete melting of Al7025 alloy matrix, degassing was carried out by using Solid Hexa Chloroethane [9], which helps to remove unwanted adsorbed gases from the melt. Once degassing is over, the preheated ceramic reinforcement particles were introduced into matrix in a novel way which involves two-stage additions of reinforcement during melt stirring. This novel two stage additions of reinforcement into matrix Al7025 will increase wettability of the matrix and ceramic reinforcement and further, which helps in uniform distribution of the particles. A continuous stirring process was carried out during addition of reinforcement into matrix. Normally for all composite preparation, stirring speed was maintained at 300rpm. After 10 minutes of continuous stirring, entire molten metal was poured into cast iron die. The prepared composites were machined and tested for micro structural studies. After revealing uniform distribution of B_4C particles in the matrix, tensile behaviour of as cast Al7025 alloy and its composites were evaluated as per ASTM standards. Figures 1 and 2 showing the cast iron die and stir casting set up and used to prepare the composites for the present study.

Specimen testing

The microstructure of the as cast Al7025 alloy and its composites reinforced with different wt.% of B_4C particulates were examined by using an optical microscope (Olympus made). The samples of as cast and Al7025-B_4C composites for microstructural study were cut from casted rods and ground by means of abrasive papers followed by rotating disc cloth polishing. Keller's reagent was used as an etching agent.

The composites and base Al7025 alloy were tested for their hardness using a Brinell hardness tester. The hardness testing was carried out in accordance with ASTM E10 standard at room temperature. A test load of 250 kg was applied to the specimens for 30s. The diameter of steel ball indenter was 5 mm. The size of the indent (d) was determined optically by measuring two diagonals of the round indent. The Brinell hardness number (BHN) was calculated for the unreinforced Al matrix and B_4C reinforced composite using equation (1). An average of ten readings was taken for each sample for hardness measurement.

$$\text{BHN}=\frac{2P}{\Pi D\left(D-\sqrt{D^2-d^2}\right)} \quad (1)$$

Where P is the applied load in kg, D is the diameter of the steel ball in mm and d is the size of the indent in mm. Each hardness value presented is an average of at least ten symmetrical indentations.

The experimental density of both unreinforced Al7025 alloy and Al7025-B_4C composites were measured by dividing the measured weight of test sample by its measured volume using an electronic weighing machine. The theoretical density of the composite was calculated by rule of mixture using formula:

$$\rho_{th}=\rho_m V_m+\rho_r V_r \quad (2)$$

Where ρ_m is the density of matrix, V_m is the volume fraction of the matrix, ρ_r is the density of reinforcement and V_r is the volume fraction of the reinforcement.

Tensile testing of the prepared samples were conducted in accordance with the ASTM E8 standard on round tension test specimens of gauge diameter 9 mm and gauge length 45 mm. Tension test was conducted by using Instron made servo-hydraulic machine,

Figure 1: Cast Iron Die.

Figure 2: Stir casting set up.

with cross head speed set at 0.280 mm/min. The experiments were conducted at room temperature. Stress versus strain graph was plotted to know the effect of B_4C particulates on tensile behaviour of Al7025 alloy composites.

Results and Discussion

Microstructural study

Figures 3a and 3b shows the optical micrographs of as cast Al7025 alloy and Al7025-6 wt.% B4C composite respectively. The grain size of the composite was much smaller than that of the alloy because particles act as nucleation sites. Figure 3b reveals good distribution of reinforcements and there is no agglomeration in the composite. From the optical microphotograph, it is clear that a good crack free bonding was formed at discrete locations between the reinforcement and the matrix alloy.

Density measurements

In the present research work, the measured densities of as cast Al7025 alloy and Al7025-6 wt.% B_4C composites are presented in the Figure 4.

It is observed that, by the addition of B_4C particles the density of the composite is slightly decreased. This decrease in density is mainly due to lower density of B_4C particles as compared to the base Al7025 alloy. Further, from Figure 4, the experimental densities for both alloy and composites are in line with the theoretical densities but slightly lesser than the theoretical densities.

Hardness

In the present work, hardness values of the Al7025 alloy and Al7025-B_4C composites have been obtained by Brinell hardness tester. The variation of hardness with Al alloy and its composite is shown in Figure 5. It is noticed that the hardness of Al7025-6 wt.% B_4C composite is more than Al7025 alloy. A notable rise in the hardness of the alloy matrix can be seen with the addition of B_4C particles. This is mainly due to the presence of B_4C particles in the matrix Al7025 alloy. Whenever a hard reinforcement is incorporated into a soft ductile matrix, the hardness of the matrix material is enhanced [10].

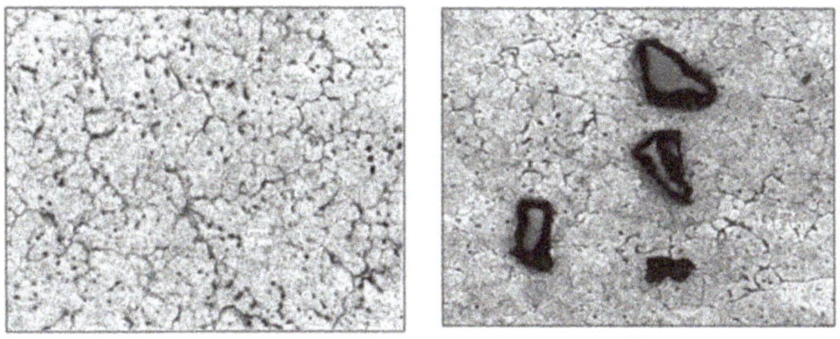
(a) (b)

Figure 3: Optical micrographs of (a) as cast Al7025 alloy (b) Al7025-6 wt. % B_4C composite.

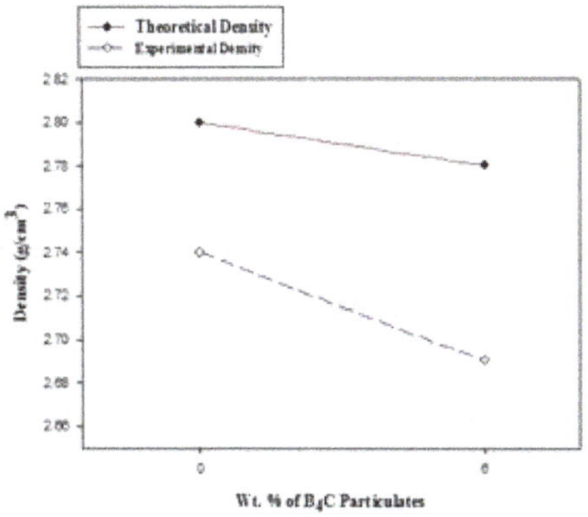

Figure 4: Comparison of theoretical and experimental densities of Al7025 alloy and its composites.

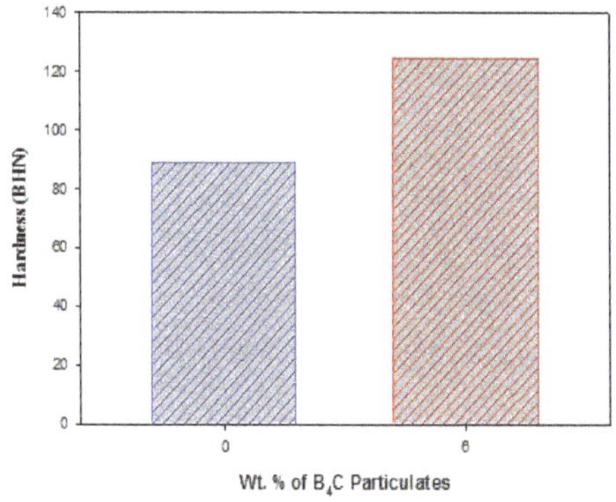

Figure 5: Variation in hardness with wt. % of B_4C particulates.

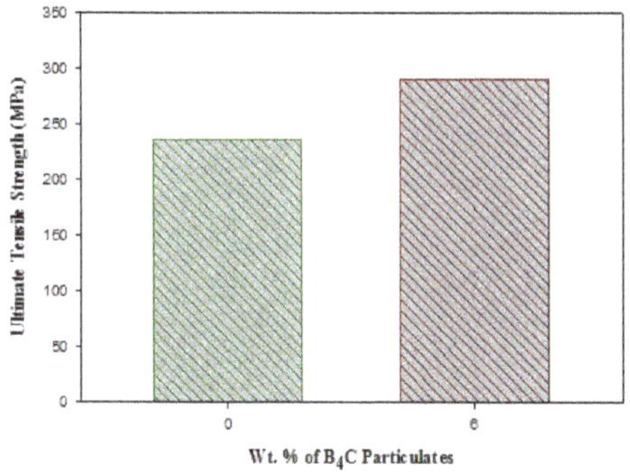

Figure 6: Variation in ultimate tensile strength with wt. % of B_4C particulates.

Figure 6 shows variation of ultimate tensile strength (UTS) with 6 wt.% of B_4C particulates. The ultimate tensile strength of Al7025-6 wt.% B_4C composite material increases by an amount of 22.42% as compared to as cast Al7025 alloy matrix. The microstructure and properties of hard ceramic B_4C particulates control the mechanical properties of the composites. Due to the strong interface bonding load from the matrix transfers to the reinforcement exhibiting increased ultimate tensile strength [11].

This increase in UTS mainly be due to B_4C particles acting as barrier to dislocations in the microstructure. The improvement in UTS may be due to the matrix strengthening following a reduction in Al7025-B_4C grain size, and the generation of a high dislocation density in the Al7025 alloy matrix a result of the difference in the thermal expansion between the metal matrix and the B_4C reinforcement.

Yield strength

Figure 7 shows variation of yield strength (YS) of Al7025 alloy

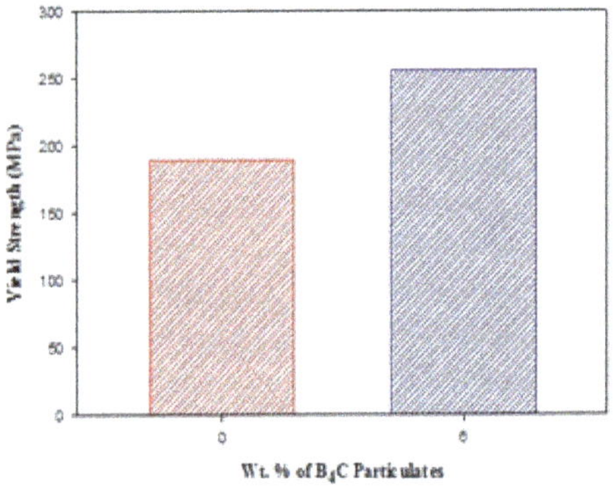

Figure 7: Variation in yield strength with wt.% of B_4C particulates.

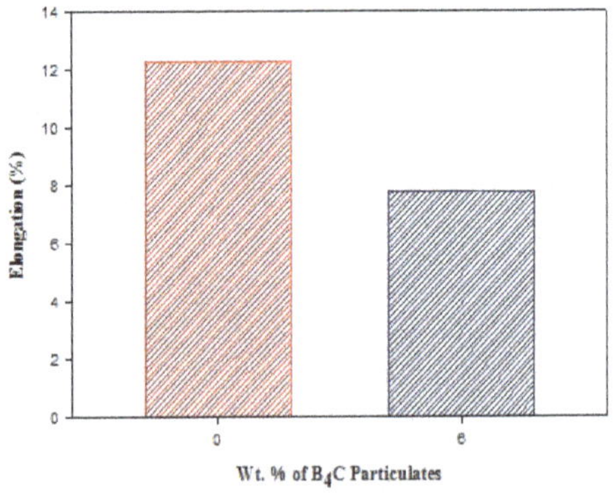

Figure 8: Variation in percentage elongation with wt.% of B_4C particulates.

matrix with 6 wt.% of B_4C particulate reinforced composite. It can be seen that by adding 6 wt.% of B_4C particulates yield strength of the Al7025 alloy increased from 189.6 MPa to 256.3 MPa. This increase in yield strength is in agreement with the results obtained by several researchers [12], who reported that the strength of the particle reinforced composites is more strongly dependent on the volume fraction of the reinforcement. The increase in YS of the composite is obviously due to presence of hard B_4C particles which impart strength to the soft aluminium matrix resulting in greater resistance of the composite against the tensile stress. In the case of particle reinforced composites, there is a restriction to the plastic flow due to the dispersion of the hard particles in the matrix, thereby providing enhanced strength to the composite.

Percentage elongation

Figure 8 is a graph showing the effect of B_4C content on the percentage elongation (ductility) of the composites. It can be seen from the graph that the ductility of the composites decrease significantly with the 6 wt.% B_4C reinforced composites. This decrease in percentage elongation in comparison with the base alloys is a most commonly occurring disadvantage in particulate reinforced metal matrix composites. The reduced ductility in Al7025-6 wt.% composites can be attributed to the presence of B_4C particulates which may get fractured and have sharp corners that make the composites prone to localised crack initiation and propagation. The embrittlement effect that occurs due to the presence of the hard ceramic particles causing increased local stress concentration sites may also be the reason.

Conclusion

The results of the study of microscopic structure and mechanical properties of the Al7025-6 wt.% of B_4C composites materials produced by stir casting are remarked as below:

- The liquid metallurgy technique was successfully adopted in the preparation of Al7025-6 wt.% B_4C composites.
- The microstructural studies revealed the uniform distribution of the B_4C particulates in the Al7025 alloy matrix.
- Density of the Al7025-B_4C composites decreased as compared to that of base alloy 7025 matrix. Further, experimental densities of base alloy and composites are in line with the theoretical densities, which show good casting procedure adopted for fabrication of composites.
- Hardness of the Al7025-B_4C composite was found to be more than base Al matrix.
- The ultimate tensile strength and yield strength properties of the composites found to be higher than that of base matrix. The improvements in UTS and YS by adding 6 wt. % of B_4C was increased by 22.46% and 34% respectively.

It was observed that the percentage elongation decreased for Al7025-6 wt.% B_4C composite as compared to base Al7025 alloy matrix.

References

1. Dora Siva P, Chintada S (2014) Hybrid composites a better choice for high wear resistant materials. Journal of Materials Research and Technology 3: 172-178.
2. Rama Rao G (2012) Padmanabhan Fabrication and mechanical properties of aluminium-boron carbide composites. International Journal of Materials and Biomaterials Applications 2: 15-18.
3. Rajmohan T, Palanikumar K, Arumugam S (2014) Synthesis and characterization of sintered hybrid aluminium matrix composites reinforced with nano-copper oxide particles and micro-silicon carbide particles Composites: Part B 59: 43-49.
4. Hamid RE, Seyed AS, Mohsen HS, Yizhong H (2014) Investigation of microstructure and mechanical properties of Al6061-nanocomposite fabricated by stir casting Materials and Design 55: 921-928.
5. Zadra M, Girardini L (2014) High performance low cost titanium metal matrix composites. Materials Science & Engineering A 608: 155-163.
6. Attar S, Nagaral M, Reddappa HN, Auradi V (2015) A review on particulate reinforced aluminum metal matrix composites 2: 225-229.
7. Cun-Zhu N, Jia-Jun G, Jun-Liang L, Di Zhang (2007) Production of boron carbide reinforced 2024 aluminum matrix composites by mechanical Alloying Materials Transactions 48: 990-995
8. Vijay R, Elanchezhian C, Jaivignesh M, Rajesh S, Parswajinan C (2014) Evaluation of mechanical properties of aluminium alloy alumina boron carbide metal matrix composites. Materials and Design 58: 332-338.
9. Nagaral M, Auradi V, Kori SA (2014) Dry sliding wear behaviour of graphite particulate reinforced Al6061 alloy composite materials 170-174; 592-594.
10. Halverson DC, Snowden WE, Aleksander J (1989) A Processing of boron carbide-aluminum composites. J Am Cerum 72: 775-780.
11. Saikeerthi SP, Vijayaramnath B, Elanchezhian C (2014) Experimental evaluation of the mechanical properties of aluminium 6061-B_4C-SiC composite. International Journal of Engineering Research 3: 70-73.
12. Shorowordi KM, Laoui T (2001) Microstructure and interface characteristics of B_4C, SiC and Al_2O_3 reinforced Al matrix composites: A comparative study 4th International Conference on Mechanical Engineering Dhaka Bangladesh 175-181.

Dynamic Analysis of Cam Manufacturing

Pham HH* and Nguyen PV

Ho Chi Minh City University of Technology – Vietnam National University of Ho Chi Minh City, Vietnam

Abstract

In cam milling process, cutting force is a variant factor during every time period and cam has a quite complex profile that leads to alternate force direction. These consequently, create machine vibration. The dynamic behaviour of machine can be predicted approximately if it is represented by a mathematical model. This paper shows result of cam cutting machine's dynamic, which used Lagrange's equation to solve. In this case, the machine vibration is surveyed only dimensions such as X and Y through using cutting condition with alloy cutting tool to mill a 10 mm thickness steel cam. The machine is modelled into the two degree of freedom vibrating system follow X and Y direction. Each of X and Y table equal to the compound: stiffness, damper and mass, which applied as constant coefficients in Lagrange's equation. On the other hand, analysing cam characteristic and milling process in detail provides the resultant cutting follow X and Y in order to become external force of previous equation. After giving data in sufficient that necessary for problem, Matlab Simulink displays the vibration of X, Y for two states tangent force factor K_t=299.3 and K_t=598.6. At the end, it gives a comparison between these states.

Keywords: Cam mechanism; Milling; Dynamics

Introduction

A cam – a part of the cam – follower mechanical system, which complex peripheral profile depends on follower movement rule is an important mechanism and used popular in automatic mechanical. Designers always expect the accuracy position of follower motion during all operating periods. It means that cam machining has to achieve a tolerance deviation. Vibration of cutting machine has a significant influence on affecting the success of machining. As an illustration, a high value cutting force ease to create a large amplitude of fluctuation that of course decrease the surface quality and dimensional tolerance [1-8]. Moreover, vibration causes the claim and machine duration to be harmful. As a result, the machine dynamic should be considered studying like a main manufacturing factor to handle issues above [9-17]. There are several documents refer to cam machining. For instance, Rothbart [6] illustrated cam manufacturing method, tolerance and errors. In addition, Altintas [17] represent the mechanics of metal cutting. Stephen and Radze [14] mentioned kinematic geometry of surface machining in 2014. They play a role just like the basic theory for this search. The paper also supplies some new acknowledge such as improving the cutting force formulations and analysing the cam machine dynamic, which others do not yet. Furthermore, it provides the essential data to evaluate and choose suitable cutting condition in order to increase the accuracy of cam.

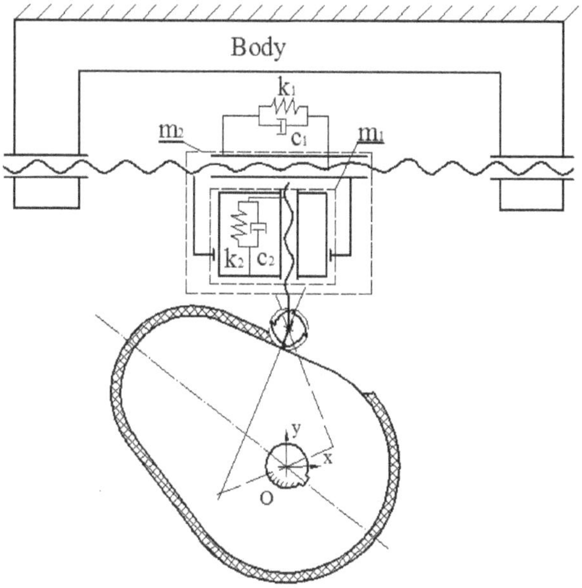

Figure 1: Model of cam machining.

Mathematical Modelling

Physical model the dynamic of cam machining

A cam machine model is explored, include table X carries milling cutting tool and table Y bears table X, both of them can only move one direction which same their name (Figure 1). Tool has clockwise spins around its axis and cam peripheral in the chosen cutting condition. In contrast, cam is fixed stationary. In mathematical model, tables X–Y become mass (m_1, m_2), damping (c_1, c_2) spring (k_1, k_2) and displacements (x, y). Likewise cutter gives the information of cutter's diameter (d) feed rate (f) (or cutting centre velocity V_{B1}), angular velocity ω_c the resultant cutting force F_c. Last of all, follower-cam displays S: follower displacement (S= (ϕ)) follower' angle rotation ϕ=ϕ (t), follower offset e, roller's diameter (D), cutting centre velocity, tangent velocity and transitive velocity of follower (v_c, v_s) (Figure 2).

As mention above, machine's kinematic depends on cam profile, so firstly it needs to determine some factors relative (Figure 2). Cam profile coordinates B at cam angle ϕ:

x coordinate of cam surface profile: x_B=x=ecosϕ-Ssinϕ

y coordinate of cam surface profile: y_B=y=esinϕ+Scosϕ

***Corresponding author:** Pham HH, Ho Chi Minh City University of Technology – Vietnam National University of Ho Chi Minh City, Vietnam
E-mail: phhoang@hcmut.edu.vn

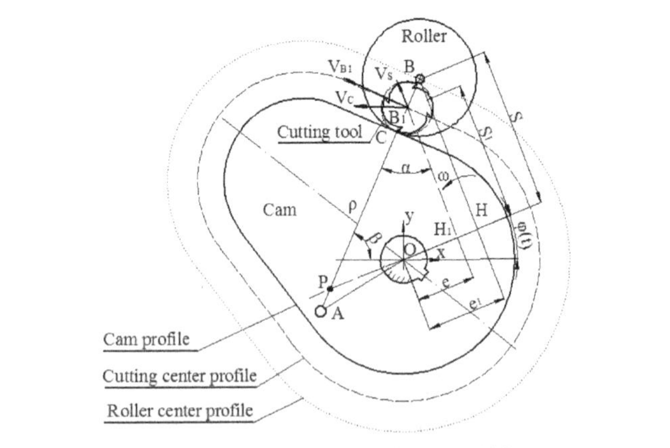

Figure 2: Analyze the kinetic of cam machining.

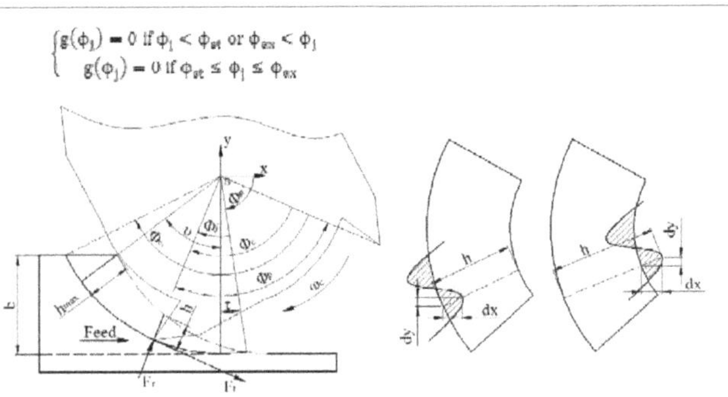

Figure 3: Cutting dynamic model.

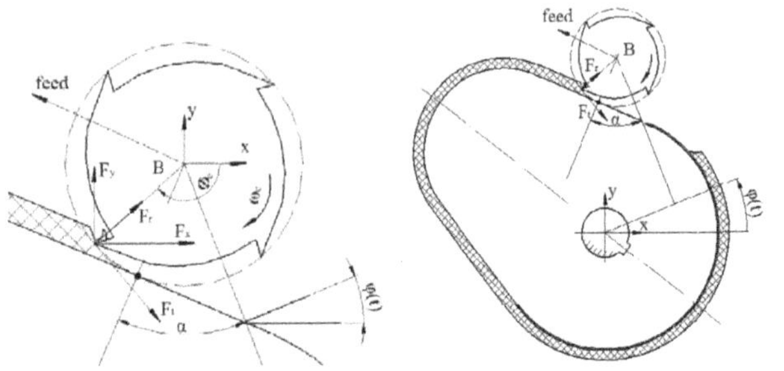

Figure 4: Cam machining dynamic.

Radius of the roller center's curvature s at B:

$$\rho= [e^2+s^2+S^2+2eS]^{3/2}/[e^2+s^2+2S^2+3eS'+SS'']\quad(1)$$

.Cam profile radius at B:

$$\rho_0=\rho-\frac{D}{2}=\left[e^2+S^2+\dot{S}^2+2e\dot{S}\right]^{3/2}/\,|\,e^2+S^2+2\dot{S}^2+3e\dot{S}+S\ddot{S}\,|\quad(2)$$

Profile radius of milling cutting centre B_1:

$$\rho_1=\rho-\frac{D-d}{2}=\left[e^2+S^2+\dot{S}^2+2e\dot{S}\right]^{3/2}/\,|\,e^2+S^2+2\dot{S}^2+3e\dot{S}+S\ddot{S}\,|-\frac{D-d}{2}\quad(3)$$

The pressure angle will be: $\alpha=\tan^{-1}[(S'-e)]/S$ (4)

The velocity of milling cutting centre: $\vec{v}_{B1}=\vec{v}_c+\vec{v}_s$ (5)

Therefore, follower's angular velocity at t is:

$$\omega=v_{B1}.\cos\alpha\,/\left(S-\frac{D+d}{2}\cos\alpha\right)\quad(6)$$

Denominate $\beta=(\overline{AB_1,OX})$ determinated by: $\beta=2\pi-\pi/2-\alpha-(\pi-\phi)=\pi/2-\alpha+$

Cam centre coordinate at A:

x coordinate of cam surface profile: $x_A=e\cos\phi-S\sin\phi-\rho\sin(\alpha-\phi)$

y coordinate of cam surface profile: $y_A=e\sin\phi+S\cos\phi-\rho\cos(\alpha-\phi)$

Dynamic of milling processing

Milling cutters can be considered to have two orthogonal degrees of freedom as shown in Figure 3. The cutter is assumed to have number of teeth with a zero helix angle. The cutting forces excite the structure in the radial force F_r and tangential force F_t, causing displacement X and Y. The dynamic displacements are carried to rotating tooth in the radial or chip thickness direction with the coordinate transformation of fdx.sin $(\varphi_1+\tau)$+dy.cos $(\varphi_1+\tau)$where $_1$ is

the instantaneous angular immersion of tooth measured anticlockwise from the cutting edge starts to contact work piece and$\tau=\sin^{-1}(0.5s/r)$. The resulting chip thickness consists of a static part h, attributed to rigid body motion of the cutter, and a dynamic component caused by the vibrations of the tool at the present and previous tooth periods. Because the chip thickness is measured in the radial direction, the total chip load [17] can be expressed by

$$h(t)=\{h+dx\sin(-\tau)+dy\cos(\varphi_j-\tau)\}g(_j)\quad(7)$$

The function $g(\varphi_j)$ is a unit step function that determines whether the tooth is in or out of cut, that is

$$\begin{cases} g(\phi_j)=0\text{ if }\phi_j<\phi_{st}\ or\phi_{ex}<\phi_j \\ g(\phi_j)=0\text{ if }\phi_{st}\le\phi_j\le\phi_{ex} \end{cases}$$

The static chip load h (7) divided into 2 period:

If $\phi_{st}\le\phi_c\le\phi_{st}+\tau+\eta$, then

$$h=s\{\cos(\tau-\phi_1)\tan[0.5\sin^{-1}\langle(\tau-\phi_1)/r\rangle]-\sin(\tau-\phi_1)\}\quad(8)$$

If, $\varphi_{st}+\tau+\eta<\varphi_c\le\varphi_{st}+\tau+\eta+\upsilon$ then

$$h=r\sin(\upsilon+\tau-\varphi_1)\,[\tan\langle(v+\tau-\varphi_1)/2+\tan(\varphi_1-\tau)].\quad(9)$$

Where $\eta=\tan^{-1}[\sqrt{2rb-b^2}-s]/(r-b)v=\cos^{-1}[(r-b)/r]$ is the angle between .perpendicular direction work plane with h_{max} and out of cutting area. In addition, start angle:

$\varphi_{st}=-\pi/2+\tau$, cutting angle: $\varphi_{st}=-\pi/2+\tau-\varphi_1$, $\varphi_{ex}=-\pi/2-\eta$, exit angle: and $\varphi_p=2\pi/Z$, .cutting pitch angle:

When mill on plane, the cutting force includes tangent and radial cutting force F_t, F_r

$$F_t=K_t th=K_t t\{h+dx\sin(\varphi_1-\tau)+dy\cos(\varphi_1-\tau)\}g(_1)\quad(10)$$

$$F_r=K_r th=K_r t\{h+dx\sin(\varphi_1-\tau)+dy\cos(\varphi_1-\tau)\}g(\varphi_1)\quad(11)$$

Project F_t, (10), F (11) into X and Y direction get F_x, F_y

$$F_x=F_t\cos[\tau-\varphi j]-F_r\sin[\tau-\varphi_j]\quad(12)$$

$$F_y=F_t\cos[\tau-j]-F_r\sin[\tau-\varphi_j]\quad(13)$$

Apply (12), (13) into cam machining in Figure 4, X and Y direction force become

$$F_x=F_t\cos[\phi-\alpha+\tau-\varphi_j]-F_r\sin[\phi-\alpha+\tau-\varphi_j]\quad(14)$$

$$F_y=F_t\cos[\phi-\alpha+\tau-\varphi_j]-F_r\sin[\phi-\alpha+\tau-\varphi_j]\quad(15)$$

Name	Value	Unit
Cam characteristic		
Follower offset e	40	mm
Roller diameter D	40	mm
Cam thickness (steel) t	10	mm
Machine parameter		
Spring k_1, k_2	537728, 296881	N/mm²)
Damping c_1, c_2	5	Ns/mm
Mass m_1, m_2	70, 100	kg
Cutting condition (up-milling)		
Cutter diameter (alloy) d	20	mm
Main spindle speed n	660	rev/m
Feed rate s	0.18	
Feed rate (*F= nSZ*)	3.96	mm/m
Thickness	10	mm
Number of teeth Z	2	tooth
Resultant cutting force F_{cmax}	383.1	N

Table 1: The value input.

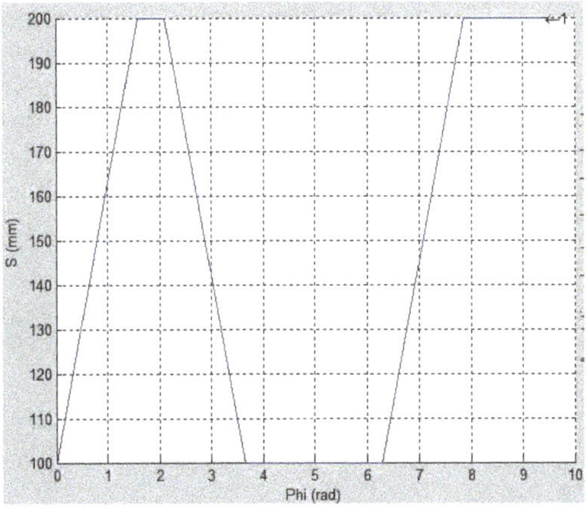

Figure 5: Follower displacement.

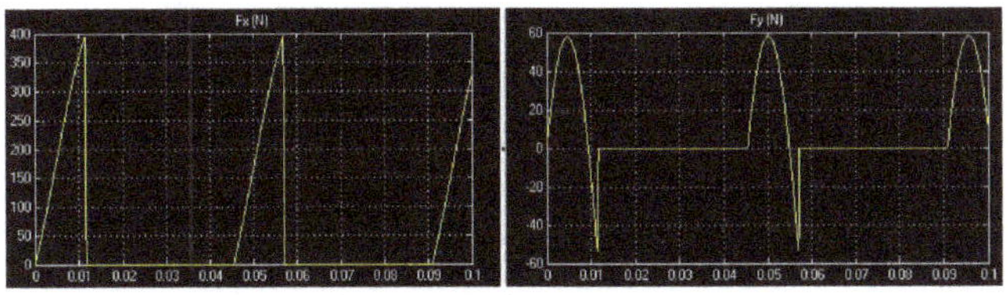

Figure 6: Cutting force diagram by K_t299.3.

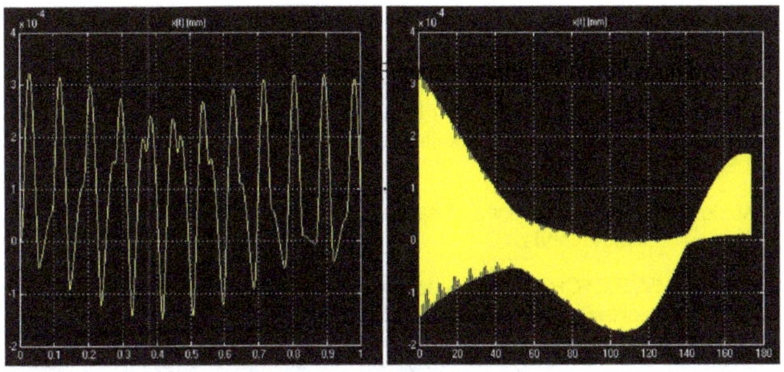

Figure 7: X table oscillate in the first second and one period (173.85 s) by K_t299.3.

Dynamics of cam cutting

To survey the oscillation of this machine, Lagrange's equation [2] is compatible to use

$$\frac{d}{dt}=\frac{d}{dt}\frac{\partial V}{\partial q}-\frac{\partial T}{\partial q}+\frac{\partial V}{\partial q}=F \tag{16}$$

F is the total external force, T [2] and V [2] are the system kinetic and potential energy, respectively. They are defined in (17) and (18)

$$T=\frac{1}{2}\sum_{1=3}^{0}[m1(x_1'^2+y_1'^2)+J_1\phi_1^2] \tag{17}$$

$$V=\sum[m_i g y_i+0.5(\Delta l)^2] \tag{18}$$

Combine equations (10) and (11) with equations (16-18) to derive the dynamic equation of cam machining (19) and (20):

$$m_1x''+c_1x'+k_1x=F_t\cos[\phi-\alpha+\tau-\varphi_j]-F_r\sin[\phi-\alpha+\tau-\varphi_j\ \tau-\varphi_j] \tag{19}$$

$$m_2y''+c_2y'+k_2y=F_t\sin[\phi-\alpha+\tau-\varphi_j]-F_r\cos[\phi-\alpha+\tau-\varphi_j\ \tau-\varphi_j] \tag{20}$$

Example Model

The input value includes machine parameters in Table 1 and follower displacement rule [4] in Figures 5 and 6.

Use the data in Table 1, the follower angle rotation to (21), (22), thus force factor K_t, K_r and follower angular rotation are:

$$\omega=F\cos\alpha/(S-\frac{D+d}{2}\cos\alpha)=3.96\cos\alpha/(S-30\cos\alpha) \tag{21}$$

$$K_t=F_c/th(t)_{max}=299, K_r=0.3K_t=89.3 \tag{22}$$

Derive milling tool angle velocity $\omega_c=22\pi$ (rad/s) $0\le\varphi 1\le\cos^{-1}0.7$ so the so cutting time $t_c=\frac{\phi_{jmax}}{\omega_c}=0.01115(s)$ cutting time:

$t_p=2\pi/(Zw_c)=1/22$ (s)

Finally, model dynamic equation will become (23) and (24), then they are solved by Matlab Simulink similarly.

$200x''+0{,}002x'+1000000x=10\{h+dx\sin(\varphi_j-\tau)+dy\cos(\varphi_j-\tau)\}g(\varphi_j)\{299.3\cos(\phi-\alpha+\tau-\varphi_j)-89.8\sin(\phi-+\tau-\varphi_j)\}$ (23)

$100y'+0{,}002+350000y''=10\{h+dx\sin(\varphi_j-\tau)+dy\cos(\varphi_j-\tau)\}g(\varphi_j)\{299.3\cos(\phi-\alpha+\tau-\varphi_j)-89.8\sin(\phi-+\tau-\varphi_j)\}$ (24)

Result and Discussion

Figures 7-11 show the table X and Y oscillation in two different stations: k_t=299.3 and k_t=598.6 during one cutting period (173.85s).

Natural circular frequency [1,2]:

$$\omega_{n1}=\sqrt{k_1/m_1}=1624.5(\text{rad/ s}) \tag{25}$$

$$\omega_{n2}=\sqrt{k_2/m_2}=2059.4(\text{rad/ s}) \tag{26}$$

Damping ratio,

$$\zeta_1=0.5c_1/\sqrt{k_1m_1}=0.00034<1;\zeta_1=0.5c_2/\sqrt{k_2m_2}=0.00055<1 \tag{27}$$

Therefore, with the aid of Euler's formula, the general solution x(t) and y(t) can be written in the form underdamping free vibration [1], derives the freedom vibrating equation of machine.

$$X=0.0023e^{-0.55t}\sin[1624.5t]\ (\text{mm}) \tag{28}$$

$$Y=0.0027e^{-1.13t}\sin[2059.4t]\ (\text{mm}) \tag{29}$$

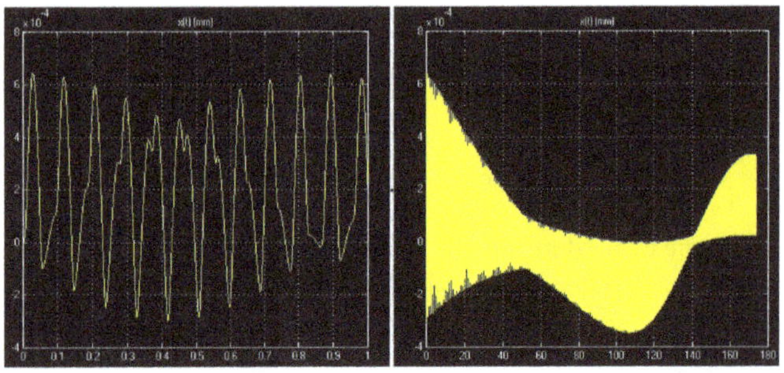

Figure 8: X table oscillate in the first second and one period (173.85 s) by $K_{t=}$598.6.

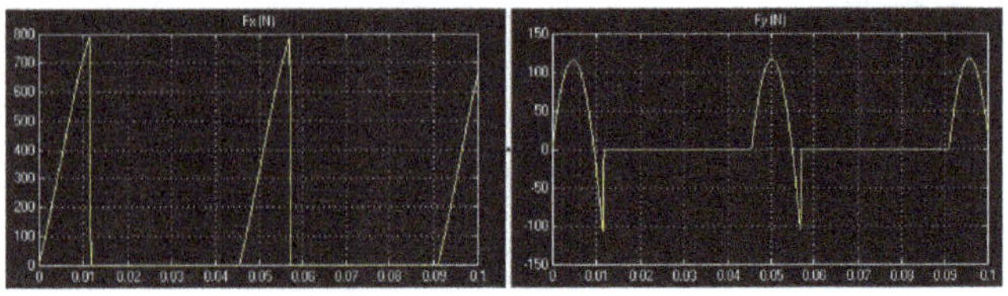

Figure 9: Cutting force diagram by $K_{t=}$598.6.

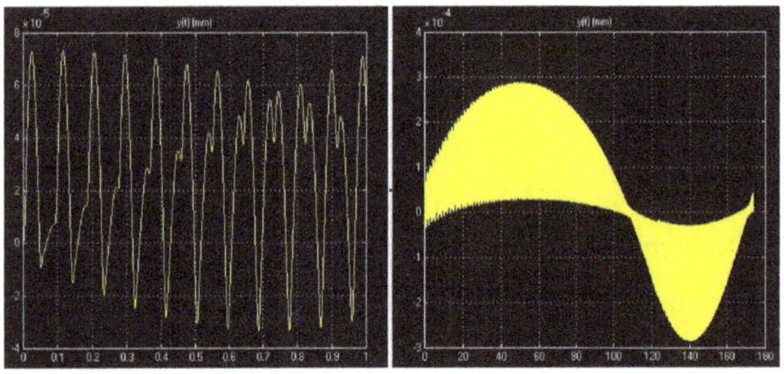

Figure 10: Y table oscillate in the first second and one period (173.85 s) by K_t=299.3.

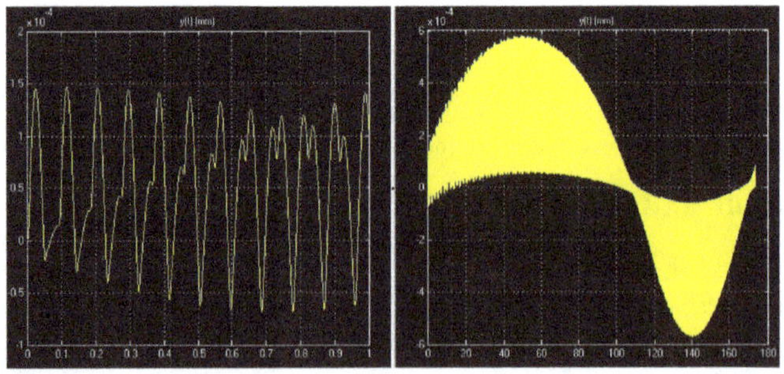

Figure 11: Y table oscillate in the first second and one period (173.85 s) by K_t=598.6.

Force F_x, F_y in Figures 6 and 8 is the result of the discontinuous cutting of milling process. They also have the variant value during each tool's revolution.

The results show the relationship between cutting force and oscillation. Assume that cutting force increase 150%, yet other conditions still unchanged. As anticipated, this boost as two times as much as the amplitude of fluctuation and displacement of X, Y. In other words, the amplitude of X change from 1.5×10^{-4} (mm) to 3×10^{-4} (mm) and other is 3.5×10^{-5} (mm) to 0.7×10^{-5} (mm). The maximum X and Y displacements also have the same trend, those rise from 3.1 $\times 10^{-4}$, 3×10^{-4}(mm) to 6.2×10^{-4} (mm), 6×10^{-4} (mm) sequentially. Their shapes alternate between F_x and F_y In contrast, the frequency responses are invariable, that means the input load and the table's physical characteristic almost affect to the table's movement. Therefore, the vibration is reduced quickly by increasing the stiffness of cutting system: tool, clamp, machine, machine's operating tightness. The rising mass and damper of tables benefit for machining in the same way.

In this situation, cam cutting machine's factors and the hardness of cam work piece are considered being unchanged. To keep machining process smoothly, it needs to decrease cutting force. Feed rate, cutting speed, thickness and width have the influence on cutting force. Tool material and abrasive cutter edge are in the same way. Those have to be controlled in strictness. The results are nearly similar as real milling models such as shoulder milling. It is useful for estimating the machining deviation.

Conclusion

The cam machining is modelled on elements: mass, springs, dampers of Lagrange's equation. Also, milling force, external force of previous equation, is also analysed thoroughly. Then the machine's vibration is completely achieved by using Matlab Simulink to solve Lagrange's equation. The results illustrate explicitly the displacements and frequencies of machine' table, those are correspondent with a rigid cutting system. Finally, this paper is a beneficial result to study the decline of cam machining deviation. In future, this model should be developed in real testing model to have more completely the evaluation of the dynamic of cam cutting machine.

Acknowledgement

The paper content is the result of project named "Development a NC milling machine for planar cams" – project code: C2015-20-03. The project is financially sponsored by the Vietnam National University of Ho Chi Minh City.

References

1. Dazzio A (2013) Fundamentals of structure dynamic. An-Najah University, Palestine.
2. Sinha A (2010) Vibration of mechanical systems. Cambridge, Britain.
3. Bisu1 CF, Zapciu1 M (2010) New approach of envelope dynamic analysis for milling process. 8th International Conference on High Speed Machining, Metz: France.
4. Myszka DH (2012) Machines & Mechanism. Prentice Hall, USA.
5. Klocke F (2011) Manufacturing process 1. Springer, Germany.
6. Rothbart HA (2004) Cam design handbook. Mc Graw-Hill, USA.
7. Yangui H, Zghal B (2010) Influence of cutting and geometrical parameters on the cutting force in milling. Scientific Research Engineering 2: 751-761.
8. Hung JP, Lai YL (2012) Prediction of the dynamic characteristics of a milling machine using the integrated model of machine using the integrated model of machine frame and spindle unit. International Journal of Mechanical, Aerospace, Industrial, Mechatronic and Manufacturing Engineering.
9. Groover MP (2010) Fundamentals of modern manufacturing. USA.
10. Farid Koura MM, Lotfy Zamzam M (2015) Simulation approach to study the behavior of a milling machine's structure during end milling operation. Turkish J Eng & Envi sci.
11. Koster MP (1973) Vibrations of cam mechanisms and their consequences on the design. Netherlands.
12. Ingalkar MV, Patil CR (2013) Computer aided kinematic analysis and simulation of the cam-follower mechanism. Int J Mech Eng and Robotics Res.
13. Petru Adrian Pop (2009) The analysis of dynamic stability at milling machine tools. ASME 2009 International Design Engineering Technical Conferences and Computers and Information in Engineering Conference.

14. Stephen P, Radze Vich (2014) Generation of surface. CRC Press, USA.

15. Wei-Hsiang Lai (2000) Modeling of cutting forces in end milling operations. Tamkang J Sci and Eng 3: 15-22.

16. You YR, Liu HB, Gao L (2013) Dynamics analysis of milling machine spindle. Applied Mechanics and Materials 345:137-140.

17. Altintas Y (2012) Manufacturing automation. Cambridge University Press, Britain.

33

Computational Study of Mixed Convection Flow with Algebraic Decay of Mainstream Velocity in the Presence of Applied Magnetic Field

Muhammad Ashraf*, Uzma Ahmad, Masud Ahmad and Sultana N

Department of Mathematics, Faculty of Science, University of Sargodha, Pakistan

Abstract

The present study concern with the two dimensional viscous, incompressible, electrically con ducting, mixed convection flows with algebraic decay of mainstream velocity $U(x)=(1-x)^{-a}$. The physical phenomenon of the problem is simulated by using the Primitive Variable Formulation (PVF) for Finite Difference Method (FDM) and Stream Function Formulation (SFF) for Local Non-Similarity Method (LNS). The physical behaviors of momentum and temperature fields are given graphically. The results obtained for skin friction and rate of heat transfer for different values of physical parameters are also compared by both methods and presented in tabular form.

Keywords: Mixed convection; Heat transfer; Skin friction; Algebraic decay of mainstream velocity

Introduction

When the surface temperature is different from the ambient fluid temperature, it gives rise to a density gradient in the thermal boundary layer region. This density gradient produces a body force in the form of buoyancy of the fluid. The buoyancy force along with some external force constitutes a mixed convection flow. There the external force can either be in the form of free stream or some moving or stretching surface. With this understanding, we formulate the problem of mixed convection flow and the possible literature survey for the problem under consideration is given in preceding paragraph.

Ostrach [1] presented the similarity solution of natural convection along vertical isothermal plate. The instantaneous action of buoyancy and induced magnetic forces leading to natural convection heat transfer was investigated by Sparrow et al. [2]. They found out that in the presence of magnetic field, the natural convection heat transfer to liquid metals may be considerably affected [3]. The mixed convection over a horizontal surface due to uniform free stream has been studied by many researchers. Gebhart [4] used an analytical technique to analyze the effect of dissipation on natural convection [5,6]. The joint free and forced convection along with uniform heat flux in the presence of strong magnetic field was studied by Lioyd el al. [7]. Dwivedi [8] examined the free convection oscillatory hydromagnetic laminar flow of a viscous, electrically conducting and incompressible fluid over a vertical infinite porous plate under the influence of a transverse magnetic field when the temperature or heat flux at the surface oscillates in magnitude but not in direction. Soundalgekar [9] studied natural convection flow along vertical porous plate with suction and viscous dissipation [10].

A particular self-similar solution of the mixed convection flow over a horizontal surface was presented by Schneider [11] for the case of specified wall temperature as an inverse square root of the distance from the top edge. Chakrobarti et al. [12] illustrated the heat transfer and flow of an incompressible electrically conducting fluid pass a stretching sheet. They presented an analytical solution for the flow and the numerical solution for the heat transfer. Under the influence of free stream oscillations, Perdikis et al. [13] examined the laminar free convection and mass transfer effects of a viscous, incompressible heat generating fluid-flow along an impulsively vertical plate with heat flux and constant suction at the plate. Wantanabe et al. [14] investigated the heat transfer in thermal boundary layer of MHD flow over a flat plate. The mixed convection boundary layer flow of an incompressible, viscous and electrically conducting fluid in the presence of transverse magnetic field along a vertical porous plate was examined by Hosain et al. [15]. In the presence of Hall current effect, Chamkha [16] expressed the MHD free convection pass a vertical plate in thermally satisfied porous medium. Chen [17] presented his study on MHD natural convection flow over a permeable inclined surface with variable wall temperature and concentration. He studied that increasing the angle of inclination decreases the effect of buoyancy force. Mostafa [18] studied the heat transfer and flow of an incompressible electrically conducting viscous fluid which passes a continuously moving vertical infinite plate with uniform suction and heat flux under the action of radiation taking into account the effects of variable viscosity. Considering a homogeneous chemical reaction of first order, Odat et al. [19] numerically examined the effect of magnetic field on the transient free convection flow of an electrically conducting fluid pass an impulsively started isothermal vertical plate. They observed that increasing the chemical reaction parameter decreases the velocity and concentration. Makinde [20] performed an analysis to examine the hydromagnetic mixed convection flow of a viscous, incompressible and electrically conducting fluid and mass transfer along a vertical porous plate with constant heat flux embedded in a porous medium.

However, when the free convective flows occur at high temperatures, radiation effects on the flow become significant. Many processes in engineering areas occur at high temperatures and knowledge of radiative heat transfer becomes very important for the design of the pertinent equipment. Nuclear power plants, gas turbines and the various propulsion devices for aircraft, missiles and space vehicles are examples of such engineering areas. The inclusion of radiation effects in the energy equation leads to a highly non-linear partial differential equation. Under the influence of thermal radiation along with variable

***Corresponding author:** Muhammad Ashraf, Department of Mathematics, Faculty of Science, University of Sargodha, Pakistan
E-mail: mashraf682003@yahoo.com

suction and thermophoresis. Alam et al. [21] numerically examined the two-dimensional steady MHD mixed convection and mass transfer flow through an inclined semi-infinite porous plate. In the presence of transversely applied magnetic field, MHD flow of an electrically conducting, viscous and incompressible fluid through a semi-infinite vertical plate with mass transfer was examined by Palani et al. [22]. Ferdows et al. [23] presented MHD boundary layer flow of a nanofluid through an exponentially stretching sheet. They also discussed and observed the heat transfer characteristics and the effects of governing parameters on the flow field. Ashraf et al. [24] studied the mixed convection flow past a magnetized vertical porous plate numerically.

In the light of above literature survey, we formulate the problem of mixed convection flow with algebraic decay of mainstream velocity in the presence of applied magnetic field.

Mathematical Analysis and Governing Equations

In this section, we formulate the bulk of mathematical equations governing the MHD convective flow along a vertical flat plate. We combine the basic momentum and energy equations into self-consistent system. This set of equations considers steady two dimensional flow of viscous incompressible conducting fluid along a semi-infinite vertical surface. The decay is only algebraic for $0<\alpha<1$ (Figure 1). A magnetic field of strength *Bo(x)* is considered to be applied parallel to the *y*-axis which is normal to the plate. Under the usual Boussinesq approximation, the flow is governed by the following boundary layer equations following by Merkin [10].

$$\frac{\partial u}{\partial x}+\frac{\partial v}{\partial y}=0$$

$$u\frac{\partial u}{\partial x}+v\frac{\partial v}{\partial y}=U\frac{dU}{dx}+v\frac{\partial^2 u}{\partial y^2}+g\beta\left(T-T_\infty\right)-\frac{\sigma B_0^2}{\rho}u \qquad (1)$$

$$u\frac{\partial T}{\partial x}+v\frac{\partial T}{\partial y}=\alpha\frac{\partial^2 T}{\partial y^2}$$

where

$$U(x)=(1-x)^{-\alpha}$$

The appropriate boundary conditions for the present problem are

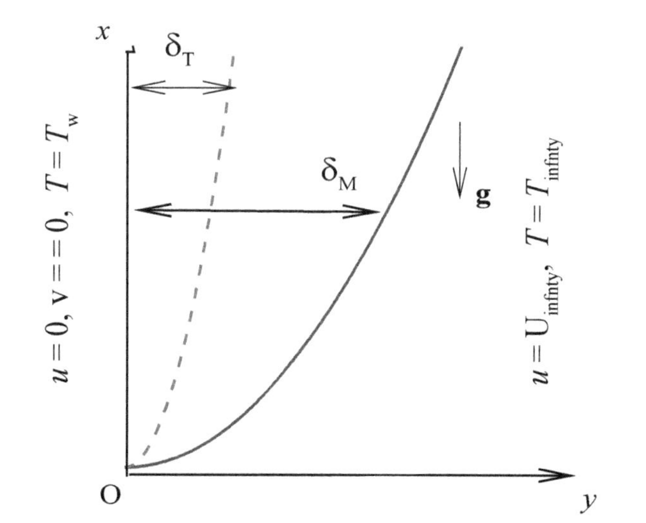

Figure 1: Boundary layer equations.

$$u(x,0)=0,\quad v(x,0)=0,\quad T(x,0)=T_w \qquad (2)$$

$$u(\infty)=U(x),\quad T(\infty)=T_\infty$$

By using the dimensionless variables

$$\bar{x}=\frac{x}{L},\quad \bar{y}=R_{ex}^{\frac{1}{2}}\frac{y}{L},\quad \bar{u}=\frac{u}{U_0},\quad \bar{v}=R_{ex}^{\frac{1}{2}}\frac{v}{U_0}$$

$$\bar{\theta}=\frac{\bar{T}-\bar{T}_\infty}{\bar{T}_w-\bar{T}_\infty},\quad \bar{U}=\frac{U}{U_0} \qquad (3)$$

The dimensionless boundary layer equations along with boundary conditions are given as:

$$\frac{\partial\bar{u}}{\partial\bar{x}}+\frac{\partial\bar{v}}{\partial\bar{y}}=0 \qquad (4)$$

$$\bar{u}\frac{\partial\bar{u}}{\partial\bar{x}}+\bar{v}\frac{\partial\bar{u}}{\partial\bar{y}}=\bar{U}\frac{d\bar{U}}{d\bar{x}}+\frac{\partial^2\bar{u}}{\partial\bar{y}^2}+\lambda\bar{\theta}-M\bar{u} \qquad (5)$$

$$\bar{u}\frac{\partial\bar{\theta}}{\partial\bar{x}}+\bar{v}\frac{\partial\bar{\theta}}{\partial\bar{y}}=\frac{1}{P_r}\frac{\partial^2\bar{\theta}}{\partial\bar{y}^2} \qquad (6)$$

The dimensionless boundary conditions are:

$$\bar{u}(x,0)=0,\quad \bar{v}(x,0)=0,\quad \bar{\theta}(x,0)=1 \qquad (7)$$

$$\bar{u}(\infty)=\bar{U}(x),\quad \bar{\theta}(\infty)=0$$

where

$$\lambda=\frac{G_{rL}}{R_{eL}^2},\quad G_{rL}=\frac{g\beta\Delta TL^3}{v^2},\quad R_{eL}=\frac{U_0L}{v},\quad M=\frac{\sigma B_0^2x^{-1}}{\rho U_0},\quad P_r=\frac{v}{\alpha}$$

the Richardson Grashof number, Reynolds number, magnetic parameter, Prandtl number and the characteristic length respectively.

Method of Solutions

We now turn to get the numerical solutions of the problem. For this purpose we will use Primitive variable formulation for Finite Difference Method and Stream function formulation for Local Non-Similarity Method.

Primitive variable formulation

We now introduce the group of transformations that is known as Primitive Variable Formulation (PVF) to transform the given system of equations into convenient form for integration

$$\bar{u}=U(\xi,Y),\quad \bar{v}=\bar{x}^{\frac{1}{2}}V(\xi,Y),\quad \bar{U}=(1-\bar{x})^{-\alpha} \qquad (8)$$

$$\bar{\theta}=\bar{x}^{-1}\theta(\xi,Y),\quad Y=\bar{x}^{-\frac{1}{2}}\bar{y},\quad \xi=-\bar{x}$$

Thus (5) and (6) becomes

$$U\xi\frac{\partial U}{\partial\xi}-\left(\frac{1}{2}YU+V\xi\right)\frac{\partial U}{\partial Y}=\frac{\partial^2 U}{\partial Y^2}+\lambda\theta-MU-\alpha\xi(1+\xi)^{-2\alpha-1} \qquad (9)$$

$$U\xi\frac{\partial U}{\partial\xi}-\left(\frac{1}{2}YU+V\xi\right)\frac{\partial\theta}{\partial Y}-U\theta=\frac{1}{P_r}\frac{\partial^2\theta}{\partial Y^2} \qquad (10)$$

The appropriate boundary conditions satisfied by the above system of equations are

$$U(\xi,0)=V(\xi,0)=0,\quad \theta(\xi,0)=1 \qquad (11)$$

$$V(\xi,\infty)=1, \quad \theta(\xi,\infty)=0$$

Discretization: Now we first discretize (9)-(20) using FDM (Finite Difference Method)

$$\frac{\partial U}{\partial \xi}=\frac{U_{i,j}-U_{i,j+1}}{\Delta \xi}$$

$$\frac{\partial U}{\partial Y}=\frac{U_{i+1,j}-U_{i-1,j}}{2\Delta Y} \tag{12}$$

$$\frac{\partial^2 U}{\partial Y^2}=\frac{U_{i+1,j}-2U_{i,j}+U_{i-1,j}}{\Delta Y^2}$$

and get discretized form

$$A_1U_{i+1,j}-B_1U_{i,j}+C_1U_{i-1,j}=D_1 \tag{13}$$

where

$$A_1=1+\frac{1}{2}\Delta Y\left(\frac{1}{2}Y_jU_{i,j}+V_{i,j}\xi_i\right)$$

$$B_1=2+\frac{\Delta Y^2}{\Delta \xi}U_{i,j}\xi_i+M\Delta Y^2$$

$$C_1=1-\frac{1}{2}\Delta Y\left(\frac{1}{2}Y_jU_{i,j}+V_{i,j}\xi_i\right) \tag{14}$$

$$D_1=-\frac{\Delta Y^2}{\Delta \xi}U_{i,j}\xi_iU_{i,j+1}-\lambda\theta_{i,j}\Delta Y^2+\alpha\xi_i(1+\xi_i)^{-2\alpha-1}\Delta Y^2$$

Similarly we discretize (10) using FDM

$$\frac{\partial \theta}{\partial \xi}=\frac{\theta_{i,j}-\theta_{i,j+1}}{\Delta \xi}$$

$$\frac{\partial \theta}{\partial Y}=\frac{\theta_{i+1,j}-\theta_{i-1,j}}{2\Delta Y} \tag{15}$$

$$\frac{\partial^2 \theta}{\partial Y^2}=\frac{\theta_{i+1,j}-2\theta_{i,j}+\theta_{i-1,j}}{\Delta Y^2}$$

and get the discretized form

$$A_2\theta_{i+1,j}-B_2\theta_{i,j}+C_2\theta_{i-1,j}=D_2 \tag{16}$$

where

$$A_2=\frac{1}{P_r}+\frac{1}{2}\Delta Y\left(\frac{1}{2}Y_jU_{i,j}+V_{i,j}\xi_i\right)$$

$$B_2=\frac{}{}+\frac{}{}U_{i,j}\ _i+\Delta Y\ U_{i',j}$$

$$C_2=\frac{1}{P_r}-\frac{1}{2}\Delta Y\left(\frac{1}{2}Y_jU_{i,j}+V_{i,j}\xi_i\right) \tag{17}$$

$$D_2=-\frac{\Delta Y^2}{\Delta \xi}U_{i,j}\xi_i\theta_{i,j+1}$$

From this, we get a tri-diagonal system of algebraic equations, which is solved by using Gaussian elimination technique. The computational domain is of size 80 × 80 mesh points, and computational is started from $\xi=0$ and march down implicitly, for this we have taken the step size Δξ=0.05 and ΔY=0.05. We calculate the effects of different parameters on coefficients of skin friction, rate of heat transfer, velocity and temperature field with the help of equations.

$$Re_x^{\frac{1}{2}}Cf_x=\left(\frac{\partial U}{\partial y}\right)_{Y=0}, \quad Nu_xRe_x^{-\frac{1}{2}}=\left(\frac{\partial \theta}{\partial Y}\right)_{Y=0} \tag{18}$$

and velocity can be calculated directly using equation of continuity (4) as shown below:

$$V_{i,j}=V_{i-1,j}-\frac{1}{2}\left(\xi\frac{\Delta y}{\Delta \xi}-Y_j\right)U_{i,j}+\frac{1}{2}\xi\frac{\Delta y}{\Delta \xi}U_{i,j-1}-\frac{1}{2}Y_jU_{i-1,j} \tag{19}$$

Stream function formulation

To get the set of equations in convenient form for integration, we define the following one parameter of transformations for the dependent and independent variables.

$$\psi=(2x)^{\frac{1}{2}}(1-x)^{\frac{1}{2}(1-\alpha)}G(x,\eta) \tag{20}$$

$$\eta=\frac{y}{(2x)^{\frac{1}{2}}(1-x)^{\frac{1}{2}(1-\alpha)}}$$

$$\theta=\frac{T-T_\infty}{T_w-T_\infty}$$

By using this group of transformations we have x and y components of velocity as

$$u=(\ -x)\quad G'$$

$$v=-\frac{(2x)^{\frac{1}{2}}(1-x)^{\frac{1}{2}(1-\alpha)}}{2x(1-x)}\left(2x(1-x)\frac{\partial G}{\partial x}+\eta(2x+\alpha x-1)G'-x(1-\alpha)G+(1-x)G\right) \tag{21}$$

Now the complete transformed form of the model of the problem attains the following form

$$G'''+2\alpha x\left(1-G'^2\right)+2(1-x)\lambda\theta-2M(G'-1)+(1-2x+\alpha x)GG''$$

$$=2x(1-x)\left(G'\frac{\partial G'}{\partial x}-G''\frac{\partial G}{\partial x}\right) \tag{22}$$

$$\frac{1}{P_r}\theta''-\theta'G(1-2x+\alpha x)=2x(1-x)\left(G'\frac{\partial \theta}{\partial x}-\theta'\frac{\partial G}{\partial x}\right) \tag{23}$$

with transformed boundary conditions

$$G(0)=0, \quad G'(0)=0, \quad \theta(0)=1$$

$$G'(\infty)=1, \quad \theta(\infty)=0 \tag{24}$$

Local non-similarity method

The solution of the set of locally non-similar equations can be obtained by using local Non- similarity method (LNS). The governing coupled Partial Differential equations are transformed into a sequence of coupled ordinary differential equations (ODE), which are then solved numerically. The local non-similarity method avoids this by deleting the stream-wise a derivative term, that is, the term with $\partial/\partial\xi$. This changes the partial differential equation (PDE) into non-linear ODE, that still has to be solved numerically. The local non-similarity method was first developed by Sparrow and Yu [6], Minkowycz and Sparrow [2] and later used by Chen [17] in his work.

To adopt this method, we first retain all terms in the transformed form with x derivative. Now the systems of equations that we have to solve by using LNS are equations (23) and (24).

Now by putting $F = \partial G / \partial x$ and $\varphi = \partial \theta / \partial x$ we have:

$$G''' + 2\alpha x\left(1-G'^2\right) + (1-2x+\alpha x)GG'' + 2\lambda(1-x)\theta + 2M(1-G') = 2x(1-x)(G'F' - G''F) \quad (25)$$

and

$$\frac{1}{P_r}\theta'' - (1-2x+\alpha x)G\theta' = 2x(1-x)(G'\varphi - \theta' F) \quad (26)$$

With boundary conditions

$$F(0)=0, \quad F'(0)=0, \quad \varphi(0)=1$$

$$F'(\infty)=1, \quad \varphi(\infty)=0 \quad (27)$$

$$F''' - 4\alpha x G'F' + 2\alpha\left(1-G'^2\right) + (1-2x+\alpha x)(GF'' - G''F)$$
$$+(\alpha-2)GG'' + 2\lambda(1-x)\varphi - 2\lambda\theta = 2x(1-x)\left(F'^2 - FF''\right) + 2(1-2x)(G'F' - G''F) \quad (28)$$

$$\frac{1}{P_r}\varphi'' - (1-2x+\alpha x)(G\varphi' + \theta') - (\alpha-2)G\theta' = 2x(1-x)(\varphi F' - F\varphi') + 2(1-2x)(G'\varphi - \theta' F) \quad (29)$$

With boundary conditions

$$F(0)=0, \quad F'(0)=0, \quad \varphi'(0)=1$$

$$F'(\infty)=1, \quad \varphi(\infty)=0 \quad (30)$$

This is the second level of truncation of the Local Non-Similarity because this approximation is made by dropping terms in second level of equations. This system of coupled nonlinear ordinary differential equation now can be solved numerically by employing the Nachtsheim-Swigert [5] iteration technique in conjunction with the shooting procedure to determine the unknown boundary conditions $f''(0)$ and $\theta'(0)$. Once we find the unknowns $f''(0)$ and $\theta'(0)$, we are able to find the coefficients of skin friction and rate of heat transfer for different effects of parameters such as algebraic decay parameter *α*, Hartmann number *M*, Richardson parameter *Ri* and Prandtl number Pr with the help of following relations:

$$Re_x^{\frac{1}{2}} Cf_x = f''(\xi, 0), \qquad Nu_x Re_x^{-\frac{1}{2}} = \theta'(\xi, 0) \quad (31)$$

Results and Discussions

In this section we will discuss the results obtained by Finite Difference Method and Local Non- Similarity Method. In this study, we explore the effects of different parameters on coefficients of skin friction, rate of heat transfer, velocity field and magnetic field. With this understanding the Figures 2a and 2b presents the effects of different values of algebraic decay parameter *α*=0.01, 0.05, 0.1 by keeping other parameters constant on momentum and boundary layer thicknesses. From these figures it is possible to see how the algebraic decay parameter *α* influences the free stream velocity distribution at Y=16-18, and there is no effects of algebraic decay parameter on temperature distribution. The reason is that if we see the equation (2) the algebraic decay is only the part of free stream velocity *U*=(1−*x*) −*α* outside the boundary so in both cases this parameter has its no effects within the boundary layer.

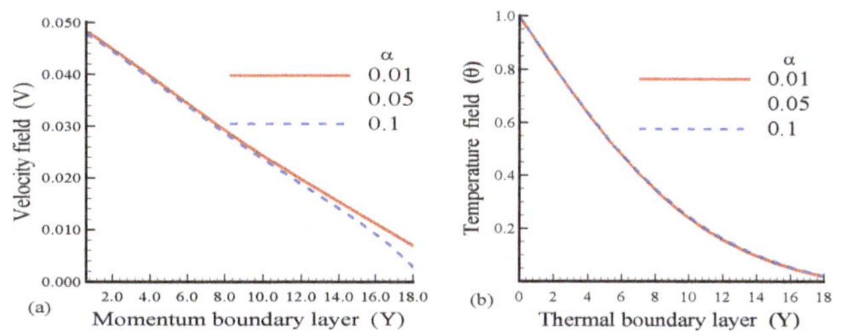

Figure 2: (a) velocity and (b) temperature field profile against boundary layer thickness Y for different values of algebraic decay parameter α=0.01, 0.05, 0.10 when P_r=0.71, M=1.0 and for Richardson number R_i=1.0.

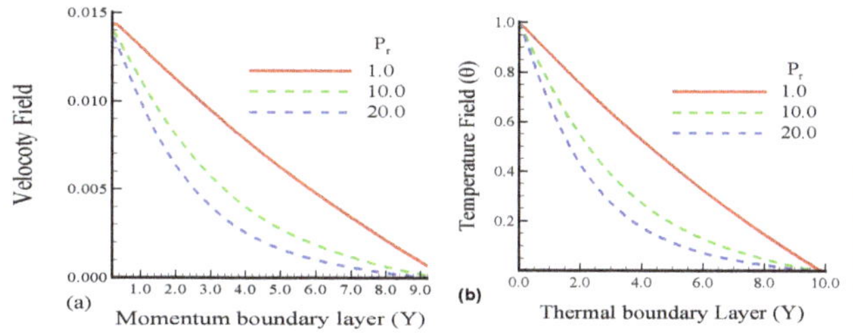

Figure 3: (a) velocity and (b) temperature field profile against boundary layer thickness Y for different values of Prandtl number P_r=1.0, 10.0, 20.0 when α=0.07 M=1.0, and for Richardson number R_i=5.0.

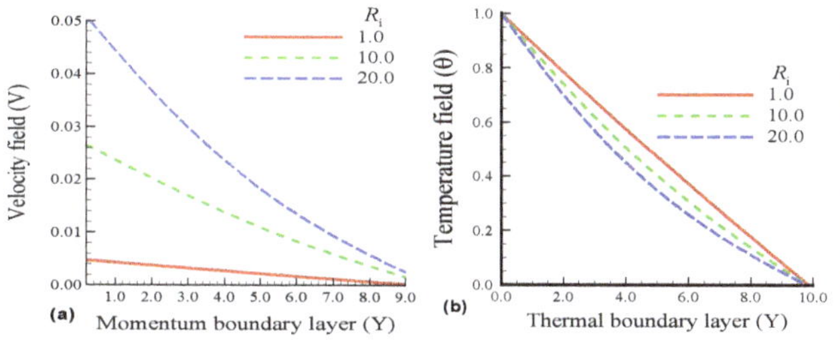

Figure 4: (a) velocity and (b) temperature field profile against boundary layer thickness Y for different values of Richardson number R_i =1.0, 10.0, 20.0 when α=0.07 M=1.0, and for Prandtl number P_r=0.71.

In Figures 3a and 3b, the influence of Prandtl number Pr (which is the ratio of kinematic viscosity to thermal diffusion) on momentum and thermal boundary layer thicknesses is shown. For increasing value of Prandtl number both the momentum and thermal boundary layer thicknesses are decreased with the reason that with the increase of Prandtl number Pr the fluid become more viscous which slow down the fluid motion, so distributions on both cases are decaying and this decay is asymptotic. Figures 4a and 4b show some results of variation of Richardson/mixed convection parameter R_i=1.0, 10.0, 20.0 on velocity and temperature profile. In this range of Richard parameter the momentum boundary layer thickness is decreased because the Richardson number acts like pressure gradient which take the responsibility of this physical phenomena but its effects in the case of thermal boundary layer is vice versa. Figures 5a and 5b shows the behavior of various values of Hartmann number *M*=1.0, 10.0, 20.0 on momentum and thermal boundary layer thicknesses. From this numerical exercise we can see the increase of applied magnetic field momentum boundary layer is decreased and thermal boundary layer is increased for maximum value of M=20.0.

Tables 1 and 2 shows some of the results obtained by Finite Difference Method and Local Non-Similarity method for different values of Hartmann number M on coefficient of skin friction and rate of heat transfer. From these tables, it is conclude that with the increase of Hartmann number M the coefficient skin friction and rate of heat transfer both are decreased. For increasing value of Hartmann number increase the magnetic energy and decrease the kinetic energy in the fluid flow domain due to this reason both quantities are decreased.

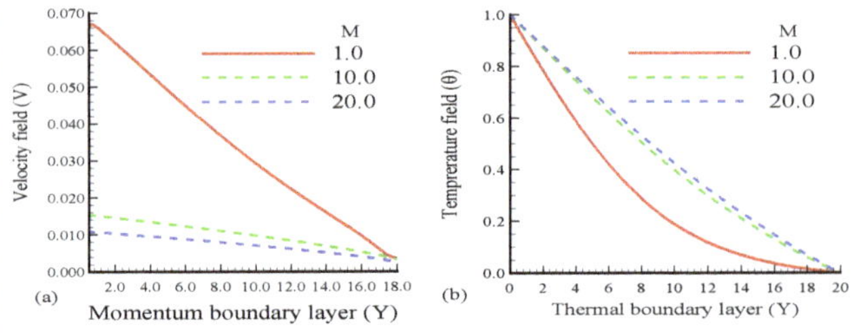

Figure 5: (a) velocity and (b) temperature field profile against boundary layer thickness Y for different values of Hartman number M=1.0, 10.0, 20.0 when α=0.07 R_i=1.0 and for Prandtl number P_r=0.71.

M	FDM	LNS
0.1	0.79478	0.79104
1.0	0.47204	0.47975
2.0	0.39717	0.37143
3.0	0.35781	0.34295
4.0	0.33172	0.33344
5.0	0.29738	0.29370
6.0	0.28504	0.27167
7.0	0.27465	0.25359
8.0	0.26571	0.22936

Table 1: Numerical values of coefficient of Skin friction $Re_x^{\frac{1}{2}}Cf_x$ obtained for different values of M when P_r=10.0, α=0.4, R_i=2.0.

M	FDM	LNS
0.1	0.58954	0.58320
1.0	0.36669	0.36761
2.0	0.31328	0.31274
3.0	0.28548	0.27415
4.0	0.26716	0.26240
5.0	0.25387	0.23197
6.0	0.24357	0.23177
7.0	0.223526	0.21319
8.0	0.22834	0.20981

Table 2: Numerical values of coefficient of rate of heat transfer $Nu_xRe_x^{-\frac{1}{2}}$ obtained for different values of M when P_r =10.0, α = 0.4, R_i = 2.0.

R_i	FDM	LNS
0.1	0.22415	0.22304
1.0	0.34958	0.34275
2.0	0.47204	0.46274
3.0	0.58357	0.57495
4.0	0.68803	0.68240
5.0	0.78664	0.78197
6.0	0.88073	0.86677
7.0	0.97108	0.91359
8.0	1.14268	1.0936

Table 3: Numerical values of coefficient of Skin friction $Re_x^{\frac{1}{2}}Cf_x$ obtained for different values of R_i when P_r =10.0, α=0.4, M=1.0.

The effects of different values of Richardson parameter on coefficients of skin friction and rate of heat transfer are shown in Tables 3 and 4. From these tables it can be seen that the skin friction and heat transfer are both increased because of the influence of the buoyancy force. The effects of various values of Prandtl number Pr on coefficients of skin friction and rate of heat transfer are shown in Tables 5 and 6.

R_i	FDM	LNS
0.1	0.28611	0.28014
1.0	0.33156	0.33975
2.0	0.36669	0.36743
3.0	0.39392	0.39214
4.0	0.41644	0.41238
5.0	0.43579	0.43184
6.0	0.45286	0.46671
7.0	0.46819	0.47347
8.0	0.49496	0.50304

Table 4: Numerical values of coefficient of rate of heat transfer $Nu_xRe_x^{-\frac{1}{2}}$ obtained for different values of R_i when P_r=10.0, α=0.4, M=1.0.

P_r	FDM	LNS
0.1	0.48354	0.48201
1.0	0.48197	0.48175
2.0	0.48039	0.48143
3.0	0.47898	0.47102
4.0	0.47772	0.462240
5.0	0.47658	0.46001
6.0	0.47554	0.45177
7.0	0.47458	0.45093
8.0	0.47368	0.45016

Table 5: Numerical values of coefficient of Skin friction $Re_x^{\frac{1}{2}}Cf_x$ obtained for different values of P_r when R_i=2.0, α=0.4, M=1.0.

P_r	FDM	LNS
0.1	0.10516	0.10404
1.0	0.14134	0.14963
2.0	0.17772	0.17702
3.0	0.20992	0.20495
4.0	0.23859	0.23209
5.0	0.26444	0.25970
6.0	0.28801	0.27653
7.0	0.30973	0.29359
8.0	0.32993	0.31936

Table 6: Numerical values of coefficient of rate of heat transfer $Nu_xRe_x^{-\frac{1}{2}}$ obtained for different values of P_r when R_i=2.0, α =0.4, M=1.0.

From these tables it is concluded that the component of skin friction is decreased while the component of rate of heat transfer is increased due to the reason that the product of density and thermal diffusion is decreased which enhance the rate of heat transfer in the fluid flow domain. At the end the numerical results obtained by both methods are found to be in good agreement.

Conclusion

In this article we have illustrated the effects of different parameters on some physical quantities such as skin friction, rate of heat transfer, momentum and thermal boundary layer thicknesses. The dimensionless local balance equations were solved numerically using a Finite Difference and Local Non-Similarity method by FORTRAN code. At the end, we summarize the following findings.

The algebraic decay parameter *a* influences the free stream velocity distribution at Y=16-18, and there is no effects of algebraic decay parameter on temperature distribution. For increasing value of Prandtl number Pr both the momentum and thermal boundary layer thicknesses are decreased with the reason that with the increase of Prandtl number Pr the fluid become more viscous which slow down the fluid motion, so distributions on both cases are decaying and this decay is asymptotic. With the increase of Richard parameter the momentum boundary layer thickness is decreased because the Richardson number acts like pressure gradient which take the responsibility of this physical phenomena but its effects in the case of thermal boundary layer is vice versa. With the increase of applied magnetic field momentum boundary layer is decreased and thermal boundary layer is increased for maximum value of M=20.0. For increasing value of Hartmann number increase the magnetic energy and decrease the kinetic energy in the fluid flow domain due to this reason both quantities are decreased. It is also noted that the skin friction and heat transfer are both increased because of the influence of the buoyancy force. It is also concluded that the component of skin friction is decreased while the component of the rate of heat transfer is increased due to the reason that the product of density and thermal diffusion is decreased which enhance the rate of heat transfer in the fluid flow domain.

References

1. Ostrach S (1952) An analysis of laminar free convective flow and heat transfer about a flat plate parallel to direction of the generating body force. NACA Technical Report 1111.
2. Sparrow EM, Chess RD (1961) The effect of magnetic field on free convection heat transfer, Int J Heat Mass Transfer 3: 267-274.
3. Sparrow EM, Minkowycz WJ (1962) Buoyancy effects of horizontal boundary layer flow and heat transfer. Int J Heat Transfer 5: 505-511.
4. Gebhart B (1962) Effects of viscous dissipation in natural convection. Journal of Fluid Mechanics 14: 225-232.
5. Nachtsheim PR, Swigert P (1965) Satisfaction of the Asymptotic Boundary Conditions in Numerical Solution of the System of Non-linear Equations of Boundary Layer Type NASA TND-3004.
6. Sparrow EM, Yu HS (1971) Local non-similarity thermal boundary layer solutions, Journal of Heat Transfer 93: 328-334.
7. Llyod JR, Sparrow EM (1970) Combined forced and free convection flow on vertical surface. International Journal of Heat Mass Transfer 13: 434-438.
8. Dwivedi RN, Dube SK (1974) Oscillatory Hydromagnetic free Convection Flow Past an Infinite Vertical Flat Plate with Suction.
9. Soundalgekar VM (1976) Effects of mass transfer on free convective flow of a dissipative, incompressible fluid past an infinite vertical porous plate with suction, Proc Indian Academy of Sciences 84A: 194-203.
10. Merkin JH (1978) On solution of the boundary layer equations with algebraic decay. Journal of Fluid Mechanics 88: 309-321.
11. Schneider W (1979) A similarity solution for combined forced and free convection flow over a horizontal plate. Int J Heat Mass Transfer 22: 1401-1406.
12. Chakrobarti A, Gupta AS (1979) Hydromagnetic flow heat and mass transfer over a stretching sheet. Quarterly of Applied Mathematics 33: 73-78.
13. Perdikis C (1986) Free convection and mass transfer effects on the flow past a vertical plate. Astro-Physics and Space Science 119: 295-303.
14. Wantanabe T, Pop I (1994) Thermal boundary layers in magnetohydrodynamic flow over a flat plate in the presence of a transverse magnetic field. Acta Mechanica 105: 233-238.
15. Hosain MA, Alam KCA, Rees DAS (1997) MHD Forced and free convection boundary layer flow along a vertical porous plate. Appl Mech and Eng 2: 33-51.
16. Chamkha AJ (1997) MHD Free Convection From a Vertical Plate Embedded in a Thermally Star- tified Porous Medium with Hall Effects, Appl. Math Modelling 21: 603-609.
17. Chen TS (1998) Parabolic systems: Local Non-similarity Method, Handbook of Numerical Heat Transfer, Chap. 5 Wiley, New York.
18. Mostafa A, Mahmoud A (2007) Variable viscosity effects on hydromagnetic boundary layer along a continuously moving vertical plate in the presence of radiation. Appl Math and Sciences 1: 799-814.
19. Al- Odat, MQ, Al-Azab TA (2007) Influence of chemical reaction on transient MHD free convection over a moving vertical plate, Emirates J for Eng Research 12: 15-21.
20. Makinde OD (2008) On MHD Boundary Layer Flow and Mass transfer Past a vertical plate in a porous medium with constant heat flux. International Journal of Numerical Methods for Heat and Fluid Flow 19: 546-554.
21. Alam MS, Rehman MM, Sattar MA (2008) Effects of variable suction and thermophoresis on steady MHD combined free-forced convective heat and mass transfer flow over a semi-infinite permeable inclined plate in the presence of Thermal Radiation, International Journal of Thermal Sciences 47: 758-765.
22. Palani G, Srikanth U (2009) MHD Flow past a semi-infinite vertical plate with mass transfer, nonlinear analysis. Modeling and control 14: 345-356.
23. Ferdows M, Md Khan S, Alam Md M, Sun S (2012) MHD Mixed convective boundary layer flow of a nanofluid through a porous medium due to an exponentially stretching sheet. Mathematical Problems in Engineering 21.
24. Ashraf M, Asghar S, Hossain MA (2010) Thermal radiation effects on hydromagnetic mixed convection flow along a magnetized vertical porous plate. Mathematical Problems in Engineering 30.

Permissions

All chapters in this book were first published in JAME, by OMICS International; hereby published with permission under the Creative Commons Attribution License or equivalent. Every chapter published in this book has been scrutinized by our experts. Their significance has been extensively debated. The topics covered herein carry significant findings which will fuel the growth of the discipline. They may even be implemented as practical applications or may be referred to as a beginning point for another development.

The contributors of this book come from diverse backgrounds, making this book a truly international effort. This book will bring forth new frontiers with its revolutionizing research information and detailed analysis of the nascent developments around the world.

We would like to thank all the contributing authors for lending their expertise to make the book truly unique. They have played a crucial role in the development of this book. Without their invaluable contributions this book wouldn't have been possible. They have made vital efforts to compile up to date information on the varied aspects of this subject to make this book a valuable addition to the collection of many professionals and students.

This book was conceptualized with the vision of imparting up-to-date information and advanced data in this field. To ensure the same, a matchless editorial board was set up. Every individual on the board went through rigorous rounds of assessment to prove their worth. After which they invested a large part of their time researching and compiling the most relevant data for our readers.

The editorial board has been involved in producing this book since its inception. They have spent rigorous hours researching and exploring the diverse topics which have resulted in the successful publishing of this book. They have passed on their knowledge of decades through this book. To expedite this challenging task, the publisher supported the team at every step. A small team of assistant editors was also appointed to further simplify the editing procedure and attain best results for the readers.

Apart from the editorial board, the designing team has also invested a significant amount of their time in understanding the subject and creating the most relevant covers. They scrutinized every image to scout for the most suitable representation of the subject and create an appropriate cover for the book.

The publishing team has been an ardent support to the editorial, designing and production team. Their endless efforts to recruit the best for this project, has resulted in the accomplishment of this book. They are a veteran in the field of academics and their pool of knowledge is as vast as their experience in printing. Their expertise and guidance has proved useful at every step. Their uncompromising quality standards have made this book an exceptional effort. Their encouragement from time to time has been an inspiration for everyone.

The publisher and the editorial board hope that this book will prove to be a valuable piece of knowledge for researchers, students, practitioners and scholars across the globe.

List of Contributors

Shah MKM, Al-Fareez Bin-Aslie, Irma Wani O and Sarifudin J
Faculty of Engineering, University of Malaysia, Sabah 88400, Kota Kinabalu Sabah, Malaysia

Sapuan SM
Department of Mechanical and Manufacturing Engineering, Universiti Putra Malaysia, 43400 UPM Serdang, Selangor, Malaysia

Kadria ML
LAMSIN, Ecole Nationale d'Ing´enieurs de Tunis, BP no 37-1002 Tunis, Tunisia

Hoshyar HA
Department of Mechanical Engineering, Bbabol Branch, Emam Sadegh University, Babol, Iran
Department of Mechanical Engineering, Sari Branch, Islamic Azad University, Sari, Iran

Ganji DD and Abbasi M
Department of Mechanical Engineering, Sari Branch, Islamic Azad University, Sari, Iran

Mohamed Larbi Kadri
Ecole Nationale d'Ingénieurs de Tunis (ENIT), Lamsin, Campus Universitaire, B.P. 37, 1002 Tunis, Tunisia

Yogita B Shinger and Thakur AG
Mechanical Department, Sir Visvesvaraya Institute of Technology, Kopargoan, Maharashtra 422102, India

Ueda S and Kishimoto T
Department of Mechanical Engineering, Osaka Institute of Technology, Japan

Banerjee MM
202 Nandan Apartment, Hill view (N), Asansol-713304, West Bengal, India

Mazumdar J
School of Electrical and Electronic Engineering and School of Mathematical Sciences, the University of Adelaide, Australia

Vijendra Kumar
Professor, Manager, Jagadguru Dattatray College of Technology, Indore, M.P, India

Viswanath T
Deputy General Manager, Kruthi Computer Services Pvt Ltd, Bangalore, Karnataka, India

Rajendra Prasad Sinha
Malaysian Maritime Academy, Advance Marine Engineering, Kuala Sungai Baru, Masjid Tanah, Melaka, Melaka 78300, Malaysia

Pragasen T Kunjambo
Malaysian International Shipping Corp, Malaysia

Bazgir AS
Department of Aeromechanics and Flight Engineering, Moscow Institute of Physics and Technology, 140180, Gagarina Street, 16, Zhukovsky, Russia

Guilin Shi, Chen Zhang, Yingguang Li, Yun Song and Ming Lu
College of Mechanical and Electrical Engineering, Nanjing University of Aeronautics and Astronautics, China

Kornel F Ehmann
Department of Mechanical Engineering, Northwestern University, USA

Tushar A Sinha, Amit Kumar and Nikhilesh Bhargava
Department of Mechanical Engineering, Thapar University, Patiala 147004, India

Soumya S Mallick
Faculty, Department of Mechanical Engineering, Thapar University, Patiala 147004, India

Zheng CM
Department of Mechanical and Electrical Engineering, Quanzhou Institute of Information Engineering, Quanzhou 36200, PR China

Junz Wang JJ
Department of Mechanical Engineering, National Cheng Kung University Tainan, 701, Taiwan

Nasreddine Benkacem, Nader Ben Cheikh and Brahim Ben Beya
Department of Physics, Faculty of sciences, Tunis ELMANAR University, Campus Universities, 2092 El-Manar II, Tunis, Tunisia

Lintu Roy and Arunabh Choudhury
Department of Mechanical Engineering, National Institute of Technology, Silchar-788010, India

Govindu N and Jayanand Kumar T
Department of Mechanical Engineering, Godavari Institute of Engineering and Technology, Affiliation with JNTU-K, Rajahmundry, Andhra Pradesh 530026, India

Venkatesh S
Department of Mechanical Engineering, Raghu Engineering College, Affiliation with JNTU-K, Visakhapatnam, Andhra Pradesh 530026, India

Martínez Concepción ER and De Farias MM
PPG, Geotechnics, University of Brasilia, DF, Brazil

Evangelista F
PECC, Structures, University of Brasilia, DF, Brazil

Wright CI
Research and Development, Global Group of Companies, Cold Meece, Staffordshire, UK

Abda AB and Khalfallah S
University of Tunis El Manar ENIT-LAMSIN, Tunisia

Jebir SK
Department of Mechanical Engineering, AL-Nahrain University, Iraq

Mahendra Kumar Trivedi, Gopal Nayak, Shrikant Patil and Rama Mohan Tallapragada
Trivedi Global Inc., 10624 S Eastern Avenue Suite A-969, Henderson, NV 89052, USA

Omprakash Latiyal and Snehasis Jana
Trivedi Science Research Laboratory Pvt. Ltd., Hall-A, Chinar Mega Mall, Chinar Fortune City, Hoshangabad Rd., Bhopal-462026, Madhya Pradesh, India

Srivastva HK
Department of Mechanical Engineering, Naraina Group of Institutions, Kanpur 208020, India

Chauhan AS
Department of Mechanical Engineering, Maharana Pratap Engineering College, Kanpur 209217, India

Raza A, Kushwaha M and Bhardwaj PK
Department of Mechanical Engineering, Naraina Group of Institutions, Kanpur 208020, India

Epifantsev K and Nikulin A
Saint Petersburg Mining University, 199106, Saint Petersburg, 21 Line of Vasilievskiy Island, Russia

Koulouris K, Konstantopoulos G and Apostolopoulos Ch
Department of Mechanical and Aeronautical Engineering, University of Patras, Greece

Apostolopoulos Alk and Matikas T
Department of Material Science, University of Ioannina, Greece

Kandasamy R and Mohamad R
Research Centre for Computational Mathematics, Faculty of Science, Technology and Human Development, Universiti Tun Hussein Onn Malaysia, Johor, Malaysia

Hopmann C and Wipperfürth J
Institute of Plastics Processing (IKV), RWTH Aachen University, Seffenter Weg 201, 52074 Aachen, NRW, Germany

Seo JK
The Ship and Offshore Research Institute, Pusan National University, Busan, Korea

Bae SY
Electric and Electronic Research Division, Korea Marine Equipment Research Institute, Busan, Korea

Shiells JE
US Government, 1401 Clearview rd, Edgewood MD, USA

Otiede ED, Odeyemi MM, Nwaguru I and Ofualagba G
Department of Electrical and Electronics Engineering, Federal University of Petroleum Resources (FUPRE)

Onuoha C, Awal SI and Ajibola O
Department of Mechanical Engineering, Federal University of Petroleum Resources (FUPRE)

Adebayo VK, Itoro E and Ejejigbe TR
Department of Marine Engineering, Federal University of Petroleum Resources (FUPRE)

Ejofodomi OA
RACETT Nigeria Ltd., 32 Agric Road, GRA, Effurun, Nigeria

Nagaral M
Aircraft Research and Design centre, HAL, Bangalore-560037, Karnataka, India

Attar S and Reddappa HN
R&D Centre, Department of Mechanical Engineering, BIT, Bangalore-560004, Karnataka, India

Auradi V
R&D Centre, Department of Mechanical Engineering, SIT, Tumkur-572103, Karnataka, India

Suresh Kumar S
Cornelius Technology Centre India Private Limited, Bangalore, Karnataka, India

Raghu S
PES College of Engineering, Mandya, 571401, Karnataka, India

Pham HH and Nguyen PV
Ho Chi Minh City University of Technology - Vietnam National University of Ho Chi Minh City, Vietnam

Muhammad Ashraf, Uzma Ahmad, Masud Ahmad and Sultana N
Department of Mathematics, Faculty of Science, University of Sargodha, Pakistan

Index

www.ingramcontent.com/pod-product-compliance
Lightning Source LLC
Chambersburg PA
CBHW080701200326
41458CB00013B/4931
* 9 7 8 1 6 3 2 4 0 8 3 2 7 *